SCANDINAVIAN CALEDONIDES

a volume in the

REGIONAL GEOLOGY SERIES

edited by

L. U. DE SITTER

EDITOR'S NOTE TO REGIONAL GEOLOGY SERIES

The aim of this series on Regional Geology is to add to the available geological literature concise descriptions of large structural units, independent of national boundaries.

It is important that the personal opinion of an author, formed by his work and experience in the structural unit he describes, comes clearly to the attention of the reader. Theorizing about the geological history of a particular kind of structure too often does not take into account the great diversity of the observed phenomena, and then generalizes in an unwarranted way. We aim to give a better basis to these general concepts and thereby stimulate a deeper understanding of the relations between different kinds of structures.

Some of the books describe classical territory where new work has brought new conceptions, others are concerned with hitherto relatively unknown regions, but always the surveys are presented from a fresh aspect.

L. U. de Sitter

SCANDINAVIAN CALEDONIDES

T. STRAND

Professor of Geology
University of Oslo

O. KULLING

Senior Geologist
Geological Survey of Sweden

1972

WILEY — INTERSCIENCE
a division of John Wiley & Sons Ltd
LONDON · NEW YORK · SYDNEY · TORONTO

Cover photo: *Reine in Moskenesøy, Lofoten (67°56′, 13°7′)*
Copyright *Widerøes Flyveselskap A/S, Oslo*

Library of Congress Catalog Card Number 77–78474
ISBN 0 471 83169 7

Printed in Great Britain by William Clowes and Sons Ltd, London and Beccles

CONTENTS

As stated on pp. 6–7, a geological map of Norway in scale 1 : 1 million, edited by O. Holtedahl and J. A. Dons, was published in 1960. The map can be ordered from Universitetsforlaget, Box 307 Blindern, Oslo 3, Norway. The price is now 30 Norwegian kroner.

Part I

THE NORWEGIAN CALEDONIDES

by

T. STRAND

PREFACE

It is no easy task to write a book on the Norwegian Caledonides in the present state of knowledge. Most, but not all, of the areas have passed the stage of reconnaissance survey and detailed investigations are under way in many districts. However, over larger regions investigations are insufficiently advanced to give a safe foundation for a regional synthesis. The author has thus felt it necessary to confine much of the text to recording the main facts.

This is one reason why the present writer did not follow the same course as Dr. KULLING, the Swedish coauthor of this volume: to describe the separate tectonic units along the entire part of the orogen. Another reason is the extent and diversity of the Norwegian Caledonides. It has thus been found preferable to divide the text into sections treating different districts separately. Comparison of the Swedish and Norwegian parts of this book will clarify the difficulty of tracing the stratigraphic and tectonic units across the border in many cases. In part this may be due to lack of detailed investigations, especially on the Norwegian side; but it may also be due to the fact that the work has been done by geologists starting from different sides of the orogen with different views and conceptions. To remedy this defect a number of short notes referring to the Swedish part of the book, or to other Swedish literature, have been included in the text.

Acknowledgements

This book ran the risk of being outdated at its appearance had not a large number of geologists, Norwegian and of other nationalities, communicated the unpublished results of their work to the author. In all cases this help was freely and generously given and the heartiest thanks are offered to them all. Naturally, the author accepts sole responsibility for the presentation of the data.

CHAPTER I

INTRODUCTORY

Norway, the western and northern part of the Scandinavian peninsula, extends from 58° to 71° 10′ N and from 4° 30′ to 31° E. The distance from south to north and northeast along the country is some 1700 km. Caledonide rocks occur along the whole of this distance, except for the southern tip of solely Precambrian rocks. The entire Caledonian orogen, including the eastern or southeastern margin, is on the Norwegian side of the Swedish–Norwegian border south of 61° 40′ and north of 68° 20′, except for a marginal strip between 68° 30′ and 69° N, most of it on Finnish territory.

As a part of Fennoscandia, Norway has its share of the old Precambrian basement rocks of the Baltic shield. They occupy large areas in the southern part of the country and in interior parts of Troms and Finnmark in the northern part, and are also found in a number of smaller areas, most of them windows in tectonic culminations. The Caledonide parts of the country include the deposits of the *Sparagmite group* and the younger Cambrian to Devonian deposits, the Caledonian intrusive magmatites and overthrust massifs of Precambrian age, and the old Precambrian gneisses which were strongly deformed and recrystallized during the Caledonian orogeny. Rocks of post-Caledonian age are: thin sediments and thick volcanic sequences with corresponding intrusive rocks of Permian age in the Oslo region, and Mesozoic sediments occupying a small area in Andøya in northern Norway.

The present land morphology of Norway is due to uplift in Cenozoic (probably Pliocene) time, accompanied by marginal faults (HOLTEDAHL, 1960d, 1960e). The recent earthquakes in Norway (stronger than in more easterly parts of Fennoscandia) were probably due to movements on the marginal fractures (KVALE, 1960b). Preliminary results of geophysical investigations now under way have shown the presence of thick blankets of young sediments on the banks around the country. The upheaval of the Scandinavian peninsula was oblique, with a maximum near the west coast and almost entirely on the Norwegian side. In the western parts of the country deep steep-sided valleys and fjords are incised into an older surface of rolling plains, or the topography is subdued, with rounded mountains (the paleic surface of HANS REUSCH), while in the east the old morphology has been preserved to a large extent. The Caledonide parts of Norway thus offer a wide spectrum of scenery, from fertile plains and broad valleys in the east to mountains of "Mittelgebirge" and the wild glaciated types in the west. The highest mountains, more than 2000 m a.s.l., are found in the resistant rocks of the *Jotun massif* northeast of inner Sognefjord, and in the *Gneiss region* around the *Jostedalsbreen plateau glacier*, northwest of the Jotun massif area. The highest mountain, *Galdhøpiggen* in Jotunheimen, is 2469 m high. Farther north, *Snøhetta*, west of the *Trondheim region* (62° 20′, 9° 15′), is 2286 m high. The greatest local relief is in the inner fjord districts north of Sognefjord, where mountains rise from sea level to 1800 or 1900 m.

HISTORICAL OUTLINE AND MAIN LITERATURE

The earliest geological investigations in Norway were made by foreign geologists travelling through the country. The best known of these is possibly LEOPOLD VON BUCH. More systematic geological work began after the foundation of the University of Oslo (then Christiania) in 1811. Outstanding in this pioneer period was B. M. KEILHAU (1797–1858). It was left to THEODOR KJERULF (1825–1888) to found modern geology in Norway, as professor at the University and from 1858 as director of the Geological Survey, begun at his initiative. The work of KJERULF and his assistants resulted in the publication in 1879 of *Udsigt over det sydlige Norges geologi* (Review of the geology of south Norway) with a geologic map. A large number of geologic maps of scale 1 : 100 000 were published in KJERULF's time, covering the Oslo region and parts of the Sparagmite and Trondheim regions. At the same time TELLEF DAHLL (1825–1893) and KARL PETTERSEN (1826–1890) carried out pioneer work in northern Norway.

KJERULF was succeeded by W. C. BRØGGER (1851–1940) as professor at the University and by HANS REUSCH (1852–1922) as director of the Survey. REUSCH wrote his classical papers on the Caledonide geology of western Norway in the 1880's. As geologists at the Survey J. B. REKSTAD (1852–1934) and K. O. BJØRLYKKE (1860–1945) widened knowledge of the Caledonides by extended reconnaissance mapping. At about the same time J. H. L. VOGT (1858–1931) was working in northern Norway and C. F. KOLDERUP (1869–1942), professor at Bergen Museum, in western Norway.

Mention must be made of the great Swedish geologist A. E. TÖRNEBOHM (1838–1911), whose work on the geology of central Scandinavia in 1896 embraced great parts of southern Norway. It is to TÖRNEBOHM's credit to have introduced the concept of overthrusts into Scandinavian geology. While BRØGGER (1893) was reluctant to accept the doctrine of overthrusts, K. O. BJØRLYKKE became an adherent, but altered his opinion before the publication of his monograph on the geology of central Norway in 1905. In the early decades of the present century there was a general scepticism towards belief in large overthrusts, but smaller ones, amounting to some tens of kilometres, were accepted. V. M. GOLDSCHMIDT (1888–1947) was an exponent of this view in his papers on Caledonide geology and petrology published between 1910 and 1921. Yet, in the same period the thrusts along the margin of the Sparagmite region (the *Vemdal nappe*) were recognized for the first time in 1902 by O. E. SCHIÖTZ (1846–1925) and described in a different area by HOLTEDAHL (1915). HOLTEDAHL (1918) also mapped and described the more magnificent thrust structures of the *Alta valley* in Finnmark. C. E. WEGMANN, who worked in Norway in the 1920's, certainly recognized the large overthrusts, but did not publish much concerning them. In a paper of 1936 HOLTEDAHL advanced the opinion that the large Jotun massif was a rootless nappe. Geologists working and publishing in later years generally recognized the reality of the large nappe structures.

Among the geologists who made contributions to Norwegian Caledonide geology after 1910, the following may also be mentioned: C. W. CARSTENS (1887–1950), S. FOSLIE (1887–1951) and TH. VOGT (1888–1958). The publications of KJERULF and TÖRNEBOHM previously mentioned are early descriptions of the geology of southern Norway. W. C. BRØGGER and HANS REUSCH wrote short presentations (in Norwegian) on the geology of Norway, and a geological description of Norway was presented in international publications by HÖGBOM (1913) (*Fennoskandia, im Handbuch der regionalen Geologie*) and by VON BUBNOFF (1930) in *Geologie von Europa*. The latter book reflects the views of V. M. GOLDSCHMIDT, who is cited as a coworker.

In 1938 HOLTEDAHL gave a lucid presentation of the geology and tectonics of the Scandinavian Caledonides (BAILEY and HOLTEDAHL, 1938). A map in this book gives a good picture of the Caledonide nappe structures. In 1953 appeared HOLTEDAHL's *Norges geologi*. This book, written in Norwegian, has appeared in a Russian translation. For the Norden Geological Congress in 1960 HOLTEDAHL edited a new book, *Geology of Norway*, with various chapters treating the Caledonides written by HOLTEDAHL (1960b, 1960c), HENNINGSMOEN (1960) and STRAND (1960a, 1960b). STRAND (1961) gives a brief review of the Scandinavian Caledonides.

A number of guide-books for the Norwegian Geological Congress excursions in 1960 appeared after the Congress as publications of the Geological Survey of Norway (DONS, 1960). There are guide-books for the Oslo and Sparagmite regions by HENNINGSMOEN and SPJELDNÆS, and by SKJESETH and HAGEMANN. A guide to Oslo and environs (not a Congress excursion guide) is that of HOLTEDAHL and DONS (1957). There are further Congress excursion guide-books by STRAND and HOLMSEN for the northeast part of the *Nappe region* and the *Oppdal district*, by KVALE for a western part of the Nappe region, by CARSTENS for the *Trondheimsfjord district*, by KOLDERUP for the Gneiss region of western Norway and by HOLTEDAHL and others for parts of northern Norway, especially Finnmark. A Swedish excursion guide partly covering Norwegian ground is KULLING's (KULLING and GEIJER (1960)) for the Ofoten district in northern Norway. A guide-book by VOKES (1960) describes a number of important Norwegian mines. The Norwegian ore deposits and all claims and mines have been listed and marked on maps in the publications of FOSLIE (1925) and POULSEN (1964).

Invaluable bibliographic references are contained in DONS (1956), which is a catalogue of all bedrock geologic maps of Norway, arranged according to Ordnance Survey map areas. A geologic bibliography of Norway (1920–1932) was published by ROSENDAHL (1933), with references to older bibliographies by HANS REUSCH. A geological map of Norway in colour at a scale of 1 : 1 million was published in 1960 to accompany the book, *Geology of Norway*. The map can be bought separately at a price of 25 Norwegian kroner. It, and other maps and publications of *Norges Geologiske Undersøkelse*, can be obtained from Universitetsforlaget, Boks 307 Blindern, Oslo 3, Norway.

STRATIGRAPHY

Beneath the oldest fossiliferous deposits of Lower Cambrian age in central southern Norway (see Fig. 102, p. 131 and Fig. 103, p. 132) are thick sequences of mainly clastic deposits with coarse arkosic sandstones as predominant constituents. The term *sparagmite* was coined for this type of rock by Professor JENS ESMARK in 1829. TH. KJERULF applied it to the corresponding stratigraphic unit, and the Sparagmite group is now recognized as a main lithostratigraphic unit. In 1900 W. C. BRØGGER introduced the term Eocambrian for a chronostratigraphic unit corresponding to the Sparagmite group. This implied that the Sparagmite deposits were closely linked stratigraphically to the overlying Cambrian deposits, thus representing a very early part of the Cambrian system. The thick sequences of sandstones and dolomites in Finnmark, the northernmost part of Norway, correspond to the Sparagmite deposits in southern Norway. A small but distinct angular unconformity is present beneath the horizon with tillites in the upper part of the group. HOLTEDAHL (1961a) in a review of the Sparagmite group and corresponding deposits in Scandinavia restricted the Eocambrian to comprise only the upper part with glacial deposits, above the unconformity in Finnmark. This part of the sequence is thus regarded as earliest Cambrian and the underlying deposits consequently as latest Precambrian. As the term Eocambrian cannot be used for the entire sequence of deposits discussed here, the term *Sparagmite deposits* is used in the text. It must be emphasized that KULLING, the coauthor, takes the term "Sparagmite group" in a more restricted sense, and does not include the younger part, here classed as Eocambrian. He also uses Eocambrian in a more restricted sense to include only the Varangian tillite horizon. The classification of the Norwegian parts follows HOLTEDAHL (1961).

The Cambro–Silurian (meaning Cambrian, Ordovician and Silurian systems) deposits show very great differences in facies in different parts of the Caledonide orogen and its foreland. It is customary to distinguish between an eastern and a western facies, corresponding to miogeosynclinal and eugeosynclinal facies types, though some of the deposits of eastern facies tend to show a foreland or shelf facies. In Norway, deposits of eastern facies are the richly fossiliferous sequence in the Oslo region, the mainly pelitic deposits beneath the nappes in central southern Norway, and all autochthonous deposits along the Caledonide margin in the southern and northern parts of the country. The deposits of eugeosynclinal facies have been divided by the writer into a Trondheim region and a Nordland facies type. The former facies type is characterized by its abundant basic volcanic rocks and by sediments of greywacke type, while thick limestones, dolomites and iron sediments are especially characteristic of the Nordland facies type, which seems to indicate more quiet and stable conditions of deposition. But volcanic rocks also occur in the Nordland sequences, especially in the lower parts, while rather thick limestones occur in the upper part of the sequence in the Trondheim region. There is perhaps no fundamental difference between the two type sequences.

The Devonian deposits are of Old Red facies and rest on metamorphosed Caledonide rocks with a pronounced unconformity.

STRUCTURE AND STRUCTURAL HISTORY

When approaching the Caledonide margin from the east or southeast one meets with autochthonous Cambrian sediments resting on a peneplained Precambrian basement. The autochthonous cover is generally thin, most of it having been sheared off by overriding thrustsheets. The Caledonian orogen has its eastern border against the peneplained floor of the stable Baltic shield, without any intervening foredeep. Nappe movement over long distances on nearly horizontal thrust planes may thus have been facilitated, and it may be said with some justification that the orogen is built of nappes. Using the same system as KULLING the nappes in Norway can be divided into a lower, a middle and an upper group. The lower nappes consist of unmetamorphosed sediments of facies largely corresponding to those of the autochthonous deposits. The group is represented by the Vemdal nappe of Eocambrian sandstone in southern Norway and by the unmetamorphosed allochthonous sandstones and dolomites of the Finnmark sequence in the north. The middle nappes consist of metamorphic rocks but the sediments involved are not of eugeosynclinal facies. In southern Norway this group is represented by the allochthonous Sparagmite deposits in the northern parts of the Sparagmite region (the *Kvitvola nappe*). It is possible that some of the allochthonous metamorphic rocks in Finnmark may be classed in this group. The upper nappes consist of sediments of eugeosynclinal facies with underlying basement rocks. To this group belongs the large charnockitic massif of the Jotun nappe in southern Norway and all the sediments in the Trondheim region, and in northern Norway the large Seve nappe or nappe complex, which extends across the border from the Swedish side. The Seve nappe is overlain by a higher nappe of highly metamorphic rocks rich in granites—the *Rødingsfjell nappe*. This upper unit does not appear to continue north of the Tysfjord culmination, where the allochthonous metamorphic rocks seem to be a continuation of the Seve nappe. In the northwestern Gneiss region in southern Norway very large recumbent folds with gneiss in the cores are met with—nappes of a Pennine type, rather different from the thrust sheets of the marginal parts.

The Vemdal nappe (group of lower nappes) shows observed lengths of transport of from 25 to 30 km. The Kvitvola nappe (middle group) in the eastern part of the Sparagmite region has its southern edge some 100 km south of a locality (west of Lake Femunden) where the autochthonous deposits on the Precambrian basement are younger than the rocks of the nappe and of a very different facies. It can only be surmised how far the upper nappes must have moved.

Distinct angular unconformities mark the end of the main Caledonian deformation, but are unknown in intermediate stages. Detailed structural studies in a number of areas have, however, given evidence of two or three phases of folding. The structures that determine the large patterns of the geologic map must be of a late origin; this was first emphasized by WEGMANN (1925). These structures fold and bend the thrust planes, which are duplicated in some places by schuppen structures. Large scale structures of this type are the "Faltungsgraben" in south Norway, the *Bergen arc structures* and the axial depression of the Trondheim region.

Tectonic unrest and upheaval of land masses is testified for by conglomerates and other coarse sediments (HOLTEDAHL, 1920; VOGT, 1928, 1945). The Trondheim orogeny or disturbance in early Ordovician times is indicated by the *Stokkvola conglomerate* in the Trondheim region. The *Trysil disturbance* may be of the same age or possibly later (in early Middle Ordovician). Other conglomerates in the Trondheim region are the *Volla conglomerate* between the Middle and Upper Ordovician, indicating the *Ekne orogeny*, and the *Lyngestein conglomerate*, probably between the Upper Ordovician and the Silurian, indicating the *Horg orogeny*. In the Oslo region an Upper Ordovician period of unrest is indicated by a conglomerate west of Oslo and by a hiatus in the sequence, largest in the northern part of the region. SPJELDNÆS (1957) claimed that faint folding took place at this time in the Oslo region. In the Bergen district KVALE found indications of a phase of metamorphism older than the Upper Ordovician sediments of the same district. Caledonian magmatism was at its strongest in Ordovician times. In a broad sense we can thus speak of a Taconic phase in Middle and/or Late Ordovician time, but no precise dating is possible. In addition Silurian sediments have been metamorphosed, folded and involved in nappe structures, and the end of the main deformation is marked by the deposition of a late Silurian (Downtonian) sandstone with a pronounced unconformity resting upon Caledonian intrusive rocks on the island of Hitra. In the Oslo region folding took place later, and here the *Ringerike sandstone* (latest Silurian–earliest Devonian) was involved in the folding, while lower Permian sediments are found above the unconformity. Data for any precise dating of Caledonian orogenies are thus not available. There is also a phase of folding later than Middle Devonian affecting the Devonian Old Red deposits (the *Svalbardian phase* of TH. VOGT.).

A limited number of age determinations on Norwegian Caledonide rocks (BROCH, 1964) have dated them at 550 to 350 m. y. old. A few of these determinations indicate metamorphism at 550 m. y., while the majority gives ages between 450 and 350 m. y. with a peak at 400 m. y. An age of 450 m. y. may indicate metamorphism in Middle Ordovician time, while ages about 350 m. y. (Devonian) may indicate not the time of congealing or metamorphism of the rock, but the time at which the rocks came near the surface and were sufficiently cooled to prevent the escape of argon.

CHAPTER II

THE OSLO REGION

The Oslo region, extending from Langesundsfjord in the south to Lake Mjøsa in the north, is a graben formed by downfaulting in Permian times, in which Cambro–Silurian sediments and Permian sediments and lavas have been spared from erosion. The Oslo graben is in addition the site of the well-known Oslo suite of deep-seated igneous rocks, which occupy the greater part of it. Cambro–Silurian outcrops are thus found in a number of separate districts within the Oslo region. Extending along the western margin of the Oslo region from south to north are the following districts: Langesund–Skien, Hedenstad, Modum, Ringerike on Lake Tyrifjord northwest of Oslo, Hadeland east of Lake Randsfjord and Mjøsa. The last district is the most northerly part of the Oslo region, where the Cambro–Silurian deposits occupy a large area on both sides of Lake Mjøsa. A further main area of Cambro–Silurian rocks is the Oslo–Asker district at the head of Oslofjord.

Apart from Jämtland in Sweden, the Oslo region is the only part of Scandinavia where a complete richly fossiliferous Cambro–Silurian sequence of eastern facies is found within the domain of Caledonian deformation.

STRATIGRAPHY

The Oslo Cambro–Silurian sequence (see Fig. 102, p. 131 and Fig. 103, p. 132) has been studied for more than a century by geologists and paleontologists of Oslo University. A review of the stratigraphy was given by Henningsmoen (1960). Kiær (1908) wrote a comprehensive monograph on the Silurian part of the sequence, while Størmer (1953) described the Middle Ordovician part. The Oslo sequence has for many years been divided into *étages*, numbered 1 to 10 and subdivided by the use of Latin and Greek letters. Division into chronostratigraphic series is a recent development.

The whole sequence comprises a maximum of 1400 m of marine deposits and up to 1000 m of a continental sandstone, the Ringerike sandstone. In most of the Oslo region the sequence commences with Middle Cambrian deposits (of the *Paradoxides paradoxissimus zone*, 1cβ) above the old Precambrian basement. The Lower Cambrian and the oldest Middle Cambrian deposits are found only in the northern part of the region (*Mjøsa district*), where they rest upon the Eocambrian Vemdal sandstone, the topmost formation in the Sparagmite group.

There are differences in lithologic and faunal facies between contemporaneous deposits in the different districts and in some cases it has been very difficult to make reliable correlations. The Oslo stratigraphers distinguish between a western facies with shallow water deposits, typically sandstones and limestones, and an eastern facies with more deep water sediments, chiefly pelitic shales and marls. There may have been a shoreline not very far to the west at some stage during the deposition.

The *Holmia series* (1a, b) comprises the fossiliferous Lower Cambrian deposits, present in the northernmost part of the Oslo region in the Mjøsa district, where they rest conformably upon the *Vemdal sandstone*. There is, however, a sharp disconformity in lithology, and thus the boundary between the uppermost (Eocambrian) part of the Sparagmite group and the overlying fossiliferous deposits is well marked (Vogt, 1924). This applies to the whole of the southern Norwegian Sparagmite region. The sediments are grey and dark shales, siltstones and fine-grained sandstones of thickness some tens of metres. Characteristic fossils are Platysolenites antiquissimus and Volborthella tenuis in the lower part (1a) and Holmia kjerulfi and Strenuella linnarssoni in the upper part (1b). In late early Cambrian time shallow seas transgressed southwards and deposited thin conglomerates or basement beds, found as far south as Hadeland (Skjeseth, 1963).

The *Paradoxides series* (1c, d) is Middle Cambrian. The oldest beds (1cα) with Paradoxides oelandicus are found in the northern parts of the region together with the underlying Lower Cambrian, whilst in the southern parts of the Oslo region, the oldest deposits (but for the thin Lower Cambrian basement beds previously mentioned) belong to the zone complex of Paradoxides paradoxissimus (1cβ). Wide areas were thus transgressed at that time. The Middle Cambrian deposits are dark shales, partly of alum shale type, with layers and nodules of black limestone.

The Upper Cambrian *Olenid series* (2a–d) is a very monotonous sequence of black carbonaceous shales (alum shales) with nodules and layers of black, bituminous limestone (stinkstone). It can be divided into a great number of biozones by genera and species of the olenid family of trilobites. The black alum shales are rich in iron sulphide and carbon (15–17% or more C). They were formerly used for the manufacture of alum. They are uranium enriched, with a maximum content of 170 p.p.m. (Skjeseth, 1958). Some layers are very rich in vanadium. The Norwegian alum shales do not yield oil or gas by distillation, unlike those of middle Sweden. The thickness of the Middle and Upper Cambrian is not more than 40–50 m in the Oslo district.

The Ordovician is divided into five series. The *Ceratopyge series* (Tremadoc) and the *Asaphus series* (Arenig) are Lower Ordovician, the *Ogygiocaris series* (Llanvirn, Llandeilo) and the *Chasmops series* (Caradoc) are Middle Ordovician and the *Tretaspis series* (Ashgill) is Upper Ordovician.

The lowermost Ordovician, the *Ceratopyge series* (2e, 3a), commences with alum shales with Dictyonema flabelliforme. There is no lithologic or faunal break with the underlying Cambrian. Above the black shales is 1 m of *Ceratopyge limestone*, a light-coloured limestone with a rich fauna with the first large asaphids (Euloma–Niobe fauna). This is probably due to a change from unaerated to aerated conditions.

The Asaphus series (3b, c) consists of a lower shale rich in graptolites and an upper limestone, the *Orthoceras limestone*, which is 10 to 15 m thick near Oslo. The Oslo Orthoceras limestone is a tongue of the thicker and more comprehensive Orthoceras limestone of the Baltic regions and has a rich fauna of asaphids and illænids. The upper part contains numerous conchs of large endoceratids. In the northern part of the Mjøsa district the Orthoceras limestone is represented by a 40 m thick impure nodular limestone (the *Stein limestone*). The thickness of the Lower Ordovician (Ceratopyge and Asaphus series) is about 40 m in the Oslo district.

The Ogygiocaris series (4aα 1–4), corresponding to the Llanvirn and Llandeilo, consists of dark shales with layers and lenses of dark limestone, with arenaceous layers at some horizons. Parts of the sequence are rich in graptolites (Didymograptus bifidus, Did. geminus, Glossograptus hincksii). In the two upper divisions (4aα 3–4) Trinucleus bronni and Tr. foveolatus are index fossils. Ogygiocaris dilatatus is a fossil characteristic of the series. Near Oslo the thickness of the series is about 100 m.

The Chasmops series (4aβ, 4bα–δ) in the district near Oslo is a sequence of marlstones about 150 m thick with layers or nodules of more lime-rich material (see Fig. 1). In 4aβ, 4bβ and 4bδ the rocks are impure nodular limestones, while the shale material predominates in the intervening divisions. The upper Chasmops limestone, 4bδ, is conspicuous for consisting of decimetre-thick layers of limestone separated by shale. The trilobite genus Chasmops is characteristic with Ch. conicophthalma in the lower part and Ch. extensa in the upper part. The Chasmops deposits show very great differences in facies, and the corresponding deposits in the Mjøsa district are not easy to correlate with those at Oslo. The *Coelosphaeridium beds* are shale and arenaceous limestone with the calcareous algae Coelosphaeridium and Mastopora as characteristic fossils. The *Cyclocrinus beds* are arenaceous shales with layers of arenaceous limestone. At the top of the sequence is the 100 m thick *Mjøsa limestone*, a massive limestone with Solenopora and corals; it is partly a reef limestone. Corresponding to the Mjøsa limestone is the 40 m thick *Encrinite limestone* in the Langesund–Skien district.

The Upper Ordovician, the Tretaspis series (4c, 5a–b) begins with a black shale, the *Tretaspis shale* (4cα) with Tretaspis seticornis. The deposits in the upper parts of 4c and in 5a are marly shales with limestone layers and nodules, similar to the Chasmops deposits near Oslo. The *Gastropod limestone*, 5a, is a rather massive, richly fossiliferous limestone, typically developed at Ringerike. The coral Sarcinula ("Syringophyllum") organa is a guide fossil. In the Oslo–Asker district the overlying 5b is a calcareous sandstone with conglomerates, containing boulders from the underlying 5a. Arenaceous deposits are common in 5b, and at Ringerike there are reef limestones. At Oslo and at Asker, west of Oslo there are indications of very weak folding before the deposition of the Silurian with a local conglomerate, 5c (Spjeldnæs, 1957). Kiær (1902) in a study of the facies of the 5b deposits at Asker found that the distribution of shallow and deeper water deposits had a trend corresponding to the trend of the folds in the same district. Upper Ordovician deposits are absent in the northern part of the Oslo region, in the Mjøsa district.

The marine Silurian sequence is divided into four series. The *Stricklandia* and *Pentamerus series* are

Fig. 1 *Middle Ordovician nodular marlstone* (Ampyx limestone, 4aβ) Sognsveien, Oslo

Lower Silurian (Llandovery); the upper and lower *Spiriferid series* are Upper Silurian (Wenlock and Ludlow). The Silurian deposits are thicker than the underlying Ordovician and Cambrian, as there was a general increase in the rate of sedimentation with time in the Oslo region. The marine Silurian sediments occur in all parts of the Oslo region, but in the north, in the Mjøsa district, the lower part is missing and the upper part was eroded away before deposition of the Permian sediments. The Silurian sediments are generally similar to the Ordovician, but contain more sandy material and limestone, while black shales are rare. There is a distinct difference between the western and eastern facies, the former having more sandy material, the latter more shale and limestone.

The Stricklandia series (6) is developed as calcareous sandstone in the west and as shale with limestone in the east. Its thickness is about 160 m in the Oslo district. In Hadeland the Stricklandia series, together with the underlying 5b, is a 100 m thick calcareous sandstone, described by MAJOR (1946). It thins to the north and is represented at Mjøsa by 10 m or less of quartzite which rests directly upon the Mjøsa limestone, the sand filling karst cavities in the limestone. Here the quartzite contains grains of microperthite of a type characteristic of the charnockitic rocks in the nappes in central southern Norway (STRAND, 1959). There is thus a break between the Ordovician and the Silurian which becomes greater to the north.

The Pentamerus series (7a–c) contains massive nodular limestones with Pentamerus laevis and P. borealis in 7a, b and marly shales in 7c. At Mjøsa 7c is a shale with graptolites. Near Oslo it is about 130 m thick.

The Lower Spiriferid series (8) with Eospirifer plicatellus is rich in shale in the lower part, while limestone predominates in the upper part. More or less reefy limestones occur at Ringerike and in the Langesund–Skien district. A limestone from this district rich in crinoid stems and corals has been much used as an ornamental stone for building. Its thickness is 100–130 m in the Oslo district.

In the northern parts of the Oslo region (Hadeland and Mjøsa districts) the deposits are calcareous sandstones and shales, indicating transport of material from the north. In the Mjøsa district the uppermost formation in the Silurian is the *Bruflat sandstone* (SKJESETH, 1963)—some 100 m of grey sandstone with ripple marks and layers of calcareous sandstone and shale, of Wenlock and Ludlow age. It is a coarse clastic facies, comparable to that of the younger Ringerike sandstone in the south. Thus the conditions for this type of facies were established earlier in the north than in the south.

The upper Spiriferid series (9) with Eospirifer elevatus is rich in limestones in the southern districts and also in the districts west of Oslo, but is more rich in shale at Ringerike. Its thickness is about 200 m in the Oslo district. At the time of deposition the belts with similar facies in the Oslo region were arranged NE–SW rather than N–S.

The uppermost marine Silurian (9g) is composed mainly of shales with mud cracks and ripple marks and has an impoverished marine fauna. There is a gradual transition into the overlying Ringerike sandstone. According to SPJELDNÆS (1966) the transition layers near Oslo are laminated red sandstones and calcarenites, indicating deposition in an estuary.

The Ringerike sandstone (10) consists of reddish brown, partly greyish sandstones of Old Red facies. It is about 1000 m thick at Ringerike and thins to about 300 m in the southern part of the Oslo region, but it is absent in the northern districts. From fossils found at some horizons (Phyllocarids, merostomes, anaspids and cephalaspids) it can be dated as youngest Silurian (uppermost Ludlow and early (?) Downtonian). Some layers in the sandstone are calcareous and contain small fragments of Silurian fossils, showing that Silurian sediments were eroded at that time in the northern districts (KIÆR, 1924). At Jeløya on the east side of Oslofjord the sandstone contains grains of feldspar which must have been derived from Precambrian rocks, indicating that such rocks were at the surface in the source areas of the sediment (VOGT, 1929).

BJØRLYKKE (1965a) made a large number of spectrometric chemical analyses of pelitic rocks from Ordovician deposits in the vicinity of Oslo. By comparing analyses of shales from 3b and 4aα 1–2 with those from 4aα 3–4, a significant increase was found in the contents of iron and magnesia in the younger deposits (the Niggli parameters *fm* and *mg* showed an increase from 38 to 62 and from 0,48 to 0,65 respectively, while the *al* parameter decreased from 46 to 29). The additional amount of ferro-magnesian material must evidently have originated from basic volcanic rocks. Its origin may be either in the volcanism recorded in parautochthonous Ordovician deposits in the most southwesterly part of Norway, or possibly in some tectonic event which made eugeosynclinal volcanic rocks victims of denudation.

Layers of bentonite interbedded with Ordovician and Silurian deposits are evidence of explosive volcanic activity in the central parts of the Caledonides. In the Middle Ordovician Chasmops deposits (4bα, lower part of the Caradoc) in the Oslo district there are more than 20 bentonite layers, most of them between 16 and 50 cm thick;

one layer measures 130 cm (HAGEMANN and SPJELDNÆS, 1955). A correlation with the rhyolitic volcanism in the lower part of the *Hovin group* in the Trondheim region is very probable. In the Silurian there are bentonites in the upper part of the Llandovery (7c) at Hadeland (SPJELDNÆS and HAGEMANN, 1955; HAGEMANN, 1966) and in the lower and middle part of the Ludlow (9c and 9f–g) at Ringerike (JØRGENSEN, 1964).

GEOLOGIC SETTING

The Caledonide structures in the Oslo region are folds in the Cambro–Silurian sediments. The folding is of *Jura type*, above a stiff Precambrian basement unaffected by the folding, and involves a *décollement* and gliding of the pile of sediments above the basement. The folding dies out along an E–W line that can be drawn some 30 km south of Oslo. Thus the southern parts of the Oslo region belong to the undeformed foreland of the Caledonides, and to this part also belong the Langesund–Skien and Hedenstad districts.

The axial trend of the folds swings from E–W to ENE and NE. The size and style of the folds differ greatly in the different parts of the sequence, depending upon the competency of the rocks. In a broad sense the thickness and competency of the folded units increase upwards in the sequence. The Cambrian and Lower Ordovician dark shales are greatly squeezed and crumpled, and certainly formed horizons of gliding during the folding. The widths of single synclines and anticlines in different parts of the sequence in the vicinity of Oslo will illustrate this point (data from maps and sections by BRØGGER (1882, 1887), BJØRLYKKE (1898) and HOLTEDAHL and DONS (1957)):

- Orthoceras limestone (Lower Ordovician, 10 m thick): 50–200 m.
- Middle Ordovician nodular marlstones: 300–700 m (Fig. 2).
- Silurian limestones: about 1000 m.
- Ringerike sandstone: large flat synclines 3–5 km wide.

The folds are of a concentric type, but are often overturned and compressed, and numerous faults cause repetition of the beds. BRØGGER considered these as fold faults formed by the squeezing out of intermediate limbs of the folds, but probably most of them are simple upthrusts (REUSCH, 1884; STØRMER, 1938).

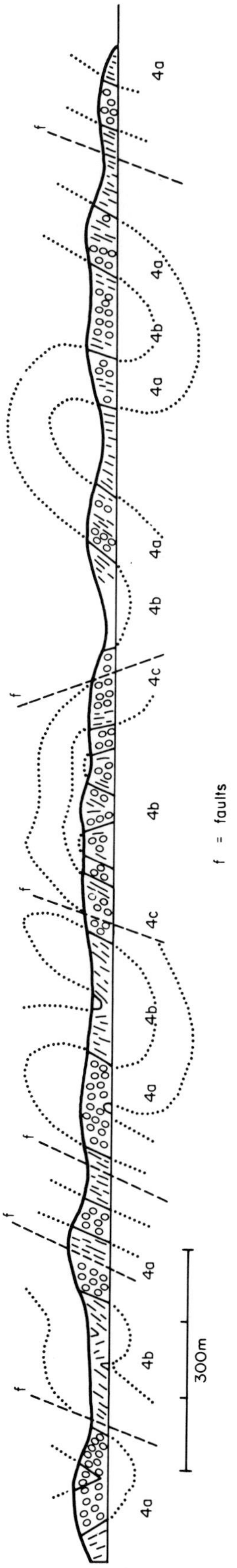

Fig. 2 *Folds in Middle Ordovician nodular limestones and shales*, section along Frognerkilen, Oslo. After K. O. BJØRLYKKE (1898)

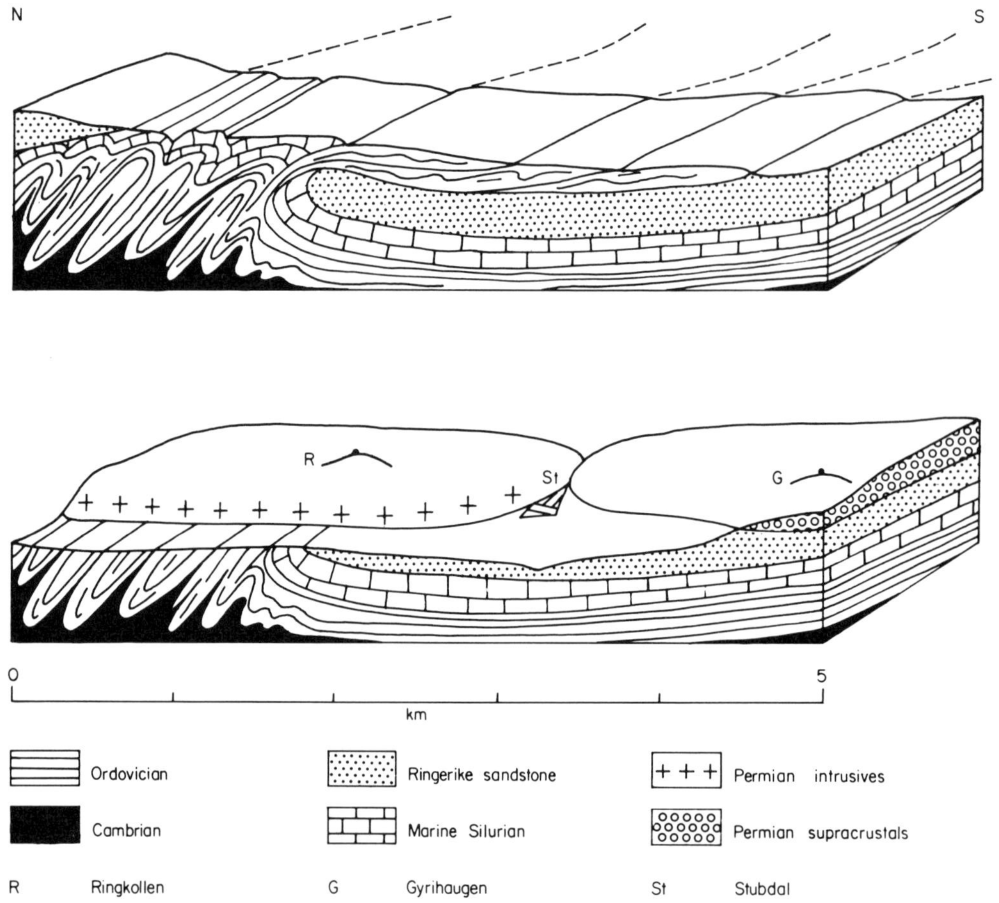

Fig. 3 *At Stubdal (St), 30 km northwest of Oslo, hornfelsed Ordovician sediments of étage 4a* were found above Ringerike sandstone (lower picture). The upper picture gives a reconstruction of the overfolding. From STØRMER (1934)

As previously mentioned the folding took place by décollement and by gliding of the folded sediments above a stiff Precambrian basement. The lowermost part of the Cambrian sequence immediately above the basement was unaffected by the folding, while the main part of the Cambrian alum shales acted as a gliding horizon, and the soft shales were smeared out during the movement of the overlying sediments. BRØGGER (1882), by measuring a carefully mapped section along Oslofjord in Asker southwest of Oslo, found shortening to 0,43 of the original width. The section displayed folds in the 10 m thick Orthoceras limestone with over- and underlying incompetent shales; it measured 970 m across the strike, while the corresponding length along the folded beds was 2250 m. Sections from the higher parts of the sequence (in which the entire folds with closures cannot be seen) indicate a smaller degree of shortening, roughly 0,6 of the original width, to

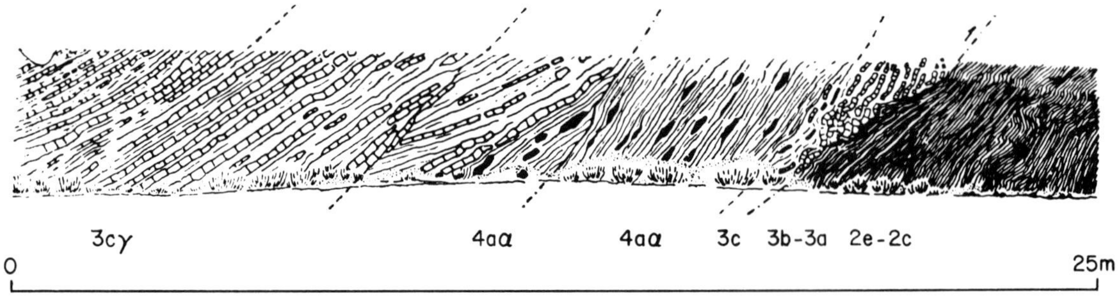

Fig. 4 *Intricately folded Cambrian and Ordovician sediments* in a road section south of Hamar, Mjøsa district. From SKJESETH (1963)

which must be added the effects of upthrusts. The large, open synclines in the Ringerike sandstone, the uppermost part of the sequence, are certainly less compressed, but this may to a great extent be compensated for by overfolds of considerable length in anticlines between the synclines, as is apparent from an observation of STØRMER (1934) at Stubdal, 30 km northwest of Oslo (Fig. 3).

In the Mjøsa district, the same relationships are found between the thickness and competency of the stratigraphic units and the size of the folds as are found in more southerly parts, but the folding is generally stronger and is more often isoclinal in the north. The 100 m thick Mjøsa limestone forms a number of synclines and synclinoria with the overlying Silurian deposits in the cores. The folds are coffer-shaped with steep limbs which may even converge upwards. The 15 m thick Orthoceras limestone forms strongly compressed and narrow isoclinal folds, while the shales in the lower parts of the sequence are intricately folded and thrusted so that very little of the original stratigraphy is left undisturbed (Fig. 4).

In the Mjøsa district the Oslo region joins the Sparagmite region without any sharply marked natural boundary (see map, Fig. 5). The Vemdal sandstone (the uppermost formation in the Sparagmite group) occurs here in an allochthonous position as the *Vemdal nappe*. It must be emphasized that the Cambrian deposits in the Mjøsa district occur in two different types of development and geologic setting. The Cambrian sequence resting directly above the Precambrian (Archean) basement has no Lower Cambrian deposits. A more complete Cambrian sequence is found above the Vemdal sandstone, containing some 15 m at least of Lower Cambrian shale and sandstone and conglomeratic layers of the ölandicus zone. Due to Permian faulting the rocks are exposed at very different erosional levels in the Mjøsa district. Just north of Gjøvik (on the west side of Lake Mjøsa) a thin autochthonous cover of Middle Cambrian alum shales above the Archean basement is

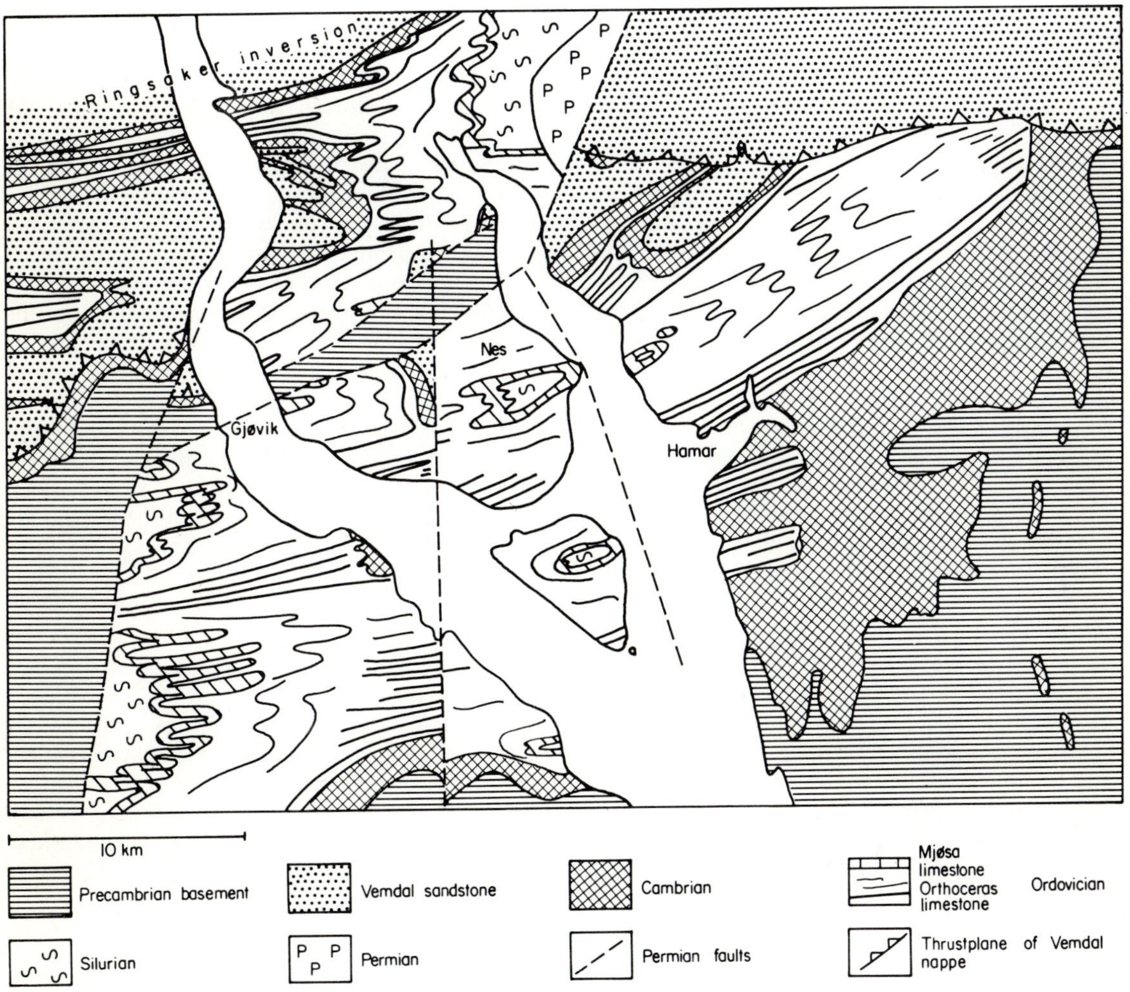

Fig. 5 *Geologic map of a middle part of the Mjøsa district.* After SKJESETH (1963)

overlain by the Vemdal nappe. In a syncline above the Vemdal sandstone a portion of the Cambro–Silurian sequence forms part of the nappe. South of Gjøvik is the Oslo graben; here, synclines of the Mjøsa limestone and overlying Silurian deposits are at the surface and the Archean basement is some hundreds of metres below. Now the observed (and minimum) length of thrusting of the Vemdal sandstone nappe is about 25 km, while the Cambro–Silurian rocks south of Gjøvik may have been moved as far as 100 km due to the shortening caused by the folding. It is thus apparent that the folded Cambro–Silurian sediments form one tectonic unit together with the allochthonous Vemdal sandstone. As the Cambro–Silurian pile of sediments was detached from the basement and folded together, the Vemdal sandstone, which was below these sediments in the northern parts, was moved at the same time and took a place above the part of the Cambrian sequence that was left on the basement (HOLTEDAHL, 1953). In the northern parts where the Cambro–Silurian sequence was underlain by the Vemdal sandstone, a nappe was formed, and in the southern parts there is a parautochthonous Cambro–Silurian sequence, folded by *décollement*. In this sense SKJESETH (1963) may be right in dividing the Cambro–Silurian sediments in the Mjøsa district into an allochthonous and an autochthonous part.

On the east side of Mjøsa and also in the Nes peninsula (south of the horst of Precambrian rocks) there are outcrops of Vemdal sandstone claimed by SKJESETH to be in a strictly autochthonous position. If this is really the case, there must be a horizon of *décollement* in the shales above the sandstone which separates the sandstone from the overlying folded Cambro–Silurian sequence. On the east side of Mjøsa allochthonous Vemdal sandstone rests above a folded Cambro–Ordovician sequence and is thus apparently in a higher tectonic position than on the west side, where the Vemdal sandstone rests on a thin cover of Cambrian shales just above the Precambrian basement.

Along great parts of the present Caledonide margin there are conditions corresponding to those displayed north of Gjøvik: a thin autochthonous sequence above the Precambrian basement which was apparently cut off by the overriding thrust. At one time there must certainly have been a brim of *décollement*-folded Cambro–Silurian along most of the Caledonide margin. It was, however, only in the Oslo graben and at Jämtland, Sweden, that this part of the orogen was spared from erosion.

Apart from overfolds of the type previously mentioned in Stubdal northwest of Oslo, no nappe structures have been found in the Oslo region. In Hadeland the Middle Ordovician deposits in two nearby areas showed great differences in facies. This suggested the presence of a nappe (STØRMER, 1944), but this has not been confirmed. At two localities in the same district 11 km apart the Middle Cambrian deposits display great differences in stratigraphy and fauna (STRAND, 1948). In one locality the deposits are strictly autochthonous above the Precambrian basement. In the other, which has a more complete sequence including the *Paradoxides ölandicus zone* (absent in the autochthonous sequence), the rocks are strongly folded, and are interpreted as being part of a nappe within the Cambro–Silurian sequence. It may well be that the transport involved in the folding was sufficiently long to account for the differences. SKJESETH (1963) explains these conditions by a local transgression in a graben or trough in the Precambrian basement, following the later Permian fault system.

The only possible unconformity registered in the Oslo Cambro–Silurian sequence is a faint one between the uppermost Ordovician and the lowermost Silurian. But there are indications of earth-movements in the central geosynclinal parts of the orogen as there are comparatively large amounts of terrigeneous material at a number of stratigraphic horizons. According to STØRMER (1965) the following five phases can be recognized in this way: Phase I (3a–3b) corresponding to the Trondheim phase; Phase II ($4a\alpha_4$–$4a\beta$) corresponding to the Pre-Bala or Pre-Caradoc phase in Wales; Phase III (5b, c–6a, b, c), corresponding to the Horg phase and to the Taconic phase; Phase IV (7c–8a, b), corresponding to the Pre-Wenlock phase in Wales; and Phase V (9g–10), corresponding to the Ardennic phase. An Ekne phase (4b–$4c\alpha$) is marked by a break in sedimentation, but not by an increased amount of material. Near Oslo, however, the contents of silt and sand in the sediments are found to have increased during late Ordovician time.

The Cambro–Silurian sequence in the Oslo region (including the Ringerike sandstone) is overlain by Lower Permian sediments with a distinct angular unconformity. The time of folding is thus known to be not earlier than early Devonian and not later than early Permian.

CHAPTER III

THE SPARAGMITE REGION IN SOUTHERN NORWAY

The Sparagmite region in southern Norway is occupied mainly by rocks of the Sparagmite group. To the south and southeast the boundary of the region is part of the Caledonian margin, where the allochthonous Vemdal sandstone rests above the thin autochthonous cover of the Precambrian basement, except in the Mjøsa district previously described. In the west the sparagmites are overlain by Cambrian and Ordovician deposits. Here there is an ill-defined boundary with the region of large crystalline nappes of southern Norway, the Nappe region. In the north and northwest the sparagmites are overlain by eugeosynclinal sediments of the Trondheim region. To the east the Sparagmite region continues on the Swedish side of the border, but in the southern part there is a sharp boundary on the Norwegian side along the Engerdal fault line (see Fig. 6). It may be appropriate to mention here that Sparagmite deposits in south Norway also occur outside the Sparagmite region, particularly in the Oppdal district west of the Trondheim region.

STRATIGRAPHY

As stated previously the thick sequences of mainly psammitic deposits in central southern Norway beneath the oldest parts of the fossiliferous Lower Cambrian are referred to as the Sparagmite group, a main lithostratigraphic unit (see Fig. 102, p. 131). There is good reason to believe that the thick series of deposits in the southern part of the Sparagmite region was laid down in troughs or basins, delineated by faults or flexures. The "Sparagmite basins", shown in Fig. 7, consist of a large western basin with deposits of the type section and a smaller eastern basin. In the areas around the basins deposition started with the very youngest Sparagmite deposits, or with fossiliferous Cambrian deposits.

In a synopsis of the stratigraphy of the Sparagmite group and corresponding deposits in Finnmark, Holtedahl (1961a) classed the upper part of the Sparagmite deposits, from the *Moelv formation* and upwards, as Eocambrian, i.e. earliest Cambrian. The older part of the Sparagmite group would therefore be latest Precambrian. (The term Eocambrian proposed by W. C. Brøgger was originally meant to apply to the time of deposition of the entire Sparagmite group.)

The stratigraphical type area of the parautochthonous Sparagmite sequence in the basins lies at the northern part of Lake Mjøsa, from which area the names of the stratigraphic units have been derived. A review of sparagmite stratigraphy in the Mjøsa district was given by Skjeseth (1963).

The *Brøttum sparagmite* is the lowermost formation exposed in the type section at Mjøsa. Due to its position in the basin the substratum cannot be seen. Near Lillehammer at the north end of Mjøsa a borehole was sunk 300 m through practically horizontal layers of the *Brøttum formation* without reaching the bottom. Taking into account the height of nearby hills, its thickness is thus at least 800 m. The Brøttum formation is a monotonous sequence of dark arkoses or greywackes alternating with shales or siltstones, commonly with graded bedding. It is widely distributed in the western Sparagmite basin where it becomes more fine grained, with shales and siltstones predominating in the northern parts. In the eastern basin it is sparsely represented and possibly absent.

The overlying *Brøttum shale* and *limestone* is thin and of local importance only.

The *Biri conglomerate* consists of well-rounded boulders of polygenous composition in a matrix of dark arkose (very similar to the arkose of the Brøttum formation) and is 150 m thick in its type area at Brøttum. Recent investigations have shown the Biri conglomerate to be a deltaic deposit occurring along the margins of the Sparagmite basins, thinning out towards the interior of the basins. In Gausdal (61° 12′, 10° 10′) Løberg (University thesis) mapped three "deltas" of Biri conglomerate within a distance of 20 km. Each "delta" was marked by a maximum boulder size

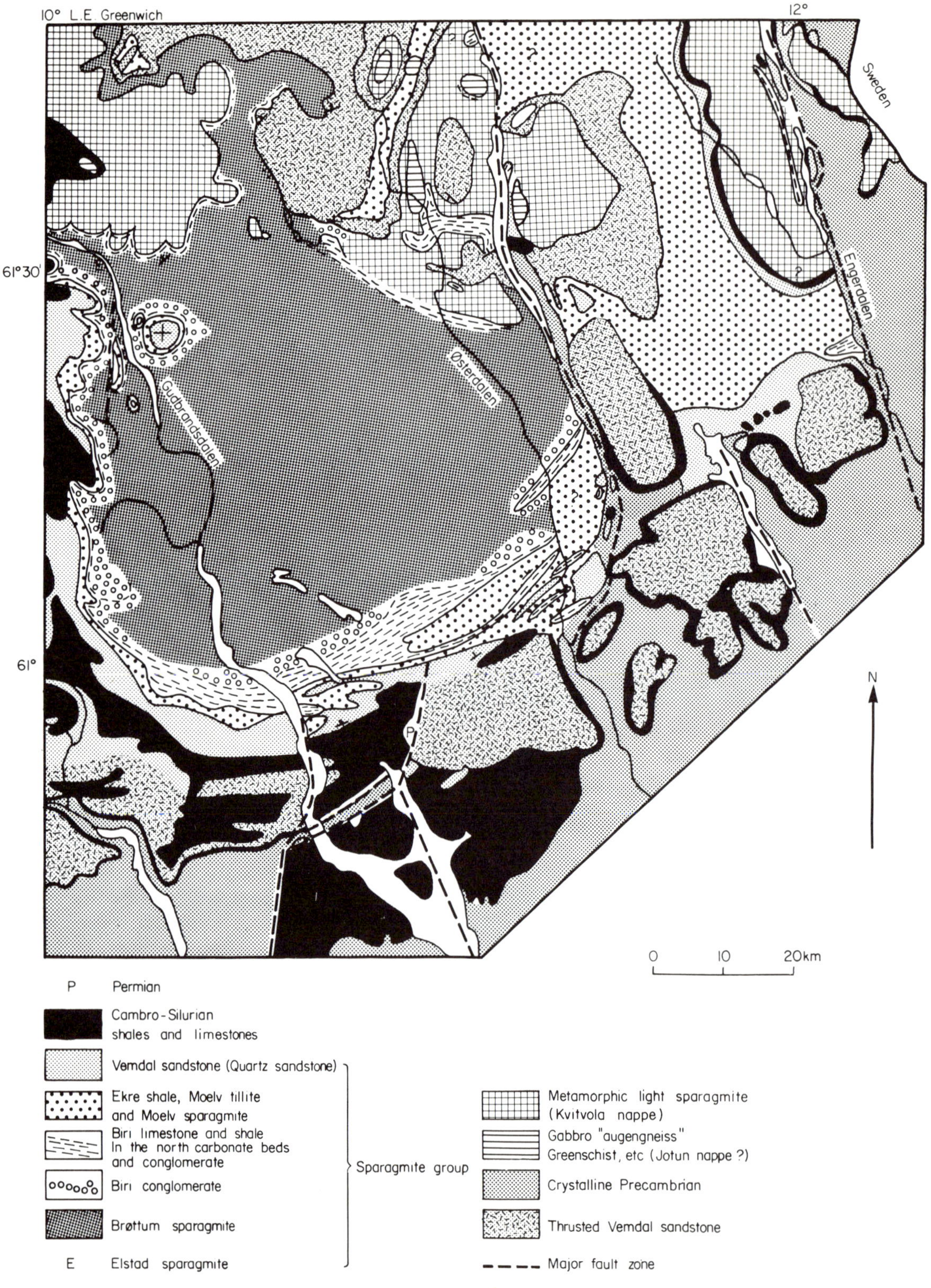

Fig. 6 *Geologic map of a southern part of the south Norwegian Sparagmite region.* By KNUT BJØRLYKKE

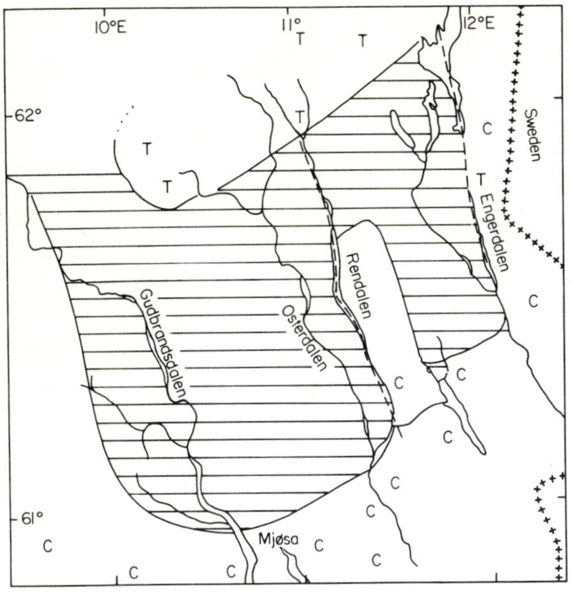

Fig. 7 *Map to show the position of the Sparagmite basins in south Norway. T—basal tillite, C—Cambrian sediments resting directly upon the Precambrian basement*

and conglomerate thickness rapidly decreasing away from the site of the maximum. The boulder material indicates transport towards the basins from the east, south and west. In a boulder of dark limestone in the Biri conglomerate, probably derived from a limestone deposited at the margin of the basin, SPJELDNÆS (1963) discovered a fossil organism of undetermined systematic position, to which was given the generic name Papillomembrana (Fig. 8).

There is generally a gradual transition from the Biri conglomerate to the overlying *Biri shale* and *limestone*, the basal layers of which are alternating shales and calcareous arkoses. The shale and limestone formation displays a large variation in facies, from black shales, partly with lenses of limestone, to impure and pure limestones or dolomites, often oolithic. In the type area at Mjøsa its thickness is more than 200 m, with shales predominating in the lower part, and limestones with intraformational conglomerates in the upper part.

The *Moelv sparagmite*, assumed to be about 250 m thick near Moelv, its type locality on the eastern side of Lake Mjøsa, consists here of coarse-grained arkoses and pebble conglomerates. Coarse grains of reddish feldspar in a light grey matrix give it a reddish colour. In the western basin the Moelv sparagmite in typical development is confined to the southern marginal parts, while in Gausdal, on the east side of the basin, its corresponding deposits are coarse-grained calcareous arkoses and limestones. In the eastern basin the Moelv formation is widely distributed and of great thickness,

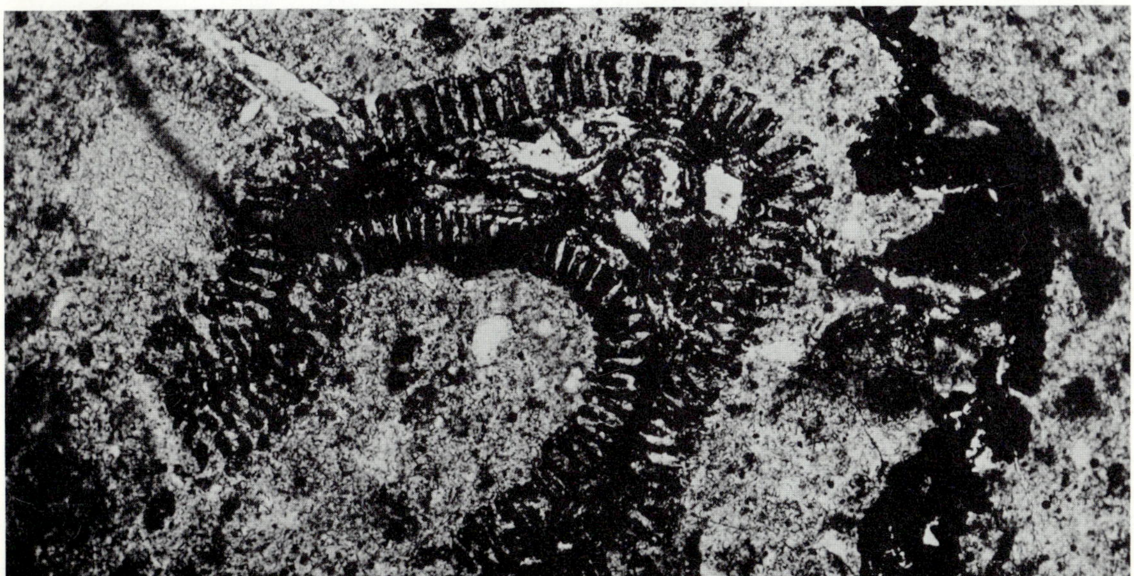

Fig. 8 Papillomembrana compta *Spjeldnæs, a fossil from a boulder of limestone in the Biri conglomerate,* Biskopåsen, northern Mjøsa district. Photo Professor NILS SPJELDNÆS

particularly near the eastern marginal fault of the basin.

The *Moelv conglomerate* which follows, fully conformably upon the Moelv sparagmite, is of tillitic character with boulders of all sizes, commonly up to 0,5 m, but up to 2 m in some localities, in an unsorted matrix lacking any sign of layering. It has a maximum thickness of more than 100 m. In parts of the eastern basin a conglomerate with well-rounded pebbles, the *Osdalen conglomerate*, intervenes between the sparagmite and the tillite. The boulder material in the Moelv tillite and in the Osdalen conglomerate contains a fair amount of Precambrian porphyries and *Dala sandstone*, indicating transport of material from the east.

In spite of its pronounced moraine-like appearance the Moelv tillite is concordant with the underlying waterlaid deposits and cannot therefore be adequately explained as a ground moraine deposited on land. It may be a water-laid moraine deposited under floating ice. Its wide distribution and lack of grading makes a turbidite interpretation less probable (cf. SPJELDNÆS, 1964). The "basal tillites", described subsequently, must have been formed in similar fashion. In some localities they rest on a substratum of arkose formed by weathering of underlying granite, which should have been removed by an advancing land ice.

The *Ekre shale* following above the Moelv conglomerate, is about 50 m thick in the type area at Mjøsa. It is a widely distributed, green or red often laminated or varved shale reaching maximum thickness in the eastern basin.

At the top of the Sparagmite group is a very characteristic sandstone occurring widely in southern Norway as well as in Sweden, to which different names have been given. The traditional name in Norway is *Kvartssandsten* ("Quartz Sandstone"). The oldest name, *Vemdal quartzite*, was given to it by A. E. TÖRNEBOHM in 1873. This will be used here as its formational name on both sides of the international border. At Mjøsa the Vemdal sandstone can be divided into two distinct parts: the underlying *Vardal sparagmite*, a rather coarse-grained arkose, and the overlying *Ringsaker quartzite*, a glassy orthoquartzite mostly dark in colour, forming benches of decimetre size. SKJESETH found Scolithus structures in the Vemdal sandstone in the Mjøsa district (Fig. 9). As far as is known pipe-rock structures of this kind have never been found in deposits older than Cambrian. In its stratigraphic setting the *Vemdal formation* is tied up with the overlying fossiliferous deposits. Unlike the older Sparagmite deposits it is not restricted to the Sparagmite basins, but extends widely outside the basins, as the younger Cambrian deposits.

According to the petrographic investigations of GRENDER (1962) on rocks from the Mjøsa type area, the sandstones of the Brøttum formation are medium-grained, poorly sorted greywackes, averaging 40% quartz, 35% feldspar and 25% micas

Fig. 9 *Pipe-rock with* Scolithus *and* Monocraterion *structures*. Vemdal sandstone, Furnes, Mjøsa district. Thickness of specimen across the layering is 13 cm. Photo Professor STEINAR SKJESETH

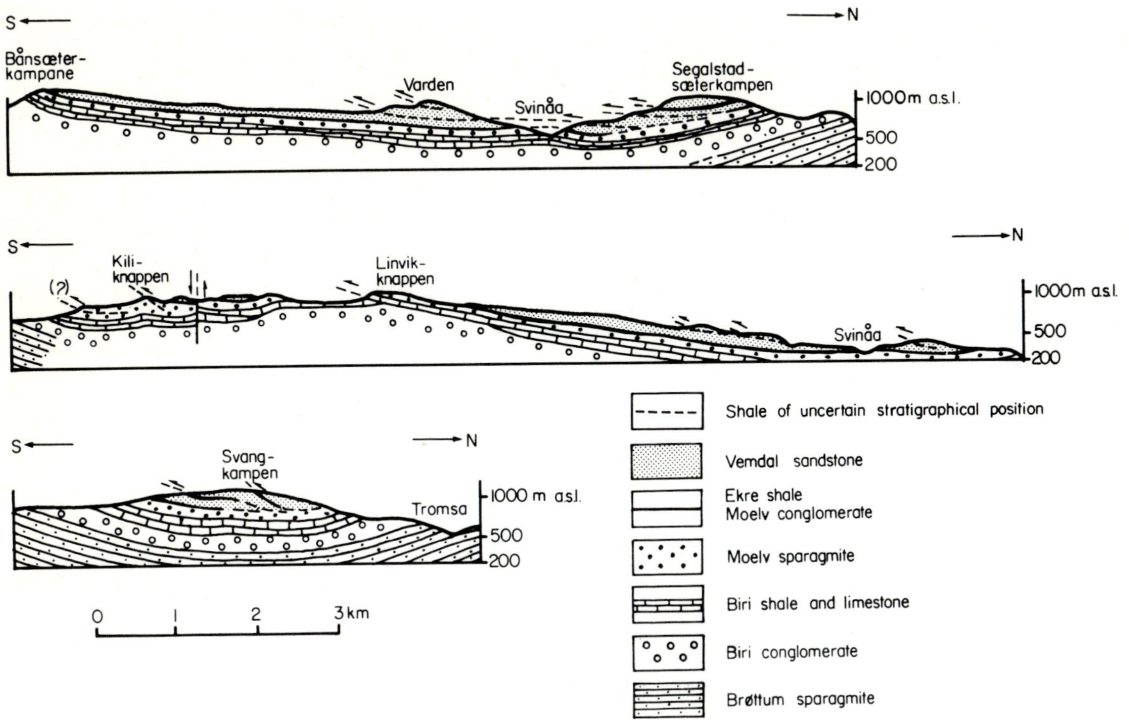

Fig. 10 *Sections across the Fåvang syncline.* By J. O. ENGLUND

and matrix. The *Moelv sandstones* are coarse-grained, poorly sorted arkoses with mean composition 55% quartz, 30% feldspar and 25% micas and matrix. The Ringsaker quartzite (the upper member of the Vemdal formation) is a coarse-grained, very well-sorted quartzite.

At Bjørånes in Østerdalen (61° 37′, 10° 54′) an autochthonous Sparagmite sequence is exposed in a window below a nappe of metamorphosed sparagmites (the Kvitvola nappe; OFTEDAHL, in HOLMSEN and OFTEDAHL, 1956; BJØRLYKKE, 1965c). Here the Vemdal sandstone (with a quartzitic Ringsaker member and an arkosic Vardal member) is underlain by Ekre shale and further by a thin tillitic Moelv conglomerate. The conglomerate has rock fragments usually not greater than 6 to 8 cm in a matrix of shale. Interpreted as an ice-dropped sediment, the conglomerate indicates a low frequency of floating icebergs and displays a facies characteristic of the interior parts of the Sparagmite basin. The conglomerate is underlain by a black shale, the *Bjørånes shale,* a deposit which can be correlated with the pre-tillite sparagmite formations.

In Gudbrandsdalen, Sparagmite deposits in a facies of the Mjøsa type section extend as far north as the Fåvang district (61° 25′, 10° 20′), which was recently mapped in detail by ENGLUND (1966) (Fig. 10). Here a formation of light grey arkoses, the *Elstad sparagmite,* occurs in an anticline below the Brøttum formation. In the Fåvang section a greywacke sandstone equivalent to the Moelv sparagmite occurs above the Biri limestone. This is overlain by a tillitic conglomerate containing boulders measuring 30 to 40 cm in an argillaceous or silty matrix, of maximum thickness 10 m.

SPJELDNÆS (1959) described boulders of Biri limestone with thin bands of recrystallized calcite in the Moelv conglomerate. The recrystallized bands were randomly orientated in the conglomerate, indicating that recrystallization had occurred before they were embedded in the conglomerate. In the same context it may be mentioned that a K/Ar determination on micas from the matrix of a specimen of Brøttum formation sandstone from the Mjøsa district indicated an age of 550 m. y. (BROCH, 1964). The micaceous matrix of the rock is recrystallized and its age should thus be that of the recrystallization. There are thus indications, certainly only vague ones, of a tectonic event before the deposition of the Moelv formation, which might possibly be correlated with the unconformity beneath the tillites in Finnmark.

The sparagmite basins are limited to the north-west by a number of windows with Archean rocks, extending as an anticlinal zone along the north-western and western boundary of the Sparagmite region. In the southern windows a tillitic conglomerate with boulders as large as 1 m rests upon the Precambrian basement, partly underlain by arkose. The "Basal tillite" (or basal conglomerate) can be correlated with the Moelv conglomerate. It is overlain by shale and sandstone representing the Ekre and Vemdal formations, and in some places also by black shales, probably corresponding to the fossiliferous Cambrian. The autochthonous sequence referred to measures from some tens of metres, up to 100 m. Similar sequences, with "Basal tillite" or younger deposits resting directly upon the Archean basement exist in the districts to the east, south and west sides of the Sparagmite basins and also in the area between the western and eastern basins.

In the Engerdal district (about 61° 40′, 11° 50′) a complex of rocks assumed to represent a northern facies of Sparagmite deposits occurs as an overthrust mass above fossiliferous Cambro–Ordovician deposits (the Kvitvola nappe, mentioned previously). HOLTEDAHL (1921) found that the lower part of the Kvitvola sequence was composed of sandstones with interlayered carbonatic sediments, both limestones and dolomites. This division was overlain by a conglomerate with boulders of quartzite, dolomite, granite and other crystalline rocks in a fine-grained, schistose matrix. The boulders were up to 0,5 m across, and often sparse and far apart. This is the *Koppang conglomerate schist*, interpreted as a tillite the boulders of which were transported by ice-rafting. The thickness of the Koppang conglomerate may be some tens of metres and at Høgberget is 15 m. Above the Koppang conglomerate are sandstones, most of them light coloured with some layers of reddish and greenish shales, together with a horizon of carbonate rocks and calcareous and dolomitic sandstones. The thickness of the whole Kvitvola nappe sequence is about 700 m.

In the Koppang district west of the Engerdal district HOLMSEN and OFTEDAHL (1956) described a sequence corresponding to that of the Kvitvola nappe. The type locality of the Koppang conglomerate schist is in the same district.

In the area between Kvam and Sjoa in Gudbrandsdalen (about 61° 40′, 9° 40′) unpublished investigations by the writer have shown the presence of a sequence of Sparagmite deposits very similar to that of the Kvitvola nappe (Fig. 11). The lowermost formation so far exposed consists of argillaceous and silty schists with interlayered quartzites, limestones and dolomites. This formation contains "conglomerate schists" with scattered boulders (maximum size 0,5 m) of the same type as the Koppang schist in the Kvitvola sequence. Overlying this formation is more than 500 m of light-coloured feldspathic sandstone. This usually has greenish streaks, due to layers enriched in muscovite. Sandstones of this lithologic type, the "light sparagmite" of Norwegian geologists, are widely distributed in the northern parts of the Sparagmite region. The "light sparagmite" in the section considered here is overlain by a formation of schists and calcareous sandstones, and above this by phyllitic schists and quartzites, the latter probably of Cambro–Ordovician age.

Still further north in the Dovre district, where the Gudbrandsdal valley transects the northern part of the Sparagmite region, the "light sparagmite" is underlain with an apparent conformity by a thick formation of pelitic schists and metagreywackes, known as the *Vardhø schists and gneisses* (STRAND, 1951a). Here the rocks are rather strongly metamorphosed and recrystallized, the typical Vardhø rocks consisting mainly of quartz, albite, epidote and micas. They must have been greywacke sediments rich in rather calcic plagioclase. Thin layers of greenschist have been found in the Vardhö formation, which thus shows the features of a eugeosynclinal type of facies.

As mentioned previously, a Cambro–Silurian sequence, largely corresponding to that of the Oslo region, is to be found above the Vemdal sandstone in the northern parts of the Mjøsa district. The same sequence together with the underlying Vemdal sandstone extends to the districts east and west of Mjøsa. The youngest deposits represented here are those of the Ogygiocaris series (4aα). When the Cambro–Ordovician deposits are traced west and northwest of Mjøsa there is an increase in the content of clastic material, and the Ceratopyge and Orthoceras limestones disappear. At Stokkbekken in Vesttorpa the deposits of the Ogygiocaris shale horizon contain layers of sandstone up to 6 m thick (MÜNSTER, 1900). To the north and west the rocks become feebly metamorphosed and the shales pass into phyllites.

The parautochthonous deposits in the northwest part of the western Sparagmite basin are overlain by Vemdal sandstone and by Cambro–Ordovician deposits, which have been investigated by H. C. SEIP in Gausdal and ENGLUND (1966) in Fåvang. Above the Vemdal sandstone are greenish shales and orthoquartzites, probably of Cambrian age. There are also black shales, measuring 120 m in Gausdal, with graptolites at two horizons. The graptolites were identified by B.-D. ERDTMANN, who found a lower horizon with Didymograpti, corresponding to 3b in the Oslo region and an upper horizon with Glyptograptus cf. teretiusculus, corresponding to 4aα. LAPWORTH (1905) described graptolites indicating the same horizons in Gausdal (from material collected by K. O. BJØRLYKKE). The graptolite shales are overlain by shales alternating with silty and sandy layers, measuring 680 m in Gausdal.

In the eastern part of the Sparagmite region, upheaval and erosion followed by the deposition

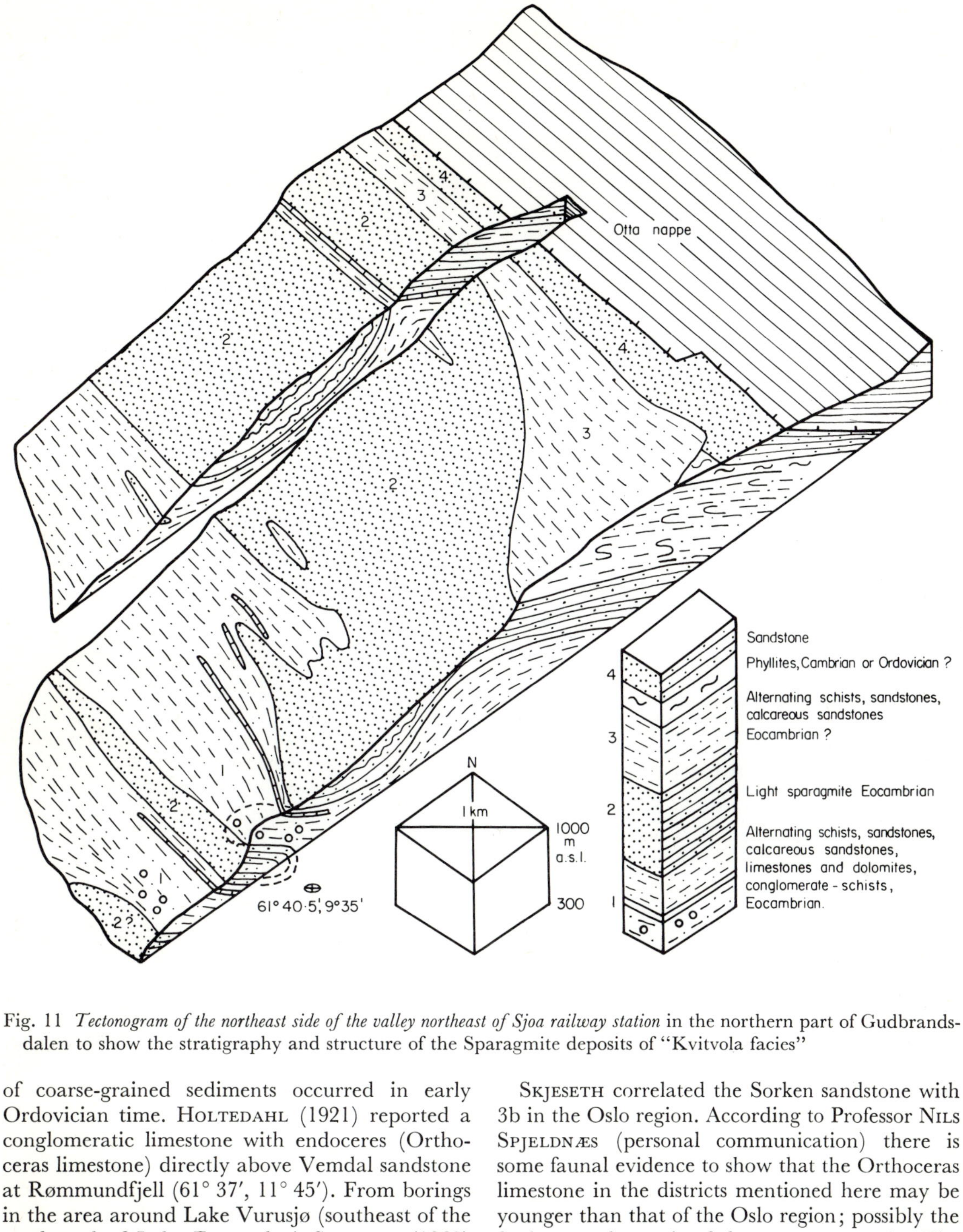

Fig. 11 *Tectonogram of the northeast side of the valley northeast of Sjoa railway station* in the northern part of Gudbrandsdalen to show the stratigraphy and structure of the Sparagmite deposits of "Kvitvola facies"

of coarse-grained sediments occurred in early Ordovician time. Holtedahl (1921) reported a conglomeratic limestone with endoceres (Orthoceras limestone) directly above Vemdal sandstone at Rømmundfjell (61° 37′, 11° 45′). From borings in the area around Lake Vurusjø (southeast of the south end of Lake Femunden) Skjeseth (1962) described a dark feldspathic sandstone up to 40 m thick with layers of conglomerate, overlain by Orthoceras limestone and shale corresponding to *étage* 4a. Black shales in the sandstone contained badly preserved graptolites. J. P. Nystuen (personal communication) has mapped the sandstone (the *Sorken sandstone*) on both sides of Lake Femunden (Figs. 12, 15).

Skjeseth correlated the Sorken sandstone with 3b in the Oslo region. According to Professor Nils Spjeldnæs (personal communication) there is some faunal evidence to show that the Orthoceras limestone in the districts mentioned here may be younger than that of the Oslo region; possibly the Sorken sandstone is of the same age as 4a. The upheaval and erosion thus recorded was called the *Trysil disturbance* by Vogt (1928). A sandstone in the area east of Lake Ossjöen (about 61° 20′, 12°) was considered as Vemdal sandstone on the basis of lithologic similarity, but Ordovician age was suspected when P. Holmsen detected an orthocer in a loose boulder of the sandstone. Boring has quite recently shown that about 100 m of the sandstone

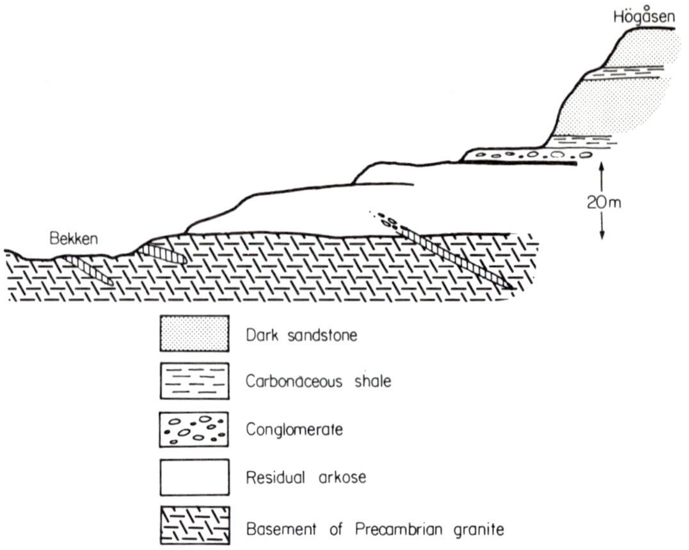

Fig. 12 *Section in Tufsingdalen west of Lake Femunden.* From G. Holmsen (1935)

rests with a normal stratigraphic contact upon Cambrian alum shales. Conglomerates in the sandstone contain fragments of feldspar and granite as large as 4 cm, and vertical movements (on faults?) must have occurred before the deposition of the sandstone (Holmsen, Skjeseth and Nystuen, 1966).

GEOLOGIC SETTING

The Vemdal nappe extends along the Caledonian margin from Hemsedal in the west to Trysil in the east. All the sparagmite rocks occurring west of the western sparagmite basin belong to the Vemdal formation (the underlying Ekre shale may also be represented). As was mentioned in the stratigraphic description, the Vemdal sandstone transgressed widely outside the Sparagmite basins together with the overlying Cambro–Silurian deposits. During the orogeny the Vemdal sandstone was detached from its basement (the underlying incompetent layer of Ekre shale may have made such detachment easy) and moved to the southeast. Outside its area of deposition, the Vemdal sandstone came to rest upon the thin cover of Cambrian shales left in place after the *décollement* and folding of the overlying Cambro–Silurian sequence (see Chapter II, Fig. 5).

The Vemdal nappe in Norway was mapped and described for the first time by Schiøtz (1902) in the district surrounding Lake Ossjøen (61° 15′, 11° 50′). Here, Vemdal sandstone extends for a distance of 60 km towards NNW at the east side of the Rendalen valley; the nappe is known to rest upon underlying fossiliferous Cambrian deposits over a distance of 30 km. West of Lake Mjøsa, in the northern part of Lake Randsfjord, the Vemdal sandstone was shown to rest upon Middle and Upper Cambrian alum shales by Holtedahl (1915). Further west, in the Aurdal area (about 60° 55′, 9° 30′), the nappe extends for more than 20 km in a NNW direction above Cambrian shales (Strand, 1954). Here the allochthonous Vemdal sandstone is overlain by Cambro–Ordovician sediments, which are themselves overlain, with a tectonic boundary, by a sandstone indistinguishable from the underlying Vemdal. Evidently there is a duplication of the nappe here.

Even though the Vemdal nappe taken as a whole forms a horizontal sheet above the Precambrian basement (separated only by the thin cover of Cambrian shales) the rocks of the nappe are seldom horizontal, but are in fact strongly folded, as the overlying Cambrian and Ordovician deposits are preserved in a number of synclines in the sandstone. Schuppen structure is very common in the sandstone and often the soft shales underlying the nappe have been squeezed in between the beds of the sandstone and detached lens-shaped pieces of it are often found in the underlying shales.

At Lake Mjøsa the Vemdal sandstone was deposited on top of a thick sequence of the older sparagmites. Here the nappe is strongly compressed together with the underlying rocks and displays a structure of isoclinal folds with steeply dipping axial planes, passing into schuppen. Thus the Vemdal nappe has a small extension in the N–S direction at Mjøsa.

The nature of the tectonics at the northern part of Lake Mjøsa is readily explained by the assumption that the rocks at this point are at the southern margin of the *Sparagmite basin* (i.e. the depression

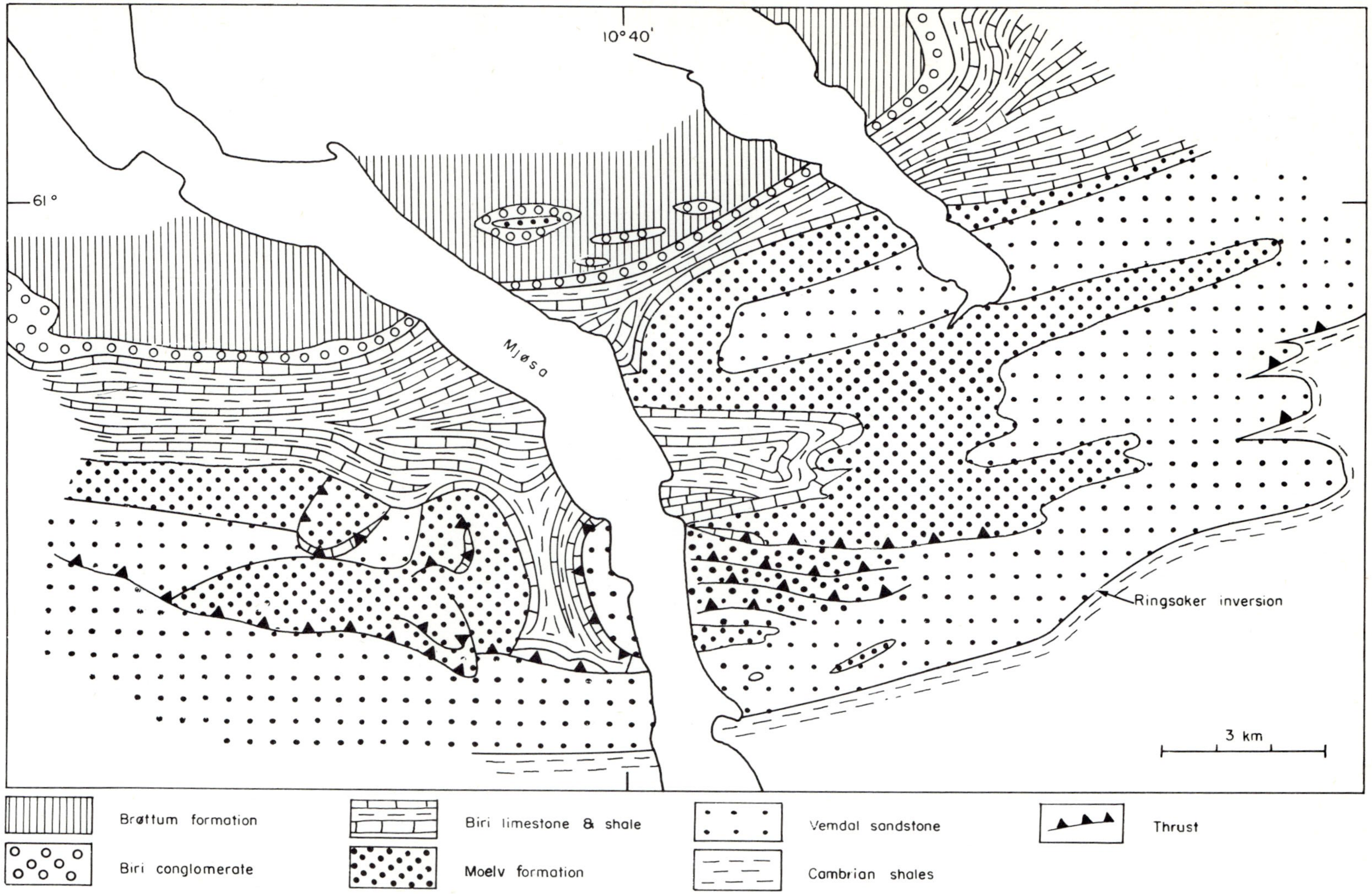

Fig. 13 *Geologic map of a northern part of the Mjøsa district*, to show the tectonics at the south margin of the Sparagmite basin. After detailed maps by D. HUSEBY (west of Mjøsa) and L. KIRKHUSMO (east of Mjøsa). The map is a continuation to the north of the map in Fig. 5

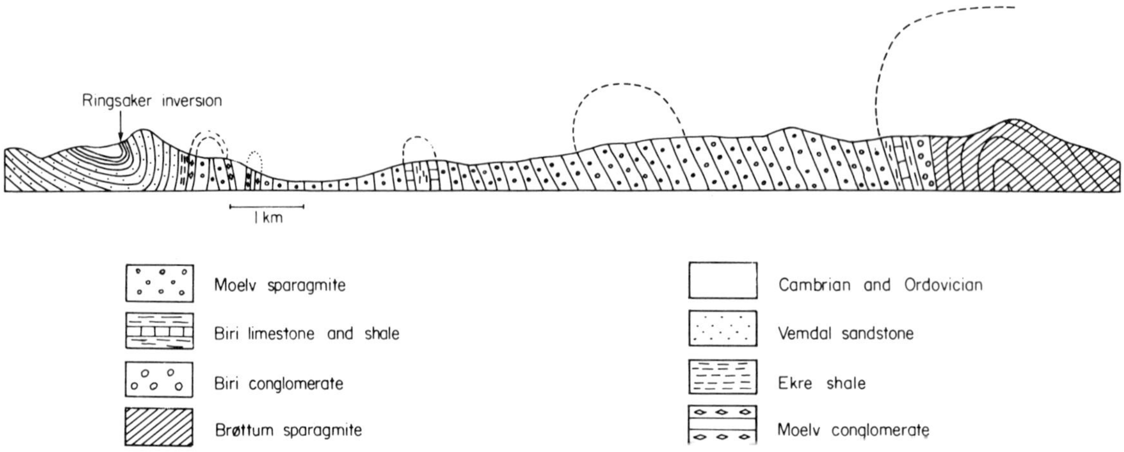

Fig. 14 *Section by* Knut Bjørlykke, *to show the folding of the Sparagmite deposits near the margin of the western Sparagmite basin. South of Rena, Østerdalen (61° 8′, 11° 22′)*

in the Precambrian basement filled with sparagmite deposits). The rocks met with an obstacle when they were moved against the margin of the basin, and were laid into sharply compressed folds. On the west side of Lake Mjøsa, a sheet of Biri limestone and Moelv sparagmite was moved above Vemdal sandstone for an observed distance of 4 km. An anticlinorium was thus formed along the southern margin of the Sparagmite basin (the *Moelv anticline* of Skjeseth (1963)). On the south side of the anticlinorium the Vemdal sandstone was inverted, together with the overlying Cambrian deposits, along the Ringsaker inversion line, extending on both sides of Lake Mjøsa (Figs. 13, 14).

The folding, so strong at the southern margin, becomes less intense towards the interior of the basin. Along the western margin of the basin in Gausdal and Fåvang (61° 25′, 10° 10′) further north, the deposits are gently folded, or almost horizontal (Fig. 10). The eastern margin of the basin along Rendalen follows a fault line, most probably of Permian age. On the west side of the fault are thick piles of Brøttum and Moelv sparagmite, and on the east side in the valley slope are Precambrian basement rocks, overlain by Cambrian shales and the Vemdal sandstone nappe. A reasonable interpretation of these conditions is that the fault-line along the valley coincides with the eastern boundary of the (western) Sparagmite basin. Skjeseth (1963) interpreted the structure as a sinistral wrench fault associated with overfolding to the south of the thick pile of older sparagmites at the west side of the valley.

In the eastern and smaller Sparagmite basin the Moelv sparagmite reaches a thickness of 1000 m and is overlain by the Osdalen conglomerate and the *Moelv tillite*. The Brøttum formation is unknown in the eastern basin, while the Biri limestone is found at Jordet in Trysil (61° 25′, 12° 10′). The eastern basin has a fault-line boundary along Engerdalen, to the east of which the Precambrian basement is directly overlain by Cambrian deposits, and in a few localities by Moelv tillite. The Moelv sparagmite in the eastern basin thins out in the west towards Rendalen. The Moelv deposits extend in a northerly direction to the area around Lake Femunden (Fig. 15). Here the depositional contact between the sparagmite and the Precambrian basement is exposed in a number of localities in upthrusted sheets of basement rock (Holmsen, 1935, 1937).

The Kvitvola nappe has its type area at Høgberget (61° 40′, 11° 50′), where Cambrian and Ordovician deposits, overlying the autochthonous sparagmites, are covered by a thick pile of light sparagmites with layers of dolomite and the characteristic "conglomerate schist" of the Kvitvola sequence, previously described.

The light metamorphosed sparagmites of the Kvitvola nappe extend in the north to the areas on both sides of Lake Femunden where they overlie much less metamorphosed red-coloured sparagmites, which are extensions of the Moelv sparagmite in the eastern basin. In Tufsingdalen west of Femunden the sparagmites of Kvitvola type overlie sediments of an autochthonous sequence (Fig. 12).

In the areas west of Lake Femunden Holmsen (1950) and Oftedahl (1950b) mapped and described thick masses of predominantly light sparagmites above the autochthonous deposits in the two large windows (with Precambrian basement rocks) (the Brydalen window in the northeast and the Atnasjø window in the southwest). "Dark sparagmites", tillites and "conglomerate schist" are also found here in the allochthonous sequence.

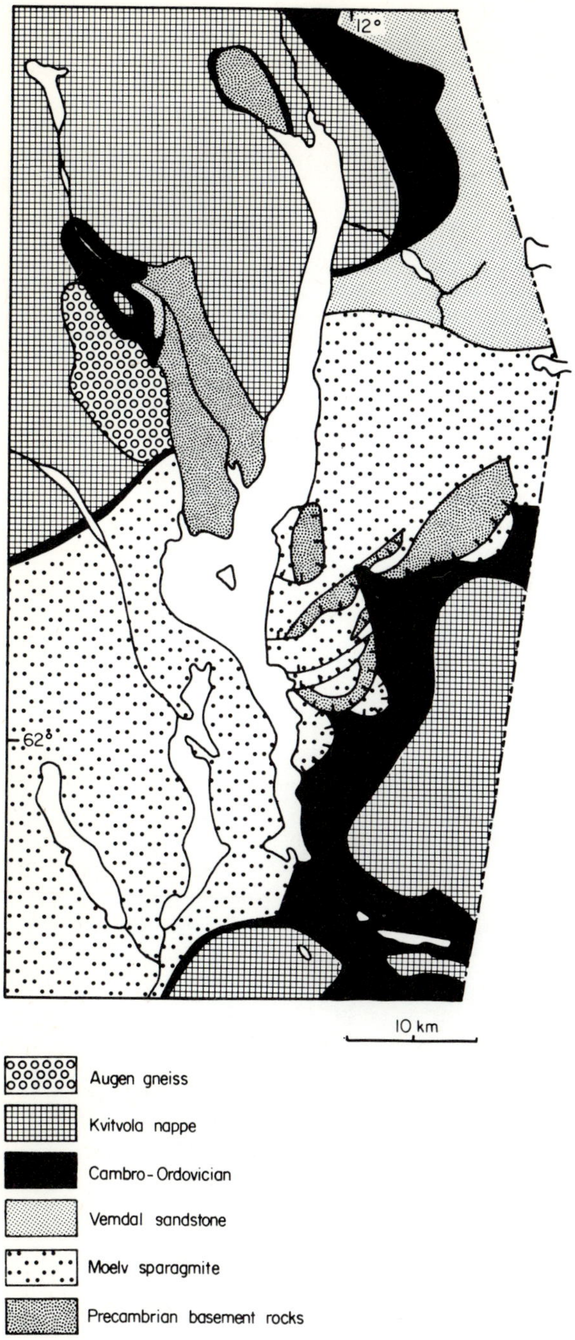

Fig. 15 *Geologic map of the surroundings of Lake Femunden*, based on the maps of G. Holmsen (1935, 1937) and recent work by J. P. Nystuen

Similar conditions exist in the *Snødøla window* (Werenskiold, 1911, Prost, 1964) where basal tillite, Ekre shale and Vemdal sandstone are the autochthonous deposits above the Precambrian basement and are overlain by dark schists (the "upper dark sparagmite" of Werenskiold) and light sparagmite; the last two are considered as allochthonous (Fig. 16). In Gudbrandsdalen, Prost (personal communication) found a tectonic boundary between the Sparagmite deposits of "basin facies" in Fåvang and the rocks of "Kvitvola facies" further north.

In the Sparagmite region one can thus distinguish deposits of "basin facies" largely in an autochthonous or parautochthonous position. The "basin facies" deposits are represented chiefly by Moelv sparagmite in the northeast and by Vemdal sandstone in the northwest. The large masses of light sparagmite and other deposits of "Kvitvola facies" are allochthonous. These rocks form the *Rondane mountains* west of the large *Atnasjø window* (Oftedahl, 1950a).

In Østerdalen (about 61° 45′, 9° 12′) four outliers of greenschist, gabbro, granitic gneiss, augen gneiss and anorthosite are found above the rocks of the Kvitvola nappe (Holmsen and Oftedahl, 1956). They are regarded as equivalents of the Otta nappe in Gudbrandsdalen.

The Kvitvola nappe is equivalent to the Särv nappe on the Swedish side of the border (Chapter XXI).

The map in Fig. 6, due to Bjørlykke (1966), gives a synopsis of the stratigraphy and tectonics of a major part of the Sparagmite region, at the present state of knowledge. A new feature shown on the map is the wide distribution of Vemdal sandstone equivalents in the northern parts of the region, in an autochthonous or parautochthonous position, beneath rocks belonging to the Kvitvola nappe. For the Østerdalen district the map is based upon the work of Holmsen and Oftedahl (1956) with reinterpretations resulting from the recent work of Bjørlykke. Asklund (1961) in a discussion of the tectonics of the Sparagmite region advanced the theory that the Precambrian rocks in the windows were allochthonous and belonged to the Vemdal nappe together with the overlying sediments. There is, however, nothing to suggest that the rocks in the large Precambrian

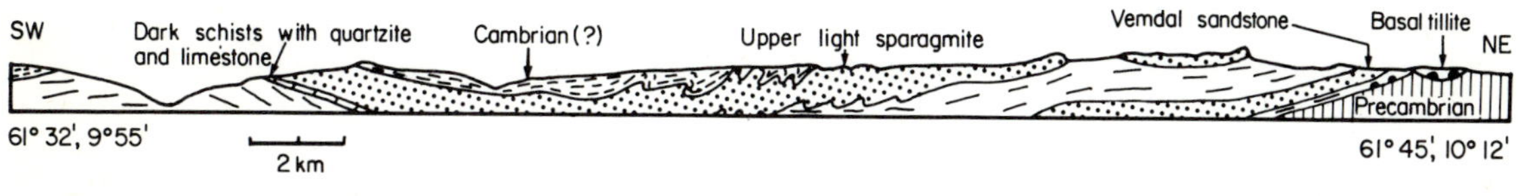

Fig. 16 *Section from the Snødøla window in the northeast to the Gudbrandsdalen valley in the southwest.* After W. Werenskiold (1911)

Fig. 17 *Augen gneiss formed by porphyroblastic growth of potash feldspar* in "light sparagmite". Høvringen, northern part of Gudbrandsdalen

a The ledge shown in the picture is some 20 cm thick

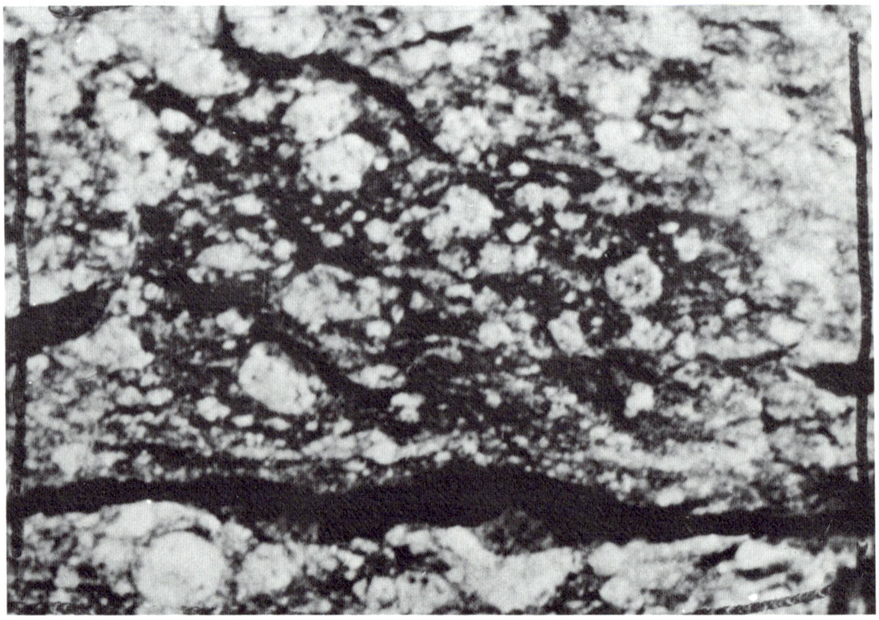

b About one-third natural size

windows of the Norwegian Sparagmite region are not autochthonous.

North of Lake Femunden the row of windows of Precambrian rocks takes a northerly direction along the international border. In the southern part of this area the basement is overlain by autochthonous Eocambrian or younger deposits (dark quartzites). There are also allochthonous sparagmite rocks. Detailed mapping was carried out in the Skardørfjell district at 62° 50′ N by SCHAAR (1962). Here a thin sequence of sparagmitic rocks with sills of diabase, most of them altered to amphibolites, rests with a tectonic contact above the Precambrian basement. SCHAAR's elaborate petrographic work on the sediments showed that they were mainly meta-arkoses with more than

25% feldspar. The same rocks on the Swedish side form part of the Särv nappe. The sparagmite rocks thin out to the north; BRYN (1959) could not find any such rocks in the area north of Skardørfjell.

A prominent tectonic feature in the Sparagmite region is the thrusting of large nappes over considerable distances (e.g. the Vemdal sandstone nappe and the Kvitvola nappe). Folding in the western sparagmite basin has been previously described. Folding on E–W or ENE axes is prevalent in the southern parts. In more northerly parts of the Sparagmite region cross folding on NW or WNW axes is the dominant type of tectonics (Fig. 11). In the northern part of Gudbrandsdalen large cross folds with axial planes dipping NE are overfolded to the southwest for a few kilometres. There is a well-marked lineation structure in the axial direction of cross folding and boulders in conglomerates are elongated in the same direction (OFTEDAHL, 1948, 1950a).

The northern parts of the region show fine examples of schuppen structures with a uniform northerly dip over wide distances, such as in the Rondane mountains (about 62°, 9° 45′) (OFTEDAHL, 1950). At Høvringen west of Rondane, STRAND (1951a) described upthrusts repeating the rocks of the Otta nappe, here involved in the schuppen tectonics together with the sparagmites underlying its thrust plane. A fine illustration of this type of tectonics is to be found near Lake Femunden, where HOLMSEN (1935, 1937) described a great number of upthrusts which had brought up slices of the Precambrian basement with overlying sparagmite deposits (Fig. 15). In some cases the upthrust sheets are inverted indicating that some folding must also have taken place. Similar upthrust tectonics have been described further south at the Mistra river (61° 42′, 11° 15′) and are also very well displayed near Andrå on Lake Storsjøen, 30 km south of Mistra. Here rocks from different stratigraphic and tectonic units—fossiliferous Cambro-Ordovician sediments, augen gneisses and Precambrian rocks—have been shuffled together like a pack of cards (HOLMSEN and OFTEDAHL, 1956).

In the southern parts of the Sparagmite region the rocks have not been affected by metamorphism, and the sandstones contain well-preserved clastic structures. Farther north the pelitic rocks are faintly metamorphosed with newly-formed sericite and chlorite, while the sandstones still preserve the clastic structures, except on zones of mylonitization. As described by OFTEDAHL (1950a), north of Rondane the sparagmites have a recrystallized matrix partly of fine-grained quartz and sericite with large grains of feldspar, partly of preserved clastic grains, and partly of newly-formed porphyroblasts. Chemical analyses published by OFTEDAHL show high potash contents and unusually low soda contents, which seem to indicate an enrichment in potash during metamorphism. Pegmatite veins in the sparagmites contain quartz, microcline (very low in soda) and hematite.

Augen gneisses formed by porphyroblastic growth of feldspar—mainly potash feldspar—in the sparagmites are common in the northern parts of the Sparagmite region and tend to be located near planes of tectonic movement. The augen, often of walnut size or larger, are single crystals or twins of microcline, or they are composites, consisting of granite-like clusters of feldspar with some quartz (Fig. 17).

SCHAAR (1962) described augen gneisses occurring as intercalations in the sparagmites of the Skardørfjell district. The rocks have a granitic, or more exactly an adamellitic composition and they are interpreted by SCHAAR as granitic rocks from the Precambrian basement, tectonically intercalated in the sparagmites. There is no reason to doubt this interpretation in this case, but it is worth stressing that such an interpretation cannot be of general validity for augen gneisses in the sparagmites. In very many cases a mere field inspection is enough to confirm that these rocks were formed by porphyroblastic growth of feldspar in a sedimentary matrix.

Like the Cambro–Silurian sediments of eastern facies, the rocks of the Sparagmite region are generally devoid of intrusive rocks. At Hummelfjell (62° 25′, 11° 15′) light sparagmites are transected by dolerites of the same type as those found in the Särv nappe in Sweden (HOLMSEN, 1950). At Høvringen in Gudbrandsdalen (61° 52′, 9° 35′) two small bodies of dolerite with partially well-preserved sub-ophitic structure were found as intrusive rocks in the *Vardhø formation* (STRAND, 1951a; see also GJELSVIK, 1952). The dolerites of Høvringen do not resemble the common type of Caledonian intrusive rocks in the same district. GJELSVIK pointed out their similarity to the dolerites in the Gneiss region.

CHAPTER IV

THE NAPPE REGION IN CENTRAL SOUTHERN NORWAY

The Nappe region in central southern Norway extends from the Sparagmite region in the northeast to the fjord districts of Hardanger and Rogaland in the southwest. It joins the Trondheim region across the northern part of Gudbrandsdalen. The name "Nappe region" is appropriate as nappe[1] structures are clearly apparent; sheets of highly crystalline rocks rest upon feebly metamorphosed late Precambrian–Eocambrian and Cambro–Ordovician sediments of eastern facies. The boundaries of the Nappe region to the northwest and southeast are erosional. The roots of the nappes are probably in the northwest.

The Nappe region can be divided into two main parts: one in the northwest extending from Gudbrandsdalen in the northeast to Hardangerfjord and the districts southeast of Bergen, and a smaller southeastern one consisting of Hardangervidda and the Rogaland fjord district.

In the northwestern area, three nappe units are distinguished in the text. In the northeast is the *Otta nappe* consisting of a basement complex of charnockitic rocks and an overlying eugeosynclinal sequence of sediments and volcanic rocks. The writer has hitherto considered the Otta nappe to be below the large *Jotun nappe*, an interpretation that may now be in need of revision (dealt with later in this chapter). The Jotun nappe, consisting of charnockitic and other crystalline rocks without any sedimentary cover, extends for 180 km, from the Otta valley in the northeast to Vossestrand in the southwest. The inner arms of Sognefjord transect the rocks of the nappe. In its southwesterly part the Jotun nappe overlies the Bergsdalen nappes which consist of sediments, volcanic and intrusive rocks of assumed Precambrian age. Some of these rocks are in contact with eugeosynclinal sediments and volcanic rocks on the northwest side of Hardangerfjord (described in chapter VI). These sediments must be considered as allochthonous, and as probably forming part of the Bergsdalen nappes. In the southeastern area there is, as far as is known, only one nappe unit, called the *Hardanger nappe* (Fig. 18).

The basement of gneisses underlying the sediments and nappes was deformed during the orogeny, probably after the emplacement of the nappes. The fairly straight line that delimits the Nappe region to the northwest follows the northwest margin of a trough in the basement, in which the rocks are bent down. The southeast margin of the trough can be observed in the Aurland window at Sognefjord. This structure was described by GOLDSCHMIDT (1912b) as the "Faltungsgraben". Outside the trough the nappes rest on a less strongly deformed basement, the ups and downs of which are reflected in the dissected outcrop patterns.

[1] The term nappe is used here in a broad sense for rock complexes of great extent moved for considerable distances above a substratum, and thus includes both thrust sheets and fold nappes. This conforms to the use of the term in Norwegian geological literature (the equivalent term in Norwegian is dekke). The term *mylonite* is likewise used in a broad sense. Many of the rocks designated thus are recrystallized and might be called mylonite schists, gneisses, or phyllonites.

STRATIGRAPHY OF THE DEPOSITS BENEATH THE NAPPES

For the most part the nappes rest upon Cambro–Ordovician deposits of eastern facies type (including the deposits of the Valdres group), but in some areas upon deposits of the Sparagmite group. As mentioned previously (chapter III) the Cambrian and Ordovician deposits at the northern part of Lake Mjøsa can be traced to the west and northwest, showing a change in facies and commencement of metamorphism. Here the Cambro–Ordovician deposits overlie the Vemdal Sandstone of the Sparagmite group.

In the *Valdres district* (about 61°, 9° 20′) (STRAND, 1938; 1951b; 1959) the Lower Cambrian deposits above the Vemdal sandstone are thicker and more coarse grained than further south, and consist of alternating shales, siltstones and sandstones of thickness perhaps 200 m. The presence of Middle

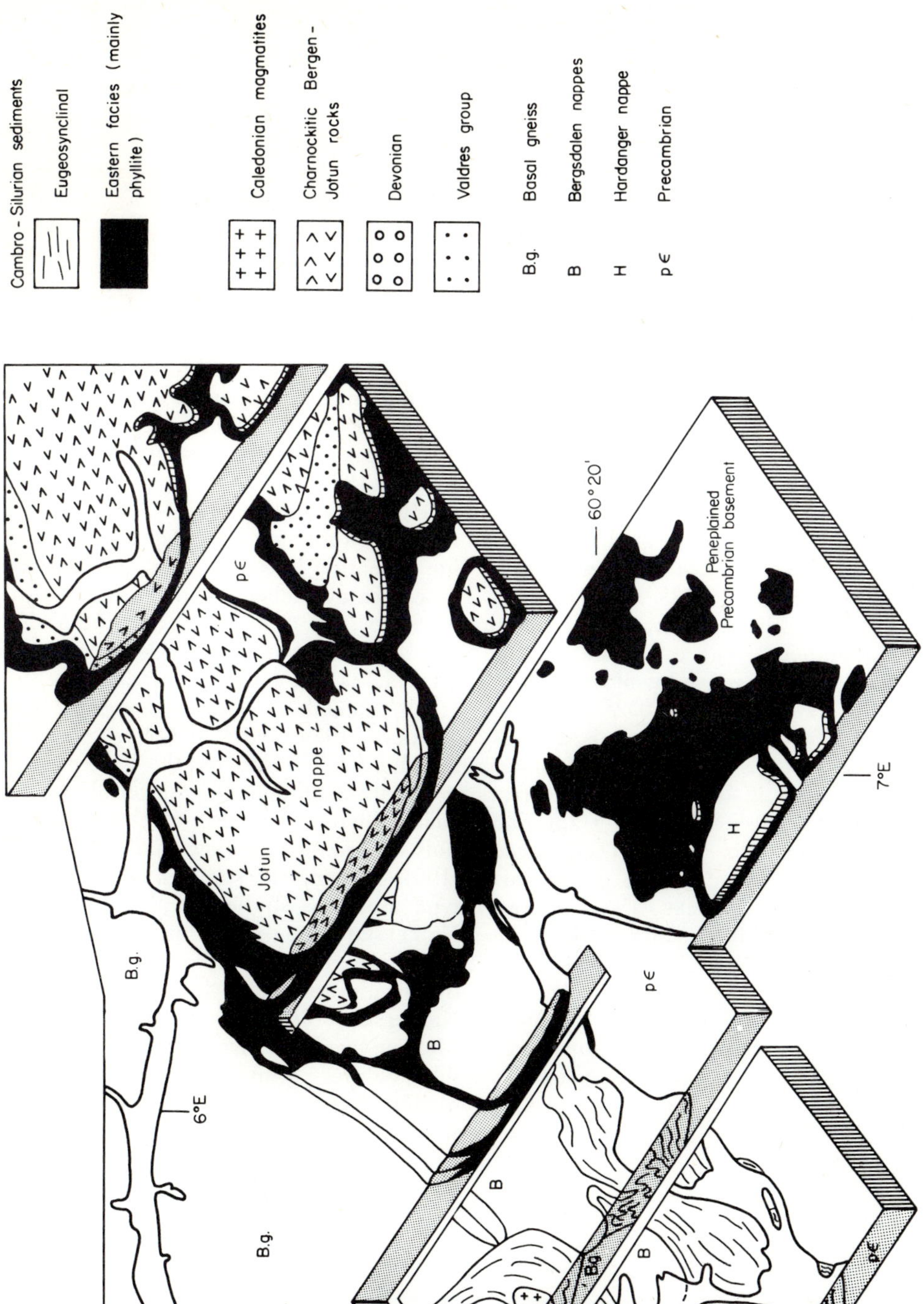

Fig. 18 *Tectonogram of a part of the Nappe region and the Sunnhordland–Bergen district.* Highly schematic

and Upper Cambrian alum shales and shales of lowermost Ordovician (3a) has been testified for by fossils. Corresponding to the graptolite shales of 3b to 4a in the Oslo region is a monotonous sequence of grey phyllites (the *Phyllite formation*), perhaps 300 m thick, which has yielded graptolite faunas with Phyllograptus and Didymograptus. The grey phyllites cover a wide area in central southern Norway, and are found at a corresponding horizon in Hardangervidda. The deposits of the Valdres group, overlying the Ordovician phyllites in Valdres, will be dealt with at the end of the section.

Cambro–Ordovician deposits in an autochthonous or parautochthonous position directly above the Precambrian basement are found in districts along the Oslo–Bergen railway (about 60° 30′, 8°) and further to the south and southwest in Hardangervidda and the districts around Boknfjord, see Fig. 103, p. 132. Thin Lower Cambrian basal deposits with Torelella lævigata and other fossils are known in localities near the Bergen railway (GOLDSCHMIDT, 1912a, 1925 and STØRMER, 1925). An early Middle Cambrian fauna with Paradoxides oelandicus (1cα) was found in the lower part of the sequence in the Jøsenfjord district (59° 18′, 6° 25′) (HENNINGSMOEN, 1952). In Hardangervidda (about 60° 10′, 7° 10′) there is, according to BRØGGER (1893), 40 to 50 m of black, carbonaceous phyllites in the lower part, corresponding to the Middle and Upper Cambrian alum shales in the Oslo region. Dictyonema flabelliforme has been found in the upper part of the black phyllites (STØRMER, 1941). Above the black phyllites is a 40 m thick dark bluish quartzite and above that a 10 m thick limestone, corresponding perhaps to the Orthoceras limestone (3c) in the Oslo region. The uppermost part of the sequence below the nappe rocks consists of 150 to 200 m of grey phyllites (Fig. 19).

Apart from thin layers of bentonite, volcanic intercalations are unknown in the Cambro–Silurian deposits of eastern facies. An exception to this is the occurrence of tuffitic (with grains of basic plagioclase) and andesitic (with traces of flow structures) layers in the phyllites in the fjord districts south of Hardangervidda as recorded by ELLEN S. KILDAL and KJELL A. TJØLSEN in unpublished theses of the University of Bergen.

The phyllites and other parts of the sequence beneath the nappes extend to the areas west and north of the Bergen railway. At Bøvertun (61° 38′, 8° 5′) on the northwest side of the "Faltungsgraben" rather thick and pure limestones occur together with dark sandstones and phyllites. It is possible that these deposits belong to the Sparagmite group, as Sparagmite deposits occur beneath the nappes northeast of Bøvertun (cf. BANHAM and ELLIOT, 1965).

The term *Valdres group* (STRAND, 1964a) was introduced to denote the deposits above the Phyllite formation in central southern Norway. In the type area at Mellene in the Valdres district (61° 5′, 9° 15′) the group comprises the Mellsenn formation and an overlying thick sequence of arkoses, greywackes and conglomerates, which BJØRLYKKE (1910) named the *Valdres sparagmite.* The writer (STRAND, 1959, 1962) in accordance with BJØRLYKKE (1905), considered the Valdres group to be in a parautochthonous position above the Ordovician phyllites and thus of Ordovician and possibly Silurian age, a view challenged by KULLING (1961). During the summer of 1965 the Mellene type area was studied in detail by Dr. R. P. NICKELSEN who worked out the structures, and LOESCHKE (1967a, 1967b) who worked out the stratigraphy and lithology. The studies of NICKELSEN (1967) give strong evidence that the Mellene section is inverted and also indicate that the Valdres group rests on the underlying phyllites with a strong tectonic unconformity

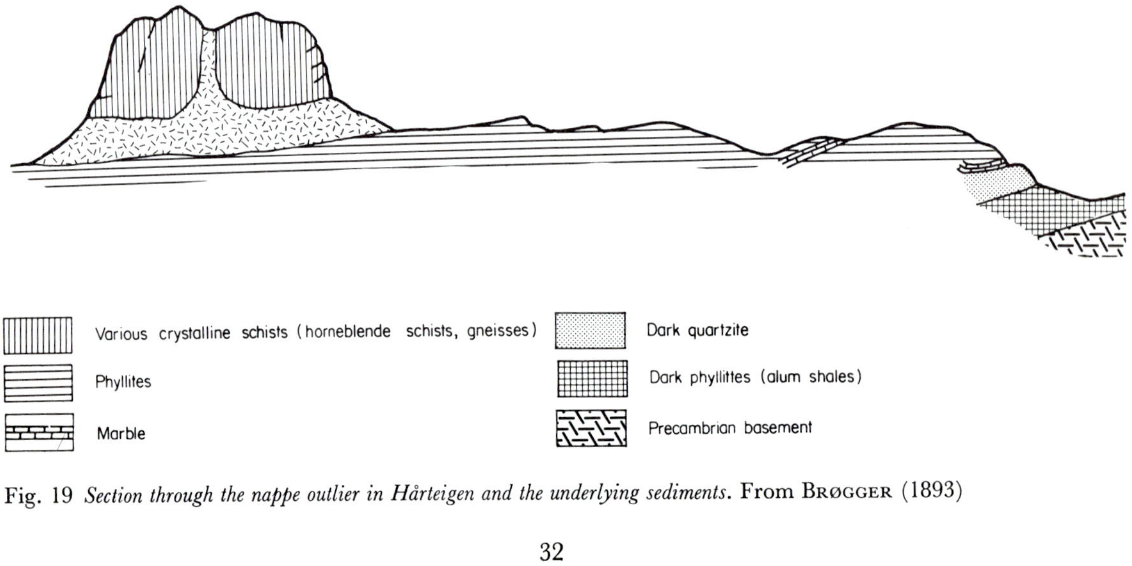

Fig. 19 *Section through the nappe outlier in Hårteigen and the underlying sediments.* From BRØGGER (1893)

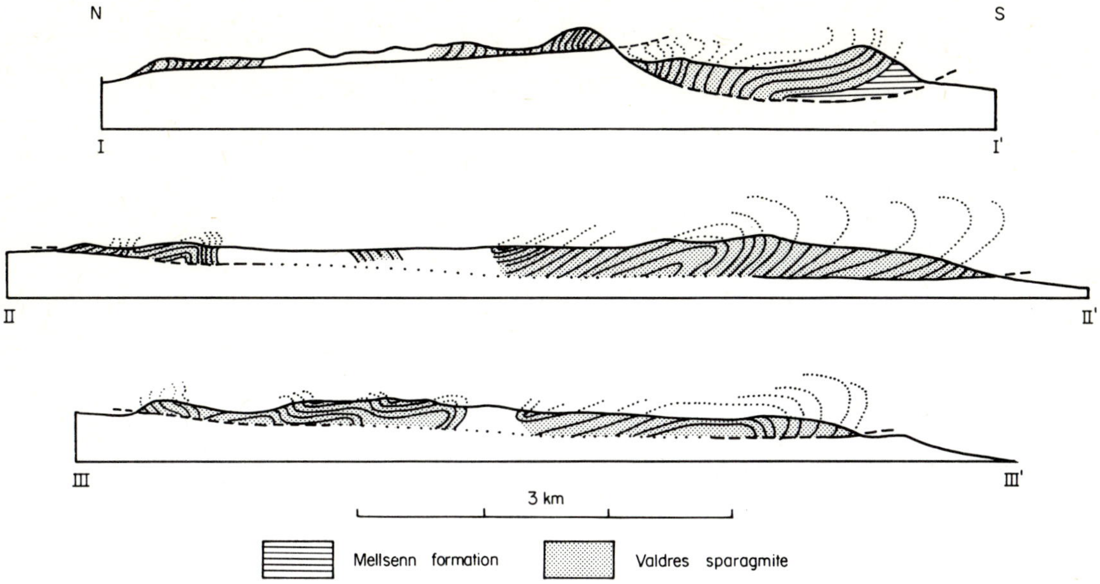

Fig. 20 *Two sections from Mellene* by R. P. Nickelsen. Dashed Mellsenn formation; Stipled Valdres sparagmite

(Fig. 20). The *Mellsenn formation*, nearly 300 m thick, is at the tectonic base of the Valdres group on the south slope of Mellene (Fig. 21). It has a (tectonically) lower member of dark and calcareous shale with Dicellograptus and Diplograptus cf.

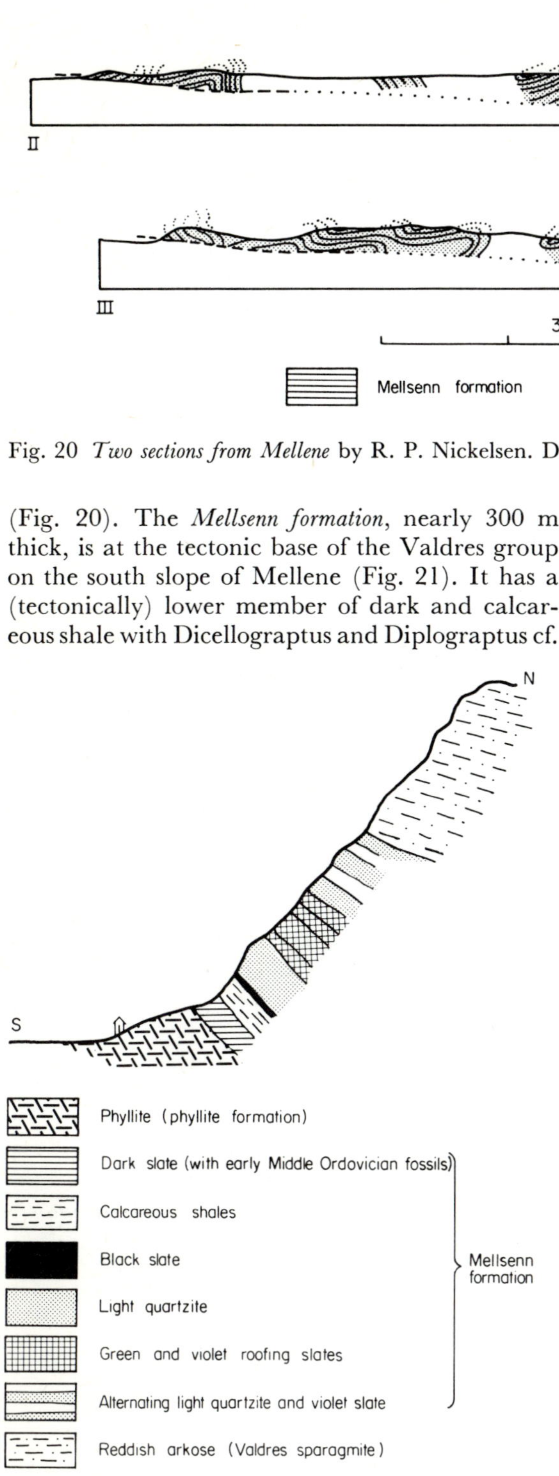

Fig. 21 *Section in the south slope of Mellene*, from K. O. Bjørlykke (1905)

teretiusculus. Loeschke found two species of brachiopods in the same shale. The fossils indicate early Middle Ordovician (Llandeilo) age. Higher up in the section, the Mellsenn formation has layers of dark sandstone and shale, a layer of green and violet roofing slates, and an upper layer of light quartzite. Higher still there is, according to Loeschke, 700 m of light feldspathic sandstone and 1400 m of coarse-grained arkosic sandstone with easily visible grains of red or violet feldspar in a greenish matrix—the typical Valdres sparagmite. There is thus an Ordovician shale stratigraphically at the top (at the tectonic base) of the Valdres group. In the absence of fossils, correlation with the other parts of the section becomes largely guesswork, but there is scarcely any doubt that the typical Valdres sparagmite corresponds to latest Precambrian or Eocambrian Sparagmite deposits, in view of its lithology. It must therefore form a nappe above the phyllites (the *Valdres nappe* of Kulling (1961)). The consequences of this on an understanding of the structure and history of the Nappe region are put forth later in this chapter. In more northerly areas the rocks are more fine grained and strongly metamorphosed and are almost identical with the "light sparagmites" in the northern parts of the Sparagmite region. In the upper part of the sparagmites are thick layers of a quartzite conglomerate (the *Bygdin conglomerate*) known for its strongly deformed boulders. Above the conglomerate is a formation of greywackes rich in basic material (the *Langsu formation*) which contains conglomerates with a corresponding matrix ("Gabbro conglomerates").

Goldschmidt (1916a) proved by petrographic investigations that the boulders in a conglomerate of the Valdres group in the Dokkvatn area (60° 10′, 9° 45′) originated from an overthrust massif of gabbroid rocks situated west of the conglomerate. In addition the arkoses and sandstones of the Valdres group contain perthite grains of a type characteristic of the charnockitic rocks in the overthrust crystalline massifs of southern Norway. Thus the deposits of the Valdres group were formed from material derived from disintegration of rocks in the nappes.

The deposits of the Valdres group cover a large area between Gausdal in the east and Valdres in the west, whence they can be traced in the west to the area around Lake Tyin. They are present in the NNE-trending Sjodalen anticlinorium and extend beneath the nappes along the northwest margin of the "Faltungsgraben" as far as south of Sognefjord. They are present most probably in the area north of Hallingskarvet (about 60° 50′, 8°) also.

THE OTTA NAPPE

The *Otta nappe* (Strand, 1951a) has its type area in the northern part of Gudbrandsdalen in the Sjoa and Otta valleys (about 61° 45′, 9° 20′) (Fig. 22). The nappe consists of an underlying crystalline massif and an overlying sequence of sediments and volcanic rocks, the youngest of which may be early Middle Ordovician. The crystalline massif is a part of the basement upon which the sediments were deposited.

The crystalline massif of the nappe (the *Rudihö massif*) consists of gneisses of varying composition, and of gabbroid and anorthositic rocks. The different types of rocks do not occur as large homogeneous massifs, but are commonly found in alternating bands (Fig. 23). The main minerals present are quartz, plagioclase, microperthite, clinopyroxene and garnet, indicating granulitic or charnockitic mineral facies. Parts of the rocks, subject to mylonitization and partial movements during the nappe emplacement, recrystallized with the formation of oligoclase, epidote, amphibole and biotite as characteristic minerals (Gjelsvik, 1946). In some areas rather large parts of the massif consist of fine-grained low grade rocks, sometimes interchanging with coarse-grained rocks. It is not certain that in all cases these rocks owe their mineral composition to retrograde Caledonian metamorphism.

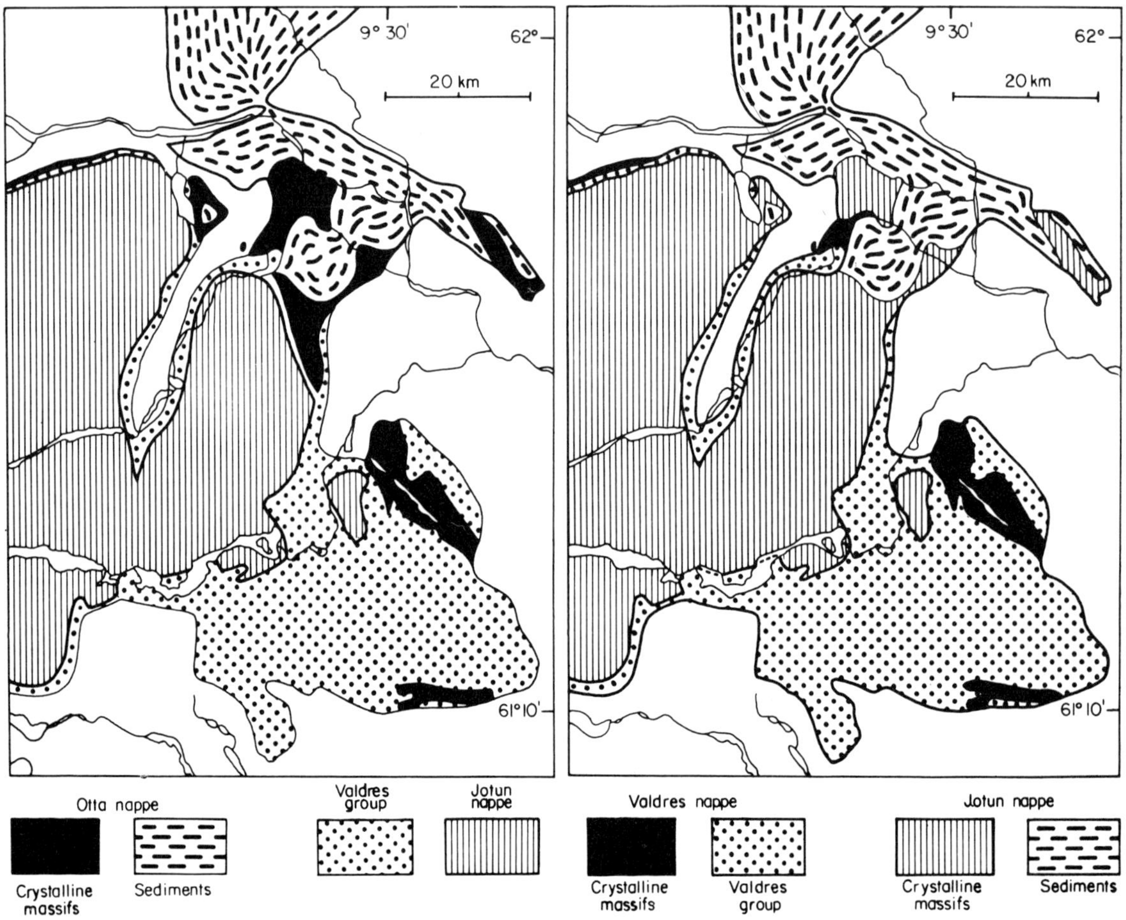

Fig. 22 *Otta nappe versus Valdres nappe.* Map showing two different interpretations

Fig. 23 *Banded meta-anorthosite–amphibolite rock of the crystalline massif of the Otta nappe. Skarvhø, Heidal* (*60° 46′, 9° 15′*)

In the lower part of the sequence above the basement complex are micaceous and feldspathic quartzites and flagstones and an overlying light-coloured garnet–mica schist, the *Heidal group* of GJELSVIK (1946).[2] The Heidal quartzites and flagstones are very similar to the "light sparagmites" and the Heidal group may well be of corresponding age. The characteristic Heidal sediments have been mapped over a large area as a basal part of the sequence, above the crystalline massif. In addition, a conglomerate at the base of the Otta nappe sequence in a more northerly locality was found to contain boulders from the underlying crystalline basement (STRAND, 1951a). From these facts it can be concluded that the Rudihø crystalline rocks form the original basement of the sediments resting upon them.

Above the Heidal group is some 2000 m of deposits of varied lithology with argillites and siltstone predominating (*the Svartkampen group*). Great parts of the sediments are of greywacke type with a profusion of minerals such as albite, epidote, chlorite and amphibole (as recrystallization products), testifying for the fact that much of the material came from igneous rocks—probably mostly volcanic rocks—without much intervention from chemical weathering. There are in addition greenschists, representing basic tuffs, and less frequently acid volcanic rocks. The volcanic rocks seldom form thick units—it is much more common to find alternations of the above-mentioned deposits with the thickness of each band varying from a few centimetres to several metres. Massive meta-basaltic greenstones about 3000 m thick are, however, found in the Otta valley west of the Otta railway station and appear to represent a volcanic centre.

In the vicinity of Otta railway station the *To conglomerate* with a meta-greywacke matrix and boulders of greenstone and hypabyssic trondhjemitic rocks is found at the top of the Svartkampen group. This conglomerate is most probably at the same horizon as the *Skardshø conglomerate* (with boulders predominantly of quartzite) from a more northerly type area.[3]

Directly above the conglomerate is the *Otta serpentine conglomerate* with well-rounded boulders (of size up to 30 cm) of serpentine. Over great parts of its area of distribution the Otta conglomerate is a pure serpentine rock. Where the serpentine has been transformed into talc, it becomes a soapstone of commercial quality. A large and rather well-preserved fauna of gastropods, cephalopods, brachiopods and trilobites has been collected from the serpentine conglomerate (HEDSTRÖM, 1930). The astonishingly good state of preservation must be ascribed to the toughness and resistance to deformation of the serpentine material. Only a small part of the fauna has at present been accurately determined. HEDSTRÖM listed species of

[2] The stratigraphic unit designated *Heidal* by GJELSVIK is taken here in a more restricted sense than was proposed by its originator.

[3] STRAND (1951a) considered the To conglomerate at Otta to be distinct from the Skardshø, but has since found the reasons for such a distinction to be invalid.

the trilobite genera Asaphus, Cybele and Pliomera common in the Orthoceras limestone (3c) of the Oslo region; these indicate late Arenig age. YOCHELSON (1963) reexamined the gastropods and found an abundance of the genus Maclurites which strongly suggests American Middle Ordovician facies, i.e. of somewhat younger age than indicated by the trilobites.

The youngest part of the sequence is the *Sel mica schist*, a grey schist often with porphyroblasts of amphibole, which is quarried as roofing material.

The fauna of the Otta serpentine conglomerate is indicative of an age corresponding to the lower part of the *Hovin group* in the Trondheim type-sequence. A correlation of the To conglomerate with the Stokvola conglomerate and of the Svartkampen group with the *Støren* and *Røros groups* is an immediate consequence.

The sediments and volcanic rocks of the nappe are in epidote amphibolite facies. There is thus a discordance in metamorphism between the crystalline basement and the sedimentary cover, indicating that the crystalline massif had acquired its high-grade mineral facies before the deposition of the sediments.

Beneath the nappe are quartzites and phyllites of eastern facies, but along the northeast side of Gudbrandsdalen the nappe rests directly upon sparagmites. The sequence below the nappe is of roughly the same age as the nappe sequence, but is of a very different sedimentary facies and of lower mineral facies. Garnet is a very rare mineral in the sequence beneath the nappe (observed in a few instances), but it is a common and characteristic mineral in the meta-sediments of the nappe. Most of the latter are coarsely crystalline, often with easily visible porphyroblasts of garnet and amphibole, and thus contrast with the fine-grained phyllites typical of the underlying sequence. Furthermore, the sediments underlying the nappe are free of intrusive rocks, in contrast to the nappe rocks themselves in which the common Caledonian intrusive rocks (saussurite gabbros or amphibolites and trondhjemites) are very common.

The crystalline massif does not extend to all parts of the nappe, which consists only of sediments for large stretches. The massif is a dome-shaped structure in the centre of the nappe, about 20 km diameter, with the sediments dipping away from it on all sides.

In its main occurrence the Otta nappe extends westwards to the east side of Lake Tesse (61° 45′, 9°), where it is represented by the crystalline basement rocks with a small remnant of the overlying sediments. A few kilometres northeast of the lake the nappe is represented by a thick pile of its sediments only. Crystalline basement rocks of the nappe can be traced along the north and northwest margin of the "Faltungsgraben" almost as far as Elveseter in Leirdalen (61° 42′, 8° 15′). Over most of this distance the nappe is only a thin sheet, but thickens to more than 500 m in the southwest. It is overlain by greywackes with conglomerates and feldspathic sandstones of the Valdres group (STRAND, 1964a). In areas near Elveseter, BANHAM and ELLIOT (1965) recorded sediments of eugeosynclinal type with volcanic greenstones in the sequence at the margin of the "Faltungsgraben". Amphibolites of this sequence were described by ELLIOT and COWAN (1966).

A number of isolated massifs of plutonic rocks have been considered as erosional remnants of the crystalline basal part of the Otta nappe. In Feforkampen (61° 33′, 9° 37′) a small sheet of gabbro lies with a tetonic contact on phyllites and quartzites. A conglomerate with boulders of feldspathic quartzite occurs immediately beneath the gabbro. The rocks in the boulders differ from the quartzites underlying the gabbro, but similar rocks are to be found in the *Espedalen massif* (DIETRICHSON, 1950). The conglomerate can be classed with the deposits of the Valdres group.

The Espedalen massif (61° 25′, 9° 35′) is elongated to the northwest and consists of anorthosite and norite in alternating bands; the two types of rocks may grade into each other. Deposits of nickeliferous pyrite, mined a century ago, occur in ultrabasic rocks of the massif. The massif rests upon underlying phyllites, but its southwest side abuts against deposits of the Valdres group with a steep boundary. In the northeast the massif is overlain by metamorphosed and deformed sediments (arkoses and conglomerates) which must be considered as part of the Valdres group (DIETRICHSON, 1945a, 1945b).

The gabbro massif of Røssjøkollene (61° 10′, 9° 40′) extends 13 km in a WNW direction. It rests with a tectonic contact upon a light-coloured sandstone that can be traced almost continuously from the type area of the Mellsenn formation, 30 km west of Røssjøkollene. The Mellsenn is at the base of the Valdres group. Above the massif are *Valdres arkoses*, and at Dokkvatn, just east of the massif, conglomerates of the Valdres group contain boulders of gabbroid rocks from the Røssjøkollene massif. Here the conglomerate is almost unmetamorphosed and undeformed, and many of the boulders show signs of weathering and a red weathering crust is preserved (GOLDSCHMIDT, 1916a).

At Skammestein in the Valdres area (61° 12′, 9°) KULLING (1961) detected a thin sheet of syenite mylonite between overlying Valdres arkose and underlying phyllites and quartzites.

OTTA NAPPE VERSUS VALDRES NAPPE

As mentioned earlier in this chapter, the question of the stratigraphic position of the Valdres group was raised anew when the manuscript of this book was near completion.

An interpretation of the Otta nappe taking the Valdres group as Ordovician is shown in the left-hand map in Fig. 22. An important point in this interpretation is that the charnockitic massifs of Espedalen and Røssjøkollene (in the southeast) and those south of the Otta valley (in the northwest part of the map), overlain by sediments of the Valdres group, are assumed to form one nappe—the typical Otta nappe—together with the charnockitic massifs and overlying eugeosynclinal sediments in Gudsbrandsdalen. Deposition of the main part of the Valdres group must thus have taken place after the emplacement of the Otta nappe and after the denudation of great parts of it. A weak point in this interpretation is that nowhere have Valdres deposits been found above, or together with, the eugeosynclinal sediments of the Otta nappe.

Figure 22 (right-hand map) gives the interpretation ensuing if the main part of the Valdres group is of late Precambrian–Eocambrian age. The Valdres group must then form a nappe together with the charnockitic massifs which underlie the Valdres sediments, and must have yielded material to them (the Valdres nappe of KULLING (1961)). There is then no reason to consider the Otta nappe (the lower Jotun nappe) as a separate tectonic unit—it must be part of the large Jotun nappe together with its cover of eugeosynclinal sediments. A further consequence is that all the allochthonous charnockitic massifs in southern Norway are Precambrian.

It is apparent that the Otta nappe has been subject to at least one phase of folding after its emplacement. The boundary between the nappe and the sparagmites on its northeast side is along a zone of folding with NW axial trend, and in some places the rocks of the nappe and its substratum are in a vertical position or even inverted locally. In the Prestberget area (STRAND, 1964b) the thrust plane of the Otta nappe and a slide plane within the nappe have been strongly folded (Fig. 24). Mesoscopic structures in this area give evidence of three phases of folding, the first probably coinciding with recrystallization and emplacement of the nappe and the second with strong folding of the thrust plane. The *Otta lunate syncline* (Fig. 25) is evidently a syncline with axial plane dipping west,

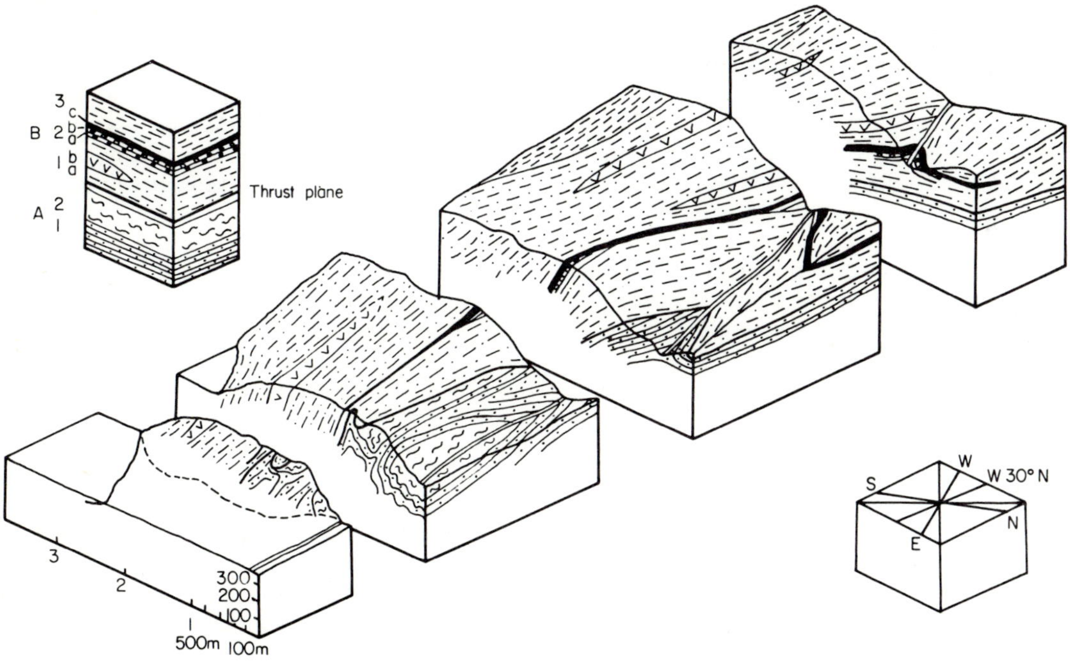

A Sediments below Otta nappe
2 Schists with carbonatic sandstones, Eocambrian and Cambro-Ordovician
1 Light sparagmites, Eocambrian

B Sediments of Otta nappe
3 Micaschist (Sel formation)
2 c To greywacke
b Otta serpentine conglomerate
a To conglomerate
1 a,b Svartkampen group, meta-greywackes and greenstones

Fig. 24 *Tectonogram of the Prestberget area, Vågåmo (61° 52′, 9° 6′)*. The left part of the tectonogram shows sharp folding of the thrust-plane of the Otta nappe

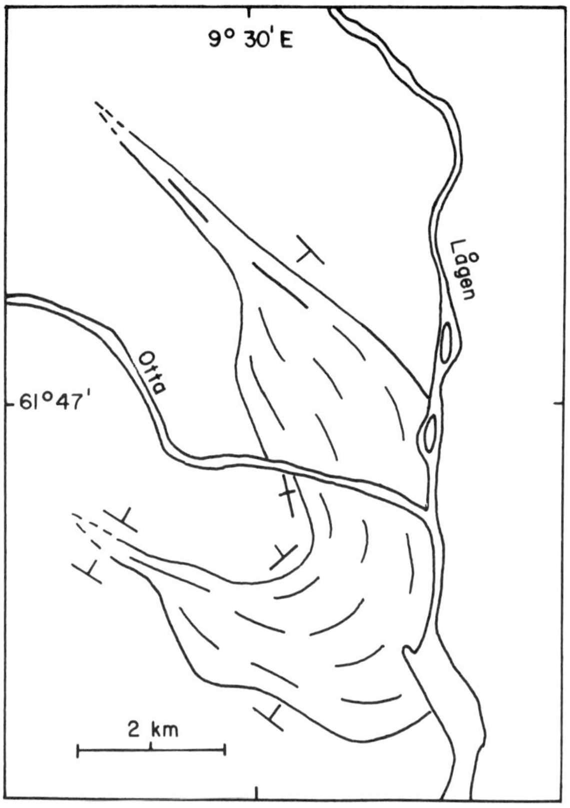

Fig. 25 *The lunate syncline of Otta*, the Sel mica schist (dashes) in the core of the fold enclosed by older deposits of the Svartkampen group

formed in connection with the nappe emplacement, which has been refolded in a NW-trending synform.

It may seem a natural conclusion to correlate the Valdres nappe with the nappes of metamorphosed Sparagmite deposits, as suggested by KULLING (1961). The Kvitvola nappe in Norway has rocks similar to those of the Valdres nappe—feldspathic sandstones of "light sparagmite" type are common to both of them. However, the Kvitvola nappe has nothing the same as the abundant basic greywackes and "gabbro conglomerates" of the Valdres nappe. Future investigations must decide whether or not the Valdres and Kvitvola nappes are in the same tectonic position.

An early Silurian sandstone in the Mjøsa district contains feldspar grains possibly derived from the Valdres nappe (STRAND, 1961). The emplacement of this nappe might thus have taken place in Ordovician time.

THE JOTUN NAPPE

The Jotun nappe, as previously mentioned, extends as a continuous massif from the Otta valley in the northeast to the Vossestrand district in the southwest, where it overlies the upper Bergsdalen nappe. This nappe appears in a small window in the Jotun nappe at about 60° 45′, 7°. The northwestern parts of the nappe are bent down into the "Faltungsgraben", while the southeastern parts rest on a flatly undulating surface and are largely separated from the main part of the nappe by erosion. Outliers also extend between those of Grønnsennknipa (61°, 8° 50′) in the northeast and Hardangerjøkulen (60° 33′, 7° 25′) in the southwest. In its northeastern parts the Jotun nappe rests upon sediments of the Valdres group. On the northwest side of the "Faltungsgraben" the Valdres deposits extend to the southwest as far as south of Sognefjord. Where the deposits are absent the nappe rests directly upon Cambro–Ordovician sediments of eastern facies (see Figs. 26–29). In the Hemsedal district the Jotun nappe lies directly above the Vemdal sandstone nappe. In the area north of Hallingskarvet (about 60° 50′, 8°) the Jotun nappe is underlain by "strongly deformed rocks, mainly supracrustals", according to the geologic map of Norway. Parts of these rocks may belong to the Valdres group.

The rocks of the great Jotun nappe were described by GOLDSCHMIDT in his monograph on the Caledonian igneous rocks of southern Norway (1916b). Together with the basement rocks of the Otta nappe, and the rocks in the central parts of the Bergen arc district, the rocks of the Jotun nappe were considered as an assemblage of magmatic rocks called *the Bergen–Jotun kindred* (*der Bergen–Jotun-Stamm*). Descriptions of the rocks of the Bergen arc district had been published earlier by C. F. KOLDERUP. The Bergen–Jotun rocks are of Charnockitic type, as indicated by the prevalence of pyroxene and garnet and the absence or scarcity of hydroxyl-bearing minerals (amphiboles and micas). The plagioclase is often antiperthitic and the alkali feldspar is a string or spindle perthite, often with very close-set inclusions. The plagioclase in the perthite inclusions is commonly not albite, but an anorthite-bearing plagioclase (oligoclase). In mineral and chemical composition the Bergen–Jotun rocks cover a wide range, from basic and ultrabasic to acidic.

Peridotites occur in the basic and intermediate rocks as inclusions ranging in size from a few metres to more than a kilometre. They consist of olivine, bronzite and clinopyroxene. The proportion of Mg:Fe in the minerals is about 0,80:1 and is thus lower than in the ultrabasic rocks in the geosynclinal sediments. The Jotunheim peridotites have been described by CARSTENS (1918) and BATTEY (1960).

Normal gabbros are relatively rare among the Bergen–Jotun rocks. Large parts of the nappe in

Fig. 26 *Bitihorn, 1608 m (61° 18′, 8° 48′), looking west.* Gabbroid rocks of the Jotun nappe above Valdres sparagmite

Fig. 27 *Northwest of Hallingskeid railway station (60° 42′, 7° 18′), looking SSW.* (Professor KVALE and members of a 1960 Congress excursion). Precambrian basement rocks (light) in foreground with overlying phyllites (dark). The front of the Jotun nappe (anorthosite) is seen in Baksafjell (underlain by a thin sheet of quartz schist of the Bergsdalen nappes). Photo J. NATERSTAD

Fig. 28 *Skagastølstindene (highest peak 2405 m), Vest-Jotunheimen (60° 27′, 7° 50′), looking northwest across the Utla valley.* Rocks of the Jotun nappe. Glacial and fluvial erosion made the youthful landforms, incised in the smoothened "paleic" surface. Copyright Widerøes Flyveselskap A/S, Oslo

Fig. 29 *Boundary between phyllite and gneiss of the Jotun nappe.* Lake Flakevatn, southwest part of Hallingskarvet. Photo Kaare Strøm

the Hemsedal district are composed of gabbros with lath-shaped plagioclases and ophitic structures. These differ from the typical Bergen–Jotun rocks. Here Bugge (1939) described a very fine-grained doleritic rock which penetrates and brecciates more coarse-grained gabbros.

Anorthosites occur in great masses and predominate in the southwestern part of the nappe. At Aurlandsfjord and Nærøyfjord (arms extending to the south from the inner part of Sognefjord) light-coloured anorthosites are exposed in the 1000 m high walls of the fjord. The Bergen–Jotun anorthosites contain a very calcic plagioclase with An 60–70. As the amounts of femic minerals (hypersthene) increase the anorthosites grade into labradorite norites. Anorthositic rocks are also found in smaller bodies as bands in gabbroid rocks. Bugge (1939) described oligoclasites—rocks very rich in albite-rich plagioclase, occurring in the Hemsedal district.

More common than gabbros are basic and intermediate rocks with microperthite and plagioclase (the *Jotun norite* of Goldschmidt and the *jotunite* of Hødal (1945)). Intermediate rocks with microperthite as the only feldspar are known as *mangerites*, and were originally described in the Bergen district by C. F. Kolderup. Leucocratic quartz-free rocks with microperthite are the *hypersthene syenites* or *mangerite syenites* (the latter name proposed by N.-H. Kolderup (1921)).

Granites of different types occur as larger massifs or form sills in the basic and intermediate rocks. They range from hypersthene granites to ordinary biotite granites. Alkaline types are also found among the granitic rocks; granites with ægirite occur north of Øye in Vang, Valdres (61° 15′, 8° 25′). Granitic gneisses without any obvious similarity to the Bergen–Jotun rocks are also found in the Hemsedal–Hardangerjøkelen areas where they predominate in the southeast part of the nappe. These rocks frequently contain microperthites of "Jotun" type.

The rocks of the Jotun nappe contain intrusive bodies of gabbro and trondhjemite of the same types as are found in the eugeosynclinal sediments. At Årdal on the inner part of Sognefjord there occurs a body of trondhjemite sufficiently large in size to show up on a small scale geologic map.

The mylonites formed at the thrust plane of the nappe have low-grade (greenschist) minerals. Other large parts of the rocks have suffered retrograde metamorphism, evidenced by saussuritization of the plagioclases and uralitization of the pyroxenes. This metamorphism may be connected with the emplacement of the nappe.

Goldschmidt (1916b) discussed the age of the Bergen–Jotun rocks without reaching any definite conclusion as to whether they were Precambrian or Caledonian. Considering the tectonic position his opinion was that the rocks had their roots in the "Faltungsgraben". Even if this were true the rocks must have been moved for some 60 km in a SE direction. Holtedahl (1936) was the first to interpret the Jotun massif as a rootless nappe. Nowhere have the rocks of the Jotun nappe been found to have an intrusive relationship to their substratum. Whatever the origin and age of the rocks may be, they must have been brought to their present position as consolidated massifs. Goldschmidt's theory of rooting of the nappe in the "Faltungsgraben" cannot be disproved, as the underside of the nappe is concealed for a minimum distance of 20 km across the graben. This theory supposes that the nappe must have moved to all sides from the fissure in the crust from which it was squeezed out, as no roots can be seen anywhere. The lower parts of the Jotun massif immediately above the underlying sediments are always strongly mylonitized. This has been observed in a great number of places and is testified for by a number of descriptions.

More detailed information on the petrography and tectonics of parts of the Jotun nappe was given by Landmark (1949), Skjerlie (1957), Hødal (1945) and Bugge (1939) for the Luster–Bøverdalen area, the Leikanger area on the north side of Sognefjord, the Vossestrand area in the southwest end of the nappe and the Hemsedal district respectively. Skjerlie recognized Valdres deposits below the nappe which were shown to consist of four thrust sheets of different petrographic composition. Hødal gave very detailed descriptions of rocks mylonitized and metamorphosed during the thrusting.

A detailed field and petrographic investigation of the Jotun rocks was undertaken by Battey (1965) in the areas northwest of the west end of Lake Bygdin (61° 23′, 8° 20′). A purple gabbro from Mjølkedøla at the west end of the lake had a coarse ophitic structure. On the east shore of Lake Tyin a gabbro of the same type was associated with ægirine granite. West of Lake Bygdin the Mjølkedøla gabbro is separated by a powerful NE-trending fault from a complex of layered two-pyroxene granulites, in which the layering is produced by variations in the proportion of dark silicates to feldspar in the granulites, and of olivine to pyroxene in the ultramafic rocks (Fig. 30). The layers vary in thickness from tens of metres to fractions of a centimetre for the ultramafic rocks

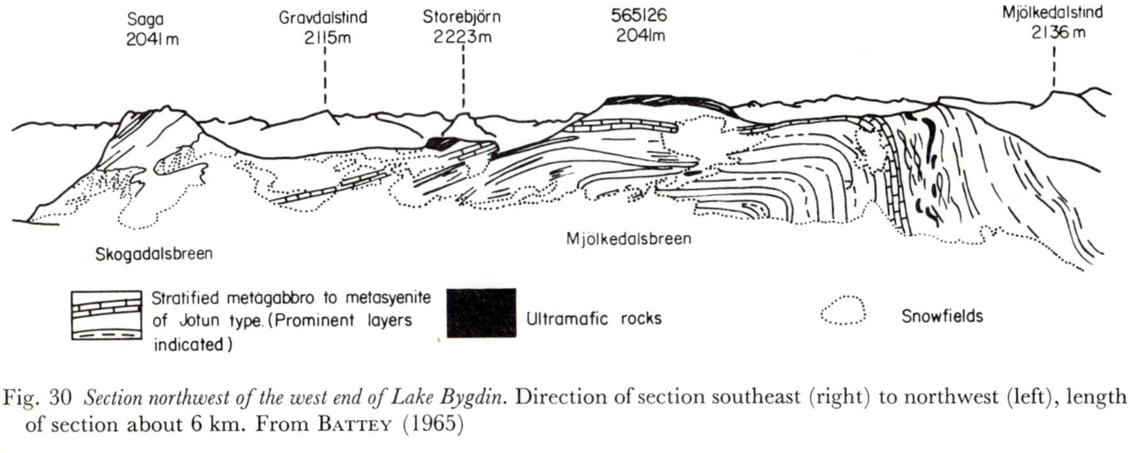

Fig. 30 *Section northwest of the west end of Lake Bygdin.* Direction of section southeast (right) to northwest (left), length of section about 6 km. From Battey (1965)

Fig. 31 *Banding in gabbroid rock of the Jotun nappe, west of Lake Tesse (61° 47′, 8° 50′)*

or the more persistent layered granulites. The strata thus produced form prominent ledges with strike northwest, broken into strips by powerful lines of dislocation trending in the same direction, along which the layers have been dragged into a subvertical attitude. In the northern part of the area mapped the banded rocks pass into homogeneous gneissic jotunites (the Jotun norites). Retrograde metamorphism occurred in the fault zone delimiting the massif to the southwest. Battey assumes a magmatic origin for the rocks and their layering, but their present mineral composition and microscopic structures are in fact due to metamorphic recrystallization. Banding and layering in rocks of the Jotun nappe are features certainly shown in many parts of the nappe outside the area studied by Battey also (Fig. 31).

In the area at the east end of Lake Bygdin (61° 20′, 8° 50′) the thrust plane of the Jotun nappe, with mylonites above the strongly deformed Bygdin quartzite conglomerate of the Valdres group, is very well exposed (Goldschmidt, 1916a; Strand, 1945 and Flinn, 1961). The boulders of the Bygdin conglomerate are strongly elongated in a NW to WNW direction, and the rocks are folded on axes plunging in the same direction. The thrust plane is thus exposed over a considerable distance. John R. Hosack has recently studied the area in great detail and has kindly given the writer access to an unpublished paper. The northwest structures in the area with conglomerate deformation belong to the first two phases of deformation recognized in the area with parallel structural planes, connected with the emplacement of the nappe. A third phase produced folds plunging east, followed by jointing.

In Grønnsennknipa (61°, 8° 50′) a nappe outlier of granitic rock borders on an outlier of Valdres deposits along a steep northwest-trending boundary (Goldschmidt, 1916a). Holtedahl (1959, 1961b) investigated the area and found tectonic contacts between the granite massif and the Valdres sediments.

Rekstad (1903, 1905) found that in the district north of Hallingskarvet (about 60° 45′, 7° 50′) the nappe had a lower part of quartzite which passed upwards into gneiss and massive granite. Askvik and Fareth, two graduate Bergen students working in the same area, found greenish metasediments with conglomerates and quartzites alternating with amphibolites above the phyllites below the nappe. The greenish sediments with conglomerates may be deposits of the Valdres group. B. Dietrichson (personal communication) found typical Valdres sediments north of Lake Djup in the same area. The occurrence of quartzites in the lower part of the nappe (or below it), also recorded at Mollnut in the Hardangerjøkelen outlier, recalls conditions in the Hardanger nappe.

A gravity survey in the Jotunheim mountains and the northeastern part of the Jotun nappe (SMITHSON, 1964) disclosed a positive Bouguer anomaly over the Jotun massif with the same trend as the "Faltungsgraben", and with a maximum local anomaly of +36 mgal. (Fig. 32). The mean density of the rocks of the Jotun massif demanded by this theory. DIETRICHSON maintained that parts of the Jotun rocks were melted and recrystallized during the thrusting. In this connection he paid particular attention to the narrow veins of dense, apparently glassy rock (pseudotachylite), the formation of which is ascribed to melting by frictional heat along planes of movement. Veins

Fig. 32 *Gravity anomaly map over a northeast part of the Jotun nappe.* From SMITHSON (1964)

(determined from samples taken at the surface) was 2,86, while the mean density of Precambrian gneisses in the region southeast of Jotunheimen was 2,74. A gravity model using a density contrast of 0,12 gave a thickness of 12 km for the nappe in the middle of the syncline, to which must be added nearly 2,5 km for the topography. A model using an assumed higher density contrast of 0,20 gave a thickness of 6,2 km (+2 km) in the middle of the syncline and 2,3 km for the plate-like extension of the nappe to the southeast.

Other interpretations of the Jotun nappe are outlined below.

DIETRICHSON (1960) divided the Jotun nappe into three different nappes: an anorthosite nappe, a granitic nappe and a jotun norite nappe. It remains to be shown by mapping whether the Jotun massif has the petrographic regularity of this material were described in the Norwegian Caledonides by GOLDSCHMIDT (1943), DIETRICHSON (1953), BRYN (1960) and SCHAAR (1962). The examples described by GOLDSCHMIDT, BRYN and SCHAAR occur in meta-sediments.

OFTEDAHL (1961) considered the Jotun nappe as an essentially autochthonous massif, admitting thrusts some tens of kilometres in length at the margins of the massif. He claimed that the Bergen–Jotun rocks originated from enormous quantities of basic magma which poured out into the bottom of the Caledonian geosyncline in Ordovician times.

THE BERGSDALEN NAPPES

In its southwestern part the Jotun nappe is underlain by the Bergsdalen nappe complex in the area northwest of Hardangerfjord. This part of the Caledonides has very intricate structures, but is also one of the best known parts, due to the work of KVALE (1946a, 1948, 1960).

The Bergsdalen nappes with the adjacent Cambro–Silurian sediments are situated in the synclinorium of the "Faltungsgraben", underlain on the southeastern side by Precambrian basement rocks at Hardangerfjord, and on the northwestern side by granodioritic gneisses with granitic veins—according to KVALE a complex of migmatites that got its structure during the Caledonian orogeny.

The Bergsdalen nappes are in contact with two different Cambro–Silurian (or Cambro–Ordovician) sequences; one consists almost exclusively of phyllites of eastern facies, and the other of western, eugeosynclinal facies with more varied types of sediments and volcanic rocks (Chapter VI).

The Bergsdalen nappes consist of a complex of supracrustal and intrusive (plutonic) rocks. Sedimentary quartzites and quartz schists are present. Some of the rocks have well-preserved sedimentary structures, e.g. cross-bedding and banding. Conglomerates with highly deformed pebbles occur. Volcanic rocks present are meta-basalts, meta-dacites and meta-rhyolites. Only the meta-rhyolites, when undeformed, show well-preserved original structures, as some of them are porphyric. The meta-dacites are fine-grained quartzdioritic gneisses rich in biotite, and their composition suggests a magmatic origin. In parts of the area the meta-basalts and metadacities occur in banded alternation. The total thickness of the sequence is not less than 3000 m (Figs. 33, 34).

Corresponding to the three types of volcanic rocks there are intrusive rocks: gabbros (mostly saussurite gabbros), quartz diorites and granites. Associated with the granites are a few bodies of granodiorite and quartz diorite, younger than the common type of quartz diorite. The intrusive rocks occur in a large number of separate plutons and form about one half of the rocks of the nappe. They are conformable and concordant with primary flow structures. The granites are younger than all the other rocks of the complex (with the exception of a few intrusive rocks). Some of the structures must have been formed during thrusting of the nappes, implying that the granite was not completely solidified at the time of the event.

It appears likely that the Bergsdalen rocks are Precambrian; a complex of Precambrian rocks of the *Telemark suite* at Sørfjord (the long, narrow arm extending south from Hardangerfjord) consists of rocks very similar to, or even identical with, the different types of rocks in the nappes.

Recent investigations have shown that charnockitic rocks (anorthosites together with intermediate and acid rocks) occur in parts of the

Fig. 33 *The mountain Vassfjøra, Hardanger (1633 m) (60° 38′, 7° 2′)*. Quartzite (lower part of mountain) and gneiss (upper part) of the upper Bergsdalen nappe. Phyllite at the foot of the mountain and between the quartzite and gneiss. Photo J. NATERSTAD

Fig. 34 *Skåndalshorgi north of the Hardangerfjord (60° 32′, 6° 37′), looking SSW.* The mountain, like the mountain in the right distance, is formed from the middle and upper sheets of the upper Bergsdalen nappe. Photo J. NATERSTAD

Bergsdalen nappes (in the middle and upper sheets of the upper nappe). The rocks are partly of gneissic appearance and are difficult to distinguish from other gneissic rocks.

The Bergsdalen nappes are divided into a lower and an upper nappe. The rocks of the lower nappe are in amphibolite, or high epidote–amphibolite facies with oligoclase; the rocks of the upper nappe are of a lower facies with plagioclase near albite. Each nappe consists of three separate sheets. Those of the lower nappe have maximum thicknesses 1000, 3000, and 5000 m, in ascending order; corresponding thicknesses in the upper nappe are 3000, 1000 and 2000 m. The nappes rest upon phyllites of eastern facies, which are also found between the different sheets of the nappes. In the lower nappe the bands of phyllite or mica schist beneath the bottom sheet, and between the others are very thin.

Due to their position in the "Faltungsgraben" the rocks in the northwest dip to the southeast, but are rather flat-lying in the central part. The tectonogram (Fig. 18) gives a simplified picture of the tectonics. A detailed map shows a much more complicated picture.

The Bergsdalen nappes must have come from the northwest, where a complex of supracrustal and plutonic rocks lay above the gneisses now found northwest of the Bergsdalen area and below Cambro–Ordovician pelites. The gneisses in the northwest contain quartz diorites and amphibolites that can be matched with corresponding rocks in the Bergsdalen nappes. The Precambrian supracrustal rocks were thrust up to the southeast together with the overlying Cambro–Ordovician mica schists. According to KVALE's interpretation the upper nappe (with a low mineral facies, indicating a relatively high tectonic level) was emplaced first and the lower nappe, which came from a deeper tectonic level, was thrust into the mica schists below the lower nappe afterwards.

KVALE paid great attention to the linear structures in the Bergsdalen nappes. Linear structures trending northeast were found in the gneisses in the basement below the nappes, in the lower part of the nappe, and also in the front of it where movement had met with obstacles. They were interpreted as having been formed by rotational movements on axes normal to the direction of rock transport. Linear structures trending SSE in the direction of transport were assumed to have been formed by laminar movements. Gradational transitions between the two directions of lineation within the nappe were not easy to account for.

No clear evidence of the time of emplacement of the Bergsdalen nappes has been found, but KVALE considers them to be of the same age as the Jotun nappe (not earlier than Middle Silurian).

Fig. 35 *The nappe outlier in Hårteigen (60° 12′, 7° 5′), in the northern part of Hardangervidda.* The sub-Cambrian peneplain with its relatively thin cover of sediments has put its imprint on the topography. Copyright Widerøes Flyveselskap A/S, Oslo

THE HARDANGER NAPPE

Hallingskarvet and Hardangerjøkelen are the most southeasterly outliers of the Jotun nappe. South of Hardangerjøkelen the paratochthonous Cambro–Ordovician sediments cover large areas of the rolling plain of Hardangervidda. Allochthonous rocks above the phyllites are met with again as an isolated outlier on Hårteigen mountain (60° 12′, 7° 5′), 40 km SSW of Hardangerjøkelen (Fig. 35). South of Hårteigen a nappe of crystalline rocks covers the phyllites in the southern parts of Hardangervidda as far as the area around Boknfjord. This tectonic unit will be called the Hardanger nappe, previously mentioned. A glance at the geologic map shows that the nappe has been strongly dissected by erosion into a large number of detached parts.

The northern part of the Hardanger nappe, north of 59° 50′, is known as a result of the publications of Brøgger (1893), Bjørlykke (1902), and Rekstad (1902, 1903), but no petrographic descriptions of the rocks have been given. According to Bjørlykke the rocks of the nappe are quartzites (meta-psammites) which are often feldspathic and quartzschists and amphibolites in alternating layers containing lenses of saussurite gabbro. There are also mica schists with quartz-feldspar nodules, gneisses of "Archean type" and porphyric granites (Fig. 36). Lenses and sheets of phyllite and quartzite from the underlying sequence are found squeezed in between the nappe rocks, all of which are very similar to the rocks of the Precambrian basement in the same area. According to Brøgger the nappe on top of Hårteigen consists of amphibolitic schists, amphibole gneisses, mica schists and pink feldspathic gneisses. All of these are supracrustal rocks. Southwest of Hårteigen a number of outliers occur with gneisses and gneissic granites directly above the phyllites. There are veins of light acid granite in these granitic rocks but nowhere do these penetrate into the underlying phyllites. In Sandskarnut on the west side of Valdalsvatnet (59° 57′, 6° 55′) the section recorded by Brøgger appears to be an unbroken sequence, as the phyllite is overlain by flagstones alternating with quartz-rich phyllites. Mica schists above the flagstone become more and more coarsely crystalline and feldspathiferous going upwards, and pass into gneisses of various types. A pink, rather coarse-grained gneiss is found at the top of the mountain, about 700 m above the phyllite of the autochthonous sequence. In many localities in this area the rock in the

Fig. 36 *Gneiss of the Hardanger nappe, Isvatn (59° 58′, 6° 50′)*. From K. O. BJØRLYKKE (1902)

lowest part of the nappe is quartzite (meta-psammite) and it passes into gneiss higher up. The same type of section is found in Molnut on the south side of Hardangerjøkulen.

The writer (STRAND) made observations at Seljestad (59° 54′, 6° 40′) (Fig. 37), where a formation of greenstones and greywackes with bands of quartzite, perhaps 100 m thick, was found above the grey phyllite. Higher up followed meta-arkoses or feldspathic psammites. In a section on the Seljestad–Røldal road the meta-arkoses were overlain by dark biotite-rich schists or gneisses and quartz-dioritic gneisses, which were probably metamorphosed greywackes and tuffs. In the same district BJØRLYKKE reported more highly crystalline rocks from higher tectonic levels at the top of the mountains. No distinct thrust plane was found by the writer within the series examined, but evidently the underlying fine-grained muscovite–chlorite phyllites are of lower metamorphic grade than the overlying rocks, which contain plagioclase, An 30–35.

In southern parts of the area the nappe rocks are of the same types as farther north, according to KALDHOL (1909). Usually there are quartzites passing upwards into gneisses and in some parts massive granites with semi-porphyric structure. The same type of granite is said to occur also in

Fig. 37 *View of the sub-Cambrian peneplain with overlying phyllites and Hardanger nappe, looking north. Seljestad, Hardanger (59° 54′, 6° 40′)*

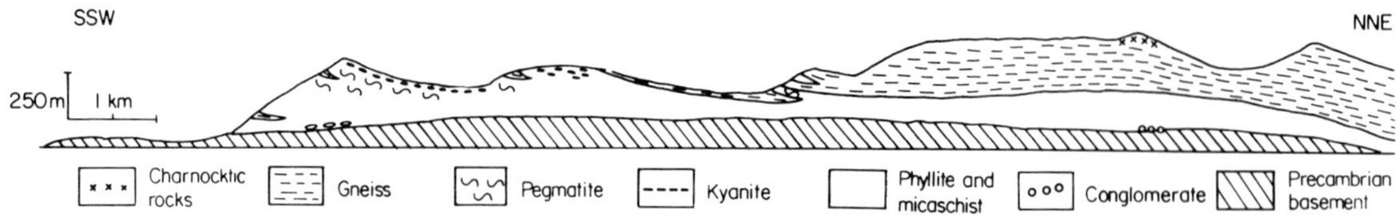

Fig. 38 *Section west of Saudafjord.* By RUTH C. SØRBYE

the northern parts of Hardangervidda, and farther north in the area around Hallingskarvet.

SØRBYE (1951, 1954) published observations made in the same areas. In eastern parts, where Caledonian deformation is relatively weak, the Cambrian basal conglomerate and sandstone rest undisturbed upon the basement, and the thrust plane above the phyllites is distinctly marked by zones of dense mylonite in the gneisses of the nappe. Of great interest is the occurrence of charnockitic material as augen and pegmatitic veins, in amounts increasing upwards, but present also at lower horizons in comparatively low-grade schists. In some sections the gneisses above the phyllites show a gradual transition into phyllites and are interbanded with them; the phyllites may pass into feldspar-rich greywackes. At still higher levels the gneisses show charnockitic features such as brown-coloured cores in the amphiboles and spindle perthites. The highest horizon, present

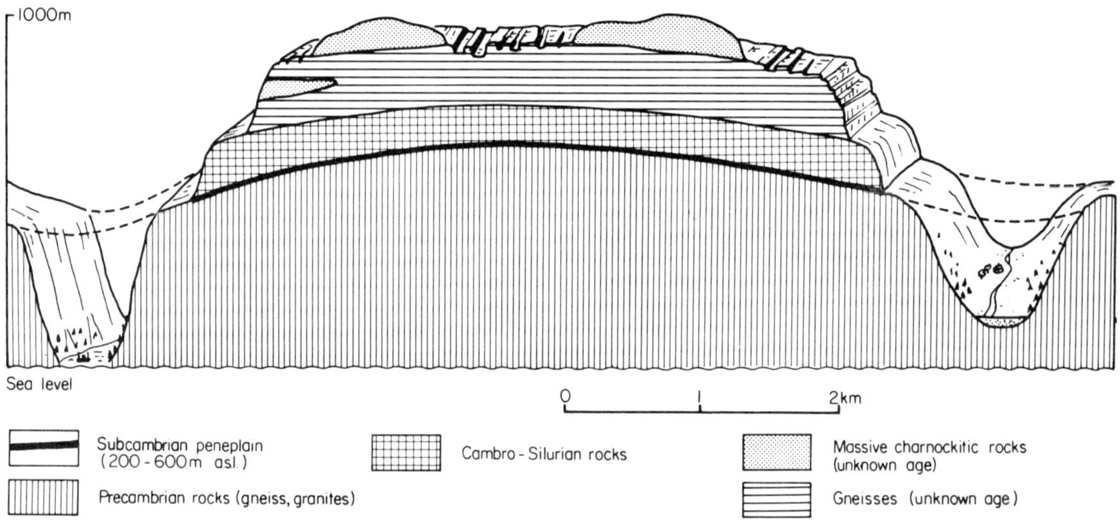

Fig. 39 *Tectonogram of the Hardanger nappe between Hylsfjord (left, north) and the Suldal valley.* By ELLEN S. KILDAL

rocks, wholly similar to the Bergen–Jotun rocks, on Napen mountain (69° 25′, 6° 40′). The rocks present are granites, diorites and syenites, one of which is a mangerite syenite with spindles of acid oligoclase in the microperthite. For the district west of Saudafjord (about 59° 35′, 6° 10′), SØRBYE very kindly made available the unpublished results of her investigations (Fig. 38). Above the Precambrian basement is the normal cover of phyllites, with black alum shale phyllites at the base and basal conglomerates preserved in a few places. While the lower part of the phyllite sequence is of greenschist grade, the metamorphic grade increases upwards to high amphibolite facies over a thickness of 400 m. The rock succession is thus phyllite–garnet–chloritoid schist–kyanite porphyroblast schist. The schists contain feldspar on Hustveitsåta mountain, contains amphibole–quartz diorite, hypersthene–quartz monzonite and potash granite, representing intermediate and acid charnockitic rocks. Large dykes of silexite are also found.

In the Ølen district north of Saudafjord conditions are similar to those west of Saudafjord, as injected schists occur beneath an overlying nappe of gneisses.

In the area between Hylsfjord and the Suldal valley (59° 30′, 6° 25′) (Fig. 39), mapped and described by ELLEN S. KILDAL (Thesis, 1962, University of Bergen) the phyllite is overlain by 300 to 700 m of gneisses, which may or may not have a well-marked tectonic boundary at the underside. The gneisses are evidently layered supracrustal rocks and both arkosic and tuffitic types are represented. Charnockitic rocks occur in the upper preserved part of the gneisses in sharp contact with them; there are no cataclastic structures at the contacts. In one instance a basic granulite (with labradorite and pyroxenes) has a 1 dm thick contact zone of well-crystallized amphibolite (with oligoclase and amphibole) next to the gneiss. As the contact rock is of a higher mineral facies than the mylonites of the nappe, the granulite must have been warm at the time of its emplacement. The charnockitic rocks vary in composition from acidic to basic, the contrasting types of rocks often forming chaotic mixtures. The oldest structure in the nappe is a pronounced northwest lineation—a large recumbent fold with axis trending NNE and overfolded to the southeast may be of the same phase. A younger phase formed flat E–W undulations accompanied by faults. Roughly similar conditions were found on Ombo island in Boknfjord, and were described by KJELL A. TJØLSEN in a thesis (University of Bergen, 1961).

The rocks in the Stavanger district belong to a sequence of micaschists and greenschists, intruded by diorites, trondhjemites and granites. In this district GOLDSCHMIDT described Caledonian trondhjemites and granites in the peninsula 20 km northeast of Stavanger, resting upon the underlying phyllites with mylonites at the underside, "as if the intrusive masses, already partly cooled, had been thrust to the southeast from their place of primary intrustion" (GOLDSCHMIDT, 1921, translated). According to BJØRLYKKE (1908) granites and gneisses occupied a position above the phyllites at Malletuva and other localities west of Stavanger. Thus it seems that the eugeosynclinal sediments with intrusive rocks form a nappe in the Stavanger district, which may perhaps be a part of the Hardanger nappe (see Chapter VI).

In northern parts of the Nappe region there is usually no problem in fixing the boundary between the crystalline rocks in the nappes and the underlying sediments. This is also the case in parts of the Hardanger nappe, for example in the northern part where granitic rocks occur directly above low-grade phyllites. In large parts of the Hardanger nappe, however, especially in the southern part, the rocks above the phyllites are sedimentary or supracrustal; in some localities they pass upwards into gneisses, and it is not always easy to find a marked thrust plane. In the same parts of the area quartzo-feldspathic deposits, not very different from the overlying gneisses, have been found to be interbedded in the phyllite sequence. It is thus possible to regard parts of the overlying gneisses as autochthonous or parautochthonous deposits in a stratigraphic position above the phyllites. BRØGGER (1893) was reluctant to accept the doctrine of overthrust (then recently introduced by TÖRNEBOHM) in the case of the Hardanger nappe and advanced a hypothesis of a massif of igneous rocks once overlying the rocks of the nappe which was later eroded away. This hot mass probably caused the upwards increase in the grade of metamorphism which led to the formation of gneisses at high levels in the sequence. This upwards increase in the sequences now seems well established for the rocks discussed here. In addition, remnants have been found of an overlying massif of charnockitic rocks which was evidently hot at the time of its emplacement. There are thus good reasons for a revival of BRØGGER's 1893 hypothesis. Certainly, BRØGGER thought in terms of magmatic intrusions at high levels, whereas most geologists today prefer to think of a hot mass of rocks which moved as a nappe.

CONCLUDING REMARKS

As previously stated, the Jotun nappe lies above the Valdres group wherever deposits of that group are present. The interpretation of the tectonic setting in the northeast part of the Nappe region depends upon the age of the Valdres group, as set forth earlier in this chapter. The Otta nappe certainly continues to the north in the southern part of the Trondheim region. There are indications in this part of the Nappe region of a tectonic event in late (?) Ordovician time.

The Bergsdalen nappes are structurally below the Jotun nappe; but as previously stated, these

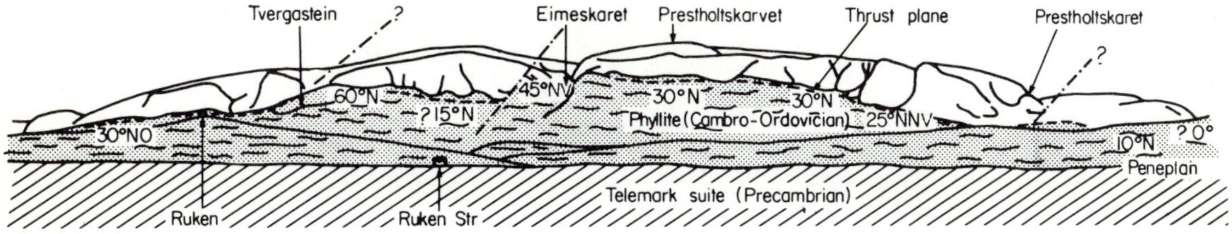

Fig. 40 *View of the eastern part of Hallingskarvet (about 6 km long) looking NNE.* From STØRMER (1964). The sub-Cambrian peneplain is at an altitude of about 1130 m, while the highest summit shown reaches 1850 m. Note the greatly varying thickness of the phyllites (shaded) and the faults affecting the nappe (white).

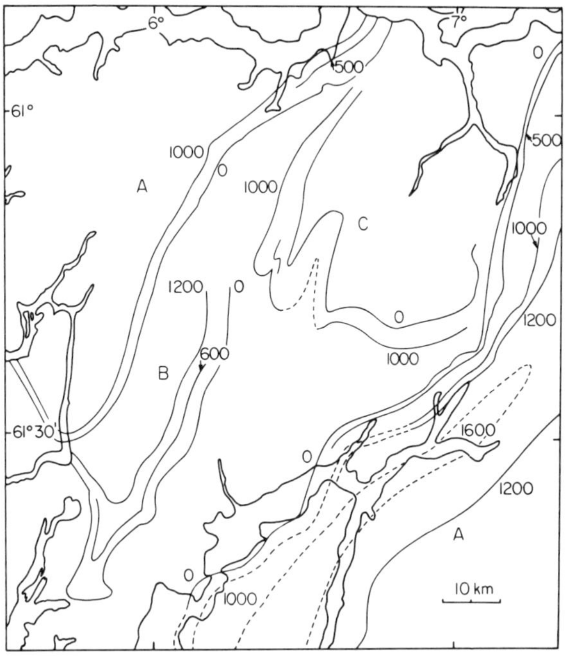

A = Surface of the gneiss basement (base of the phyllites)
B = Top of the lower Bergsdalen nappes
C = Base of the Jotun nappe

Fig. 41 *Structural contour map of a southwest part of the "Faltungsgraben"*. After unpublished maps by Professor ANDERS KVALE

nappes are considered to have been emplaced in the same phase, i.e. not earlier than the Middle Silurian.

The Hardanger nappe is less well known, and the problems connected with it are unsolved. The rocks in the Jotun nappe north of Hallingskarvet and in the Hallingskarvet and Hardangerjøkulen outliers show some similarity to those of the Hardanger nappe, and a connection between them is thus possible.

The thrust planes below the nappes are surfaces of discontinuity and sites of very large horizontal movements. However, partial movements within the different rocks may also have accumulated to considerable amounts. Conspicuous examples are the elongation of conglomerate boulders in the general direction of movement of the nappes (SE or ESE). The sediments beneath the nappes have been greatly influenced, particularly in the more westerly areas. The original stratigraphy has been disturbed in many instances and the sediment thickness may vary from practically nothing to some 200 or 300 m within a short distance, as for example at Hallingskarvet (STØRMER, 1964) (Fig. 40). The parts of the autochthonous sequence best protected against these disturbances are the basal conglomerates or sandstones, but even these are often sheared away in more westerly districts. The upper parts of the Precambrian basement have also been affected and have become foliated parallel to the surface

Fig. 42 *Aurlund at Aurlandsfjord (60° 53', 7° 10') looking northeast.* The Precambrian basement rocks of the Aurland window with overlying phyllites and Jotun nappe slope to the northwest (left in picture) into the "Faltungsgraben". Photo J. NATERSTAD

of contact with the overlying sediments. The following examples have been taken from an unpublished paper kindly placed at the disposal of the writer by Professor ANDERS KVALE. At Finse and Hallingskeid on the Bergen railway there are no signs of Caledonian deformation beyond the uppermost 1 or 2 m, but the zone of foliation is 15 to 30 m thick in localities at Hardangerfjord, where amphibole in the gneisses has been altered to biotite. In the Aurland window, south of the inner part of Sognefjord, a conformable foliation with growth of feldspar augen has developed from 10 to 30 m below the contact. At Sneisagavlen, west of Sørfjord (KVALE, 1946b) Caledonian lineation has been impressed on quartz schists of the Precambrian basement as far as 500 m below the contact with the overlying phyllites. Similar development of foliation in Precambrian basement gneisses has been reported on the northwest side of the "Faltungsgraben" northeast of Sognefjord, by LANDMARK (1949) and BANHAM and ELLIOT (1965).

It is apparent that there was a phase of deformation later than the emplacement of the nappes, which resulted in bending of the thrust planes and the underlying rocks, including the Precambrian basement. The structures in the northwest affecting the Otta nappe have been described in a previous section (see Fig. 24). Probably the formation of the "Faltungsgraben" and the Bergen arc structures and also the deformation of the Precambrian basement of southern Norway occurred in the same phase. The structural contour map (Fig. 41) (due to KVALE) of a southern part of the "Faltungsgraben" and part of the Bergen arcs shows slopes between 16° and 35° at the margins (Fig. 42). The Jotun nappe is of maximum thickness (12 km as deduced from gravimetry) near the northwest margin of the "Faltungsgraben" and requires a slope of about 45°. No faults have been observed along the margin of the "Faltungsgraben", which must have been formed by bending of the basement.

CHAPTER V

THE TRONDHEIM REGION

The Trondheim region is a large synclinorium of axial depression occupied by more or less metamorphosed Cambro–Silurian sediments of western, eugeosynclinal facies with many intrusive rocks. The narrow southwestern part of this region is connected with the Nappe region, where the sediments of the Otta nappe pass into the Trondheim region. On the southeast side the rocks of the Trondheim region lie above the sparagmites of the Sparagmite region, while north of 62° 40′ they rest upon sparagmites and underlying Precambrian basement rocks in the anticlinoria extending along the international border. In the north this eastern boundary almost coincides with the international border. On the west and northwest sides the sediments lie above gneisses of the Basal gneiss region, but fairly thick layers of sparagmitic rocks intervene between the gneiss and the Trondheim sediments. Here the *Surnadal synclinorium* protrudes to the WSW.

Northeast and east of Trondheimsfjord the rocks of the Trondheim region occur in two synclinoria, separated by the intervening *Tømmeras anticlinorium*. The synclinoria terminate in the northeast by a rise of their axes towards the gneiss area of the *Grong culmination*. The northwesterly of the synclinoria, the *Snåsa synclinorium*, continues southwest as the *Fosen synclinorium* on the northwest side of Trondheimsfjord. The islands of Hitra and Smøla, southwest of the mouth of Trondheimsfjord, consist mainly of intrusive rocks, and are part of the continuing trend southwestwards from the Fosen synclinorium.

STRATIGRAPHY

The Trondheim region contains a Cambro–Silurian sequence of western, eugeosynclinal facies (see Fig. 196, p. 272). The stratigraphic relationships are clearest in the northwestern parts of the region, where the area between the Orkla and Gaula river valleys (Orkdalen and Gauldalen) is of a special importance to the stratigraphy. Here the rocks are practically unmetamorphosed; thus a number of formations have been dated by fossils, and most of the names of the groups and formations have been derived from this district (Kiær, 1932; Vogt, 1945). Professor Chr. Oftedahl has kindly placed unpublished information concerning this area at the disposal of the writer.

With regard to the relation of the oldest part of the Trondheim region sequence to its substratum, the sequence is known to be normally above Precambrian rocks in the northeastern part of the region. In the Surnadal synclinorium in the northwest, the sequence is apparently conformable with the underlying deposits of assumed Eocambrian and late Precambrian age.

The lowest part of the Trondheim region sequence is a group of mainly pelitic schists beneath the volcanic deposits of the *Støren group*, which is assigned to the *Røros group*. Objections can be raised to the use of the term "Røros group" for a stratigraphic unit in this part of the Trondheim region, and it might be more appropriate to use the name "Trondheim group", as the deposits beneath the greenstones west of Trondheim were referred to as "*de ældste Trondhjemslag*" (oldest Trondheim deposits) by Th. Kjerulf. The rocks of the Røros group are mica schists and phyllites, partly of normal pelitic composition, partly of schists rich in plagioclase and other lime-bearing silicates.

The basaltic volcanic rocks of the overlying Støren group are 2500 m thick in the section along the River Gaula. Very thick piles of massive meta-basalts are present in the area around Løkken mine. Pillow structure is common in the greenstones of this and other regions. The massive basaltic rocks of the Løkken district probably represent a centre of volcanic activity. At Vassfjellet 20 km south of Trondheim there is another volcanic centre with massive meta-basalts (Figs. 43–45). In many other areas a major part of the greenstones are less massive and much more schistose, indicating that

Fig. 43 *Meta-basaltic breccia or agglomerate, Støren group. Road north of Hjerkinn (62° 16′, 9° 37′).* The match serves as a scale

Fig. 44 *Meta-basalt with pillow structure (large pillows), Støren group. North of Løkken mine (63° 8′, 9° 42′)*

Fig. 45 *Meta-basalt with pillow structure* (smaller pillows of fine-grained greenstone). Støren group. Locality as in Fig. 44

they were tuffs deposited at some distance from a volcanic centre. The Støren volcanism is not exclusively basaltic, as rhyolitic volcanites are also found interbedded with the greenstones, in modest quantities compared with them, however.

Associated with the greenstones are deposits of iron sulphide, pyrite and pyrrhotite; these are distinctly layered and banded rocks, which differ from the epigenetic sulphide deposits of the same area in being practically free of copper. They are assumed to have been deposited by bacterial action under anaerobic conditions, during periods when the sea water was particularly rich in sulphur and iron compounds originating from volcanic exhalations. Another type of deposit interbedded with the greenstones is jasper, a tough chert coloured red by finely disseminated hematite. With increasing metamorphism the jasper becomes a black quartz–magnetite rock.

In some areas limestones and other sediments are interbedded with the greenstones of the Støren group.

The Stokvola conglomerate above the Støren greenstones marks the boundary with the overlying Hovin group deposits. Red jasper is conspicuous among the boulders of the conglomerate, besides greenstones and trondhjemitic intrusive rocks. The conglomerate does not form a continuous layer, but occurs over a wide area in the northwestern part of the Trondheim region, with the type locality on Stokvola mountain about 40 km northeast of Trondheim. In some localities the Stokvola is a sedimentary breccia with sharp-edged fragments. Thus at Høidalskammen east of Løkken mine sharp-edged blocks of jasper more than 1 m across are found, indicating that transport must have taken place by sliding, under conditions of considerable local relief. Conglomerates also occur in other parts of the Trondheim region, probably at a corresponding stratigraphic level.

The sediments above the Stokvola conglomerate are assigned to the Hovin and Horg groups, and they occur in the wide northeast-trending Horg syncline or synclinorium, obliquely cut by the Gaula valley (Fig. 46). The Hovin group, immediately above the Stokvola conglomerate, is divided into a lower part with fossiliferous formations below the polygenous Volla conglomerate and an upper part above this conglomerate. The Hovin sediments were deposited under unstable conditions—partly under conditions of considerable local relief with a certain amount of penecontemporaneous erosion in places.

In the Hølonda area (63° 7′, 10°) (Vogt, 1945) and in the area immediately to the west (Chadwick and others, 1963) the *Hølonda formation* lies directly above the Stokvola conglomerate. It contains deposits of contrasting types. Prominent in the western parts of the area are sandstones and greywackes, often with graded bedding and high plagioclase contents, and conglomerates and breccias with small fragments of red jasper (the *Gaustadbakk breccia* of Th. Vogt). Interbedded with the above-mentioned deposits are fossiliferous sediments of "normal" type, namely shales and impure limestones, which predominate in the eastern part, in Hølonda. The *Bogo shale* member, just above the base of the Hølonda formation, consists of 60 m of black shale with well-preserved graptolites (Dichograptus, Didymograptus, Tetragraptus), indicating a horizon in the middle part of the Arenig, corresponding to the upper part of 3b in the Oslo region (Blake, 1962). In the Hølonda type area, the impure limestones of the formation contain a rich and tolerably well-preserved fauna of brachiopods, gastropods, cephalopods and trilobites. The trilobites belong to the genera Illænus, Niobe, Pliomera, Gonotelus and Calliops (?), indicating old Middle Ordovician (Llanvirn, Llandeilo) or older (Arenig) age. According to a written communication from Professor Nils Spjeldnæs the fauna may belong to more than one faunal horizon and may represent a comparatively great span of time.

The *Hølonda porphyrites*, of andesitic composition, are distinguished by their large insets of plagioclase. They mostly occur as concordant intrusive bodies or dikes, but pyroclastic fragments of Hølonda porphyrites have been found in *Hølonda limestones*, and the suite of rocks seems to have resulted from volcanic and subvolcanic activity contemporaneous with the Hølonda formation. This formation has an upper member of sandstone, overlain by a conglomerate (the *Sundsetkjølen conglomerate*) and above this by the *Hareklett formation*, consisting of alternating layers of black shale and rhyolitic tuff or tuffite. A large boulder that must have come from the Hareklett formation has yielded a number of fossils, particularly brachiopods and trilobites. Caradoc age is indicated by the presence of the trinucleid trilobite Cryptolithus.

Above the Hareklett formation is the Volla conglomerate, marking the base of the upper part of the Hovin group. The polygenous Volla conglomerate occurs widely in the northwestern parts of the Trondheim region. Its boulders, generally well rounded, are composed of sediments, volcanic and intrusive rocks of Caledonian types, and gneisses and granites probably derived from the Precambrian substratum.

Between the Volla and the overlying Lyngestein conglomerates a relatively thin formation of rhyolitic tuff, the *Grimsås rhyolite*, represents the deposits of the upper part of the Hovin group on the northwestern side of the Horg syncline.

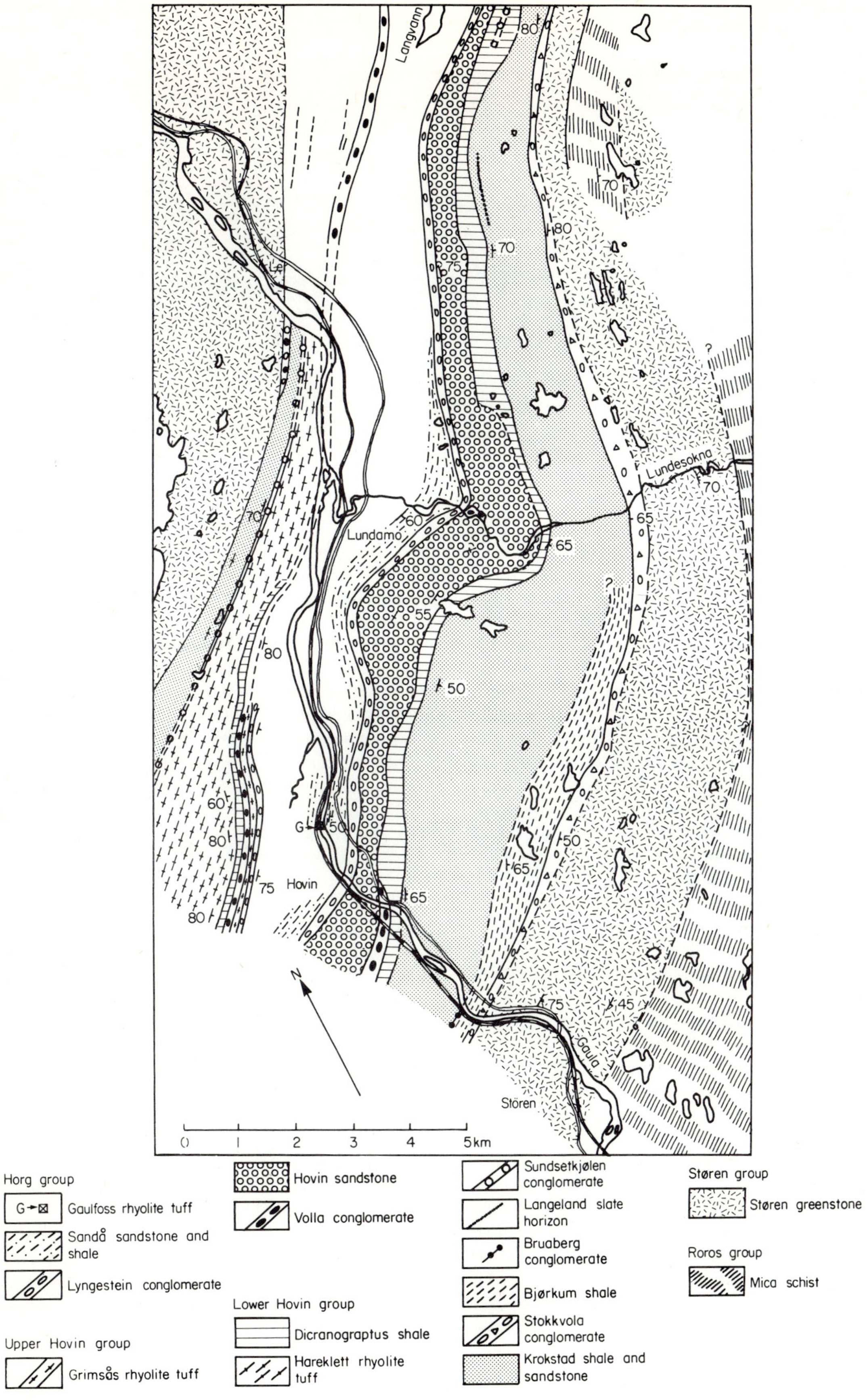

Fig. 46 *Geologic map of the Horg syncline*. By Professor CHR. OFTEDAHL

Fig. 47 *The Lyngestein conglomerate, road section near Hovin, Gauldalen.* From O. HOLTEDAHL (1953)

A rather different sequence is found on the southeast side of the syncline. Above the Støren greenstones and the Stokvola conglomerate is the *Bjørkum shale*, a greenish-grey argillite, very rich in chlorite in some parts. A slate in the same stratigraphic position as the Bjørkum shale, in the northern part of the Horg syncline, has yielded a specimen of a Trinucleus, T. forosi, indicating late Arenig age. The *Krokstad formation* overlying the Bjørkum shale has a maximum thickness of 600 m. It consists of alternating 1 cm to 1 dm thick layers of shale and sandstone, and includes a conglomerate (the *Bruaberg conglomerate*) with boulders of jasper and large fragments of white marble. It is overlain by the *Dicranograptus shale*, a 200 m thick formation of black shale containing numerous layers of sandstone to the north. It contains graptolite assemblages indicating the zones of *Dicranograptus clingani* and *Pleurograptus linearis*, of Middle Ordovician (Caradoc) age.

The Volla conglomerate lies above the Dicranograptus shale and is itself overlain by the *Hovin sandstone* which represents the upper part of the Hovin group on the southeast side of the syncline. The Hovin sandstone, 500 m thick, has alternating layers of greywacke sandstone, siltstone and shale, and often contains fragments of shale. A tuffitic horizon in the upper part most probably corresponds to the Grimsås rhyolite.

The uppermost part of the preserved sequence in the Horg syncline (Horg group) is the Lyngestein quartzite conglomerate (Fig. 47) and the overlying *Sandå formation* of sandstones and shales, also with rhyolitic tuffites. The deposits are of the same type as occur in the underlying formations.

It is apparent from the previous section that the deposits assigned to the lower part of the Hovin group, between the Stokvola and Volla conglomerates, contain a number of fossiliferous formations ranging from late Lower Ordovician to Middle Ordovician. The underlying deposits in the Røros and Støren groups must thus be Early Ordovician and older, and the overlying upper Hovin and Horg groups must be Upper Ordovician and possibly younger. The Lyngestein conglomerate has been correlated with the *Vojtja conglomerate* in Västerbotten, Sweden, and has thus been assumed to mark a horizon at the Ordovician–Silurian boundary. Consequently Silurian age has been assumed for the Sandå formation, lying above the Lyngestein conglomerate.

The sandstones in the Horg syncline and adjacent areas are of greywacke type, and this type of facies with alternation of sandy and more fine-grained sediment is similar to that of the Alpine Flysch (Figs. 48, 49). The conglomerates give evidence of vertical movement during deposition; this is particularly true of the Stokvola conglomerate with its large angular boulders of jasper, which in some localities measure more than 1 m across.

Even though the fossiliferous formations cannot be traced far outside their type areas, the Stokvola and Volla conglomerates are widely distributed and can be recognized over large areas in the northwestern part of the Trondheim region. This is also true of the volcanic deposits.

Outside the Hølonda–Horg district, fossiliferous deposits with graptolites have been found at Lånke, about 20 km southeast of Trondheim. According to a personal communication from Professor NILS SPJELDNÆS, it may be possible to recognize Glossograptus and Dicellograptus, which might indicate Llandeilo age.

The Kalstad limestone is absent in the Hølonda–Horg area, but occurs about 30 km WSW of the

Fig. 48 *Siltstone with sandy bands, from the lower part of th Hovin group. Fættenfjord, a branch of Trondheimsfjord, 30 km ENE of Trondheim.* (The lobe protruding from the sandy layer is probably a secondary feature)

area in the Orkla valley, resting directly upon Støren greenstones. Here the older Hovin sediments, if they were ever present, must have been eroded away before the deposition of the limestone. The Kalstad limestone contains rich and well-preserved flora and fauna, particularly calcareous algae and corals, which correspond closely to the flora and fauna of the Mjøsa limestone and indicate late Middle Ordovician age.

In the area on the inner part of Trondheimsfjord and further northeast along Lake Snåsavatn, the stratigraphy has been clarified by the work of Carstens (1960a) and Peacey (1964). East of the inner part of Trondheimsfjord the Tømmerås anticlinorium of Precambrian rocks extends northeastwards and reaches the southeast shore of Lake Snåsavatn (Fig. 50). According to Peacey it consists of a complex of leptites (fine-grained quartzo-feldspathic rocks of volcanic derivation) with granitic gneisses and a band of sediments. Along the south, west and north sides the Tømmerås anticlinorium is normally overlain by Eocambrian (?) and Cambro-Ordovician deposits, and, remarkably, the outcrops of the lithologic units within the massif are substantially parallel to the erosion surface upon which the younger sediments were laid down.

The oldest deposits above the Tømmerås massif are those of the *Leksdalsvann group*, overlying the southwestern part of the massif. This group has a lower unit of dark fine-grained sandy siltstone with thin pale bands or lenses of calcareous material, and an upper unit of arkosic sandstones. Primary sedimentary features such as cross bedding and grading are well preserved. From a maximum thickness of about 500 m, the Leksdalsvann deposits thin and wedge out to the north and south; correlation of the grain size and sorting with the thickness shows this to be a primary feature. The lithologic facies of the Leksdalsvann group recall the Sparagmite deposits and suggest a corresponding age.

The Cambro-Ordovician deposits can be seen to rest upon the rocks of the Tømmerås massif on the southeast shore of Lake Snåsavatn, on both sides of the gap in these deposits where the Tømmerås rocks reach the shore. A basal conglomerate with boulders of the underlying leptites is present. The overlying sequence is a basal quartzite, a member consisting of greenschists alternating with amphibole-bearing mica schists, and a mica schist member, partly with amphibole. Above these deposits is the *Steinkjer conglomerate* with boulders of jasper and other rocks, directly overlain at this point by the *Snåsa limestone*. At the

Fig. 49 *Banded argillite and siltstone, Hovin group. Heirsane near Levanger (63° 45′, 11° 20′).* Photo Fr. Chr. Wolff

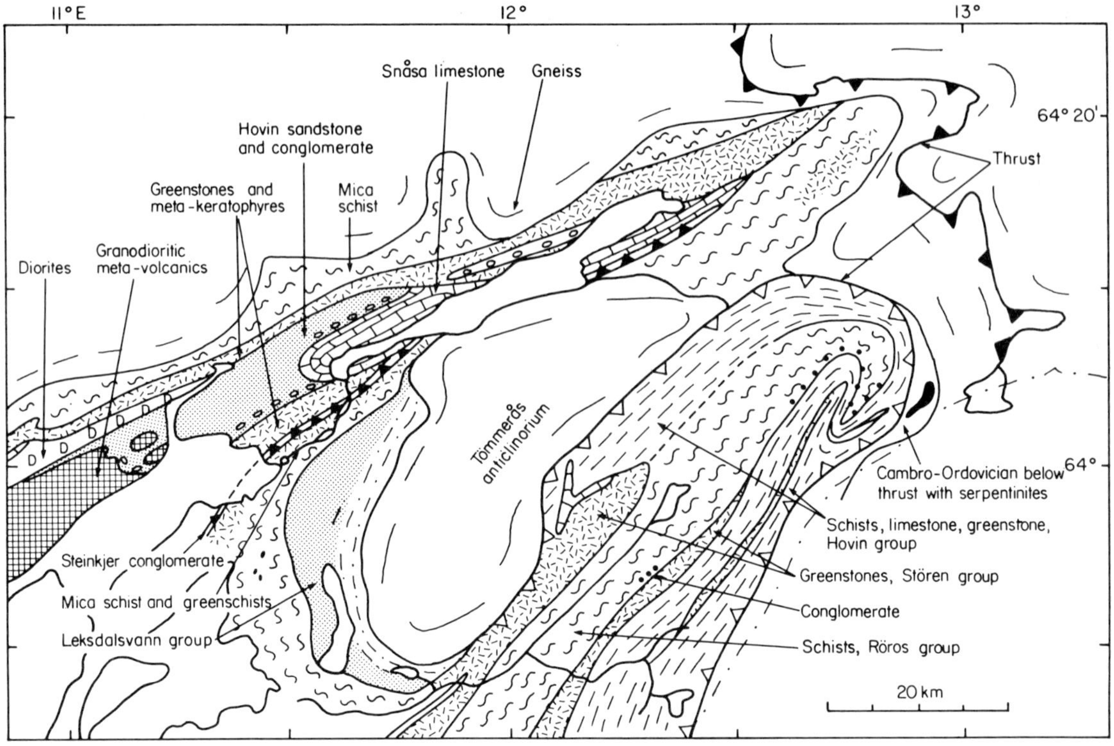

Fig. 50 *Geologic map of the Tømmerås anticlinorium and surroundings.* Based on maps by H. CARSTENS, J. S. PEACEY and FR. CHR. WOLFF

gap on Lake Snåsavatn the sequence is thin and incomplete, measuring about 350 m, but it thickens rapidly to roughly 2000 m within a distance of about 20 km, to the southwest and northeast.

On the inner parts of Trondheimsfjord, in the Steinkjer–Malm district, the Steinkjer conglomerate, previously mentioned as outcropping on Lake Snåsavatn, is a correlate of the Stokvola conglomerate horizon and separates mica schists and greenschists corresponding to the Røros and Støren groups, from the overlying deposits of the Hovin group. North of Steinkjer the Steinkjer conglomerate is overlain by mica schists and further by a sequence of alternating greenschists and meta-keratophyres, interpreted as a volcanic pyroclastic succession. Interlayered in this sequence are oxydic iron deposits, worked at Fosdalen mine on the innermost part of Trondheimsfjord. Above the greenstone–keratophyre division follows the Snåsa limestone, some 100 m thick. This has yielded badly preserved fossils, including gastropods of Ordovician types (recalling Hormotoma and Maclurites) and hemispherical fossils, which make Middle Ordovician age probable. Granodioritic gneisses—shown by CARSTENS (1960b) to be metamorphosed pyroclastic rocks, a more coarse-grained facies of the rhyolites—are members of the same sequence also.

The greenstone–keratophyre succession, the Snåsa limestone, and the granodioritic gneiss of volcanic origin, are considered as equivalent to the lower part of the Hovin group. They are overlain with an erosional unconformity by a polygenous conglomerate and a sandstone and shale division which can be correlated with the Volla conglomerate and Hovin sandstone in the upper part of the Hovin group.

A limestone with indeterminable fossils on Tautra island, 20 km northeast of Trondheim, is possibly equivalent to the Snåsa limestone, as may also be the case with other limestone occurrences in the area between Tautra and Snåsa.

Southwest of the mouth of Trondheimsfjord are two large islands, Hitra and Smøla. In the southern part of Smøla the southwesterly of the two islands, the *Smøla limestone* has yielded tolerably well-preserved fossils, mostly silicified gastropods of genera Eunema, Lophospira, Helicotoma, Hormotoma and Maclurites (STRAND, 1932). Dr. ELLIS L. YOCHELSON (written communication) after a study of the writer's paper expressed the opinion that the fauna were not older than Middle Ordovician. Associated with the limestone are basaltic–andesitic and rhyolitic lavas and a sandstone with conglomerates that may be equivalent to the Hovin sandstone. According to CARSTENS (1955) the lavas on Smøla correspond to the greenstone–keratophyre succession of the Snåsa syncline. The two districts seem to have many similarities in stratigraphy and the Snåsa and Smøla limestones may well be contemporaneous. Sandstones, probably corresponding to the Hovin sandstone, are also known from the Ørlandet peninsula at the mouth of Trondheimsfjord (RICHTER, 1958).

In the eastern part of the Trondheim region, a zone of greenstones and other volcanic rocks of the Støren group can be traced along the south-eastern and eastern boundaries of the region, from Folldal in the south, further along the north-west side of the Glomma river, and across the Gauldalen, Tydalen and Meraker valleys to Skjækerhatten (64°, 12° 30′) in the northeastern part of the region. The zone has been known for a long time, its northern part having recently been mapped by WOLFF (1960, 1964). In Tydal the volcanic zone has been mapped and described in great detail by KISCH (1962). Here there is a sequence with steep dip and N–S trend which consists of alternating basic and acidic volcanic rocks and sediments: oligoclase amphibolites and blastoporphyric meta-basites, meta-rhyolites and rhyodacites, pelitic schists and gneisses. The intimate and regular banding, with bands of amphibolite of thickness down to 1 cm, leaves no doubt as to the volcanic nature of the igneous rocks. To the south of the area described by KISCH, in the upper part of Gauldalen (62° 55′, 11° 35′), a black schist occurs which contains Dictyonema flabelliforme, first found by J. H. L. VOGT in the 1880's (STØRMER, 1941). VOGT (1940) showed that a porphyric greenstone, corresponding to the blastoporphyric metabasites of KISCH, occurred in the Dictyonema-bearing schist. (VOGT interpreted all the basic igneous rocks in the zone under consideration here—his "Fongen injection zone"—as intrusive rocks, but this is certainly not correct.) He also pointed out that black carbonaceous schists, possibly equivalent to the Dictyonema schist, are common within the greenstone and amphibolite zone. The occurrence of a volcanic greenstone in the Dictyonema schist proves that the volcanic rocks in the eastern zone are of the same age as the Støren greenstones in the west, as the latter are overlain by the Bogo shale with early Ordovician graptolites. The older parts of the Støren volcanic group are thus at least as old as earliest Ordovician.

To the west of the zone with volcanic rocks are found pelitic schists, the *Drøia schists* of VOGT. A quartz conglomerate intervenes between the volcanic zone and the schists west of it; this is known as the *Bukhammer conglomerate* (KISCH) in Tydalen and the *Gudå conglomerate* in Meraker to the north (WOLFF, 1964). The pelitic schists in question are part of a mass which intervenes between the zones of Støren greenstone in the east and west. The zone is some 50 km wide in the section along Gauldalen, but shrinks to some 10 km in the section along the Meraker valley due east of Trondheim. The schists are often black and carbonaceous and are staurolite schists (the millstones of Selbu, formerly quarried) and andalusite schists in zones of higher metamorphism, while the lime silicate schists and gneisses in Gauldalen are derivates of calcareous schists. In the section along the Meraker valley the eastern zone of greenstones has rocks of the Hovin group on the east side and the schists just mentioned on the west side. Provided that the boundaries are not wholly of a tectonic nature, the schists in the west must be older than the Støren greenstone and must form a central anticlinal zone. WOLFF (1964),

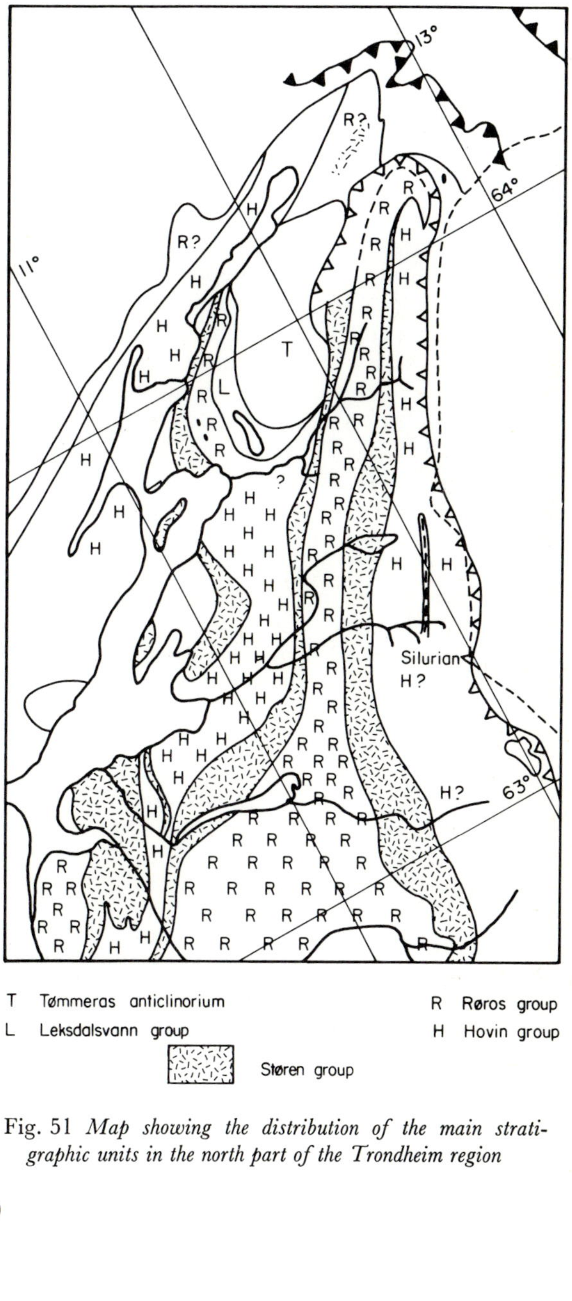

Fig. 51 *Map showing the distribution of the main stratigraphic units in the north part of the Trondheim region*

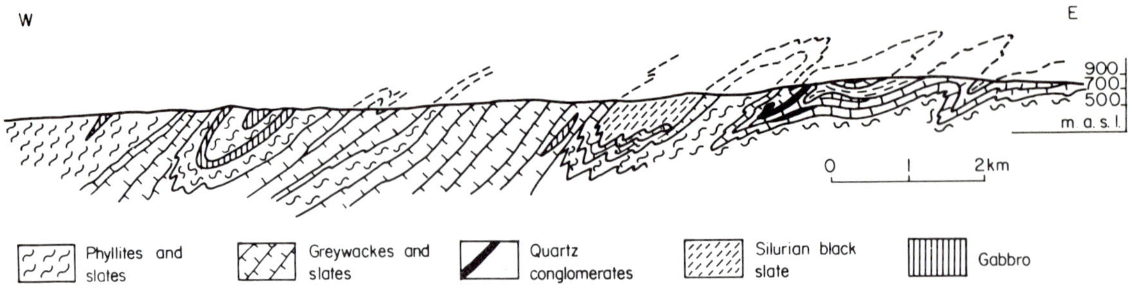

Fig. 52 *Section of the sediments of the Hovin group north of the Meraker valley.* By DR. ANNA SIEDLECKA; east end of section at the international border

in accordance with CARSTENS (1921), but in opposition to VOGT (1940), considered the schists in the central zone (the *Gula schists* of TH. KJERULF) as belonging to the Røros group (Fig. 51).

East of the eastern band of Støren greenstones in the Meraker section is a grey phyllite, locally with a limestone, followed to the east by a zone of volcanic greenstone. This part of the sequence represents the lower part of the Hovin group. According to Dr. ANNA SIEDLECKA, who has recently mapped them in detail (Fig. 52), the younger rocks further east have a lower division of greenish to grey phyllites with subordinate layers of greywacke, a higher division with greywackes predominating and with subordinate amounts of phyllite, and a top division of black argillites. The same black schists yielded early Silurian graptolites (GETZ, 1890) at Kjølhaugene north of the Meraker valley. According to a review of the fauna by Dr. G. L. ELLES (KIÆR, 1932) there are species of Monograptus and Rastrites, indicating the zone of *Monograptus convolutus* in the upper part of the Lower Llandovery (*étage* 7 in the Oslo region). In the greywacke below the Silurian argillite are quartz and quartzite pebble conglomerates, which may correspond to the much coarser and thicker Lyngestein conglomerate to the west in Gauldalen. Some of the schists in the sequence, containing biotite porphyroblasts are very similar to the "Stuedal schists", a type of rock common in the areas south of Meraker. It thus seems probable that the Hovin group deposits extend for a considerable distance south of the Meraker valley.

MAGMATISM

The oldest phase of magmatic activity in the Trondheim region is that connected with the volcanism of the Støren group. That intrusive rocks were formed at the same time is well testified for in the Løkken district (CARSTENS, 1951), where gabbroid and trondhjemitic intrusive bodies occur plentifully in the rocks of the Røros and Støren groups, but are virtually absent in the younger Hovin rocks. The *Hølonda porphyrites* occurring east of Løkken in the Hølonda formation are very easy to distinguish from the other intrusive rocks. Similar conditions are found in the Otta district in Gudbrandsdalen (STRAND, 1951a) where nearly all the intrusive rocks occur in the lower part of the sequence, corresponding to the Røros and Støren groups. The basic volcanic and intrusive rocks of this old phase were described by GOLDSCHMIDT (1916b) as his *Stamm der grünen Laven und Intrusivgesteine*. The ultramafic dunites and serpentinites must also belong to the old phase, as they have not been found in rocks younger than the Støren group in any area where a reliable stratigraphy has been established. A possible exception must be made for some ultramafic rocks which occur as parts of larger massifs of gabbroid rocks. The dunites occur as lens-shaped bodies, varying in size from a few metres to some kilometres. They contain olivine very rich in Mg (Fo 90), most of them having been altered to serpentine and sometimes to talc. The larger bodies contain deposits of chromite. Ultrabasic rocks from the Trondheim region were described by CARSTENS (1918).

The gabbros occur in bodies of all dimensions from lenses measuring a few metres to large phacoliths several kilometres across. The small bodies are commonly completely saussuritized and uralitized. Gabbroid intrusive rocks forming sills are commonly recrystallized to amphibolites. The larger bodies are not entirely free of similar alterations, but in parts may be of fresh, unaltered rock. The gabbro massifs of Tronfjell and Fongen (CARSTENS, 1920 and HOLMSEN, 1950), consist of olivine gabbro and norite and also contain peridotitic lenses and rocks tending towards anorthosites with about 80% plagioclase. The *Hyllingen gabbro* in Tydal, south of Fongen, was described in great detail by KISCH (1962). One part of the massif is of gabbros with or without olivine, whilst another part is meta-gabbro with hornblende or cummingtonite, cut by veins of pegmatite. Parts of the gabbro show compositional layering with bands of thickness down to 1 cm.

The light-coloured trondhjemites or "oligoclase granites" occur as medium-grained intrusive rocks, as porphyrites, and as aplites and pegmatites. This last type of rock intrudes and injects the older part of the sequence (Fig. 53). A varied assemblage of intermediate and acidic rocks from the Tydal district were described by KISCH as mafic hornblende quartz diorites, a group of iron-rich granites and granodiorites, and typical trondhjemites. Some of the rocks were gneissic.

A special type of acid intrusive rock of trondhjemitic affinities is the felsite of the Løkken area described by C. W. CARSTENS. This is a dense albite–quartz rock that occurs as lens-shaped bodies, the larger ones measuring some hundreds of metres across. Some coarse-grained trondhjemitic rocks occurring in the Støren group are strongly gneissose and were called *protogine granites* by TH. KJERULF. Some of these rocks may be of volcanic (pyroclastic) origin, as indicated by H. CARSTENS for rocks of this type which form a band in the Støren group greenstones west of Trondheim. Another example is the "granulite" of Folldal, about 1 km wide and 12 km long, which has distinct layers of varying composition, as described by VOGT (1890).

Volcanic activity in middle and late Ordovician times is testified for by the acid volcanic rocks in the Hovin and Horg groups of the type area. There are no intrusive equivalents to these rocks in the same areas, but the intrusive rocks of the

Fig. 53 *Trondhjemitic intrusions in mica schists of the Röros group. Langfredagsnes, Meraker valley (63° 28′, 11° 30′).* Photo FR. CHR. WOLFF

Snåsa and Fosen synclinoria, and on the islands of Hitra and Smøla, intrude into sediments of the Hovin group and must thus be of late Ordovician or younger age. The rocks in question have some similarity to those of GOLDSCHMIDT's *Opdalit–Trondhjemit-Stamm*, typified by the rocks of the Oppal–Innset massif, situated near the west margin of the Trondheim region at 62° 40′, which contains a differentiated series of rocks ranging from norites to trondhjemites. The norites, with inclusions of peridotites and pyroxenites, are very coarse-grained rocks found in the central parts of the massif. They pass into quartz–biotite norites. More acidic rocks present are diorites with hypersthene and mica, and common hornblende-bearing diorites. A pyroxene-bearing quartz monzonite, with a chemical composition near to the calculated average for igneous rocks, was called *opdalite* by GOLDSCHMIDT. The narrowing western part of the Oppdal–Inset massif is composed of trondhjemites ("oligoclase granites") and similar rocks with considerable amounts of potash feldspar. Another large body of intrusive rocks occurs in the southeastern part of the Trondheim region, where it extends for 35 km ENE of Dombås. In the west it consists of trondhjemite and in the east of gabbro; intermediate rocks are also present.

The Ørlandet peninsula which protrudes northwest of the mouth of Trondheimsfjord is the most southwesterly part of the Fosen synclinorium. (RICHTER, 1943, 1958.) The Hovin sandstones and shales which extend along the southeast part of the peninsula are intruded by a red gneissic granite which interfingers with the sandstone over a distance of 1 km. The granite contains inclusions of the sandstone as large as small hillocks; these inclusions of sandstone in the granitic gneiss are also found further to the northeast. A similar migmatitic rock is found in the northwestern part of Ørlandet, where a colour photo reproduced in RICHTER's paper of 1943 shows the granite brecciating a greenish sedimentary rock. The same migmatites are also found northwest of Bjugnfjord and on Tarva island, northwest of Ørlandet. The middle part of Ørlandet is occupied by a light massive granite or granodiorite, younger than the reddish gneissic granite, but not post-Caledonian, as boulders from the granite are found in the Devonian conglomerates.

On the islands of Hitra and Smøla the rocks are known as a result of the publications of SCHETELIG (1913), CARSTENS (1924b), RICHTER (1943, 1958) and KOLLUNG (1963). SCHETELIG characterized the rocks as a differentiated suite ranging from ultrabasic and basic to granitic, with diorites, trondhjemites and granodiorites predominating. In many cases the more acid, younger rocks break through the older more basic ones and brecciate them. There are gabroid and quartz-dioritic dykes, and swarms of granite pegmatites break through all the other rocks.

The diorites, which form a major part of the rocks, range from dark types near gabbro to light quartz-bearing rocks. The granodiorites are red-coloured in some places, an unusual feature among the intrusive rocks of the Trondheim region.

In the southeast part of Smøla, where dioritic intrusive rocks border on nearly unmetamorphosed sediments, C. W. CARSTENS found large inclusions of limestone in the diorite, which had exerted a strong contact metamorphism on the limestone. It is of considerable interest to note that one of the limestone fragments could be seen to have been folded before it was immersed in the magma.

In the northeast part of Hitra, according to KOLLUNG, a large massif of diorite has been intruded by younger trondhjemites and granodiorites, which in the northern part of the island have intruded and injected a suite of strongly metamorphic sediments, evidently extending from the Fosen synclinorium in the northeast. Micaceous gneisses and augen gneisses were thus formed by injection and metasomatism. Frøya, the large island north of Hitra is composed of granitic rocks, most of them gneissic (REUSCH, 1914). REUSCH found inclusions of limestones, amphibolites, schists and gneisses in the granites; large fragments of limestones were interwoven with granitic material. HOLTEDAHL (1944) reported massive and partly very coarse-grained granites in the northern part of Frøya and on the island groups of Froan and Halten northeast of Frøya.

GEOLOGIC SETTING

As previously mentioned the sediments of the Otta nappe pass into the southwestern part of the Trondheim region; this part of the region belongs to the same tectonic unit as the Otta nappe. The boundary between the sparagmites and the overlying Trondheim sediments in the southeast coincides with the thrust line of the nappe, at least for some distance ENE. Here, two lenses of the nappe with parts of its crystalline basement rocks are preserved between schuppen of sparagmites on the southeast side of the Trondheim region boundary. On the west side, immediately north of the Otta valley the thrust line of the Otta nappe is known to be present at the boundary also.

The southern part of the Trondheim region, between Hjerkinn in the West and Folldal in the southeast, has a synclinal structure with the older sediments along the margins and the younger ones in the middle part (GEIS, 1961a).

In the Hjerkinn district (STRAND, 1951a) and in the Kongsvoll district north of Hjerkinn (DU DRESNAY, 1950) the light sparagmites and flagstones in the west are overlain by gneisses and augen gneisses (of sedimentary origin) and further to the east by mica schists and greenschists with a body of serpentinite. In the Folldal district (MARLOW, 1935) mica schists, phyllites, and amphibole-bearing quartz schists and greenschists occur along the southeast margin of the Trondheim region, together with a large number of serpentinite bodies. Mica schists, often with a brownish hue, predominate between Hjerkinn and Folldal. They are probably equivalent to the Sel mica schist in the Otta nappe sequence and to the lower part of the Hovin group, in accordance with their position in the middle of the supposed synclinorium.

The zone with serpentinites along the southeast boundary of the Trondheim region extends to the northeast as far as the Røros district, but serpentinites are not found north of Lake Aursunden (62° 40′). According to TÖRNEBOHM (1896) the sparagmites east of Røros are overlain by dark mica schists or phyllites; these he assigned to the *Brek schist* group. Bands of greenschist are interlayered in the mica schists around Røros (J. FÆRDEN, personal communication) and in the area south of Røros (HOLMSEN, 1950).

The Skardørfjell–Sylene district is the Norwegian part of a window with Precambrian basement rocks along the international border between 62° 40′ and 63° 10′; in the southern parts sparagmites are found immediately above the basement (BRYN, 1959; SCHAAR, 1962). On the west side of the window the Precambrian and Sparagmite rocks are overlain, with a tectonic contact, by a sequence of biotite-bearing quartz schists, biotite porphyroblasts–mica schists (a petrographic type called "Stuedal schist") and garbenschists with amphibole. The "Stuedal schist" has large visible porphyroblasts of biotite and is often laminated. At the east side of the window a small part of an amphibolite massif, resting above the basement rocks in the Sylene mountains, lies on the Norwegian side of the border. Lower in the massif the amphibolite occurs as thin bands in a plagioclase-bearing mica schist, but upwards it becomes the dominant rock. At Øyfjellet west of Lake Essandsjø, on the west side of the window, a mass of amphibolite of Sylene type rests with a tectonic contact upon the underlying garbenschists and "Stuedal schists", and thus apparently forms a western outlier of the Sylene amphibolite massif. Similar outliers of amphibolite occur in the areas to the north, zones of mylonitization also being recorded (BJØRLYKKE, 1963). In the Tydal section KISCH (1962) described diaphthoritic and probably mylonitized rocks along the boundary between the volcanic and intrusive rocks in the west and the underlying "Stuedal schists" in the east. It thus seems that the volcanic and intrusive rocks were thrust upon the tectonically underlying "Stuedal" and other schists to the east in the areas under consideration.

The investigations of STRAND and DU DRESNAY in the southwesterly part of the Trondheim region have previously been reported. Further north in Drivdalen the N–S direction of the boundary in this part of the Trondheim region is due to a sharp flexure in the E–W-trending axes of earlier folds, bent down to the east with an axial plunge of up to 45° (WEGMANN, 1959). In the Oppdal district (HOLMSEN, 1955) low-grade rocks of the Trondheim region (green-coloured greywackes with interlayered meta-basalts, conglomerates and breccias) rest above higher grade rocks of the Gneiss region in the west, on a tectonic boundary with mylonites at the base.

The central, more southern parts of the Trondheim region are inadequately known; the pelitic and silty meta-sediments contain no marker horizons and little or nothing can be said of their stratigraphic and structural relationships.

Those parts of the Trondheim region situated south of Trondheimsfjord and northwest of a line drawn in a northeasterly direction through Støren (63° 2′, 10° 20′) are the stratigraphic type areas. Along the boundary to the underlying basement complex, the lower (Røros) part of the sequence rests upon sediments of probable Eocambrian age with an apparent conformity. The tectonic conditions seem uncomplicated, with the lower part of the sequence lying along the boundary to the Basal gneiss, and the younger (Hovin and Horg) sediments in synclines or synclinoria above the Støren greenstones. In addition, the metamorphic grade of the rocks makes a regular pattern with comparatively high-grade metamorphism (oligoclase–epidote) in the structurally (and stratigraphically) lower rocks just above the basement, and practically unmetamorphosed rocks in the younger groups of rocks in the synclines.

Apparent on the map is the offshoot from the Trondheim region towards southwest in the Surnadal strip, a syncline in the Basal gneiss. According to HERNES (1956a, 1956b) the syncline is overturned to the northwest, very strongly so at Stangvik on Todalsfjord. Southwest of Stangvik the Trondheim sediments disappear due to a rise in the axis of the syncline, but appear again on islands in the fjord off Molde (BUGGE, 1934; HERNES 1956a) and still further west on the north coast of Hareidlandet (GJELSVIK, 1953). The Surnadal strip can thus be traced over a distance of 170 km.

Between the Orkla and Gaula valleys the Hovin–Horg deposits form the generally NE-trending *Horg synclinorium*, 25 by 15 km large, centring about 63° 5′, 10° 0′. Detailed maps of parts of this area were published by VOGT (1945), CHADWICK and others (1963) and CHALOUPSKY (1963). The area mapped by CHADWICK and associates in the northern part of the synclinorium shows rather complicated tectonics with the axial trend of the folds varying from east to northwest. Faults, partly with horizontal displacements, are also of importance. VOGT in the Hølonda area (east of CHADWICK's area) and CHALOUPSKY in the area west of Støren in the southern part of the synclinorium mapped a number of strongly compressed folds within the synclinorium. Most of the folds are overturned towards the southeast, but there are also folds (less strongly) overturned in the opposite direction. The southeastern part of the synclinorium continues to the northeast as a simple syncline, obliquely cut through by the Gaula valley north of Støren. The syncline widens in the area south and southeast of Trondheim, where the Hovin sandstone with its rhyolitic volcanic rocks forms a 6 km wide syncline or synclinorium at the head of the fjord due east of the city (CARSTENS, 1960). Wide synclines and anticlines formed by rocks of the Støren and Hovin groups, the latter with rhyolitic volcanic rocks, are also found further north on the east side of Trondheimsfjord. At Forbordfjell (63° 30′, 10° 55′) greenstones of the Støren group form a 5 km wide synformal structure, underlain on all sides by the younger deposits of the Hovin group, according to a map of CARSTENS (1960). This structure must be either a mushroom-shaped fold or the inverted limb of a large recumbent fold. The degree of metamorphism in this district is low and traces of fossils are preserved in the limestones.

In the Tømmerås anticlinorium the Eocambrian (?) and Cambro–Ordovician deposits are autochthonous in relation to the underlying Precambrian rocks. The Snåsa syncline extending along Lake Snåsavatn northwest of the Tømmerås anticlinorium seems to be of an uncomplicated nature, with the younger deposits in the middle part. The rocks of the Snåsa syncline extend in the southwest to the outer part of Trondheimsfjord, where they are folded, together with the underlying gneiss, into a number of synclines and anticlines with NE axial trend (RAMBERG, 1944).

The *Grong culmination*, a small part of which is shown on the map (Fig. 50), is composed of quartz porphyrites and granitic gneisses of Precambrian age, rather strongly influenced by Caledonian mylonitization. Tectonically the Precambrian rocks of the Grong culmination fall into two parts—an upper one in the periphery of the area and a lower one in the centre of the area; the latter is the Olden massif of Swedish geologists and is regarded by ASKLUND as a part of the Olden nappe. A sequence of limestone, quartzite and phyllite—the autochthonous cover of the Olden massif—is found between the two units in the southern parts of the Grong culmination, near the Swedish border. OFTEDAHL (1956) has drawn the boundary between the two units, considered as the thrust line of the Seve nappe, on the basis of FOSLIE's geologic map.

In the synclinal depression on the southeast side of the Tømmerås anticlinorium the stratigraphic units show a symmetrical arrangement with the east side the rocks of the Hovin group lie at the base of the nappe. On the international border near Storlien, due east of Trondheim, they rest with a well-marked tectonic boundary upon strongly mylonitized rocks, which, according to the geologic map of Sweden, are Precambrian; the overlying Vemdal sandstone appears in the windows along the border (Fig. 54). Recent investigations by ROBERTS have shown the presence of an older set of similar folds with axial trend NE and a younger set of concentric folds trending roughly NNE (Fig. 55).

As mentioned previously the stratigraphical and structural units in the northeast synclinorium can

Fig. 54 *The thrust front at the eastern boundary of the Trondheim region. Looking southwest from the Swedish side of the border at the Meraker road (63° 20′ N).* Photo FR. CHR. WOLFF

older rocks of the Røros group in the middle part. The rocks of the synclinorium have been shown to form a nappe, resting on a thrust plane marked by strong mylonitization above the Precambrian rocks of the Tømmerås anticlinorium (PEACEY, 1964; WOLFF, 1964). The thrust plane can be traced around the synclinorium; on the east side, where it runs along the international border, it lies above Cambro–Ordovician rocks, which are apparently in an autochthonous position above the Precambrian basement of the Seve nappe. Further south the overthrust is directly above strongly mylonitized Precambrian rocks in the windows along the border. In the northeastern part of the synclinorium a quartz conglomerate between the Røros and Støren groups forms a pattern on the map indicative of a fold with a double closure, which is interpreted as a recumbent anticline overfolded to the east and refolded in a NNE-trending synform. The nappe can thus be interpreted as a large recumbent anticline, complicated by folding on a smaller scale. On the be traced to the south. The Hovin group sediments in the east are probably represented by the "Stuedal schists" south of the Meraker valley. The band containing schists of the Røros group between bands of Støren greenstone widens to the south. The eastern band of Støren greenstone can be traced to the southwestern part of the Trondheim region, where there seems to be a synclinal structure with the younger deposits in the middle part. Detailed mapping is required to unravel the true structure.

The distribution of zones of regional metamorphism in a major part of the Trondheim region is shown in the map, Fig. 56. In the northwest the grade of metamorphism increases downwards, and is largely concordant with the stratigraphic and tectonic position. The high degree of metamorphism in the central part of the region also affects rocks in a low stratigraphic position (Røros group), according to the interpretation of the stratigraphy previously set forth. In some cases an increase in the metamorphic grade is apparent in the neigh-

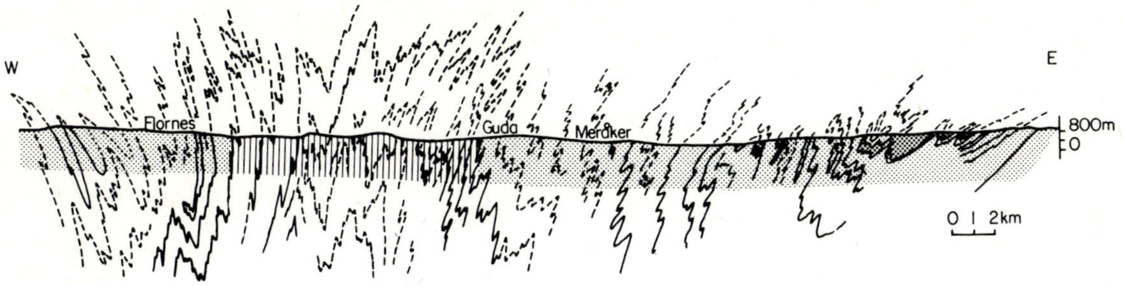

Fig. 55 *Section giving an interpretation of the folding tectonics along the Meraker valley.* By DR. DAVID ROBERTS (left, west; right, east)

bourhood of large intrusive bodies. On his map GOLDSCHMIDT marked a zone of contact aureole around the Oppdal–Innset massif, but he has not described the phenomenon. In the eastern part of the region fine-grained phyllites pass into andalusite-bearing mica schists within a few kilometres on approaching the gabbro massif at Lake Øyungen (62° 45′, 11° 5′) in an area which has been mapped by TORE BIRKELAND, an Oslo graduate student.

There can be no doubt that all rocks of the Trondheim region are allochthonous, and have been moved for considerable distances from their place of formation. In the southwest the rocks of the region are a part of the Otta nappe, and in the northeast the sediments with part of their underlying Precambrian basement rest on a thrust plane above the Olden massif and its sedimentary cover. This thrust plane is the same as that forming the base of the Seve nappe complex of Sweden. Along the east and southeast sides of the Trondheim region the boundary to the underlying rocks is of a tectonic nature. In the northwest, on the other hand, the boundary to the gneiss or to the

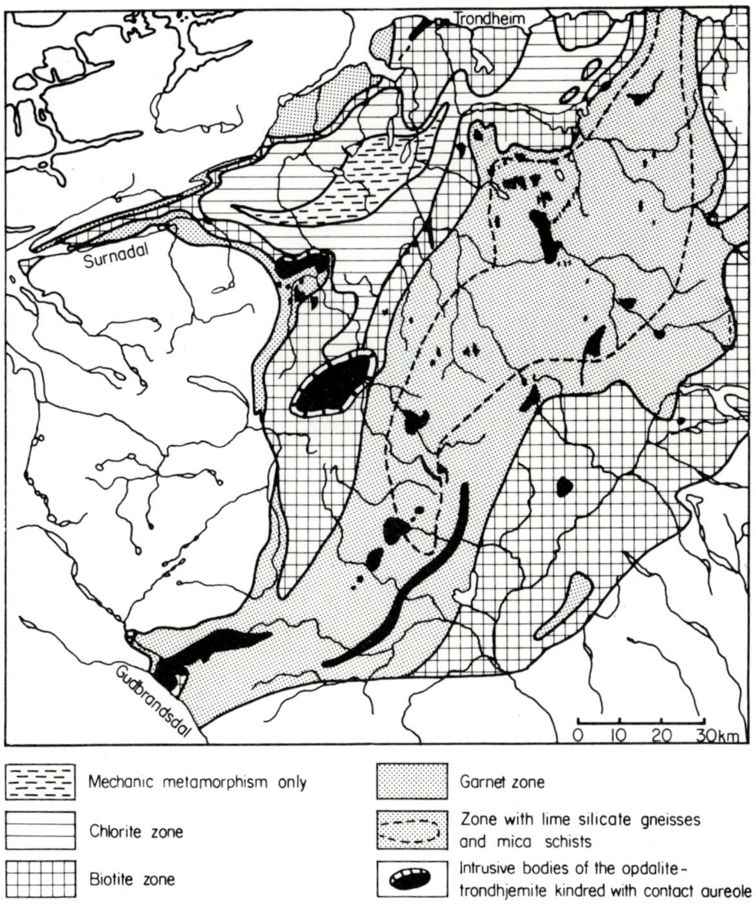

Fig. 56 *Zones of metamorphic grade in the Trondheim region.* From GOLDSCHMIDT (1915)

next underlying deposits (of Eocambrian to late Precambrian age?) has been proved or indicated to be one of normal superposition. The nappe in the northeastern synclinorium is one of a lower order, as here sediments have been moved in relation to other sediments within the region. The thrust zone, which can be traced on the Norwegian side as a continuation of the thrust below the Sylene amphibolite massif, is of unknown tectonic significance at present.

It would be premature in the present state of knowledge to attempt any synthesis of the tectonics of the Trondheim region. In particular, the relationship between the Cambro–Silurian sediments, and the underlying rocks of the Gneiss region in the west and northwest, presents unsolved problems.

On the Swedish side of the border, the Tännfors district in Jämtland (Chapter XXII) may be a continuation of the Trondheim region; possibly there is an outlier of the nappe present on the Norwegian side.

Addendum. TORSKE's paper (1965) gives a detailed description of the area north of Lake Selbusjöen (southeast of Trondheim). Here a sequence of phyllites, meta-greywackes and quartz schists (Røros group) are overlain by greenstones of the Støren group, including a large proportion of pyroclastic deposits. Younger deposits present are conglomerates overlain by andesitic and dacitic tuffs and tuffites (the lower part of the Hovin group).

Mapping during the summer of 1967 by officers of the Geological Survey and graduate students of Oslo University has produced new, important results. In the eastern part of the Trondheim region, about 62° 50′N, the greenstones of the Støren group are in an inverted position below overlying older beds. Probably the greenstones in question are in the inverted limb of a large fold-nappe thrust above the younger rocks east of them. The rocks in the eastern part of the Trondheim region, from the Meraker valley southwards at least as far as the Røros district, belong to the Hovin group (as was indicated on a map by C. BUGGE (1954)).

CHAPTER VI

AREAS WITH CAMBRO–SILURIAN DEPOSITS ALONG THE WEST COAST OF SOUTHERN NORWAY

A great part of the area along the west coast of southern Norway is underlain by Cambro–Silurian deposits. There are three main districts: a southern one comprising Jæren, the Stavanger area and Karmøy; a middle one, the Sunnhordland–Bergen district; and a northern one between Sognefjord in the south and Nordfjord in the north. The first two districts join the Nappe region and pass into it without any marked boundary.

STRATIGRAPHY

Most of the Cambro–Silurian deposits (see Fig. 103, p. 132) display eugeosynclinal facies and can briefly be characterized by their general resemblance to the deposits of the Trondheim region. Volcanic rocks, both basic and acidic, comprise considerable parts of the sequence (Fig. 57). There are in addition many intrusive rocks of the types known from other parts of the Caledonides: ultramafics, gabbros, diorites, trondhjemites and granites. A fair number of sulfide deposits of Trondheim region type are present in the area.

Over most of the area there is little or nothing known about the age and order of deposition of the sediments. Around Os, 25 km south of Bergen, it is possible to distinguish between an older and a younger part of the sequence, the latter having late Ordovician and early Silurian fossils. At the top of the older division is a polygenous conglomerate, the *Moberg conglomerate*, which can be correlated with the Volla conglomerate in the Trondheim region. An impure fossiliferous limestone follows above the Moberg conglomerate. This limestone also occurs on Stord island, where the fossils are very well preserved; Kiær (1930) described 17 species of corals. The fauna agrees very closely with that found in the Gastropod limestone (5a) in the Oslo region. In the Os area there are in addition a quartzite, a quartzite conglomerate (comparable to the Lyngestein conglomerate in the Trondheim region and to the early Silurian sandstone in the northern parts of the Oslo region), and schists with calcareous nodules containing Stricklandia lens and other fossils, indicating the horizon of *étage* 6c (early Llandovery) in the Oslo region. Schists with early Silurian graptolites are found both in the Os area and on Stord. Three other localities (Karmøy, Bømlo, and the northern district) have yielded more or less well-preserved fossils of Late Ordovician or Silurian types.

It is hardly a mere coincidence that fossils have been preserved in the upper part of the sequence only. According to unpublished investigations by Professor Anders Kvale the older parts of the sequence show stronger metamorphism and deformation than the younger parts with the fossils, thus indicating a phase of metamorphism prior to

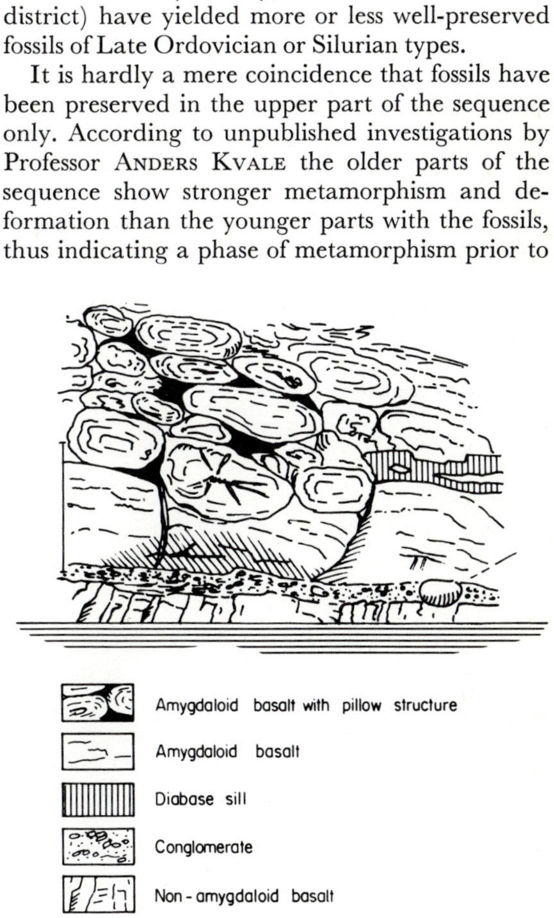

Fig. 57 *Section near Finnås in the middle part of Bømlo* (*69° 45′, 5° 15′*). From Reusch (1888). The scale to the left is 1 m.

the deposition of the younger parts of the sequence. In the Bergen district all the intrusive rocks present are found in the older part of the sequence—not in the younger part. It may also be emphasized that the faunas and lithofacies of the younger deposits are very similar to those of the Oslo region deposits, and must have been laid down under conditions very different from those prevailing during the deposition of the older parts of the sequence.

The stratigraphy of the area northwest of Hardangerfjord has been worked out by NATERSTAD (1965, and personal communication) who divided the sequence into five groups (I–V). Group I consists of gneiss with interbedded quartzite. At Ljones (60° 15′, 6° 10′) the gneiss, which is strongly mylonitized in the lower part, rests upon mica schist. Here the mica schists overlie the Precambrian basement gneiss in an assumed autochthonous or parautochthonous position. Groups II and III both contain volcanic rocks, basic and intermediate as well as acidic, together with greywackes, mica schists, quartzites and green polygenous conglomerates. A quartz porphyry with an overlying marble at the top of group II serves as a widely distributed marker horizon (Fig. 58).

Group IV commences with a green polymict conglomerate with boulders formed from rocks from the underlying sequence (and possibly from Precambrian rocks also). Higher up are sandstones with conglomerates, a grey impure limestone, and calcareous phyllites. Group V consists mainly of black phyllites and dark quartzites, with intermediate and acidic volcanic rocks in the lower part.

The polymict conglomerate at the base of group IV can be correlated with the Moberg conglomerate in the Bergen district. As in the Bergen district, the rocks above the polymict conglomerate (groups IV and V) are less metamorphosed and lack the intrusive rocks that occur in the lower parts of the sequence. The effusive rocks in group V contain unchanged original minerals, such as pyroxene and biotite. Thus groups I–III should be Middle Ordovician and older, while groups IV and V should be Late Ordovician; the upper parts are possibly Early Silurian.

Fig. 58 *View to the northwest from Tveitakvittingen (1299 m) on the northwest side of Hardangerfjord. (60° 20′, 5° 55′).* Area covered by Cambro-Silurian sediments and volcanic rocks; the light rocks in the foreground consist of quartz keratophyre interbedded in greenschists. The front of the Bergsdalen nappes lies in the left distance. Photo J. NATERSTAD

GEOLOGIC SETTING

Jæren, the low coastal district south of Stavanger, is heavily covered with Pleistocene deposits, and little can be seen of the solid rock. According to BJØRLYKKE (1908) there are phyllites in the eastern part of the district along the boundary with the Precambrian gneiss (possibly there is a fault along this boundary). In more westerly parts there are highly metamorphosed rocks: gneisses, hornblende schists, mica schists and crystalline limestones.

The Stavanger district, composed of the islands north of the city, was described by GOLDSCHMIDT (1921). Above the Precambrian basement is a sequence of low-grade phyllites, evidently an extension of the phyllite sequence below the nappes in the districts further north. Greenstones, mica schists and other sediments above the phyllites, intruded by granites and trondhjemites as well as by more basic rocks, display a higher grade of metamorphism, and many of the mica schists have been transformed into plagioclase porphyroblast schists and augen gneisses. GOLDSCHMIDT interpreted the rock series phyllite–porphyroblast schist–augen gneiss as formed by a process of progressive metasomatism connected with the intrustion of acid magmas. KALSBEEK (1964) studied zircons from the different types of sediments and found a marked difference between the zircons of the phyllites (small in size and well rounded) and those of the overlying plagioclase-bearing schists (larger, more angular and elongated). He concluded that the two types of sediment might have been different originally, and that the process of metasomatism favoured by GOLDSCHMIDT was open to doubt, though not disproved. As stated in Chapter IV, there are indications that the sequence of greenstones, mica schists and associated intrusive rocks forms a nappe above the phyllite sequence.

Just off the coast northwest of Stavanger lies the island of Karmøy. The northeast part of the island (and a narrow strip on the mainland in the east) is underlain by amphibolites and greenstones of partly intrusive and partly volcanic origin (GEIS, 1961b, 1962). In the southern and western parts occur gneissic rocks with conglomerates, interpreted by GOLDSCHMIDT (1921) as arkoses formed from trondhjemitic material. The rocks are probably of pyroclastic origin, corresponding to the granodioritic gneisses in the northern parts of the Trondheim region. In the southeast corner of Karmøy a sequence of sediments, comprising a polygenous conglomerate and sandstones and shales with late Ordovician or Silurian fossils, is inserted as a tectonic wedge between greenstones in the east and gneisses in the west (BROCH and others, 1940).

The Sunnhordland–Bergen district is composed of the following: the Sunnhordland island group with the larger islands of Bømlo, Stord and Tysnesøy; an area along the northwestern side of Hardangerfjord; and the Bergen area with the Bergen arc structures. The geology of the district can be divided into the following three parts: a strip with Cambro–Silurian sediments in the southern part of the Sunnhordland islands and along the northwestern side of Hardangerfjord; the Sunnhordland massif of igneous rocks in the northern parts of the Sunnhordland islands; and the Bergen district.

On Stord island pyrite deposits of sedimentary origin are mined and the ore is associated with greenschists and black phyllites (SKORDAL, 1948). Further north, the thick Moberg conglomerate and overlying fossiliferous late Ordovician and Silurian sediments occur beneath the overthrust Sunnhordland igneous massif. A 2 km wide band of hard felsitic quartz porphyry is found in the southern part of Tysnesøy and on adjacent Huglo island (KOLDERUP, 1929). Parts of the areas with Cambro–Silurian sediments along the northwestern side of Hardangerfjord were mapped and described—a southwestern part by FOSLIE (1955) and a northeastern part by KVALE (1946). J. NATERSTAD (personal communication) mapped the intervening area and worked out the stratigraphy. As mentioned previously the gneiss in the lower part of the sequence (group I) rests above a phyllite of eastern facies with a tectonic boundary. The Cambro–Silurian sediments of eugeosynclinal facies are thus allochthonous and form a nappe below the Bergsdalen nappes or incorporated in them (cf. KVALE, 1960a) (Fig. 59). To the northeast the sediments of eugeosynclinal facies disappear beneath the rocks of the Bergsdalen nappes, and 20 km further to the northeast at the inner part of Hardangerfjord there occurs a Cambro–Ordovician sequence of eastern facies (BRØGGER, 1893) corresponding to the sequence in Hardangervidda and other areas beneath the great nappes.

The Cambro–Silurian sediments in the southern part of the Sunnhordland islands and at Hardangerfjord have a dominant NE trend. According to NATERSTAD there are structures of at least two ages: an older set of lineations and folds directed NW, and a younger set of structures generally directed NE. The large folds follow the latter direction, and some of them are recumbent and strongly overturned to the west or northwest. As they also affect the rocks of the overlying Bergsdalen nappes they must belong to a late phase—the phase during which the nappes in the northwest part of the "Faltungsgraben" got their present position with dips towards SE.

The Sunnhordland igneous massif contains intrusive, as well as effusive, rocks, of both acidic and basic composition. It forms the northern parts of the large Sunnhordland islands and the smaller islands north of them, and sends offshoots into the Bergen district. The boundary between the massif and the sediments to the south on Stord and Tysnesøy is a tectonic line with a N–E trend along which the igneous massif was pushed to the southeast, against the sediments. This tectonic boundary forms a distinct feature of the landscape (KOLDERUP, 1931).

On Bømlo the boundary between the igneous massif and the sediments in the south is not marked topographically as on Stord and Tysnesøy. The basaltic rocks of the massif in the middle part of Bømlo show very well-preserved features, such as pillows, agglomerates and porphyric and amygdaloidal structures (REUSCH, 1888). The part of the massif situated on Stord was mapped and described by KVALE (1937). The basic rocks are saussurite gabbros intimately associated with meta-basalts. There are granitic rocks of two types, a granodioritic one and a less common granitic one with dominant microcline. The granites are younger than the gabbroid rocks and penetrate them in a large number of dykes in different directions. Corresponding to the two types of granite, effusive quartz porphyries form layers several hundreds of metres thick, and there are also volcanic conglomerates or agglomerates 190 m thick. On Møkster, an island north of Bømlo, the igneous rocks of the massif contain inclusions of limestone and dolomite, with skarn minerals (KOLDERUP, 1924).

A large gabbro massif offshoots towards the northeast from the Sunnhordland massif; it is 35 km long and 8 km wide and lies between the charnockitic massif and the Cambro–Silurian sediments in the southern part of the Bergen district. Parts of it consist of fresh, unmetamorphosed

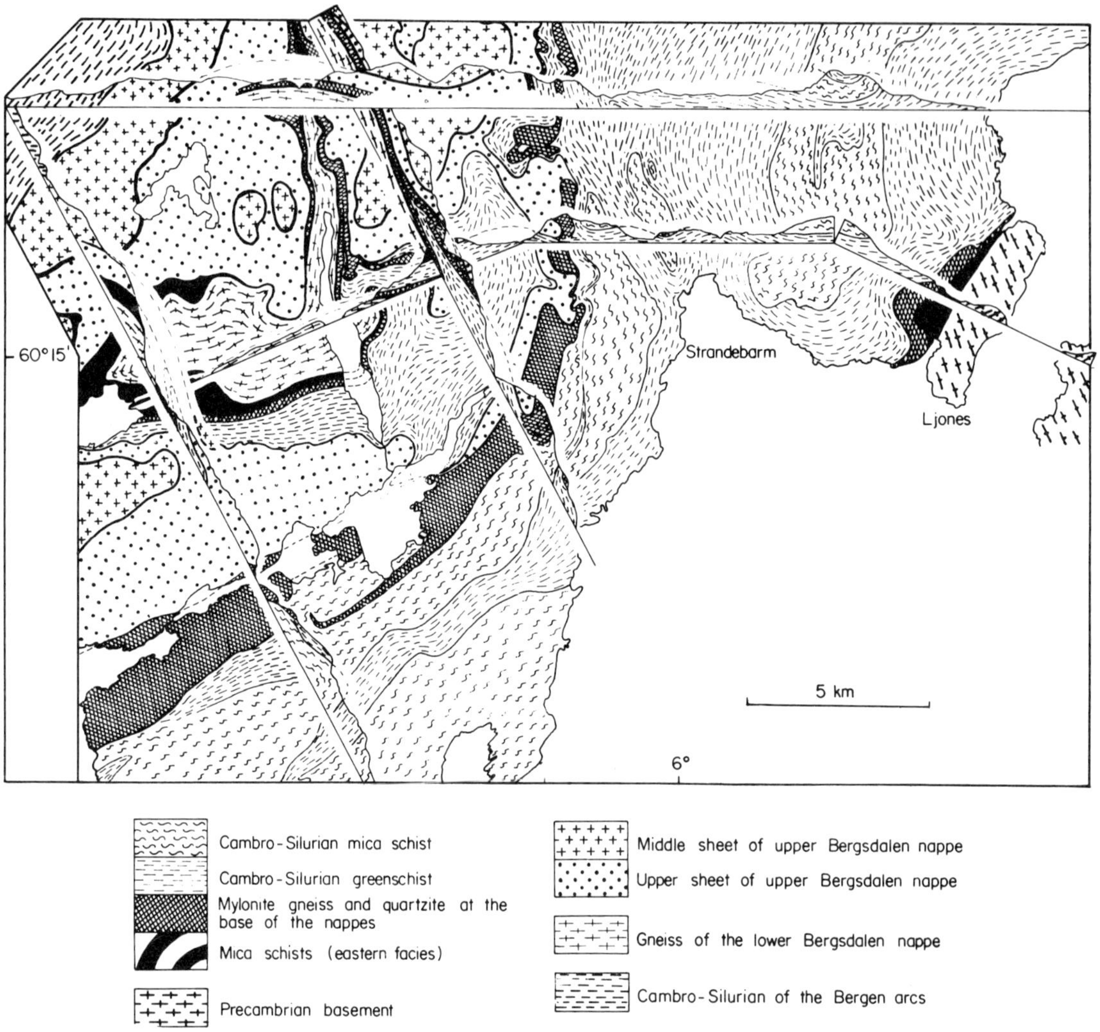

Fig. 59 *Geologic map with sections of the Strandebarm district on the northwest side of Hardangerfjord.* By J. Naterstad

olivine gabbro which shows banding in parts. The main part of the mass is, however, saussurite gabbro, parts of it transformed by tectonization into a schistose rock scarcely distinguishable from a greenschist. The gabbro and the adjacent sediments are intruded by numerous large sills of trondhjemite.

The Bergen district with the Bergen arcs (Kolderup and Kolderup, 1940) displays a synclinal structure with the synclinal axis curved in an arc convex to the east. The sediments at the sides of the synclinorium are underlain by gneisses and overlain by a massif of charnockitic rocks in the core of the synclinorium. The strips with the Cambro–Silurian sediments form the Bergen arcs; a smaller inner arc in the west, where the city of Bergen is situated, and a larger outer arc in the east. The gneisses east of the outer arc sink beneath the surface to the south, at the apex of the triangular anticlinorium between the northern part of the outer Bergen arc in the west, and the margin of the "Faltungsgraben" with the Bergsdalen nappes in the east. The sediments in the southern part of the outer arc are separated from the sediments on the northwest side of Hardangerfjord by the rocks of the Bergsdalen nappes.

In the inner arc the Cambro–Silurian sediments are not immediately adjacent to the charnockitic rocks, but are separated from them by an intervening zone of sedimentary gneiss with layers of quartzite and conglomerate. The inner arc is very strongly curved and describes a semi-circle; the gneiss massif west of Bergen seems to have been pushed vigorously to the east, the sediments generally dipping between 60° and 80° to the east. In the outer arc also the sediments have been strongly compressed into a vertical or steep attitude. The strip of sediments gets very thin and pinches out to the north, but widens in the southern part, in the Os–Samnanger area with fossiliferous Upper Ordovician and Silurian sediments and the large gabbro massif.

The charnockitic rocks in the Bergen district ("rocks of the anorthosite kindred" of the Bergen geologists) cover a wide range of chemical and mineralogical compositions, and are generally of the same types as the rocks in the Jotun nappe. Many of these rock types (such as mangerite) were described for the first time in the Bergen district. Eclogites occur as segregations in anorthosite gabbros. Most of the rocks occur in more or less rapidly alternating bands with trends conformable to the arcs. Parts of the rocks have been metamorphosed into ordinary amphibolites and gneisses, which are hardly distinguishable from the underlying gneisses. The Bergen charnockitic massif can best be interpreted as a part of the Jotun nappe.

The Bergen arc structures must have been formed in a late tectonic phase, probably at the same time as the "Faltungsgraben" structure. Where the rocks of the Bergsdalen nappes are adjacent to the arcs, it can be seen that the linear structures of the nappes have been bent and folded. However, the arc structures are not older than the Devonian sediments in the northern part of the Bergen district, as these sediments have transgressed above rocks stamped with the Bergen arc structures (Kvale, 1960).

In the northern district between Sognefjord and Nordfjord Cambro–Silurian sediments occur in four synclinal depressions, connected under the sea to islands off the coast. The four synclinal areas are, from south to north: the Solund area off the mouth of Sognefjord, and the Kvamshesten, Håsteinen and Hornelen areas, all with Devonian deposits.

In a westerly part of the Solund area, on islands facing the ocean, a massif of greenstones or dolerites and gabbros appears beneath the Devonian sediments. All the rocks display a structure of lath-shaped plagioclases, and there is usually no distinct boundary between the coarse-grained and the fine-grained rocks, which are assumed to form a volcanic core.

Of special interest is the Kvamshesten area, where a nappe of mangerite syenite, a rock type originally described in this area by Kolderup (1921), occurs above Cambro–Silurian (?) sediments; the nappe is, however, overlain by a thick quartzite, phyllites and a limestone with badly preserved corals of Upper Ordovician or Silurian age. According to Kolderup (1960) the nappe is inserted between the two series of Cambro–Silurian strata. The sediments beneath the mangerite syenite nappe in Kvamshesten will also be dealt with in Chapter VII.

CHAPTER VII

THE GNEISS REGION IN SOUTHERN NORWAY

The Caledonide rocks in the Nappe and Trondheim regions are limited to the west and northwest by areas underlain mainly by gneisses, generally of the same lithologic types as those occurring in the Precambrian basement on the east and southeast side of the Caledonides. The gneisses in the northwestern areas were looked upon as Precambrian rocks and were marked thus on the geologic map of Norway published in 1915. However, the gneisses in the northwest, in contrast to the Precambrian basement rocks in the southeast, in most cases show conformable relationships to the overlying deposits which are known to be of latest Precambrian–Eocambrian or Cambro-Ordovician age; in some cases it is difficult even to fix a sharp boundary between these deposits and the underlying gneisses. Boundary relationships of this kind have been observed and described, e.g., in the Oppdal district, by BJØRLYKKE (1905).

The introduction of the concepts of migmatization and granitization into geological thought certainly brought about a change in the understanding of the history and age of the northwestern gneisses. WEGMANN (1935) claimed that a migmatite front was present beneath Caledonide sediments in Lesja. HOLTEDAHL (1936, 1938) expressed the view that the northwestern gneisses had lain deep-down during the Caledonian orogeny and had been too strongly influenced by Caledonian processes to be regarded simply as Precambrian rocks. In papers dealing with the Oppdal area HOLTEDAHL (1938b) and BARTH (1938) maintained that the gneisses underlying flagstones or micaceous quartzite, corresponding to Sparagmite deposits in that area, were products of transformation of deposits normally underlying the sparagmitic flagstones. In a later paper (1944) HOLTEDAHL again expounded this view and coined the term "Basal gneiss", a term that can be generally applied to any gneiss complex conformably underlying a suite of sediments. In southern Norway we thus speak of the *Basal gneiss region* or, for short, the *Gneiss region.*

In the main the ideas thus advanced have been generally accepted; no-one doubts today that the Gneiss region contains gneisses that are derived from supracrustal rocks of the Caledonian cycle.

The rocks that lie conformably below the oldest sediments of definite Cambro-Ordovician or Eocambrian age have recently been investigated along the western margin of the Trondheim region. These investigations have shown that the rocks in question are largely meta-arkoses and meta-greywackes, and possibly also volcanic deposits. Worth mentioning are the publications of HERNES (1965a, 1965b) concerning the Tingvoll–Raudsand district. It seems probable that parts of the gneisses in the Gneiss region are late Precambrian eugeosynclinal deposits, partly contemporaneous with the Sparagmite deposits in the eastern parts. In the northern part of the Sparagmite region the Vardhø formation has previously been mentioned as showing a eugeosynclinal type of facies. OFTEDAHL (1964) maintained that in large parts of the Gneiss region Precambrian deposits were in an undisturbed position at the beginning of Cambrian time. Most probably, however, the main part of these are late Precambrian—younger than the orogeny and metamorphism which were shown, by a great number of age determinations on Precambrian rocks from the southeastern parts of Norway, to have occurred about 900 m. y. ago (BROCH, 1964).

A number of writers have considered feldspathization and granitization to be of great importance in the Caledonian transformation of the rocks of the Gneiss region. The evidence for this is largely from field observations of porphyroblastesis of potash feldspar in many kinds of rocks and gradation of these rocks into granitic or granodioritic gneisses.

An elaborate petrographic and geologic investigation of a suite of rocks said to have undergone potash metasomatism was given by ROSENQVIST (1944), who described rocks from the Oppdal district, varying from greenstones and other gabbroid rocks (very poor in potassium) to coarse augen gneisses of a quartz-monzonitic or granodioritic composition. All gradations could be found between the end members of the series. Augen gneisses with large porphyroblasts or heteroblasts of feldspar or granitic material are spectacular rocks, very common in parts of the Gneiss region.

Dunites and peridotites occur widely in many parts of the Gneiss region. They are olivine rocks (dunites) and olivine–enstatite rocks (saxonites), some of them serpentinized. They form lenticular or flat bodies, varying in size from very small ones to up to 2 by 10 km. The dunites and peridotites of the Gneiss region are exact counterparts of the corresponding rocks in the eugeosynclinal Cambro–Ordovician sediments in other parts of Scandinavia. It is thus highly probable that the ultrabasic rocks in the Gneiss region are of the same age, and have the same stratigraphic relationships, as those in the other parts of the Scandinavian Caledonides. The ultrabasites in question are known to be of Early Ordovician or older age, wherever a reliable stratigraphy has been established.

Some of the larger dunites contain inclusions or segregations of garnet peridotite, with garnet and pyroxene of eclogite mineral facies (O'Hara and Mercy, 1963; Mercy and O'Hara, 1965). There is, however, no gradation in chemical and mineralogical composition from the garnet peridotites to the eclogites proper.

The eclogites occur in similar fashion to the dunites and peridotites, as rounded or lenticular bodies in the gneisses, of all sizes up to about 100 m in their longest dimension. The type of eclogite described by Eskola (1921) consists mainly of garnet and jadeitic clinopyroxene and is of gabbroid bulk composition. Lappin (1960) described eclogites with quartz and kyanite. Alteration of the eclogites into amphibolites is common at the boundary of the bodies and along joints. An alteration product of the eclogite pyroxenes, useful for the recognition of diaphthoritic eclogites, is a myrmekite-like diopside–plagioclase symplectite. The gneisses surrounding the eclogites are in amphibolite or even lower facies. If the eclogites did crystallize at their present sites, it is necessary to assume that the mineral facies of the surrounding gneisses is a result of diaphthoritic processes (Kolderup, 1955). Rosenqvist (1956) found string perthites characteristic of the granulite facies and also garnet and orthopyroxene in gneisses from the Lesja–Sunnmøre districts. Many of the rocks were of the same type as the Bergen–Jotun ones. O'Hara and Mercy considered the eclogites to have intruded as crystalline masses from deeper horizons by tectonic processes. They found a lack of structural conformity between the eclogites and their country rocks.

Anorthositic rocks are characteristic of the Gneiss region. They are meta-anorthosites—more or less pure andesine or labradorite rocks. Some of them occur as integral parts of banded rock complexes which in some cases display charnockitic facies. Kolderup (1928) described and mapped rocks of the Bergen–Jotun type in the area between Sognefjord and Nordfjord. It may be possible to look upon these rocks as counterparts of the charnockitic rocks of the Nappe region. O'Hara and Mercy (1963) described anorthosites from Sunnmøre as concordant sill-like masses several hundred feet thick, extending for many miles along the strike.

Anorthosites may also occur as intrusive bands or lenses in the country rocks, which in some cases are sedimentary and may contain inclusions of anorthosite (Eskola, 1921; Inge Bryhni, personal communication).

A type of intrusive rock peculiar to the Gneiss region is a gabbro with lath-shaped plagioclases, described in the Sunnmøre area by Gjelsvik (1952) as a dolerite. This dolerite occurs as lens-shaped bodies of size varying from a few metres to some 2 km. The boundaries to the gneiss country rock are usually tectonically disturbed and the dolerites have been amphibolitized at the margins. Carstens (1957) described gabbros of the same type in the Gneiss region north of Trondheimsfjord. Gjelsvik discussed the problem of the age of the dolerites and compared them with similar rocks, described by Strand (1951), at Høvringen in the northern part of Gudbrandsdalen, which are intrusive in rocks of the Vardhø formation, a part of the Sparagmite group. This was taken by Gjelsvik as an indication that his dolerites were of Caledonian age and were not old Precambrian rocks.

A tentative interpretation of the stratigraphy and history of the Gneiss region is that geosynclinal sedimentation began in late Precambrian time. Equivalents of the Sparagmite deposits are present as arkoses, plagioclase-bearing greywackes and volcanic rocks. The dolerites described by Gjelsvik, not known to occur in Cambro–Ordovician deposits, may be a manifestation of the magmatism at that time. During a quieter interval in Eocambrian (?) times a micaceous and feldspathic sandstone was deposited over large parts of the area. This is the "sparagmitic flagstone" of the Oppdal district of "light sparagmite" type, present over large parts of the Gneiss region as well as in northern parts of the Sparagmite region. The overlying Cambro–Ordovician (and younger?) deposits were also largely feldspar-rich arkoses and greywackes alternating with basaltic and acidic volcanic rocks. The old basement rocks, deeply buried beneath the sediments, were strongly deformed and parts of them were subject to granitization and anatexis; the unconformities once present were thus obliterated.

A number of K/Ar determinations have been carried out on rocks from the Gneiss region (Broch, 1964). A group of seven determinations in the interval 425 to 367 m. y. indicates an event of Caledonian metamorphism at about 400 m. y. Two determinations gave ages of 582 and 590 m. y., while an Re/Os determination on molybdenite gave an age of 540 m. y. Hernes (1964) interpreted these determinations as indicative of an early Caledonian event of orogeny and metamorphism in latest Precambrian time. A few determinations on rocks from other parts of the Caledonides gave results in the same time interval, among them an age of 550 m. y. for the sample of Brøttum sparagmite previously mentioned.

REGIONAL DESCRIPTIONS

It may be convenient to commence the description of the Gneiss region with a part of it in which the Precambrian basement rocks are present in a largely unaltered state. Such an area is found along the northwest margin of the "Faltungsgraben" northeast of Sognefjord. Here the upper parts of the basement rocks have been mylonitized to conform with the overlying sediments, in the same way as at the southeast margin of the "Faltungsgraben". The unconformity is, however, preserved in deeper parts of the basement and the overlying sediments can be followed for long distances with an undisturbed stratigraphy (Holtedahl, 1944; Landmark, 1949). Banham and Elliot (1965) found similar conditions in Hestbrepiggene (61° 45′, 8° 5′). The massif with Precambrian rocks probably extends to the northwest in the area around the Jostedalsbreen plateau glacier. Inge Bryhni (personal communication) crossed this area and observed massive granitic gneisses without conspicuous structures.

At the northwest margin of the "Faltungsgraben" there occur (according to Landmark and Skjerlie, 1958), large synclinal and anticlinal structures formed by a deformation of the basement, and also schuppen structures by which parts of the overlying mica schists were pinched between wedges of gneiss. The structures indicate a direction of movement to the northwest. A similar synclinal structure was found by Elliot and Shouls near Sotaseter (subsequently described).

South of Sognefjord is a triangular area of gneisses, delimited by the margin of the "Faltungsgraben" to the southeast and by the Bergen arcs to the southwest. There are also gneisses west of the Bergen arcs. The rocks along the margin of the "Faltungsgraben", underlying the Bergsdalen nappes, were described by Kvale (1946a) (as recorded in a previous chapter) whilst the rocks adjoining the Bergen arcs were described by Kolderup and Kolderup (1940). The area is a complex of gneisses also including nearly massive granites, gabbros and amphibolites. The prevailing opinion is that the gneisses in this part consist mainly of Precambrian material, influenced to a high degree by Caledonian tectonics and recrystallization.

The *Haugesund peninsula* between the outer part of Hardangerfjord in the north and Boknfjord in the south contains a massif of gneisses and other crystalline rocks, overlain on parts of its circumference and interior by Cambro–Silurian phyllites or mica schists. It is delimited in the east by a sharp flexure and is transected by N–S running faults. The mica schists and phyllites overlying the gneisses are transected and migmatized by quartz-dioritic intrusive rocks (Sørbye, 1949, 1964). Pegmatites, which are the youngest rocks within the massif and were earlier suspected as being of Caledonian age, were proved to be Precambrian by a K/Ar age determination, which gave an age of 900 m. y.

In the areas between Sognefjord in the south and Nordfjord in the north are the Cambro–Silurian sediments described in a previous chapter, in addition, Devonian sandstones and conglomerates occur in four separate areas; from south to north these are Solund, Kvamshesten, Håsteinen and Hornelen. The relationship between the Cambro–Silurian sediments and the underlying gneisses is one of the main problems of the area, which has been described in a number of publications by N. H. Kolderup (1928, 1950, 1955, 1960a, 1960b); whilst latterly Finn J. Skjerlie and Inge Bryhni worked in the area. The writer is indebted to all three for personal communications on the geology of the area.

According to Kolderup (1960b): "The gneisses . . . vary within wide limits. We find massive granitic rocks and analogous rocks with basic rock fragments forming a breccia—the variety of migmatite known as agmatite. But we also find veined gneisses, augen gneisses and schistose gneisses. Some of the last-named varieties may resemble the schists closely and in many places pass into the schists by gradations. There are gneisses passing into mica schists, quartz schists, and hornblende and chlorite schists. Some pegmatite dykes and veins are also present in the gneiss area." The schists referred to in the preceding quotation are conformable with the Cambro–Silurian sediments of the area and contain bodies of dunite and serpentinite. It is difficult or impossible to find clear and sharp boundaries between these sediments and the underlying gneissic migmatites, and there are thus strong indications that the migmatites are of Caledonian origin and that large portions of them were formed from deposits of the Caledonian cycle. However, SKJERLIE has recently made observations which may not be in harmony with such an interpretation. As mentioned previously, the greywackes and greenstones with serpentinites that underlie the mangerite syenite nappe in Håsteinen extend to the east in a narrow syncline. The sediments in the syncline are devoid of the pegmatites and granitic veins which occur in the underlying gneisses, and thus either the migmatization of the underlying gneisses was a Precambrian event or the granitization front did not reach up to the level of the sediments in that part of the area. KOLDERUP (personal communication) tentatively indicates that a number of granites (or granodiorites and quartzdiorites) of intrusive or palingenic origin formed centres of Caledonian granitization and migmatization.

Gabbroid rocks occur in large massifs in the southern parts of the district. These rocks surround parts of the Solund syncline, their most northerly occurrence being at Førdefjord between the Kvamshesten and Håsteinen synclines. In the most southerly-lying body the rocks present are olivine norites with orbicular structures, but otherwise they are mainly amphibolites, often with garnet. A giant crystal of garnet from Gjølanger on Dalsfjord measuring 90 cm across was described by KOLDERUP and ROSENQVIST (1950). Some of the rocks in the complex are in eclogite facies. Small deposits of titaniferous iron ore occur in the basic complex. The basic bodies are surrounded mainly by more or less granitized gneisses and are intersected by granitic and gneissic rocks which are hard to explain as magmatic intrusive bodies. The surrounding gneisses generally contain inclusions or fragments of basic rocks in various mineral facies.

A tectonic boundary runs along the south side of the Håsteinen area with Cambro–Silurian and Devonian deposits, and separates the southern area with amphibolites and eclogites from the northern area with anorthosites.

The northern area south of the Hornelen Old Red basin has been investigated by BRYHNI (1962),

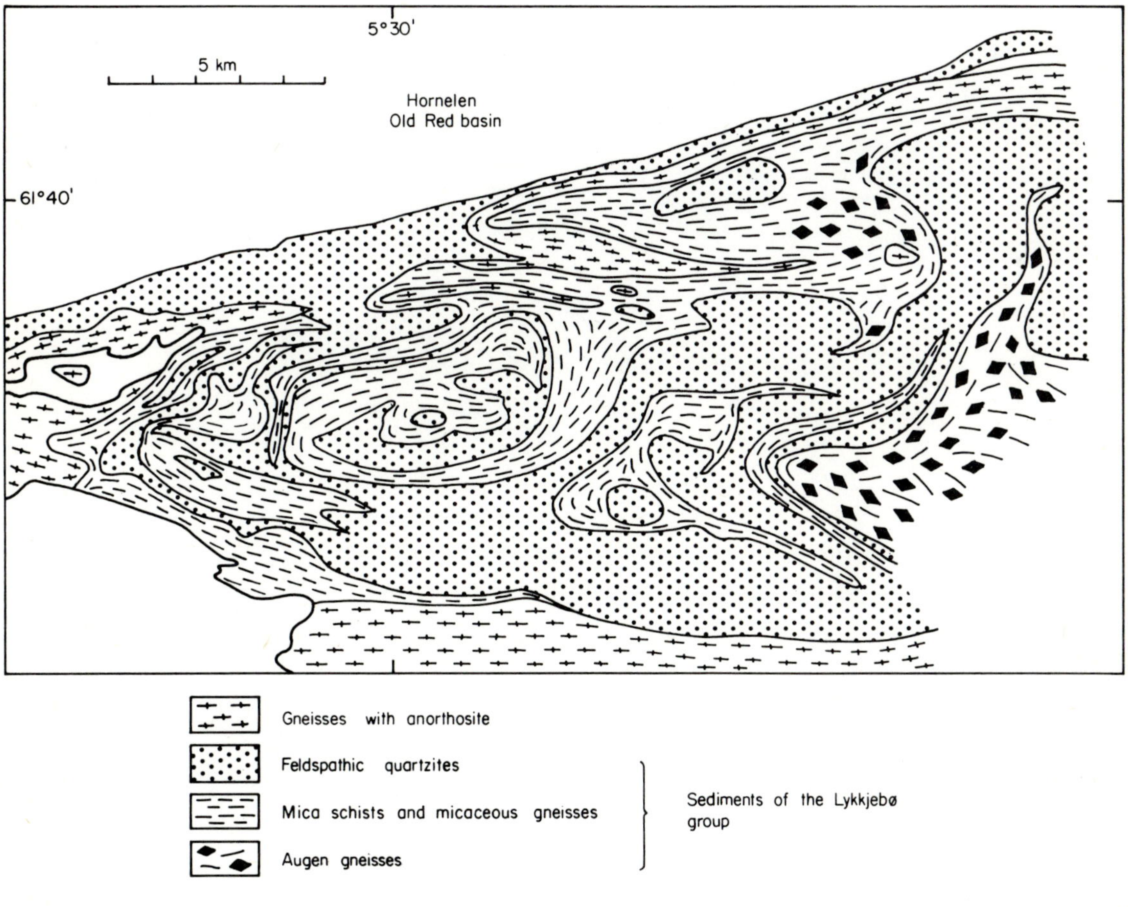

Fig. 60 *Geologic map of an area on the south side of the Hornelen Old Red basin.* After I. BRYHNI, unpublished map

while KOLDERUP (1950) described the Gloppen area around Nordfjord, to the west of the basin. The present account is also based on an unpublished manuscript and the personal communications of BRYHNI. Here is a rock complex containing bands of anorthosites and meta-anorthosites as parts of a banded complex of amphibolites (with some massive relics) and gneisses containing biotite and epidote. The gneisses contain relics of string perthites, indicating that they are diaphthoritic after granulite gneisses. North of the complex with anorthositic rocks are rocks of the *Lykkjebø group* with light feldspathic quartzites as their main constituent. The Lykkjebø quartzites are very similar to the "light sparagmites" in the northern parts of the sparagmite region. In addition to the quartzites the Lykkjebø group contains mica schists, dolomites and gneisses. These rocks display a very complicated map pattern with strongly bent trend lines, possibly due to recumbent folding in a main axial direction WNW (Fig. 60). BRYHNI found gabbros as intrusive rocks in the Lykkjebø group similar to the dolerites described by GJELSVIK in more northerly areas.

Sunnmøre, the coastal and fjord district between Nordfjord in the south and Romsdalsfjord (62° 40′) in the north, is known for its numerous occurrences of dunites and eclogites. Between the two fjords mentioned lies Storfjord with its many branches.

A reconnaissance survey of Sunnmøre was carried out by GJELSVIK and CHR. C. GLEDITSCH (GJELSVIK, 1951). Apart from limestones, lime silicate gneisses and psammitic flagstones the rocks are mainly fine-grained to coarse-grained granodioritic gneisses commonly banded with dark layers rich in amphibole and biotite, or alternating quartzitic and mica-rich layers. The gneisses are transected by numerous small sills and veins of granite and pegmatite, which have migmatized parts of the gneisses. The main structural trend in the area is in an E–W direction. It is possible

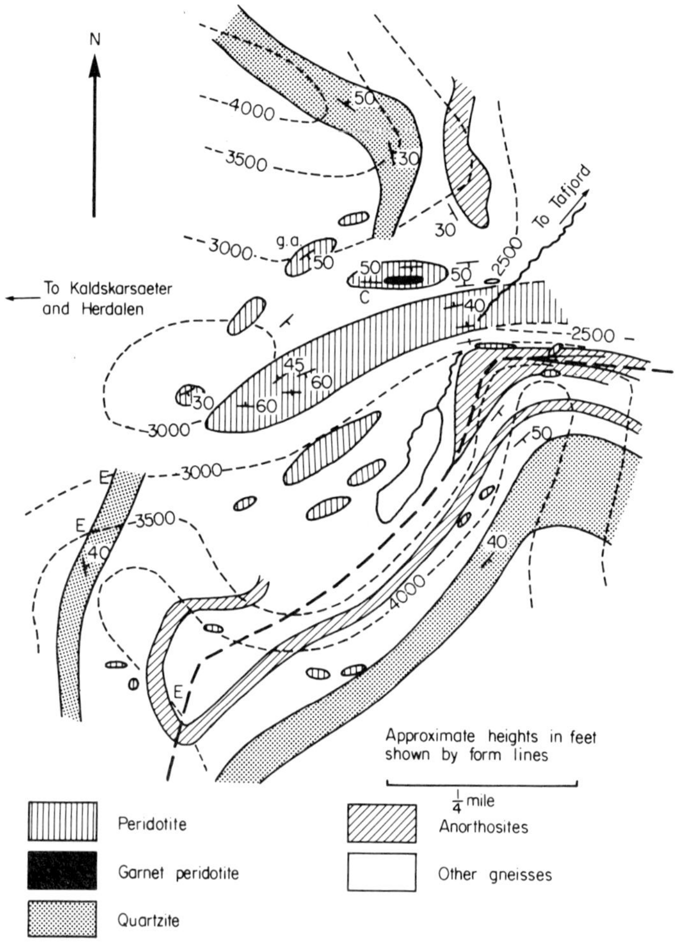

Fig. 61 *Outcrops of peridotite (vertical shading) and garnet peridotite (solid) among quartzites (stippled), anorthosites (oblique shading) and other gneisses (plain). Recumbent fold with axis plunging 30° SE. 4 km southwest of Tafjord (62° 14′, 7° 25′).* From O'HARA and MERCY (1963)

to distinguish between two different parts of the district, an area in the south where the rocks contain inclusions of anorthosites, dunites and eclogites, and a smaller area north of Langfjord, to which are confined all larger bodies of the dolerites described by GJELSVIK (1952).

In the Tafjord area (62° 15′, 7° 25′) O'HARA and MERCY (1963) found an upper division of pelitic and semipelitic gneisses with amphibolites and hornblende schists and with conspicuous horizons of quartzite. Every occurrence of dunites and anorthosites within the area is found in this group of rocks, while eclogites are rare. At lower horizons in the sequence are layered gneisses, also of sedimentary origin, intruded and injected by granite. They become increasingly frequent downwards. Small bodies of eclogite are relatively frequent at the lower levels. The gneisses of the upper division with interlayered quartzites and anorthosites lie in large recumbent folds with axes plunging east at moderate angles, the axial planes striking NNE and dipping ESE (Fig. 61).

In the southern parts of the district, north of Nordfjord, INGE BRYHNI (personal communication) found sedimentary gneisses with inclusions or intrusions of anorthosite. Together with the anorthosites occur massive rocks containing relics of minerals showing charnockitic facies. The major parts of the rocks in Sunnmøre are in amphibolite facies, but ROSENQVIST (1956) recorded granulite facies rocks in the Stadt peninsula and other western areas.

As appears from the preceding, the greater part of Sunnmøre is underlain by supracrustal rocks of the Caledonian cycle. Previously no basement to these deposits was recognized in the district.

To the northeast of Sunnmøre is the Oppdal district, situated west of the Trondheim region between 62° 20′ and 62° 50′ N (Fig. 62). At the boundary to the Trondheim region, directed N–S along the axial flexure previously mentioned, the rocks of the Oppdal district are overlain by the more low grade rocks of the Trondheim region with a tectonic boundary. As mentioned previously it was the work of HOLTEDAHL (1938) and BARTH (1938) in the Oppdal district which gave an impetus to commencement of work in the Gneiss region. Later contributions to the geology and petrography of the district were those of ROSENQVIST (1941, 1944), HOLTEDAHL (1950) and HOLMSEN (1955, 1960, 1965). It is possible to distinguish between a basement, the basal gneiss, and an overlying series of sediments, particularly in the eastern parts of the Oppdal district. The basal gneiss, exposed in the large Lønset anticlinorium, is a banded granodioritic gneiss. At Nonshø in the northwest part of the district an almost massive quartzdioritic rock forms the basal massif. Later workers in the Oppdal district interpreted the basal gneiss as a Precambrian basement complex with an uncomformable relationship to the overlying sediments. The unconformity is not clearly apparent in the easily accessible localities along the road in the northern part of the Lønset anticlinorium, but there is a sharp boundary to the overlying "Sparagmitic flagstones" with a conglomerate just above the base. Above the sparagmitic rocks are mica schists and hornblende schists with bodies of serpentinite. They can be correlated with the sedimentary and volcanic rocks in the lower part of the Trondheim region sequence (of Cambrian and early Ordovician age) and are called "Trondheim schists". The sediments are set through by lenses, dykes and sills of amphibolite, and also by a large body of saussurite gabbro at Grahø.

Large parts of the Oppdal rocks consist of coarse augen gneisses. They have been formed from sparagmitic rocks as well as from rocks belonging to the Trondheim schists. Some of the augen gneisses belong to rocks of undetermined stratigraphic position (Figs. 63, 64). ROSENQVIST's petrographic work on these rocks was dealt with in a previous section. Meta-anorthosites (albite zoisite rocks) are found in Drivdalen in the southeast part of the district as schlieren or sill-like bodies (or layers) in augen gneisses and hornblende schists. In other parts of the area they occur in rocks interpreted as basal gneisses.

The tectonics of the district is dominated by arc-like trends of the rock units, indicating a complicated structure which is certainly due to more than one phase of folding. Nappe structures occur with basal gneisses situated above the sediments. On Hornet mountain west of Oppdal a large mass of sparagmitic flagstones lies in a large synform, underlain on the north, east and south sides by "Trondheim schists". This structure has been interpreted as a large recumbent fold. H. HOLTEDAHL, who mapped the northwest part of the district in detail, interpreted the structure in a similar way (Fig. 65).

MURET (1960) made a reconnaissance survey covering the wide areas between the Oppdal district in the northeast and Sunnmøre in the southwest (Fig. 66). On a small scale map and a diagrammatic section he drew a structure of very large recumbent fold nappes of Pennine type with E–W axial trend having basal gneiss in the cores and envelopes of quartzites, overlain by Cambro–Ordovician sediments. MURET's picture may be slightly exaggerated as regards the very large size of the folds. However, structures with a similar style and orientation are known in the Oppdal district as well as in Sunnmøre.

South of the Oppdal district the rocks of the Trondheim region are underlain by thick masses

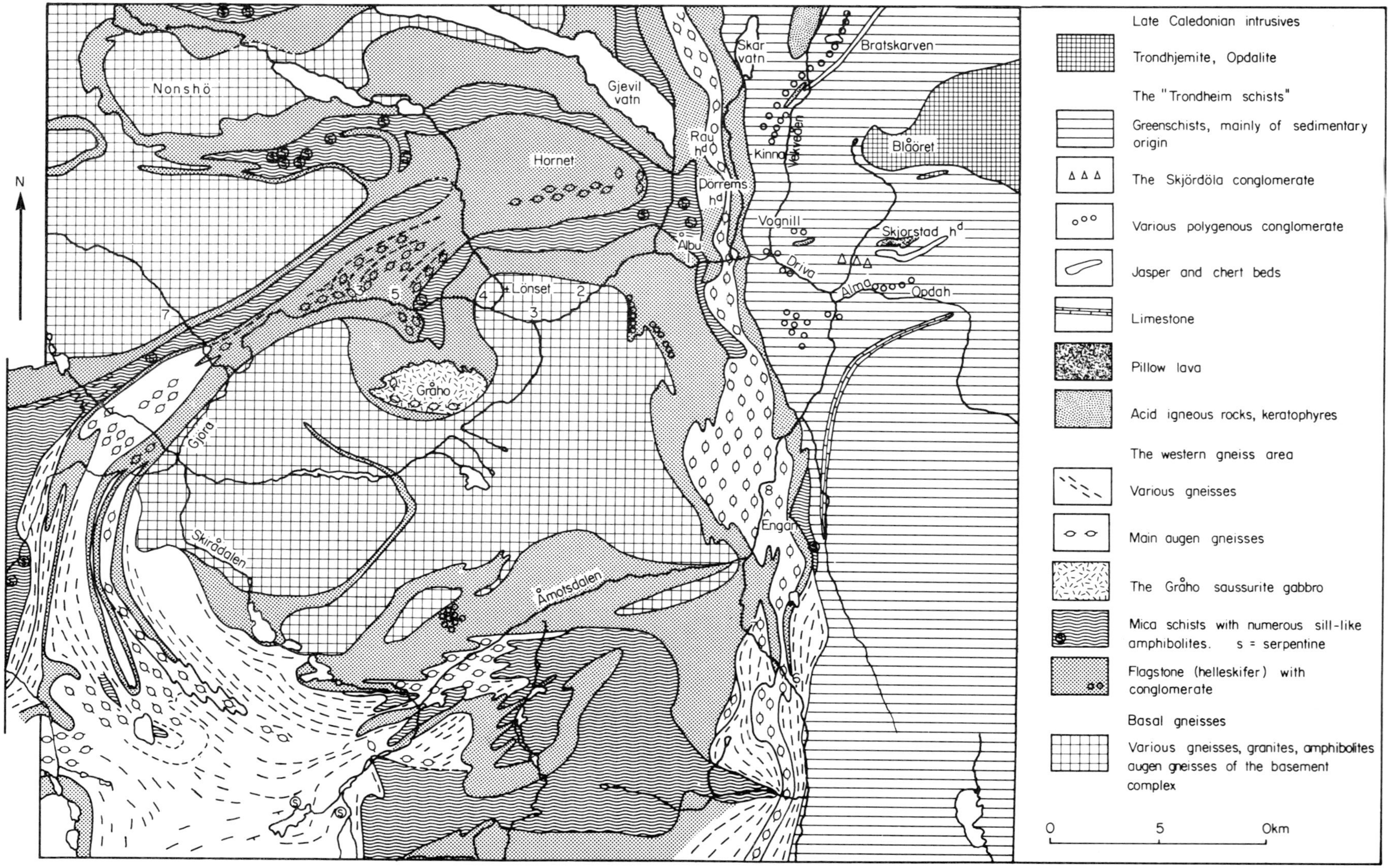

Fig. 62 *Geologic map of the Oppdal district.* From P. HOLMSEN (1960). The Lønset anticlinorium is in the centre, and Nonshø with massive quartz-dioritic rock in the northwest corner

Fig. 63 *Alternating sparagmitic flagstone and mica schist with small augen of feldspar.* Hornet mountain at Lake Gjevilvatn, Oppdal district. From O. Holtedahl (1938)

Fig. 64 *Coarse augen gneiss with layers of light granitic rock.* Near Engan railway station, Oppdal district. From O. Holtedahl (1938)

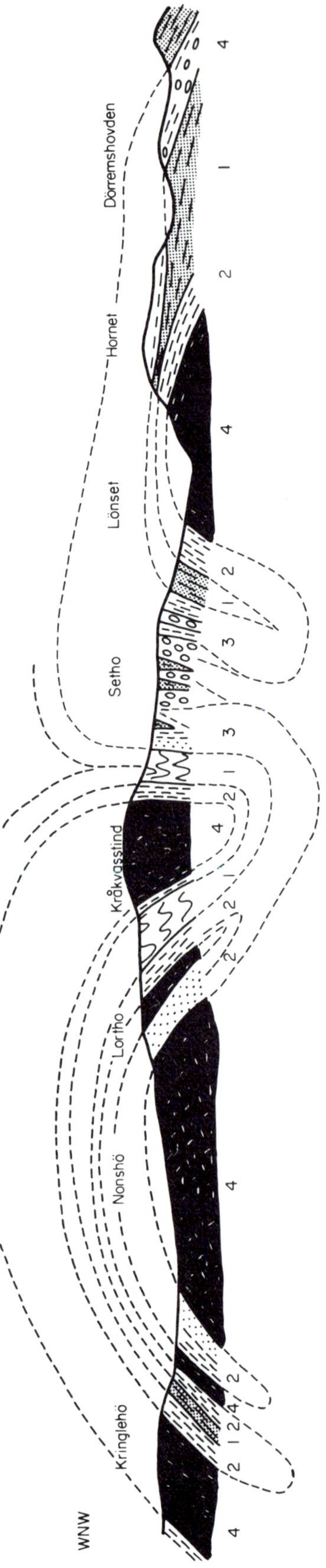

of sparagmitic rocks in the Snøhetta massif (62° 20′, 9° 15′). According to investigations by Du Dresnay (1950) and Strand (1951) there is no sharp boundary between the Trondheim region sediments and the schists, gneisses and sparagmites underlying them in the west. Thus conditions seem to be different from those prevailing in the Oppdal district. A small area at the eastern boundary of the Gneiss region in Lesja (62° 12′, 8° 55′) has been investigated by William H. Scott, a graduate student of Yale University, to whom the writer is indebted for a written report on the area. As in the Oppdal district a psammitic formation of "light sparagmite" type is overlain by a formation of mica schists and amphibolites with large bodies of serpentinite (the *Sjong formation*). In the northern part of the area the "light sparagmite" is overlain by gneissic meta-arkoses, differing considerably from the Sjong formation overlying the sparagmite in the south. Below the "light sparagmite" are deposits of greywacke type: quartz schists and biotite–epidote gneisses, the lowest unit of the area being a banded gneiss with alternating amphibolitic and granitic bands. There is no structural evidence that any of the lower units are parts of an old basement, unconformable with the overlying deposits.

The main structures in the area are due to two phases of folding, both of which produced folds of great dimensions. A first phase formed attenuated folds with axial trend ENE, plunging NE. These folds were refolded by second generation folds with axial planes striking NNW, dipping NE and with axial plunge NE. A first generation anticline was bent into a curve by a second generation fold as shown in Fig. 67. To the east the area is delimited by a thrust upon which sparagmitic rocks were brought above the previously described rocks with a pronounced tectonic unconformity.

A small area near Garmo in the Otta valley (61° 51′, 8° 50′) has recently been investigated by Wilhelm Ruud, a graduate of Oslo University (Fig. 68). The area is at the "tectonic corner", at the junction of the Nappe region and the "Faltungsgraben" with the Trondheim region. Here the "light sparagmite" is overlain by the Otta nappe on a well-marked thrust. The upper part of the "basal gneiss" below the "light sparagmite" is of dark greywacke or arkose and is underlain by quartz-dioritic gneisses with bands of amphibolite, resting on a massive granitic gneiss which may belong to an old basement complex. No unconformity can be detected in any part of the

Fig. 65 *Generalized section from the Oppdal district* (see map, Fig. 62). From H. Holtedahl (1950). 1 "Trondheim schists"; 2 Sparagmitic flagstones (stippled); 3 Various rocks ("Sethø complex"); 4 Basal gneiss (solid)

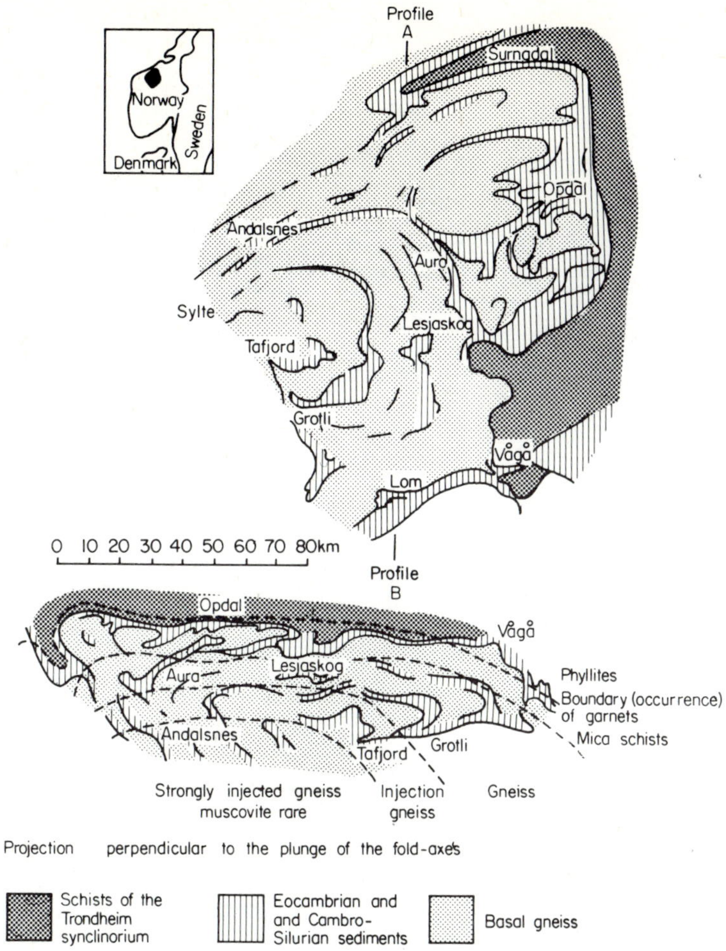

Fig. 66 *Map and section of the Gneiss region between Sunnmøre in the southwest and the Trondheim region.* From MURET (1960). The section is constructed perpendicular to the fold axes, plunging 10° to 50° to the east.

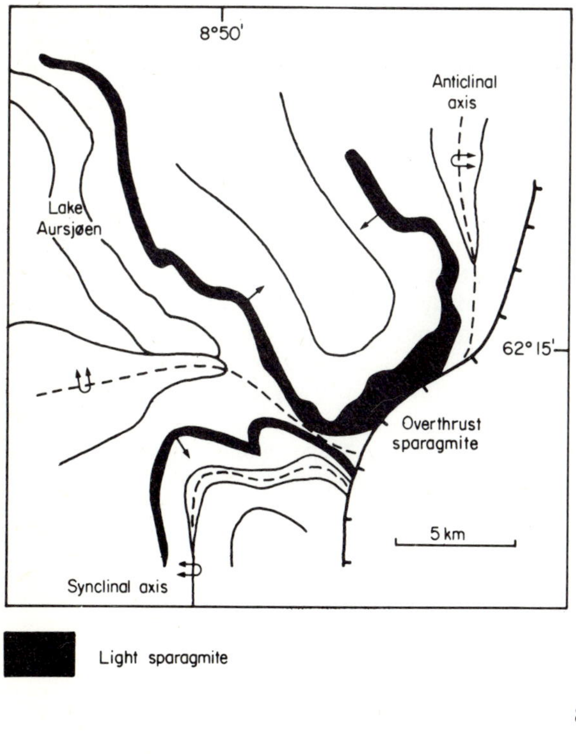

Fig. 67 *Structural map of an area east of Lake Aursjøen in Lesja.* After an unpublished map by WILLIAM H. SCOTT. The arrows point stratigraphically upwards from the "light sparagmite" (solid)

sequence. The deposits beneath the "light sparagmite" are thus very similar to those found by SCOTT in Lesja. As previously reported, quite different conditions were found by BANHAM and ELLIOT (1964) in Hestbrepiggene, 40 km WSW of Garmo, where there is a distinct unconformity between a Precambrian gneiss and overlying Cambrian (?) sediments. In the Garmo and Lesja areas there are evidently late Precambrian Sparagmite deposits not present in Hestbrepiggene. The rocks form an anticlinorium, tightly folded on axes trending ESE and plunging in the same direction. Some of the arkoses have been transformed into massive augen gneisses with closely packed feldspar porphyroblasts.

In Hestbrepiggene, there occurs part of the Precambrian basement gneiss at the northwest margin of the "Faltungsgraben" which extends

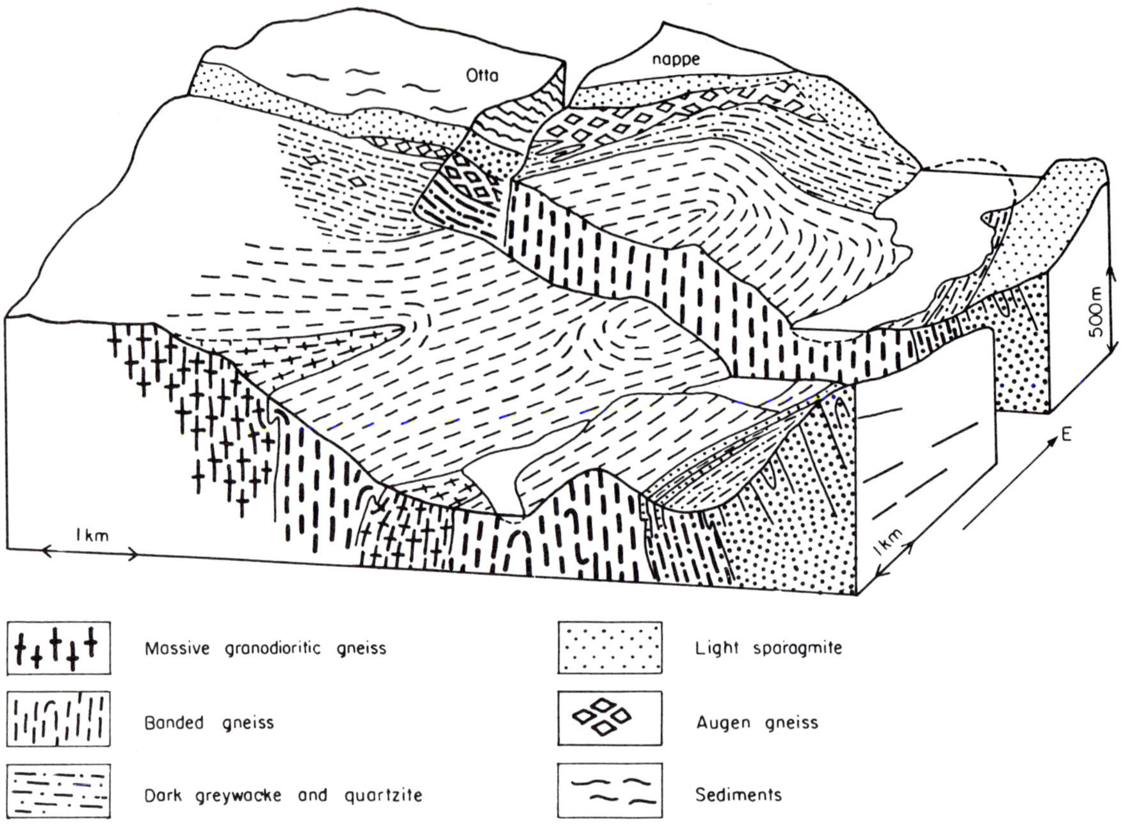

Fig. 68 *Tectonogram of an area around Garmo (61° 51′, 8° 50′)*. By Wilhelm Ruud

northwards in the Skjåk district in the upper part of the Otta valley. In the same district psammitic rocks of sparagmitic type and other sediments have been reported by a number of writers (Holtedahl, 1941; Strand, 1949; Gjelsvik, 1953). Detailed investigations have recently been made by R. B. Elliot and M. M. Shouls northeast of Sotaseter (61° 50′, 7° 45′), 30 km northwest of the margin of the "Faltungsgraben". The writer is indebted to Dr. Elliot for a written report on the area.

A complex of rocks consisting of anorthosites together with sediments with ultrabasic bodies lies down in a syncline in the underlying Precambrian gneiss (Fig. 69). The syncline trends NNE with axial plane dipping west. The sediments present are greywackes and quartzites of "light sparagmite" type. The rocks form two separate thrust sheets above the gneiss, which is highly sheared and recrystallized at the boundary with the overlying rocks. The latest structure found in the area is a vertical sinistral tear fault striking west with a displacement of up to 6 km. The anorthosites and amphibolites with associated sediments, evidently thrust above the gneiss basement, can be interpreted as a continuation of the nappes in the "Faltungsgraben", particularly of the Otta nappe.

In the environs of Grotli (62°, 7° 40′) (Strand, 1966) banded gneisses occur, with bands varying from amphibolites to granitic and quartz-dioritic gneisses, having anorthosites as a conspicuous constituent. A sequence of sediments, consisting of light quartzites, mica schists, meta-greywacke and amphibolites rests above the gneiss massif with tectonic contracts. The rocks are folded in reclined folds with dip of axial planes and plunge of axes to the east.

Between Romsdalsfjord and the outer part of Trondheimsfjord is a marginal part of the country with a dominant NE structural trend, as shown by the direction of the coastline and some of the fjords, and by geologic structures such as the Surnadal syncline. Hernes (1954, 1956a, 1956b, 1965a, 1965b) has investigated part of this district and has established a stratigraphical column (Figs. 70, 71).

In the core of the Surnadal syncline, greenstones and underlying mica schists of the Støren and Røros groups of the Trondheim region sequence occur. Comformably beneath these rocks are micaceous gneisses with layers of quartzite. The dark micaceous gneisses pass into augen gneisses and more massive granodioritic gneisses by porphyroblastesis and metasomatism (?)

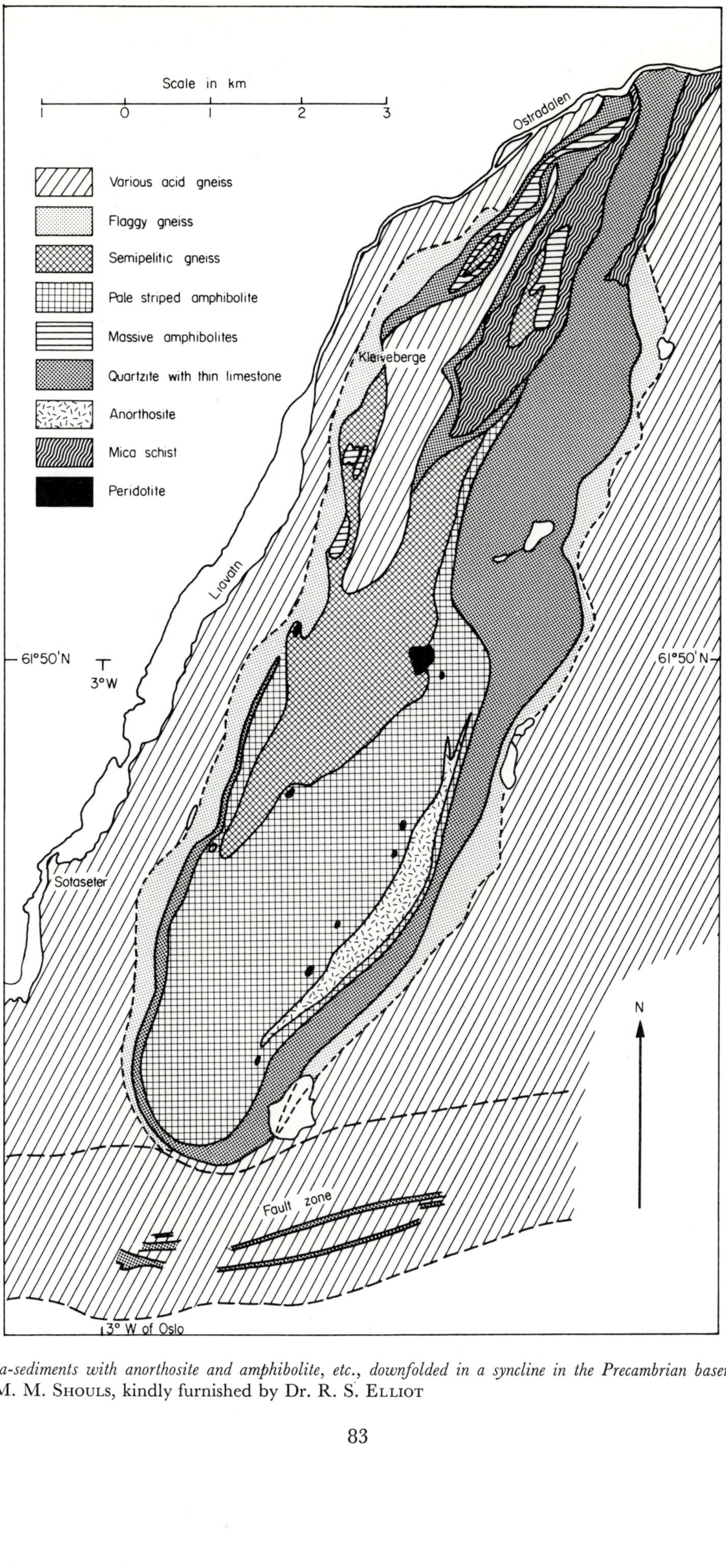

Fig. 69 *Meta-sediments with anorthosite and amphibolite, etc., downfolded in a syncline in the Precambrian basement gneiss.* Map by M. M. SHOULS, kindly furnished by Dr. R. S. ELLIOT

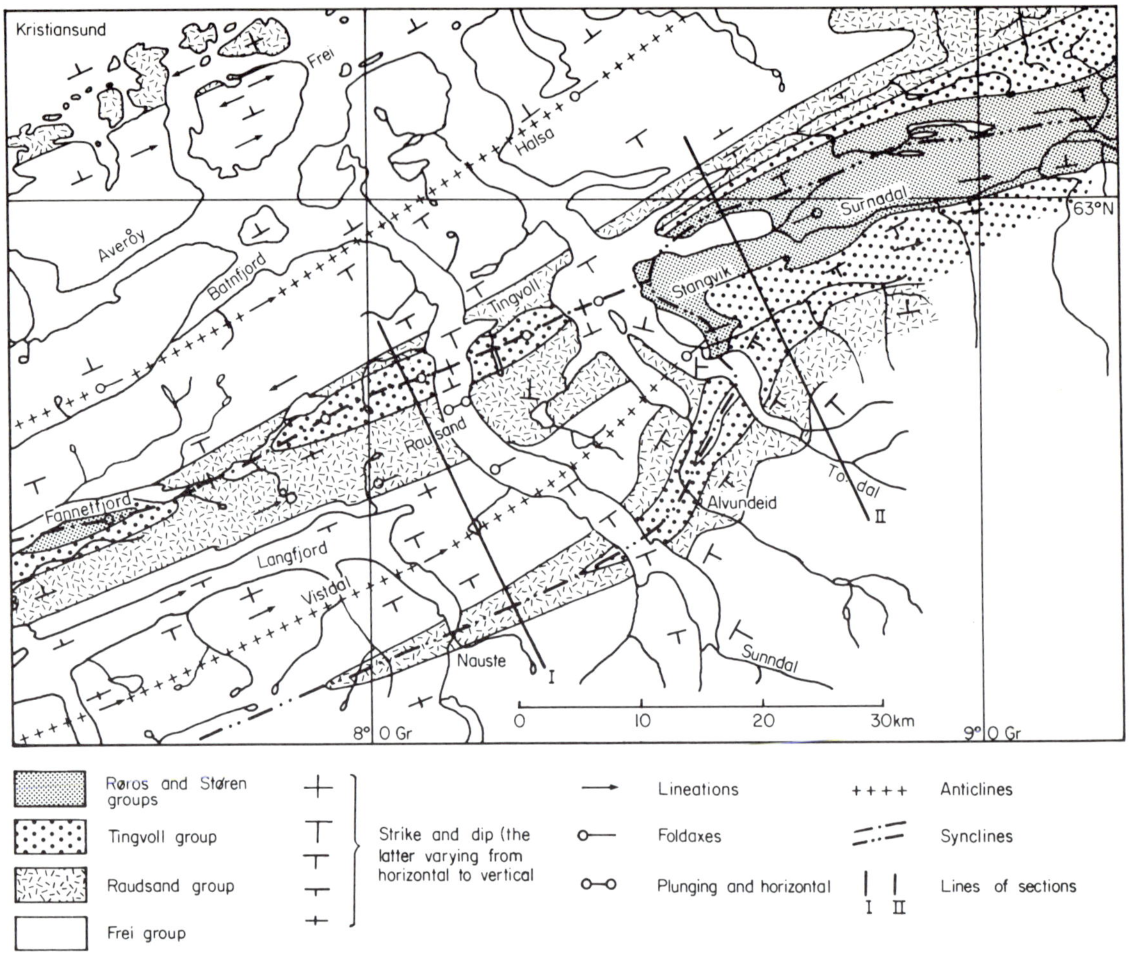

Fig. 70 *Geologic map of the Surnadal syncline and surroundings.* From HERNES (1965)

(STRAND, 1953). These rocks constitute the *Tingvoll group* of HERNES, considered to be of late Precambrian age. The underlying *Raudsand group* includes more massive and homogeneous gneisses, grey granodioritic and red granitic rocks, considered as being of volcanic origin, and in the lower part amphibolites also interpreted as volcanic rocks. Northwest of the Surnadal syncline on the peninsula northeast of Romsdalsfjord (the *Kristiansund peninsula*) there are more heterogeneous and generally less massive rocks than in the Raudsand group: micaceous gneisses with bands of mica schist and flaggy gneisses of quartz-dioritic composition. Pure limestones of great thickness occur in the coastal parts of the area, and also to the southwest in Sunnmøre and to the northeast on both sides of Trondheimsfjord. HERNES unites these rocks in the *Frei group*, stratigraphically below the Raudsand group. Other writers (RAMBERG, 1944; GJELSVIK, 1953) considered the limestones in question to be of Cambro-Ordovician age. Massive gneisses in the coastal area at Kristiansund are considered to belong to the Tingvoll group. The rocks of the Frei group must accordingly form an anticlinorium between the Surnadal syncline and the coast.

Serpentinites are rare in the outer districts considered here, while metadolerites of the type described by GJELSVIK are common. In Sunnmøre, as mentioned previously, dolerites are common in the area north of Langfjord, in the continuation of the trend in this district. The amphibolite-eclogites described by HERNES in the Kristiansund peninsula contain plagioclase and amphibole together with eclogite facies minerals, and are interpreted as diaphthoritic rocks. They were originally gabbroid rocks of the area which were transformed into eclogites.

RAMBERG (1944) described the rocks at the outer part of Trondheimsfjord and later mapped the areas between outer Trondheimsfjord and Hemnefjord, 35 km to the southwest, in great detail. The writer had access to a manuscript map and written communication from Professor RAMBERG.

The rocks at the outer part of Trondheimsfjord,

described by RAMBERG in 1944, occur in a number of synforms and antiforms trending northeast. They consist of mica schists, limestones and lime silicate rocks, with greenstones and amphibolites of the Trondheim region sequence conformably underlain by gneisses (Fig. 72). (According to OFTEDAHL (1964) there is no original unconformity between the gneisses and the overlying sediments in this area.) The gneisses are strongly influenced by Caledonian migmatization and parts of them have been derived from Sparagmite deposits. The grade of metamorphism varies greatly within different parts of the folded complex, apparently due to post-metamorphic folding of the isograde surfaces. Similar conditions are met with in the areas southwest of Trondheimsfjord mapped by Professor RAMBERG, a short account of which has now been published. Rock types here (in Vestranden) are "various quartzo-feldspathic gneisses alternating with strongly metamorphosed sparagmites and Trondheim schists, including metabasalt (greenstone) and marble. A characteristic red potash gneissic granite occurs often as anticlinal cores robed by flagstone (sparagmite), mica schists and/or amphibolites, which may be traced continuously to demonstrable Cambro–Silurian strata in the Trondheim area. Some of the red granitic gneisses appear to derive from flagstone or a fine grained leptitic rock whose origin is unknown. Granodioritic gneiss, often carrying hornblende, is also positioned in the core of anticlines" (RAMBERG, 1966). The tectonics is dominated by folding on axes trending NE and plunging 10 to 20° in the same direction. A significant structural feature is an intensive stretching in the axial direction.

The *Gangåsvann area*, adjacent to the Trondheim region on the northwest side of Orkdalen, was mapped by PEACEY (1964). Below the lower part of the Trondheim region sequence (Røros group) are quartzites, semi-pelites and pelites, with calcareous lenses at some horizons (the *Sognsjø group*). The *Våvann group* lower down in the sequence is a variable complex of grey plagioclase-rich gneisses with thin bands of quartzites, mica schists and amphibolites. The rocks of the two groups are cut by meta-dolerites with remnants of ophitic structure, not found in the overlying Trondheim region sequence. The rocks were folded in two main phases, an early phase producing isoclinal recumbent structures, which were refolded in more simple, open folds with axial trend northeast.

The migmatites on the northern part of Hitra and on Frøya were described in the section on the Trondheim region, in this part of the Gneiss region migmatization occurred in connection with large bodies of Caledonian intrusive rocks. These circumstances are unknown in other parts of the Gneiss region.

A northern part of the Gneiss region lies north of Trondheimsfjord; in the east the Precambrian rocks of the Grong culmination pass into this northern part of the Gneiss region, where they grade into augen gneisses. The northern districts are known as a result of a reconnaissance survey by BIRKELAND (1958). A basement complex of granodioritic and granitic gneisses is overlain by

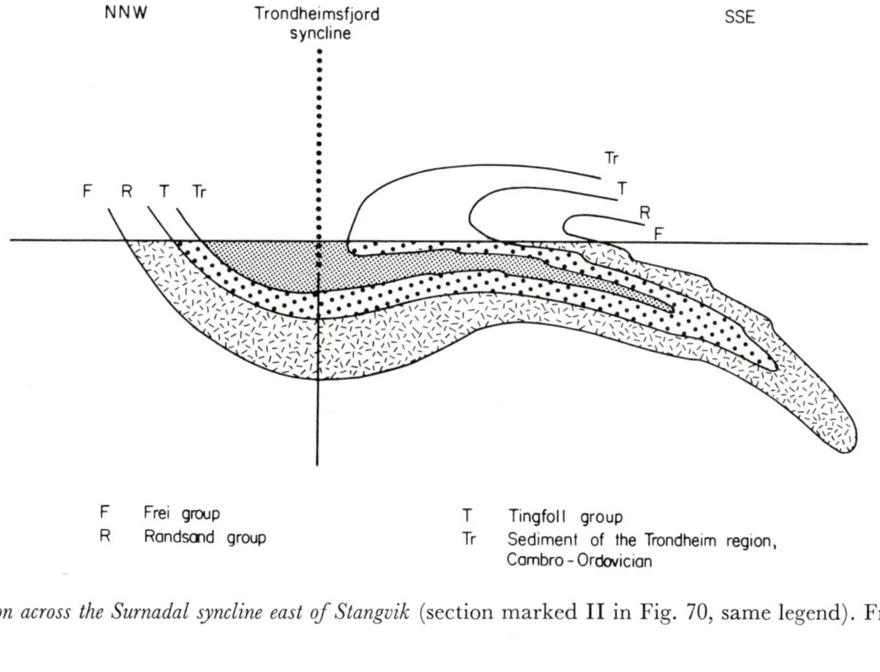

Fig. 71 *Section across the Surnadal syncline east of Stangvik* (section marked II in Fig. 70, same legend). From HERNES (1965)

meta-sediments of various sorts, namely quartzites, quartz schists and mica schists, micaceous gneisses, limestones and lime silicate gneisses. The predominant trend of the structures in the area is NE, but N- and NW-trending structures are also found in the central part of the area. A few bodies of serpentinite are known in the area and gabbroid rocks corresponding to GJELSVIK's dolerites at Sunnmøre occur (CARSTENS, 1957a).

The information available on the northern districts indicates a history and structure similar to that of more southerly parts of the Gneiss region.

Addendum. After completion of the manuscript of this book Professor HANS RAMBERG kindly let the writer see his 1966 paper on centrifuged dynamic models and their bearing on the structure of the Scandinavian Caledonides, particularly of the Gneiss region. It is not appropriate to discuss the merits of the theory of buoyant gravity tectonics here, but RAMBERG's admirable experiments have certainly produced models reminiscent of structures in the Gneiss region. He particularly stresses the *Möre culmination* in the south (emphasized by GJELSVIK (1953) as the *Skjåk–Tingvoll culmination*) and the *Grong–Namsos culmination* in the north as areas of characteristic curving strikes of "mantled dome" type, contrasting with the homoaxial tectonics of other parts of the Gneiss region.

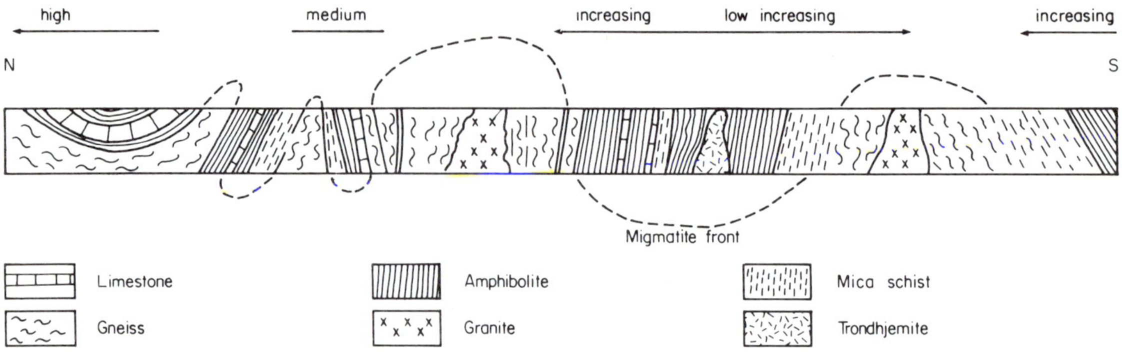

Fig. 72 *Section across the Fosen peninsula on the northeast side of the outer part of the Trondheimsfjord.* After H. RAMBERG (1944). Length of section about 32 km

A paper by BANHAM (1966) describes faults and fault vein mineralization in Hestbrepiggene. Two papers by LAPPIN (1966) and BRYHNI (1967) deal with the petrography and structure of parts of the Gneiss region with emphasis on ultrabasic rocks, eclogites and anorthosites. LAPPIN considers the eclogites as being derived from mantle material, while BRYHNI considers them to be products of Caledonian metamorphism. KILDAL (1966) records observations on the gneisses in islands northwest of Bergen.

CHAPTER VIII

THE GRONG–HATTFJELLDAL DISTRICT

The *Grong–Hattfjelldal district* is separated from the northeastern part of the Trondheim region by the intervening *Grong culmination* and extends as a rather narrow strip along the international border. To the west the district is limited by a thrust line which brings highly metamorphic rocks of the Rødingsfjell nappe above the more low grade rocks of the district. There is a natural boundary to the north, at the north end of Lake Røsvatn (65° 55′ N), where the Rødingsfjell thrust line turns to the east onto Swedish territory. The *Store Børgefjell basal massif* divides the district into a southern and a northern part.

STRATIGRAPHY

The Børgefjell massif in the southern half of the district consists of schists, quartzites and granitic gneisses. The autochthonous sediments of the massif are best known in its northern part (Strand, 1955 and unpublished report; Zachrisson, 1964), where they consist of a lowermost arkose with overlying schists and quartzites. Some of the schists are black and highly carbonaceous. In the southern part of the Børgefjell massif (Foslie and Strand, 1956) the *Dærgafjell quartzite* lies just above the Precambrian rocks of the massif. At least part of the rocks mapped as Dærgafjell quartzite belong to the autochthonous sequence of the massif, according to Zachrisson (1964).

The *Grong district* is a continuation of the Trondheim region north of the Grong culmination, with the same general types of rocks. The stratigraphic relationships appear rather clearly in the area around Lake Limingen (Foslie, 1926; Foslie and Strand, 1956; Strand, 1963). Above the Dærgafjell quartzite is the *Rørvik group*, a sequence of phyllites and mica schists, some dark and carbonaceous, others calcareous. Interbedded with these sediments are greenstones and greenschists and also acid (quartz keratophyric) volcanic rocks. It may be reasonable to assume that the Rørvik group is of Early Ordovician age, corresponding to the Røros and Støren groups. Interbedded in the Rørvik group is a quartz conglomerate, the *Portfjell conglomerate*, and a limestone, the *Bjurbekk limestone*. According to the investigations of Nilsson (1964) on the Swedish side of the border, the Portfjell conglomerate comes between a lower member of alternating basaltic and quartz keratophyric volcanic rocks and an upper member of basaltic to andesitic volcanic rocks. The Bjurbekk limestone is interlayered in the upper member of the sequence. Above the Rørvik deposits, in the middle of a large syncline, lies the *Brakkfjell formation* of calcareous mica schists, rich in plagioclase.

On the Norwegian side of the border the *Liming group* of calcareous mica schists, sandstones, feldspathic sandstones and conglomerates lies above the Rørvik group with a profound erosional unconformity. The conglomerates of the Liming group contain boulders of all the igneous intrusive rocks in the underlying formations. The deposits of the Liming group can be correlated with the Volla conglomerate and the upper part of the Hovin group in the Trondheim region, and with the *Gillik group* in Västerbotten, Sweden. Deposits younger than the Liming group have not been found in the Grong district.

The *Hattfjelldal district* lies north of the Børgefjell massif and extends along the border to Lake Røsvatn (65° 50′ N). Here the lowest formation in the allochthonous sequence above the Børgefjell massif is a quartzite, correlated by Strand (1955) with the Dærgafjell quartzite on the south side of the massif and shown by Zachrisson (1964) to correspond to the *Fjällfjäll quartzite* on the Swedish side. It is overlain by a group of alternating phyllites, often black and carbonaceous, mica schists and calcareous mica schists, quartz schists, quartzites and greenschists, which can be correlated with the Rørvik group in the Grong district. Zachrisson (1964) gave a highly detailed map of the *Remdalen district* on the Swedish side, adjoining the most southerly part of the Hattfjelldal district. In addition to the Fjällfjäll quartzite, his

Tjoapsi formation and *Lasterfjäll group* extend across the border and are represented in the deposits assigned to the Rørvik group.

Above the schists and greenschists of the Rørvik group lies a group of calcareous mica schists and the Liming group with conglomerates. In the northern areas the Liming group also contains conglomeratic limestones and meta-greywackes rich in gabbroid material. Above the Liming group lies the *Hattfjelldal limestone* but its boundary is a tectonic one and the stratigraphic position of the limestone is thus undetermined.

GEOLOGIC SETTING

The *Grong* and *Liene areas* have been mapped in great detail by the late state geologist, Steinar Foslie. The maps are now printed by the Geological Survey, but so far only one has been published with a description (Foslie and Strand, 1956). Oftedahl (1956, 1958a) has given short reviews of the structure of the area in two papers.

As stated previously the Precambrian rocks of the Grong culmination consist of an underlying unit, the *Olden massif*, and an overlying unit forming the lower part of the large Seve nappe which is overlain by the sediments of the nappe.

East of the Grong culmination and south of the Grong area lies the Liene area. Here the supracrustal rocks of the Seve nappe consist mainly of greenschists and mica schists with several bodies of serpentinite, and smaller intrusive bodies of trondhjemite. Some of the mica schists are strongly injected with granitic material.

The Grong area (Fig. 73), north of the Grong culmination, is underlain in its western part by

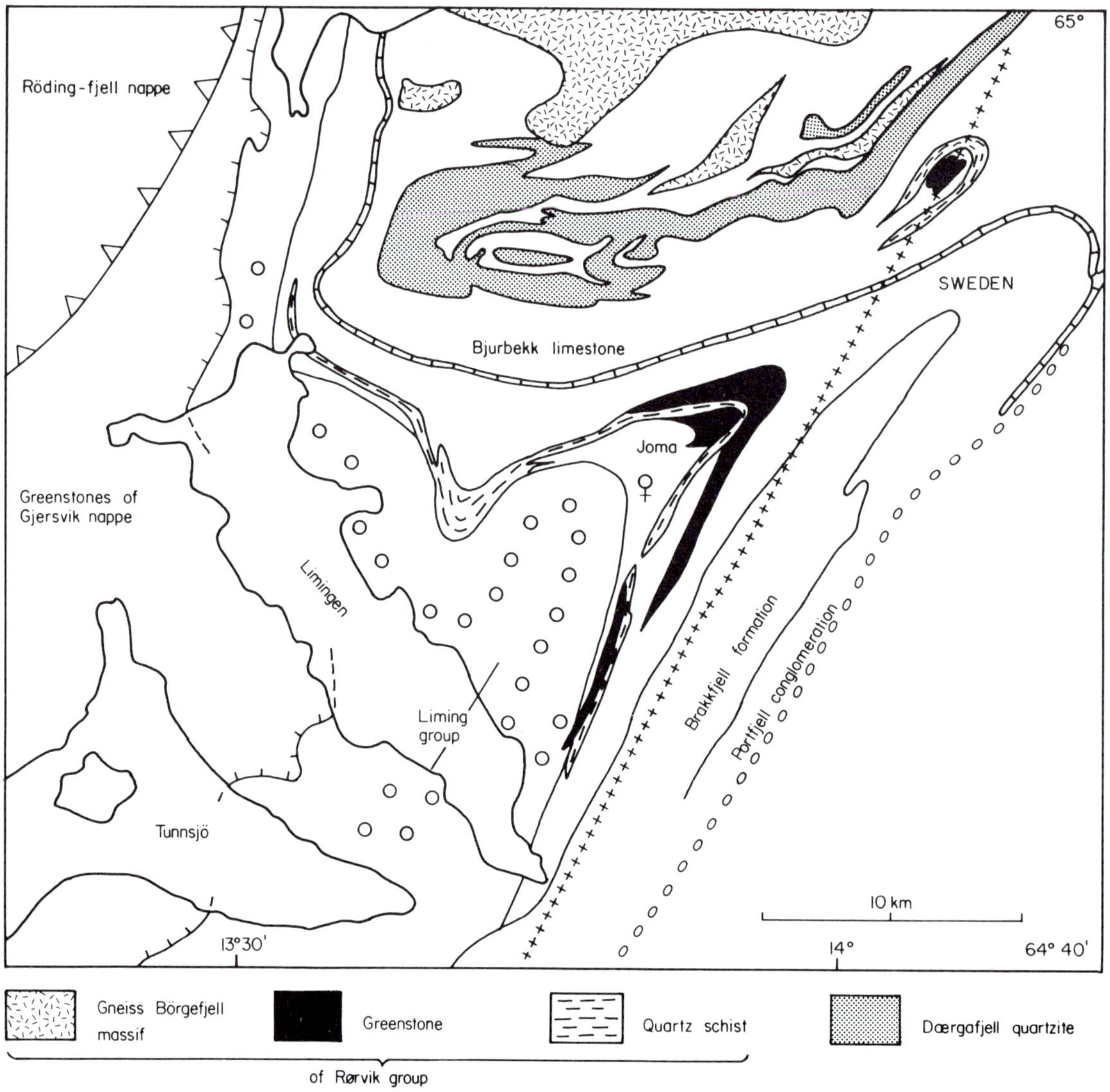

Fig. 73 *Simplified geologic map of a northern part of the Grong district.* Based on the maps of Foslie, and on the map of Nilsson (1964) for the Swedish part

massive greenstones with numerous rather large intrusive bodies of trondhjemite. In this complex occur the large deposits of cupriferous pyrite ore of Gjersvik and Skorovatn. The greenstones cover a wide area in the south, but wedge out north of Lake Limingen (almost exactly at 65° N). At Lake Limingen the greenstones can be seen to rest upon the rocks east of them with a tectonic boundary. They thus form a separate tectonic unit—the *Gjersvik nappe* (Oftedahl, 1956; Foslie and Strand, 1956).

The sequence described in the stratigraphic outline underlies the eastern parts of the area. On the south side of the Børgefjell basal massif the Dærgafjell quartzite occurs in several bands with an east to northeast trend and is bordered to the south, west and north by phyllites. The quartzite may lie in the core of a large recumbent anticline—in any case the structure seems highly complicated. On the southern and western sides of this structure the Bjurbäck limestone, which crosses the border from the Swedish side, extends continuously for nearly 40 km in a westerly and northerly direction and describes a curve convex to the southwest. On the Swedish side (Nilsson, 1964) the limestone makes a still sharper curve convex to the NNE. On the east side of the curve the limestone disappears, but the Portfjell quartz conglomerate (just below the Bjurbekk limestone) continues to the SSW and crosses the border onto the Norwegian side east of the south end of Lake Limingen. The stratigraphic horizon marked by the limestone and conglomerate is thus continuous for a distance of nearly 100 km. On the northeast side of Lake Limingen the rocks of the Rørvik and the overlying Liming groups form a number of wide and open antiformal and synformal structures. Some of the rock units mapped show a pinching out in two directions, indicative of the lunate outcrops produced by folding in two directions more or less normal to each other (Oftedahl, 1958a). Whatever they may be interpreted as, the structures invite detailed investigation.

The Store Børgefjell basal massif consists of mica schists, quartzites and coarse-grained and massive granitic or granodioritic gneisses. The overlying Caledonide sediments have tectonic boundaries with the rocks of the massif and show no signs of granitization or feldspathization at the contact. The Børgefjell massif must thus consist of unchanged Precambrian rocks, but may have been affected by Caledonian tectonics.

The conglomerates and calcareous sandstones of the Liming group strike along the west side of the Store Børgefjell massif and continue to the north in the western part of the Hattfjelldal area, where they extend to the south shore of Lake Røsvatn. To the west the rocks of the Liming group are overlain with a tectonic contact by the *Hattfjelldal limestone*. The limestone is dark, fine grained and of a very low grade of metamorphism, but parts of it have been altered into hard, sometimes light-coloured dolomitic rocks by a secondary process of dolomitization. In the eastern part of the area the sequence of the Liming group and underlying deposits shows folding in both NNE and WNW directions and contains a great number of serpentinite bodies, some of them of considerable dimensions (Strand, 1955, 1958).

Rocks of the same general type as in Hattfjelldal occur further north on the east side of Lake Røsvatn. A gabbro massif on Krutfjell mountain in this area has exerted strong contact metamorphism on the underlying sediments (I. B. Ramberg, personal communication).

On the Swedish side of the border the Storfjäll nappe complex (Chapter XXII) extends from 65° 31′ to 66° 18′ with thrust planes traceable to the border, the southern point being in the area under consideration. In the present state of knowledge, nothing can be said about the extension of the Storfjäll tectonic unit on the Norwegian side.

CHAPTER IX

THE BINDALEN–VEFSEN DISTRICT

The *Bindalen–Vefsen district* is limited to the east by the thrust line of the Rødingsfjell nappe, which separates it from the Grong–Hattfjelldal district. The most southerly part of the district is in the west, adjacent to the northern part of the Gneiss region. Further north, it is delimited to the west by the ocean. Its northern boundary can be drawn along Ranfjord and along a line elongated in the northeasterly direction of this fjord, towards the southern end of the *Nasafjell basal massif* (66° 30′, 15°). In a geological sense the Bindalen–Vefsen district is part of northern Norway, with the highly metamorphic sediments and granitic rocks characteristic of that part of the country, but the southern part of the district belongs to southern Norway (in an administrative sense).

STRATIGRAPHY

Nothing is known of the order of deposition of the sediments over most of the area, with the exception of the *Rana district* in the north. The sediments present are: mica schists, micaceous gneisses, limestones and dolomites and, more subordinately, quartzites. Also present are mica schists with kyanite and staurolite, representing pelitic sediments formed from strongly weathered material, and also micaceous gneisses rich in plagioclase, representing sediments of greywacke type. Some of the concordant amphibolites occurring among the layers are possibly of volcanic origin. Sedimentary iron deposits typical of the Nordland facies occur; they are of maximum thickness and commercial value in Dunderlandsdalen in the Rana district, where J. A. W. Bugge (1948) described a sequence about 1000 m thick of four named formations—the *Lasken*, *Nevernes*, *Ørtvann* and *Stillvann*, in ascending order. The Lasken formation consists of dolomite and limestone, the Nevernes of mica schist with several layers of iron ore, and the Ørtvann and Stillvann consist of mica schists, carbonate rocks and iron ores. The mica schists are partly calcareous (with lime silicates formed at the expense of carbonate originally present in the rocks) and partly kyanite schists. Bugge published seven chemical analyses of the schists and pointed out that some of them at least were of extremely residual type, indicating thorough weathering of the material before deposition. The presence of iron sediments also shows strong geochemical differentiation during deposition.

GEOLOGIC SETTING

The thrust line of the Rødingsfjell nappe, delimiting the Bindalen–Vefsen district to the east, is marked by a strong lithologic contrast and by a zone of heavy mylonitization (Fig. 74). It has been traced in the south as far as west of Lake Tunnsjö at 64° 50′ N, but becomes less clearly marked further to the south. A well-marked thrust plane appears again on the north side of the Sanddøla valley (64° 28′, 12° 50′) where greenstones and greenschists extending from the Grong district are overlain by a large massif of trondhjemite with a zone of heavy mylonitization at its base. This is probably the continuation of the Rødingsfjell nappe thrust plane. The trondhjemite mentioned together with gabbro composes a large massif of gabbro and trondhjemite, which should thus form part of the Rødingsfjell nappe. Oftedahl (1956) does not share this view and considers the massif in question to belong to a separate tectonic unit, namely the Gjersvik nappe mentioned previously. The thrust plane can be followed to the west across the Namsen valley and further north and northwest to the area north of Foldenfjord. Along this line the large massif of *Bindal granite* rests upon highly metamorphic mica schists and basaltic and dacitic metavolcanic rocks as described by Birkeland (1958). The latter rocks probably correspond to the sediments and volcanic rocks of the Grong district.

Fig. 74 *Front of the Rødingsfjell nappe in Hattfjelldal (65° 34′, 13° 58′)*, looking north, Lake Røsvatn in the distance to the right. Granitic gneisses of the nappe above Hattfjelldal limestone (growth with birches)

BIRKELAND describes a fault with intense mylonitization at the lower side of the Bindal granite and interprets it as evidence of the upthrust of an eastern block above a stationary western block. The writer has visited this region and has observed the zone of strong mylonitization at the lower side of the Bindal granite massif as far north as east of Foldereid (64° 58′, 12° 10′). To the west the thrust plane of the Rødingsfjell nappe apparently passes into the air. If this is really the case, the east to west extension of the nappe should be about 70 km.

The island of Leka, northwest of Foldereid (BIRKELAND, 1958), consists largely of a serpentinite massif nearly 15 km long. The serpentinite is surrounded by saussurite gabbro which is partly foliated. On a narrow strip along the southeast coast lies a polygenous conglomerate together with sandstone and some layers of phyllite and limestone. It is tentatively suggested that the rocks on Leka belong to the rock complex beneath the Bindal granite massif which extends from the Grong district.

The area north of the 65° N line is covered by four 1:250 000 Geological Survey maps with descriptions (REKSTAD, 1917a, 1925a, 1925b; HOLMSEN, 1932). Detailed investigations of particular parts of the area have been described by BIRKELAND (1958), FOSLIE and STRAND (1956), GUSTAVSON and GRØNHAUG (1960) and NISSEN (1965). There are also the printed but as yet unpublished 1:100 000 maps of Trones and Sanddøla by S. FOSLIE. HÅKON LIEN has contributed in an unpublished university thesis (1956) and the writer is also indebted to SIGBJØRN KOLLUNG of the Geological Survey for a written communication on the southwestern part of the area.

According to REKSTAD (1917a), definite volcanic rocks occur on the island of Vega (65° 40′, 11° 55′) and on the small islands to the west of Vega. Also present are greenstones interpreted as tuffs, which are developed in some localities as conglomerates and agglomerates. Some of the rocks are vesicular, the cavities being filled with calcite.

Gneisses cover large areas according to the geologic maps available. The gneisses are partly (probably for the most part) injection gneisses associated with the granites and partly rather homogeneous gneisses rich in microcline. GUSTAVSON and GRØNHAUG (1960) describe rocks of this group as fine to medium grained and grey or light red in colour, often banded and containing bands of amphibolite and crystalline limestone, suggesting a sedimentary origin. They contain lenses and veins of pegmatite and are interpreted as sediments metasomatically changed by the addition of potash; this is indicated by the fact that the potash feldspar occurs in particularly great profusion near planes of tectonic movement.

A glance at the geologic map will show that granites cover large parts of the area. In the southwest lies the large massif of Bindal granite. According to KOLLUNG the dominant rocks within the Bindal massif are light grey to reddish granodiorites; a typical rock is porphyric granodiorite with large insets of microcline. There are in addition mostly small bodies of darker quartz diorites and diorites and monzonites and syenites poor in quartz. Most of the Bindal rocks have massive structures or only faint foliation. Light granodiorites and trondhjemites penetrate the other rocks and are thus the youngest rocks in the massif. All rocks of the massif, however, have a border zone of migmatite against the adjoining meta-sediments. The granitic rocks in Hattfjelldal (65° 30′, 15° 30′), known to the writer from field observations, are light-coloured granites or granodiorites occurring in oblong massifs with strong gneissic schistosity concordant with the structures of the country rocks. They commonly contain inclusions of the adjoining sediments (Fig. 75), apparently with undisturbed geometry, and are bordered by an injection zone with numerous granitic sills and lenses in the sediments, so that often no marked boundary exists between granite and country rock. This type of injection gneiss or migmatite was described by OXAAL (1919) in Dunderlandsdalen.

Of a more obvious magmatic nature are granodiorites and trondhjemites (with idiomorphic, zoned plagioclases) occurring as phacolithic bodies (FOSLIE and STRAND, 1956). It is of interest to note that dykes from these rocks cut through the granitic veins in adjacent injected micaceous gneisses. A large massif of trondhjemite and gabbro measuring 20 by 20 km lies on the north side of the Sanddøla river, as mentioned previously.

The available information shows that the granites of the district under consideration are of the intrusive Nordland type and are not basal massifs, as is the case with most of the granitic rocks of the Svartisen–Salta district to the north.

A group of intrusive rocks differing from those hitherto considered were briefly described by VOGT (1897), in two areas; one at Velfjord (65° 20′, 12° 35′), the other just north of Mosjøen (65° 55′, 13° 20′). The massif north of Mosjøen contains gabbroid, banatitic and syenitic rocks, some of them with porphyric structures. They break through the structures of the Nordland rocks and are surrounded by well-marked contact aureoles with minerals such as wollastonite and lime garnet. VOGT considered the rocks to be younger than the regional metamorphism. Possibly related to these rocks are the diorites and syenites with alkali pyroxenes described by VOGT (1916) in the Horta island group (65° 12′, 11° 25′). The rocks contain calcite and are assumed to have assimilated limestone. An ultrabasic rock with microcline dominating over plagioclase among the feldspars which occurs in this area is called Hortite, after the island group name.

Not much more is known about the tectonics of the district other than the fact that there is generally a N–S axial trend. In the area mapped by GUSTAVSON and GRØNHAUG, between 65° 10′ and 65° 20′ N and east of 13° 10′ E, the rocks can be divided into two tectonic units: a western nappe with rocks trending NNE and dipping steeply west, and resting upon it an eastern nappe on a steeply dipping thrust plane with strong mylonitization. The eastern nappe, which rests on the main Rødingfjell thrust plane, has a less regular structure with its main axial trend to the northeast. The western nappe shows stronger metamorphism and granitization than the eastern one. There are two massifs of trondhjemite and granodiorite. The intrusive rocks are younger than the schistosity of the micaceous gneisses and contain rotated fragments of the latter, yet dykes from the intrusive rocks were folded in a later phase.

In the area around Bleikvassli mine RAMBERG (1967) mapped two complexes of rocks (mica schists, lime silicate schists, limestones and amphibolites, in amphibolite facies) separated by a slide (Fig. 76). During an early phase the rocks were folded in tightly compressed similar folds with a NNE or west to WNW axial trend. In a later phase the rocks were laid in large concentric folds, affecting the slide between the two complexes as well as the early folds, the axial planes of which were bent. A complex of highly metamorphosed

Fig. 75 *Granitic gneiss of injection type with inclusions of mica schist. Nellifjell, Hattfjelldal (65° 34′, 13° 57′).* Photo O. JØSANG

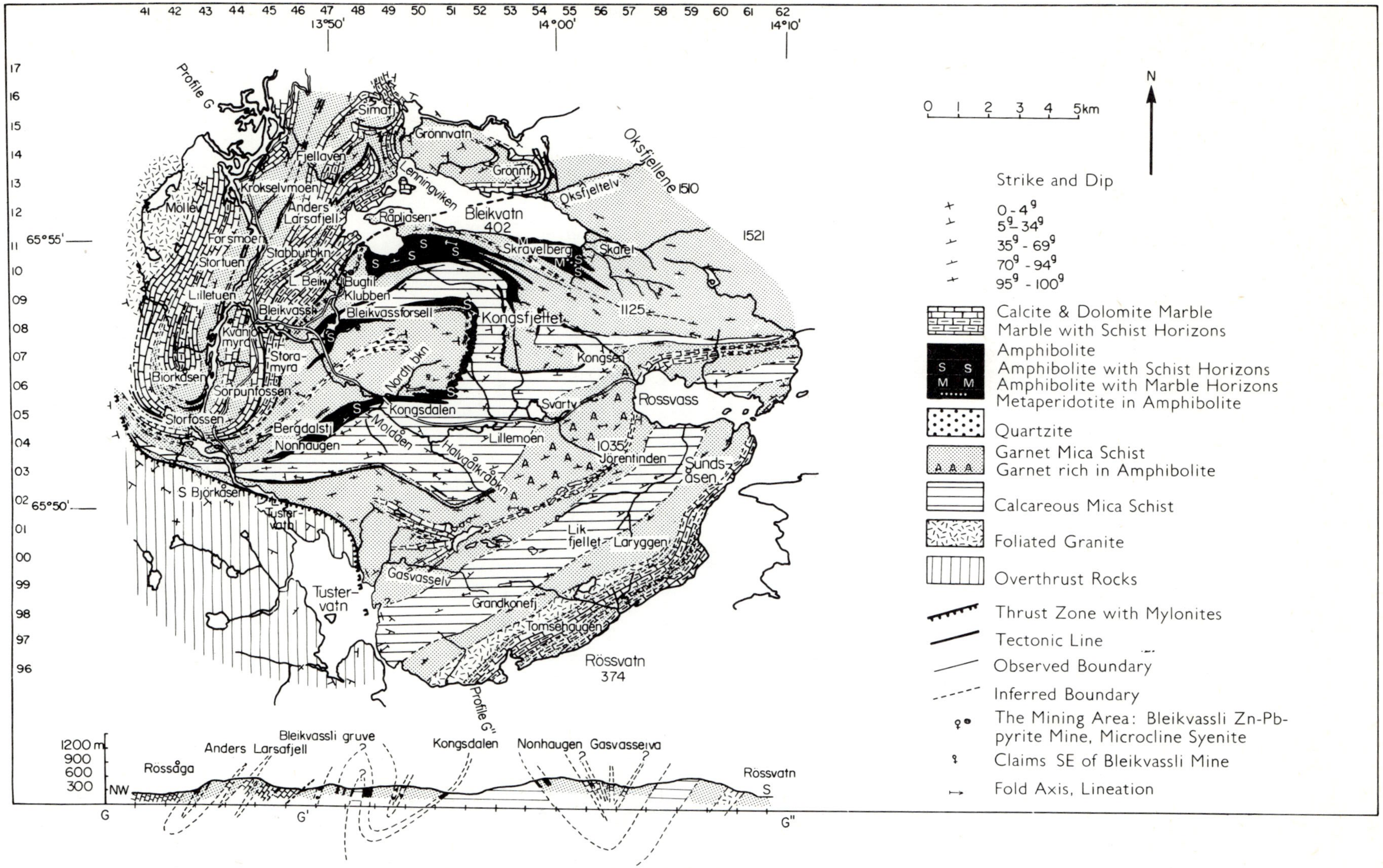

Fig. 76 *Geologic map of the Bleikvassli area.* By Ivar B. Ramberg

sediments, differing from those previously mentioned in being rich in granitic intrusions (in the southwestern part of the mapped area, Fig. 76), rests upon the other rocks on a well-marked thrust plane with structures indicating movement to the ESE.

RAMBERG also mapped the thrust plane of the Rødingsfjell nappe from the international border westwards to Lake Røsvatn. This thrust plane separates the highly metamorphic rocks of the Bleikvassli area from much less strongly metamorphosed phyllites, quartzites and greenschists which extend from the Hattfjelldal area. Here there is a persistent stratigraphy on both sides of the thrust plane, and the cataclastic rocks at the thrust plane have been metamorphosed and folded. (The persistence of stratigraphy at the thrust plane is reminiscent of conditions beneath the equivalent of the Rødingsfjell nappe in the Sulitjelma district.) The thrust in the southwestern part of the Bleikvassli area contains thick zones of mylonite and rests above the underlying rocks with a distinct tectonic unconformity. A characteristic rock in this upper nappe consists of quartzites alternating with amphibolites, which extend some 30 km to the north according to REKSTAD's (1924) map. RAMBERG suggests that this thrust plane in the Bleikvassli area is the continuation of the thrust plane which extends from the south and which has been considered as a continuation of the Rødingsfjell nappe thrust plane. According to this interpretation, the large nappe of highly metamorphic rocks and large granite massifs in the southern parts of northern Norway must be an independent tectonic unit, and the Rødingsfjell nappe in a restricted sense does not extend south of Lake Røsvatn. In its type area at 66° N the nappe occurs in a tectonic depression.

In the Dunderlandsdalen area where large deposits of sedimentary iron ore occur (OXAAL, 1919, J. A. W. BUGGE, 1948) the rocks of the sequence of limestones, dolomites and mica schists form a series of folds with the axes curved in an arc convex towards the southeast. On the east and south sides of the arc there are massifs of granite and injection gneisses; the arc arrangement of the axes is explained by the fact that the sediments were squeezed between the resistant massifs. BUGGE's detailed map of the iron ore deposit shows a very complicated structure that is probably due to repeated folding, as has been proved by British workers in the areas to the north (see the following section).

The *Langfjell injection zone* trends northwest and lies on the southwest side of the Dunderlandsdalen arc. Here, schists several hundred metres thick are intruded and injected by granodioritic and trondhjemitic veins. The rocks are folded on axes pitching 20 to 25° to the west of southwest, assumed to be in the direction of differential movement of the rocks. This differential movement took place over a considerable thickness in connection with the injection of acid magmas. As will appear from the following section, this tectonic zone has been interpreted by RUTLAND and NICHOLSON as marking the lower side of the Rødingsfjell nappe, the main part of which should end towards the north here. But, as emphasized by BUGGE, the tectonic zone of Langfjellet was formed at a rather deep horizon in the crust with plastic behaviour of the rocks, under quite different conditions from those commonly prevailing at the thrust planes beneath the nappes.

CHAPTER X

THE SVARTISEN–SALTA DISTRICT

The *Svartisen–Salta district* (Fig. 77) lies north of the Bindalen–Vefsen district. The boundary between the two districts lies along Ranafjord and continues along a line in the direction of the fjord to the southern end of the Lønsdal basal massif. The area on the Norwegian side of the border south of the Lønsdal massif is included in the area under consideration. To the north the district has a rather irregular boundary with the Tysfjord culmination, where rocks of the large *Tysfjord basal massif* come up beneath the sediments.

STRATIGRAPHY

The *Nasafjell* or *Lønsdal basal massif* (the latter name is used for the Norwegian part), is divided into two almost equal parts by the international border. The rocks of the massif are definitely Precambrian, as has been shown in the eastern Swedish part. Fine-grained light-coloured gneisses, common on the Norwegian side, may correspond to the meta-arkoses which according to Du Rietz (1949) form a large part of the massif. The rocks in the northern part of the massif in Saltdal were described in detail by Steenken (1957) as gneisses with bands of amphibolite and lime silicate rocks, of assumed supracrustal origin. They pass into massive granitic rocks towards the interior of the massif (at least in the limited area mapped). In the western part Caledonian rejuvenations may have taken place, as intrusive contacts between sediments and underlying gneisses have been recorded (Oxaal, 1919).

In 1956 the writer examined a section from the gneisses of the massif to the base of the Rødingsfjell nappe in the area south of the Lønsdal massif at Krokstrand (66° 27′, 13° 8′). Nearest to the basal massif are strongly deformed quartzites and black carbonaceous schists. Further up follows a formation of feldspathic quartzschist or greywacke, a limestone (about 300 m thick) and above this a division of greenschists. Mylonitic garnet–mica schists of the overlying Rødingsfjell nappe were found at the foot of Kjerringfjell. The rocks of the sequence are of low grade metamorphism. According to Oxaal (1919) the limestone of the sequence extends to the north along the western margin of the Lønsdal massif and to the southeast into Swedish territory. A corresponding sequence on the Swedish side was described by Marklund (1952). The limestone of this sequence (the *Jullega limestone*) was correlated by Kulling (1955) with the *Pieski limestone*. (See the description of the Aivo–Sarvas area in Chapter XXII.)

On the northeast side of the Lønsdal massif, along Junkerdalen (Rekstad, 1917b), the granitic rocks of the massif are overlain by 100–200 m of phyllitic schists, partly black and pyrite-bearing with some calcareous layers also, and then by a limestone, which strikes to the east across the border. On the Swedish side this limestone has been traced in connection with the Pieski limestone in its type area at Lake Pieskijaurre (67°, 16° 30′). The limestone in Junkerdalen is overlain by schists rather like those underlying it, but further upwards more coarse-grained and highly metamorphosed schists occur.

At the northern apex of the Lønsdal massif in Saltdalen, Steenken (1957) found a similar sequence immediately above the massif: a formation of carbonaceous schists with quartzites and a formation of calcareous mica schists with a limestone corresponding to the Pieski limestone. A higher-lying unit of staurolite gneisses and amphibolites is of higher metamorphic grade than the underlying rocks and is separated from them by a thrust plane.

Thus on the Norwegian side the Lønsdal massif is overlain by a sequence consisting of black carbonaceous schists with quartzites in the lower part and a limestone corresponding to the Pieski limestone at a higher horizon. According to the stratigraphy established by Kulling on the Swedish side, the Pieski formation is of early Ordovician age. The black schists may be Cambrian alum

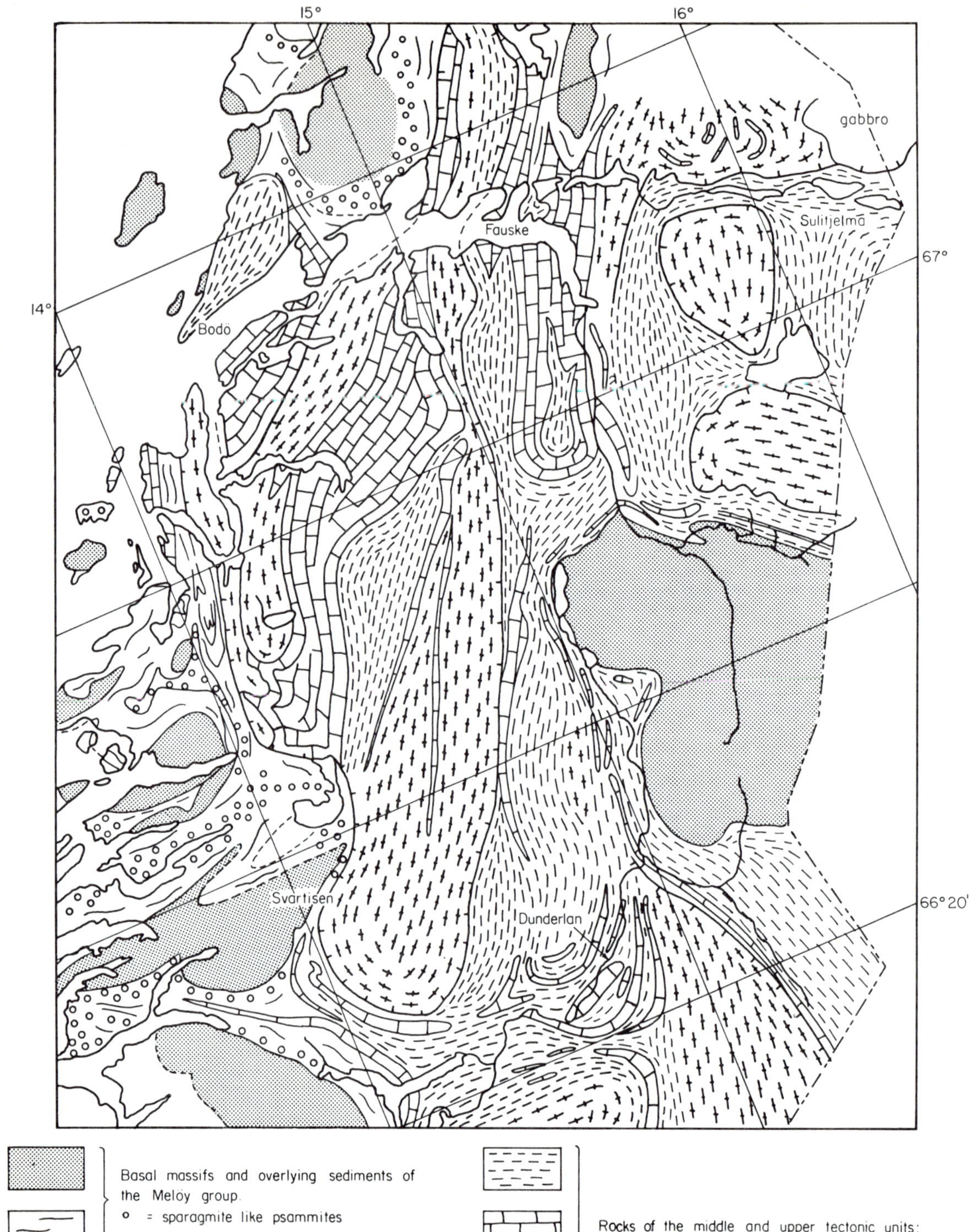

Fig. 77 *Geologic map of the Svartisen–Salta district*, based especially upon maps by Rutland and Nicholson (1965), Steenken (1957) and upon unpublished material furnished by Professor J. A. W. Bugge (Sulitjelma area).

shales, which according to KULLING (1962) are commonly found in the autochthonous or parautochthonous covers of the windows, while the Pieski limestone belongs to the Seve nappe complex.

A sequence of low grade sediments above the Pieski limestone extends along the border on the Norwegian side as a 10 km wide strip going along Lake Balvatnet to the Sulitjelma gabbro mountains, as described by VOGT (1927). The same sequence on the Swedish side was studied by KAUTSKY (1953). Above the Pieski limestone is the 1500 m thick *Muorki schist formation*, consisting of quartz–albite–muscovite–chlorite schists having a green colour due to the coarse grains of chlorite. Thin layers of dolomite occur in the upper part. The overlying *Furulund schist formation* (1100 m) is of a more varied lithology. In the lower part there are greenschists similar to the Muorki rocks and also black carbonaceous schists, limestones and albite porphyries; the last rock mentioned occurs in one locality as boulders in a highly deformed conglomerate or agglomerate. The upper part of the formation consists of banded biotite-bearing calcareous schists (with calcite and clinozoisite). In the lower part of the Furulūnd formation is a horizon with fossiliferous limestones extending on both sides of the border at 67° 5′ N. Besides encrinites, the limestones have yielded a fossil determined as the bryozoan Dianulites petropolitanus. Professor NILS SPJELDNÆS (personal communication) questions this identification, and considers the fossil to be a coral or a coarse-tubed bryozoan of late Middle Ordovician or Upper Ordovician age. According to G. KAUTSKY, the deposits east of Vastenjaurre in Sweden, which yielded a Halysites, are at a corresponding stratigraphic horizon. KULLING (1962) considers the deposits here in question to be of Upper Ordovician and Silurian age.

Along the western margin of the district a number of granitic massifs have been shown to be basal massifs, probably of Precambrian age, overlain by a sequence of sediments, known as the *Meløy group* (HOLLINGWORTH and others, 1960; RUTLAND and NICHOLSON, 1965). Although the boundary to the overlying Meløy sediments is generally tectonically disturbed, there is a sharp petrographic contrast between gneiss and overlying sediments and the same detailed stratigraphy is found above all the massifs. In the lower part of the Meløy group the most conspicuous rocks are banded or unbanded generally massive psammites associated with semi-pelitic schists, mica schists and variable hornblendic schists. In the upper part of the group pelitic mica schists and variable hornblendic and calcareous schists are about equally represented. Hollingworth and others correlate the lower part of the Meløy group, on the basis of lithologic similarity, with the Sparagmite schists of the Oppdal district and the upper part of the group with the Cambro–Ordovician "Trondheim schists" of the same district.

At Fauske in the Salta district (67° 15′, 15° 20′) the Fauske limestone with a thickness of more than 1000 m is composed of various layers of calcite and dolomite marbles, many of which are exploited for their beauty and quality. Structurally at the top of the limestone is the *Øynes conglomerate*, up to 200 m thick, with boulders of quartzite and limestone in an argillaceous matrix. There seems good reason to correlate the conglomerate and limestone at Fauske with the *Evenskjær conglomerate* and the *Evenes limestone* in Ofoten, but if this is correct the Fauske sequence must be inverted.

The limestones of the Fauske sequence extend to the south along Saltdalen, towards the northern part of the Lønsdal massif, where the Pieski limestone strikes from the east. Earlier writers, such as VOGT (1927), correlated the two limestones in the west and the east and assumed a field connection between them along Saltdalen and Junkerdalen, but this has not been convincingly demonstrated. In the western part of the district the limestones of the *Gildeskål* and *Saura groups*, with conglomerates corresponding to the Øynes conglomerate at Fauske, have been correlated with the Fauske limestone by British workers, while the *Bodø schist group* has been compared to the Furulund schists of Sulitjelma. The deposits overlie rocks of the Meløy group, but are separated from them by tectonic glide planes or slides. The highly metamorphic schists and limestones in the higher tectonic units of the district are of wholly unknown age and stratigraphic position.

GEOLOGIC SETTING

The western part of the district has been investigated by a group of British workers (RUTLAND, 1959; HOLLINGWORTH and others, 1960; RUTLAND and others, 1960; ACKERMANN and others, 1960; NICHOLSON and WALTON, 1960; RUTLAND and NICHOLSON, 1965; HOLMES, 1966); the second last paper gives a summary of the tectonic relationships. There are three main structural elements, of which the basal massifs with their parautochthonous cover of Meløy group sediments is the lowest. The middle main unit contains the Fauske limestone and other sediments assumed to be in a corresponding stratigraphic position, as mentioned previously. The upper main unit contains highly metamorphic rocks, partly injected by granitic material and granitized. Various groups of pelitic and hornblendic schists, marbles and other rocks

Fig. 78 *Limestone transected by granite. Kapplandsøya, Salten (67° 12′, 14° 38′).* Photo J. Rekstad

have been distinguished, among them the *Sokumfjell marble group* and the *Harefjell gneissic group.* The latter rock unit is situated around Sokumvatnet (65° 55′, 14° 15′) and consists of quartz monzonitic and dioritic gneisses. The dioritic gneisses can be proved to be of volcanic origin by means of interbedded meta-sedimentary bands that can be followed over long distances. In the microcline-bearing quartz monzonitic gneisses, on the other hand, the compositional banding is not parallel to the original layering shown by the bands of meta-sediments, and the formation of these rocks is ascribed to a soda and potash metasomatism undergone by the original dioritic rocks. Granites occur in the gneisses of the Harefjell group as well as in the underlying Sokumfjell marbles as a network of sheets and veins, the emplacement of which was evidently connected with the granitization of the dioritic gneisses. Granitic veins are also found in the basal gneiss massifs and in other rocks, but much more sparsely than in the Harefjell and other rocks of the upper unit, the *Beiarn nappe* (Fig. 78).

Detailed structural investigations have given evidence of three phases of folding; it may, however, be sufficient to distinguish between an early phase which produced isoclinal and recumbent folds, the larger of which were fold nappes of a Pennine type, and a later phase which produced more open synforms and antiforms, largely responsible for the present geometry as shown on the geologic map (Fig. 79). The early folds have variable axial trends, the average being in an E–W direction while the later folds are predominantly in a N–S direction.

The metamorphism in the district reaches a maximum in the almandine amphibolite facies. Evidence of the metamorphic history, best preserved in the pelitic rocks, shows that muscovite was the first mineral to be formed during an early phase of folding, while kyanite, staurolite and garnet grew later, and were rotated during later episodes of folding. Sillimanite occurs only in connection with granitization.

In the southwestern parts of the district are found the basal gneiss massifs of *Høgtuva, Svartisen* and *Glomfjord,* besides a number of smaller detached parts of the massifs. The gneisses of the massifs have taken part in all phases of folding together with the overlying sediments. The early phase produced recumbent folds of nappe dimensions with the gneiss in anticlinal cores, and refolding during later phases produced synforms which can be shown to be drooping anticlines. A good example is the *Fondal synform* which curves around the Svartisen massif. Folds of this type, with gneiss in the core of the synform, were mapped by Skjeseth and Sørensen (1953) at Holandsfjord, west of Svartisen. The largest of the early isoclines, the *Krokvatn fold* with its N–S axial trace just east of the 14° meridian, is a reclined fold with axis plunging SE, refolded by a younger synform overturned to the west. It has an amplitude of several kilometres; a possible continuation to the north might increase the amplitude to some 40 km.

The *Ørnes peninsula* (Figs. 80–82) shows structural relations similar to the southern districts, the basal massifs being involved in large recumbent folds.

The *Heggmovatnet massif* north of Skjerstadfjord (67° 25′, 15°) is regarded as an anticlinal culmination with a core of basal gneiss. Much of the massif is made up of rocks of sparagmite type. Reitan (in Holtedahl and others, 1960) records meta-arkoses with 50% or more quartz, and the general

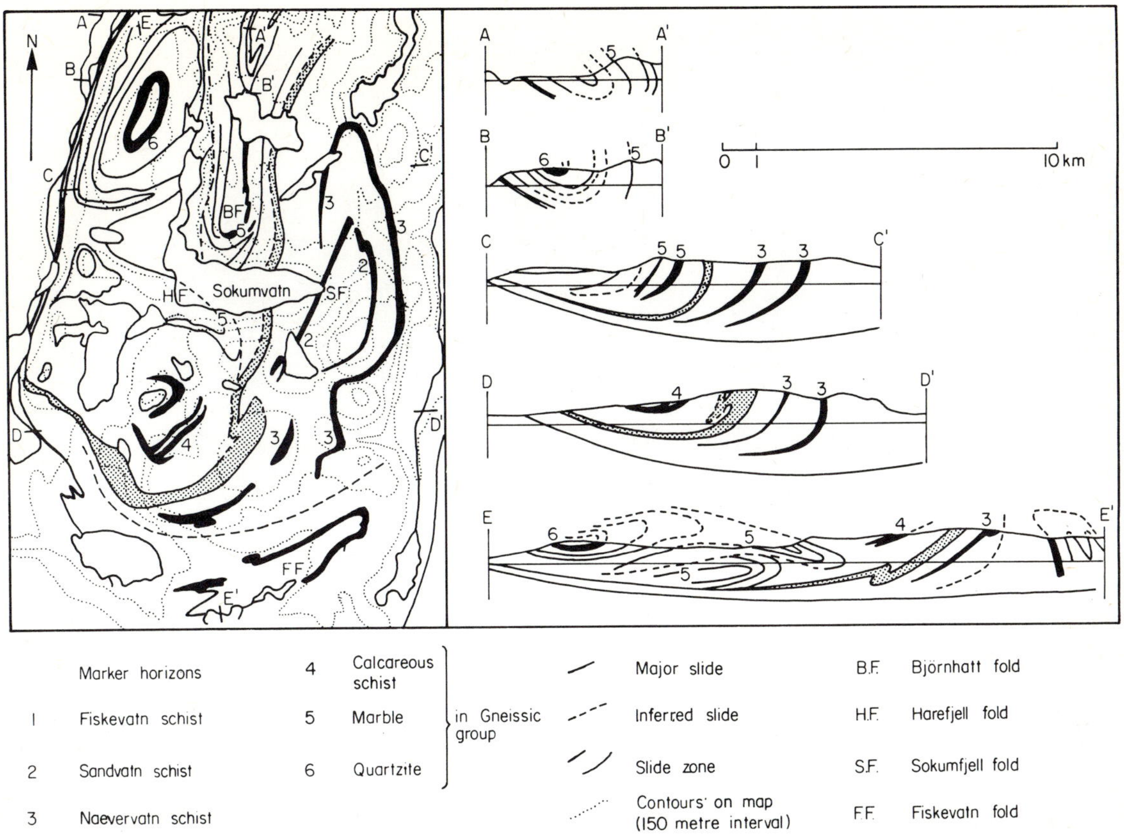

Fig. 79 *The area around Lake Sokumvatn (66° 65′, 14°15′).* A number of marker horizons show the style of folding. From Rutland (1959)

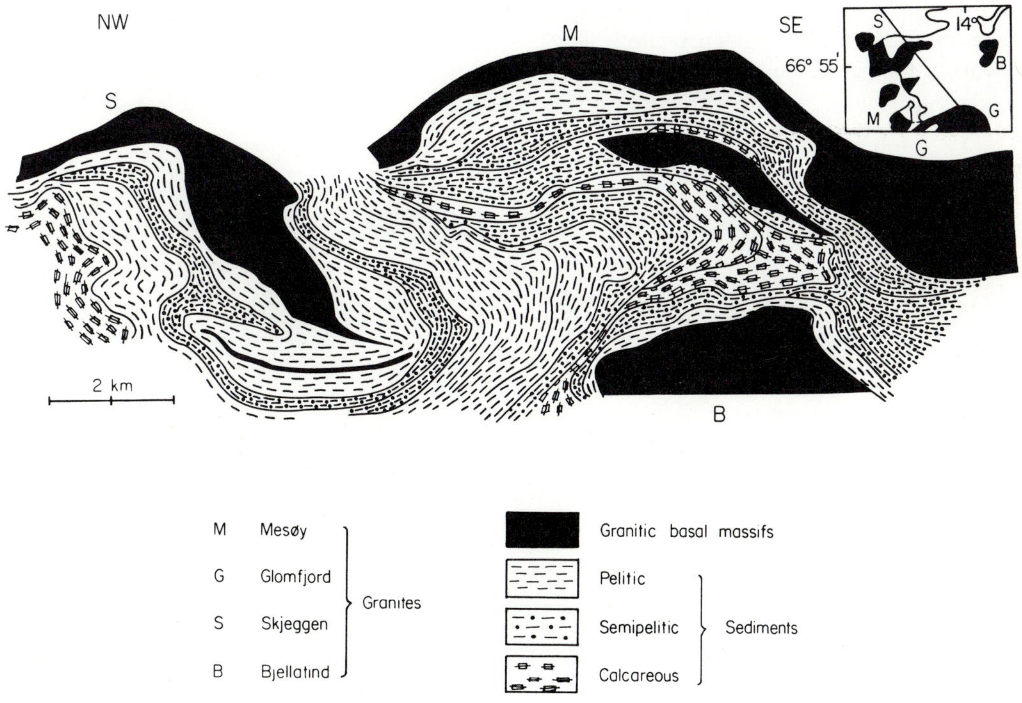

Fig. 80 *Section across the Ørnes peninsula (map in upper right corner), projected on plane normal to regional fold axis, plunging 17° SW.* After M. Holmes (1966)

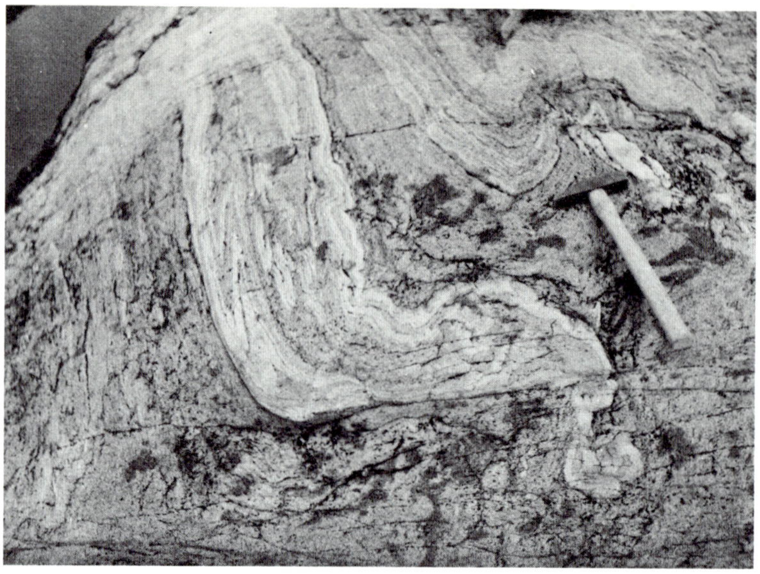

Fig. 81 *Disharmonic isoclinal F_1 folds in semipelitic and pelitic schists, refolded by F_2 shear folds. Steffodalen (66° 50′, 13° 50′), looking south.* Photo M. Holmes

succession outwards from the core is like that of the other massifs. The underlying gneisses, however, as far as they are known, are much coarser than those of the southern massifs and usually contain porphyroblasts of potash feldspar—they may be granitization products of underlying sediments.

As emphasized by the British writers, the western parts of the Svartisen–Salta district display structures characteristic of a deep tectonic level, different from the conditions in marginal parts of the orogen, where the nappes were moved above a stiff basement. They speak of fold nappes of *conjunctive* type in the former case and of *disjunctive* type in the latter case. The upper tectonic unit with highly granitized rocks, the Beiarn nappe of Rutland and Nicholson, is still assumed to have its root outside the area and to be an equivalent of the Rødingsfjell nappe in the southern districts. They suggest that the Rødingsfjell nappe ends in a synformal closure west of the south end of the Lønsdal massif. The Langfjell injection zone south

Fig. 82 *Disharmonic F_1 folds in semipelitic schists and amphibolites.* Locality as in Fig. 81, looking southwest. Photo M. Holmes

of the Dunderlandsdalen iron ore district is assumed to form the northwest boundary of the nappe. Further investigations are needed to prove or disprove this hypothesis. In any case the sharply marked thrust plane under the east side of the nappe shows conditions radically different from those displayed by the Langfjell tectonic zone.

The most southerly of the basal massifs recognized by the British workers is the Högtuva massif, previously mentioned. Southwest of this massif, separated by a thin intervening zone of gneiss, lies the *Sjona granite massif* north of Sjonafjord (66° 25′, 13° 15′). According to HOLMSEN (1932) the intervening gneiss dips beneath the Högtuva massif; the Sjona massif is thus (apparently) at a deeper tectonic level than the Högtuva massif. From the relationships between the two it seems reasonable to regard the Sjona granite as a basal massif also.

The Norwegian Sulitjelma district lies between the Saltdal valley in the west and the international border in the east. In the south it has a natural boundary to the Lønsdal massif, while its northern part extends to the high Sulitjelma mountains. The district was described by SJÖGREN (1900), VOGT (1927) and KAUTSKY (1953). The writer is indebted to Professor J. A. W. BUGGE for information on unpublished work in the district and further to K. J. HENLEY and R. J. MASON, who gave access to manuscripts on tectonics and metamorphism, and on the Sulitjelma gabbro, respectively.

Three main structural units can be distinguished in the Sulitjelma district. The lower one is the sequence of the Muorki and Fūrūlūnd formations, while the upper one consists of various high grade sedimentary schists, with limestones in parts of the district. Between these two main units is a zone with basic volcanic rocks as the main constituent (greenstones, chlorite–albite schists). A thin band of granitic rock above the greenstones, which extends all around the Baldoaivve synclinorium, is of volcanic origin. The same is true of the Fūrūlūnd "granite" in the northern part of the district. The middle zone also contains a considerable variety of other rocks, such as staurolite- and garnet-bearing schists, pegmatite and quartzite. The rocks have suffered strong tectonization combined with chloritization. There is lensing and brecciation on a large scale and according to SJÖGREN's map a "chloritized and brecciated amphibolite" is widely distributed. The middle zone is the site of the large bodies of cupriferous pyrite worked by the Sulitjelma mining company.

In the Sulitjelma district the valley in which lies Lake Langvatnet is excavated along an anticlinal structure trending WNW. Here the rocks of the lower unit are at the surface, and the valley separates a northern mining area from a southern mining area in the *Baldoaivve synclinal depression.* Still further to the south is the *Skaiti synclinal depression* with rocks of the upper unit. Structures trending E–W are prominent in the district; for example in such features as mineral lineation, elongation direction of the ore bodies and axes of folding. The E–W folds are isoclinal in the lower units, but only tightly compressed in the upper unit. A younger set of folds trends roughly N–S and is overturned to ESE.

The grade of metamorphism varies greatly within the district. In the eastern part near the border the rocks of the lower unit are nearly unmetamorphic (with fossils), while the overlying rocks of the upper unit are of high metamorphic grade. This difference disappears, however, towards the west where the metamorphic grade in the lower units increases from greenschist to amphibolite facies within a distance of some 15 km.

Granitic intrusive rocks occur among the rocks of the upper unit, particularly in the Baldoaivve synclinorium where granitic and trondhjemitic rocks are widely distributed. A swarm of doleritic intrusions are found in the schists of the Muorki formation in the eastern part of the district.

The *Sulitjelma gabbro*, occurring on both sides of the border, rests with a tectonic contact upon greenstones of the middle zone, while it has intrusive and hornfelsed contact with the rocks of the upper unit. It is an olivine gabbro displaying rhythmic banding as a result of the concentration into bands of large crystals of clinopyroxene. In the lower parts of the body the rock has been changed into a hornblende meta-grabbro. (VOGT (1927) also considered the underlying greenstones as metamorphic derivates of the gabbro.)

KAUTSKY (1953), who mapped the Swedish part of the Sulitjelma district, classed the three units of the Norwegian part of the district as belonging to his *Pieski*, *Vasten* and *Gasak nappes*, in ascending order. KULLING (1960, 1962) considered KAUTSKY's Gasak nappe as part of the Rødingsfjell nappe. On the Norwegian side the Rødingsfjell nappe, together with the Sulitjelma gabbro, should thus comprise the upper unit. In the eastern part of the district these rocks have a much higher grade of metamorphism than the underlying rocks; there is in addition a zone with intense tectonization and chloritization in the middle zone below the upper unit. On the other hand, the ore-bearing rocks of the middle unit extend for a considerable distance, apparently with undisturbed stratigraphy, a fact which seems to throw doubt on the presence of a plane of large tectonic movement at the same horizon.

As appears from the description in the Swedish part of the book (see Chapter XXII) KULLING has revised the tectonic scheme of KAUTSKY and introduced new names for the tectonic units. KULLING includes TH. VOGT's Fūrūlūnd formation of the Norwegian Sulitjelma district in his Virihaure nappe. The lower parts of the Norwegian Sulitjelma sequence should thus belong to the Seve–Köli nappe complex, as delimited by KULLING.

The large massif of gneissic granites at Nordfolla (67° 50′, 15° 30′) lies above the surrounding sediments, according to the sections of REKSTAD (1929), and is thus not a basal massif. The boundary between the sediments of the district under consideration and the basal gneiss in the Tysfjord culmination to the north owes its geometry to large folds affecting the basement, as was commented on by KAUTSKY (1953), who compared the granitic rocks of the basement massif with the mantled domes described in other areas.

As may be apparent from the foregoing descriptions, three main tectonic units can be distinguished in the Svartisen–Salta district. To the lower unit belong the basal massifs with parautochthonous covers of sediments, known in western parts of the district. The middle unit consists of the *Fauske limestone sequence* in the west, and of the Sulitjelma sequence of the Pieski, Muorki and Fūrūlūnd formations, with their equivalents. There is in addition an upper unit of various high grade rocks, often highly injected by granitic material. This includes the Beiarn nappe of RUTLAND and NICHOLSON and the upper unit in the Sulitjelma district, and must be considered as part of the large Rødingsfjell nappe, which has no known counterpart north of the district under consideration. The boundaries between the middle and upper units are, however, still unknown over large parts of the area.

It is apparent that the western part of the district with the basal massifs represents a zone of tectonic culmination. A synclinal zone contains the Beiarn nappe south of Skjerstad, and a second one occurs in the Sulitjelma district.

Addendum. A paper by NICHOLSON (1966) gives details of the stratigraphy of the fossiliferous sediments and associated volcanic rocks in the eastern part of the Sulitjelma area.

CHAPTER XI

CRYSTALLINE BASAL MASSIFS IN THE TYSFJORD DISTRICT AND ON THE ISLANDS ALONG THE COAST FROM LOFOTEN TO RINGVASSØY

In the *Tysfjord district* (about 68° N) an axial culmination brings to the surface a massif of granite beneath the Cambro–Silurian sediments. Granites, gneisses and other plutonic rocks compose the islands of the Lofoten and Vesterålen groups and also the large islands of Hinnøy, Senja, Kvaløy and Ringvassøy with smaller adjacent islands, thus extending northeast to about 70° 20′ N. In general the rocks in these districts are in a similar position to the Basal gneisses in southern Norway.

The main part of the Tysfjord district has been mapped and described in great detail by Foslie (1941, 1942). The *Tysfjord granite*, here forming a basal massif, is a greyish or faintly reddish coarse granite, almost massive or with faint schistosity. It is a potash granite rich in microcline with dark minerals such as lepidomelane and a hastingsitic amphibole, and having fluorite and epidote-orthite as common accessories. Above the coarse-grained Tysfjord granite, at the boundary to the overlying sediments, are fine-grained gneissic granites, which are otherwise, in their chemical and mineralogical composition, wholly similar to the coarse Tysfjord granite. In parts of the area a thin band of mica schist or quartzite comes between the coarse granite and the fine-grained gneiss, and in other parts of the area the fine-grained granites apparently intrude the sediments and may be found as several thin bands alternating with the mica schists (Fig. 83). The sediments overlying the granite are not at the same stratigraphic level in all parts of the area, representing higher parts of the sequence in the north than in the south. Because of the similarity and the intimate connection between the fine-grained gneissic and the coarse-grained main Tysfjord granite, Foslie considered the whole granite to be of Caledonian age. A fact not in harmony with this interpretation is the occurrence in the *Rombak window*, east of Ofotfjord, of coarse-grained granites very similar to the Tysfjord granite. The Rombak window is an area of largely unchanged Precambrian rocks, as is shown by the presence of autochthonous conglomerates and quartzites above the basement, which are present in the Norwegian part of the window also. East of Æfjord only 15 km of intervening sediments separate the Tysfjord granite from a similar granite in the Rombak window, where even the presence of fluorite is common to both granites. For this reason Vogt (1942) considered the Tysfjord granite as Precambrian and its fine-grained gneissic facies as having been formed from palingenetic magmas formed by remelting of the granite at deep levels. In this connection Vogt mentioned the *Lødingen granite*, a red potash-rich granite occurring on both sides of Ofotfjord. On the north side of the fjord it has an injection contact with the overlying sediments, and on the south side it passes into the Tysfjord granite without any marked boundary.

A different interpretation of the gneissic granite immediately below the sediments in the Tysfjord area has been given by Swedish geologists. In its extension onto Swedish territory the gneissic granite proves to be in an allochthonous position above sediments of the "Hyolithes zone" and below the sediments of the great Seve nappe. Thus Kautsky (1953) and Kulling (1964) interpreted the gneissic granites in the Tysfjord area as corresponding to the nappe of similar rocks on the Swedish side. (The two writers had somewhat different views on the tectonic relationships of the rocks in question on the Swedish side.) The concept of Kautsky and Kulling of tectonic emplacement seems at least as reasonable as the concept of intrusion of anatectic magma. On the Norwegian side recrystallization can be assumed to have obliterated all traces of mylonitized tectonic boundaries. Oftedahl (1966) commented on this matter and on the overthrust tectonics in north Scandinavia.

On the northwest side of the large and open Vestfjord, the islands of the Lofoten group and the southwestern parts of Hinnøy and Langøy in the Vesterålen group are underlain by very massive,

Fig. 83 *Stolotindan east of the Hellemofjord (67° 54′, 16° 10′).* Tysfjord granite overlain by mica schists with bands of gneissic granite. From REKSTAD (1919)

charnockitic rocks, known as the *Lofoten rocks.* Rocks of the same type are also found on Hamarøy on the southeast side of Vestfjord. KOLDERUP (1899), VOGT (1907) and VOGT (1910) described the Lofoten rocks and interpreted them as a differentiated suite of magmatic intrusive rocks. Limestones and other sediments, including layered iron ores found as inclusions in the Lofoten rocks, were interpreted as xenoliths. At present there seems good reason to consider the main parts of the Lofoten rocks as old Precambrian supracrustal rocks metamorphosed under granulite facies conditions (Fig. 84).

The rocks on the island of Langøy (68° 45′, 15°)

Fig. 84 *Reine in Moskenesøy, Lofoten (67° 56′, 13° 7′) to show the morphology typical of the charnockitic Lofoten rocks.* The strandflat, an abrasion surface of Pleistocene age, forms a narrow brim at the foot of the mountains. Copyright Widerøes Flyveselskap A/S, Oslo

in the Vesterålen group have recently been mapped and a detailed petrographic description of them given by HEIER (1960). The eastern part of Langøy and parts of Hinnøy east of Langøy are underlain by veined gneisses in amphibolite facies of granodioritic composition with bands and lenses of amphibolite. The amphibolite facies gneisses in the eastern part of Langøy are overlain by charnockitic rocks in the west on a thrust plane. In spite of this sharp, tectonic boundary the difference between the gneisses in the east and the charnockitic rocks in the west is due solely to an increase of the metamorphic grade from amphibolite to granulite facies conditions, shown by the fact that biotite in the former rocks is replaced by hypersthene in the latter. The so-called "charnockite border series" forms a transition between the amphibolite facies gneisses and the charnockites proper. In the border series, as in the eastern gneisses, it is easy to distinguish between the amphibolitic and acidic bands, but this is difficult in the charnockites proper, due to the uniform dark colour of all rocks in this group. Modal and chemical analyses show that the rocks in the eastern gneiss complex and in the border series tend to separate into a group poor in silica (50% or less SiO_2) and another group with 60% or more SiO_2. The banded charnockites, on the other hand, evenly span the whole range of silica composition. The mineral composition of the charnockitic rocks shows no regular variation with the silica contents, contrary to what might be expected in a series of magmatic rocks.

A number of bodies of augen gneiss, granite, monzonite and gabbro occur within the banded complex. The amphibolite facies banded gneisses in the southeast part of Langøy pass into coarse red augen gneisses. There are granites of two types: "old red granites" surrounded by a migmatite aureole and "young red granites" with sharp concordant boundaries to granulite facies rocks, yet the contact is gradational in the sense that increasing amounts of small pegmatitic and granitic veins are found on approaching the granite body. In the southwestern part of Langøy the banded granulites pass into granulitic monzonite with large porphyroblasts of potash feldspar. An intrusive body of gabbro with monzonite occurs in the northwestern part of Langøy. The gabbro contains segregations of ultrabasic rocks and titaniferous iron ore. Parts of the gabbro are banded with alternating dark and light bands.

As emphasized by HEIER, the Lofoten–Vesterålen rock complex includes old Precambrian massifs and must have a long and complicated history with Precambrian as well as Caledonian periods of metamorphism. Molybdenites from the district gave an age of more than 2000 m. y. by the Re/Os method. Later K/A determinations gave 640 m. y. for "young red granite" from Langøy, 650 m. y. for a Lofoten granite and 575 m. y. for an augen gneiss from Langøy, while three determinations on biotite from pegmatite, grey gneiss and mica schist on Langøy gave 423, 430 and 450 m. y. (BROCH, 1964).

According to Professor J. A. W. BUGGE (personal communication) the monzonitic charnockites in the southwestern part of Hinnøy contain bands of quartz-banded iron ore, different from the iron ores of "Nordland type" found in the Cambro–Silurian sequence, but similar to a well-known type of Precambrian sedimentary ore. Pure magnetite ores on Austvågøy, occurring as bands in granitic gneisses with remnants of porphyryic structure also conform to a Precambrian type of ore. Some iron ores (of skarn type), however, are of "Nordland type"; they occur together with limestones and graphite schists and extend for considerable distances with structures conformable to the surrounding charnockitic rocks, and cannot be inclusions or xenoliths in magmatic rocks. Lime silicate gneisses in the same assemblage have a field appearance like the charnockitic gneisses (HEIER, 1960).

On Flakstadøy in the Lofoten group, CARSTENS (1957b) described an iron-rich gabbro with bands of magnetite pyroxenite at the base; the gabbro passed upwards into anorthosite. Anorthosites occur in other parts of Lofoten and Vesterålen also, but are not quantitatively important. The gabbro on Flakstadøy is connected with a large massif of monzonite on Vestvågøy, the island adjacent to it to the northwest, according to a personal communication from Dr. CARSTENS. Here the core or dome of monzonite is bordered by gneisses of supracrustal origin with banded iron ores.

Southwest of Lofoten are two groups of smaller islands, those of Værøy and Røst, consisting of gneisses, micaceous gneisses, crystalline limestones and mica schist with kyanite; the massive Lofoten rocks do not occur here.

Some facts about the rocks on Hinnøy, Andøy and the islands further northeast are known as a result of the pioneer work of KARL PETTERSEN and from more casual later observations, a compilation of which was given by HOLTEDAHL (1953). The rocks are mainly gneisses and granites, but crystalline limestones, micaceous gneisses and other rocks apparently representing Cambro–Silurian sediments occur in concordant interlayering with the gneisses and also underlie relatively large parts of the islands.

At Kasfjord on Hinnøy, west of Harstad (68° 50′, 16° 20′) FOSLIE (in BUGGE and FOSLIE, 1922) gives a section of the boundary between granite and overlying sediments, dipping about 45° southeast. In the sediments just above the granite there are concordant intrusions of aplitic granite which led to the formation of skarn masses with magnetite and a little arsenopyrite in the overlying calcareous sediments. The relationship between the granite and the overlying sediments seems to be the same as in the Tysfjord district.

The sedimentary iron ores in these areas are of "Nordland type" (of Cambro–Silurian age) according to Professor J. A. W. Bugge (personal communication). On Bjarkøy north of Harstad (69°, 16° 10′) Kautsky (1946) reported a sedimentary iron ore, half of which was surrounded by a metamorphosed sediment, while the other half occurred in a granitic rock—this might show that Caledonian granitization was of importance here.

The city of Tromsø is situated on an island of the same name. The northern part of the island is composed of amphibolites which are overlain by sediments in the southern part of the island. The amphibolite is cut through by a network of dykes of oligoclasite, of thickness varying from a few centimetres to 5 or 10 metres (Landmark, 1951b). Later studies by Landmark (personal communication) have shown that the oligoclasite and associated light-coloured trondhjemitic and granitic rocks occupy the northeastern parts of Kvaløy and the eastern parts of Ringvassøy, the two large islands west and north of Tromsø. The complex of magmatic rocks is interpreted by Landmark as a nappe thrust into the sediments at a high stratigraphic horizon (below the *Tromsø limestone*).

CHAPTER XII

THE OFOTEN–TROMS DISTRICT

The *Ofoten–Troms district* comprises the areas covered by Cambro–Silurian sediments north of the Tysfjord culmination. To the northeast it extends to the boundary of the Finnmark region, where psammitic deposits of late Precambrian and Eocambrian age predominate.

STRATIGRAPHY

The Caledonide margin lies on the Norwegian side in parts of the Ofoten–Troms district, where deposits of Cambrian age are found in an autochthonous position above the Precambrian basement. These deposits, known for many years as the "Hyolithes zone", will be described in the following chapter on the Finnmark region. Separated from the underlying deposits of the "Hyolithes zone" by a thrust plane or a plane of tectonization are the deposits of the Storfjell group (GUSTAVSON, 1965), of feldspathic quartz schists with layers and lenses of dolomite. According to GUSTAVSON the Storfjell group may be a parautochthonous deposit in a normal stratigraphic position above the "Hyolithes zone" or it may form a lower nappe of sediments of late Precambrian–Eocambrian age. The corresponding group of rocks on the Swedish side is the *Rautas nappe complex* (KULLING, 1964) and the complex of sediment with dolomites found by HAUSEN (1942a) at the Caledonide margin in Enontekiö, Finland. The *Jerta nappe* of SKJERLIE and TAN consists of similar rocks in a corresponding tectonic position (Chapter XIII).

The main part of the sediments in the district consists of more or less strongly metamorphosed deposits of Nordland facies, separated from the above-mentioned deposits by a well-marked thrust plane. In western parts of the district, autochthonous sediments are absent and the metamorphosed sediments rest directly upon rocks of the basement complex.

The most complete section of the deposits of Nordland facies is in the Narvik–Ofoten district at Ofotfjord, with the following main units, according to VOGT (1922, 1942) and FOSLIE (1941, 1949), in descending order:

- Niingen schist group
- Bogen group
- Evenes limestone group with Evenskjær conglomerate at the base
- Narvik group with Melkedal limestone and Ballangen schists in the upper part
- Rombak group with Rombak limestone in the upper part

The rocks of the *Rombak group* come above the Archean substratum in the Rombak window east of Narvik, separated, however, by autochthonous quartzites of assumed Early Cambrian age. The rocks are predominantly mica schists with the *Rombak limestone* in the upper part. Above the Rombak limestone is a horizon with amphibolites, present in the mountains north of Narvik and present also at a corresponding horizon on the Swedish side. The nature of the amphibolites, i.e. whether they are intrusive or extrusive, is uncertain.

The *Narvik group* is composed mainly of mica schists with the *Djupvik quartzite* in the lower part. Higher up the mica schists become more strongly metamorphosed and injected by granitic veins.

In the upper part of the Narvik group is the 30 m thick *Melkedal limestone*, traceable over a distance of 70 km, and also a horizon of iron ore and the *Ballangen schists*, with several horizons of carbonaceous schists, rich in pyrite.

The *Evenes limestone group* is exposed in the Håfjell syncline on the south side of Ofotfjord as 1200 m of predominantly calcite marble, but with some horizons of dolomite, mica schists and quartzite. At the base of the Evenes group is the *Evenskjær conglomerate*, 75 m thick in the Håfjell section, with boulders of quartz, diorite, trondhjemite and quartzite in a matrix of calcareous mica schist. As emphasized by FOSLIE and KULLING the Evenskjær conglomerate is similar to the

Liming conglomerate in the Grong district and to the Gilliks conglomerate on the Swedish side of the border. If this lithologic similarity has a stratigraphic significance, the Evenskjær must be of late Middle Ordovician or early Late Ordovician age.

The *Bogen group* consists of alternating mica schists, quartzites, dolomites and limestones and two horizons of iron ore. The thickness of the sequence above the Melkedal limestone in the Håfjell syncline is nearly 5100 m. Taking the thickness of the deposits below that horizon into account, the thickness of the whole of the Ofoten sequence cannot be less than 7000 m.

Above the Bogen group on the north side of Ofotfjord is the *Niingen schist group* consisting of mica schists with granitic injections.

The Ofoten sequence can be regarded as a type section of the stratigraphy of the Nordland facies, at least insofar as many other districts in northern Norway have a sequence predominantly of mica schists in the lower part and thick limestones in the upper part, as was recognized many years ago by VOGT (1897).

In the Tysfjord district south of Ofoten (FOSLIE, 1941, 1942) the 500 m thick *Reppi calcareous schist formation* has a stratigraphic position corresponding to the lower part of the Narvik group, while the overlying 1500 m thick *Gicce formation* of granite-injected schists probably corresponds to the granite-injected schists in the Narvik group at Ofoten. Both below and above the Reppi schist are horizons, 300 and 900 m thick, with concordant bands of amphibolites, often alternating with trondhjemitic bands and mica schists. It is likely that these rocks are of volcanic origin (OFTEDAHL, 1959). As the rocks in question are associated with talcified dunites, it is tempting to correlate them with the volcanic horizon of the Støren group in the Trondheim region. In the Linnajavvre district, on the international border at 67° 35′ N, the Reppi formation is more calcareous than in the north and contains limestones.

In the areas north of Ofoten and south of the 69th parallel, GUSTAVSON (1967) has traced the stratigraphic units from Ofoten. The Narvik group contains acid Na-dominant gneisses interpreted as volcanic rocks. In northern parts of the area it is not possible to distinguish between the Evenes and the Bogen groups, which are joined to the Salangen group by numerous limestones.

In the area around the city of Tromsø and further to the south as far as the Målselv valley (69° N), as yet unpublished investigations have been done by K. LANDMARK. The sediments in this area can be divided into three parts. The lower part (unnamed) consists chiefly of mica schists and micaceous gneisses rich in plagioclase and contains a formation of feldspathic quartzite. The mica schists contain bands of amphibolite, probably of volcanic origin. The *Balsfjord group* in the middle part of the sequence contains low grade mica schists. The *Balsfjord limestone* occurs in the middle of the group and is directly underlain by a conglomerate with boulders of quartzite and limestone. Correlation with the Evenskjær conglomerate and Evenes limestone in Ofoten is probable. The upper unit, the *Tromsø group*, has a lithology and metamorphic grade similar to that of the lower group. There is a quartzite formation at the base of the group and higher up is the *Tromsø limestone*, between two formations of mica schist.

In the Birtavarre district east of Lyngsfjord (about 69° 25′, 21°) PADGET (1955) described a sequence of thickness some 3000 to 4000 m, as listed below in descending order:

Quartzite (unnamed)
Store Borsejok schists
"Schists-with-thin-limestones"
Green beds
Ankerlia schists
Guolas limestone
Nitsim Varre schists and quartzite
Big limestone

The sequence, assigned to the *Birtavarre group*, is apparently conformable with underlying sparagmitic deposits; of special interest is the occurrence of basic volcanic rocks. The *Guolas limestone* contains greenstones with pillow structures. In the *Ankerlia formation* there are rocks rich in amphibole and zoisite which are probably sediments mixed with volcanic rocks, while the overlying "Green beds" are amphibolites of extrusive origin.

In the *Lyngen district* B. A. O. RANDALL (written communication) mapped 3000–4000 m thick piles of sediments on the east, west and southwest sides of the *Lyngen gabbro massif*. The different units present are mica schists, limestones and greenstones, a few of which have preserved structures typical of lavas. The stratigraphic relationships between the different parts are not clear (Fig. 87).

The *Vaddas district* (about 69° 50′, 21° 30′) is the most northeasterly part of the Ofoten–Troms region (VOGT, 1927; H. BARKEY, personal communication). Here the lower part of the sequence, the *Lilleelv group*, consists mainly of more or less feldspathic quartzites and is assumed to be of late Precambrian or Eocambrian age, corresponding to the psammitic rocks which underlie the districts to the east and northeast. The overlying *Vaddas group* consists of mica schists with quartzite in the middle part while the *Oksfjord group* of mica schists and biotite-bearing quartz schists lies at the top of

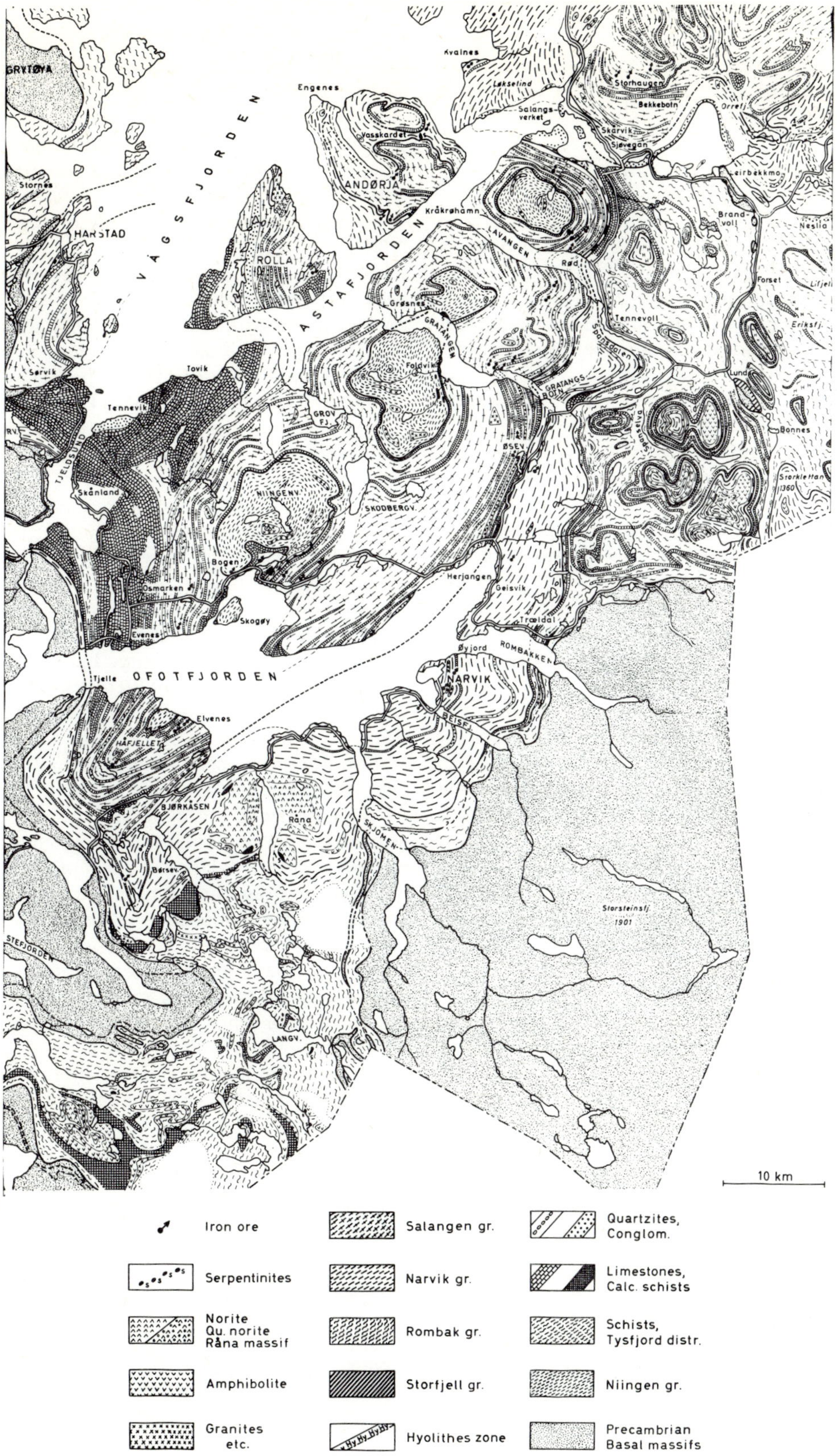

Fig. 85 *Geologic map of a southern part of the Ofoten–Troms district (between 68° and 69° N and about 16° 20′ and 18° 20′ E).* Based on maps by S. FOSLIE and TH. VOGT (southern part) and by M. GUSTAVSON (northern part). Drawn by M. GUSTAVSON, Geological Survey of Norway

the sequence. The Lilleelv and Vaddas groups contain amphibolites, assumed to be of volcanic origin.

GEOLOGIC SETTING

The geologic conditions at Ofotfjord have been dealt with by TH. VOGT (1942), who is also responsible for the geologic map of Narvik (published without a description).

In the Rombak window of Precambrian rocks inside Ofotfjord the basement is overlain by a basal quartzite and conglomerate, probably of Cambrian age. In Bjørnefjell near the international border (about 68° 30′) sharp and narrow folds with axial direction NNE have affected the rocks of the basement and have produced synclines containing the basal quartzites and conglomerates. The Caledonian deformation of the peneplained sub-Cambrian surface of the Rombak massif (structural contour map in VOGT, 1942) shows a synclinal depression directed N–S along the international border. To the northwest the massif sinks down beneath the overlying sediments with a slope of 15°.

On the north side of the Rombak window the sediments of the Rombak group with limestones and concordant amphibolites form a number of basin-shaped short synclines, clearly under the influence of folding in two axial directions. North of Ofotfjord occurs a very wide syncline or synclinorium which is a continuation of the Håfjell syncline on the south side of the fjord (Fig. 85).

The parts of the district lying south of Ofotfjord have been mapped and described in great detail by FOSLIE (1941, 1942, 1949). Here the basement is folded together with the overlying sediments on axes trending NNE and WNW or west. The strongly compressed *Kjøpsfjord syncline* extends for a considerable distance to the south. The lower part of the sequence above the basement is strongly and intricately folded in both the axial directions, while the folding seems to be less strong in the upper part above the Melkedal limestone. The folds with axes trending NNE are always overfolded to the west, while the folds with axes trending WNW or west are overfolded to the south. The combination of folding in two directions gives rise to synclines closed at both ends, and to anticlinal structures convex in all directions. At Håfjell on the south side of Ofotfjord the upper part of the sequence with the thick Evenes limestone forms a 10 km wide syncline with its axis plunging down to the NNE.

The sediments contain many intrusive bodies, predominantly basic ones. They occur in great profusion in the mica schists both below and above the Reppi schist, while they are more sparse above the Melkedal limestone. Above the Evenskjær conglomerate (at the base of the Evenes limestone) there are few intrusive rocks, only basic ones. The distribution of the intrusive rocks between the different horizons, compared with conditions in other parts of the Caledonides, should be indicative of Middle Ordovician or younger age for the Evenes limestone and may thus support the correlation previously mentioned.

Common intrusive rocks are coarse-grained and light-coloured gabbros, diorites and quartz diorites with a gneissic schistosity. They are evidently connected with the trondhjemites of the area, which occur as concordant sill-like bodies persisting over large parts of the area. Trondhjemites and granites also occur as smaller veins and lenses with structures varying from aplitic to pegmatitic. At some horizons the material is injected in very numerous and small veins, or as patches and porphyroblasts, thus taking the form of granitization.

Differing from the other basic intrusive rocks are the norites of the large *Råna massif* (68° 20′, 17° 5′). These norites are wholly unmetamorphosed and have exerted some contact metamorphism on the surrounding rocks. The whole massif has an exterior selvage of olivine norite with ultrabasic segregations and a central core of quartz norite. It is possible that the norite body occupies an inverted syncline and that the distribution of the rocks within the massif is a result of gravitative differentiation.

In the districts north of Ofoten (GUSTAVSON, 1965 and 1967), the allochthonous metamorphic sediments rest above autochthonous deposits at

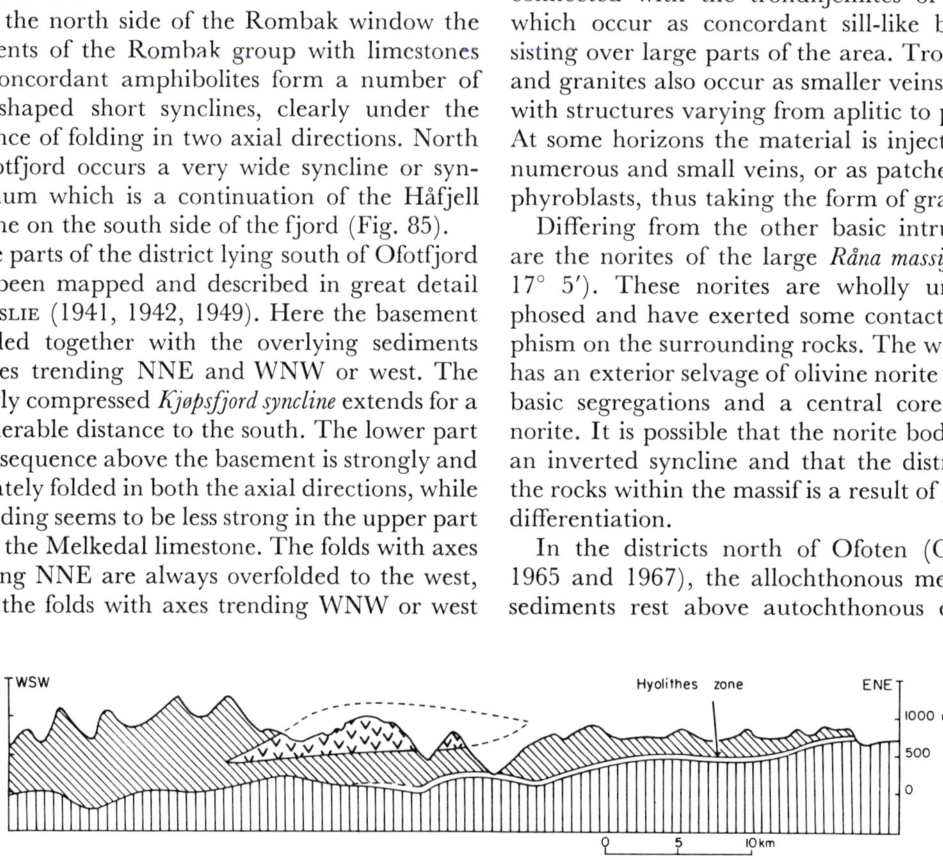

Fig. 86 *Section through the nappes (with an amphibolite lens) with underlying sediments of the "Hyolithes zone" and the Precambrian basement, north of Altevatn, about 68° 45′, 19° 30′.* By B. ROTHÉ, A/S Sydvaranger, Oslo, based on magnetic measurements

the Caledonide margin and also in windows in the interior as much as 80 km distant from the Caledonide margin (Fig. 86). In this part of the region the Rombak group is separated from the overlying Narvik groups by a thrust plane, partly with a tectonic unconformity and discontinuity in metamorphism, at a maximum in the Narvik group. This tectonic boundary becomes less marked towards the south and west. The Niingen group has a similar position as in the southern areas: the supposed tectonic boundary under the group is usually covered by scree. The grade of metamorphism varies from epidote amphibolite to high amphibolite facies.

The rocks were folded in two main phases. An early phase, more or less contemporaneous with the metamorphic recrystallization, produced folds of a plastic style and of recumbent type with axial trend NE, usually overturned to the SE. Probably of the same age are cross-folds of a similar style with NW-trending axes, overturned almost equally often to SW and to NE. Later folding produced flexure-like folds with axial trend NNE and a less pronounced cross-folding with axial trend WNW. The geologic map gives a picture of a simple structure with nearly flat-lying layers. Large folds of recumbent type seem to be rare, or at least are difficult to detect, but evidence of large folds of this type has been found in the eastern part of the area.

A recent contribution to the geology of the area was made by LUND (1965).

The area around the city of Tromsø and in Målselvdalen south of Tromsø has been investigated by KÅRE LANDMARK of the Tromsø Museum. A geologic map of the southern part of the area has been printed, but until now not much has been published (LANDMARK, 1951a, 1951b); the following account is based on written communications from LANDMARK. At Mauken in Målselvdal (69° 5′, 18° 40′) granodiorites and amphibolites of the Precambrian basement are exposed in a window; here also parts of the autochthonous sediments of the "Hyolithes zone" are preserved (about 80 km west of the Caledonide margin). The overlying sediments of the large nappe contain interlayered gneisses and the highly metamorphic rocks in the lower and upper part of the sequence are rich in plagioclase; the feldspar also occurs as porphyroblasts or in veins, schlieren and pegmatites. Dominant structures are large basin-shaped synclines (or synforms) with a NW axial trend. The Tromsø limestone forms a large synform in the peninsula south of Tromsø. The west limb of the synform pinches tectonically out to the west and is overlain by gneiss. Here there is certainly a nappe structure, perhaps a recumbent fold with gneiss in the core.

The Lyngen peninsula between Lyngsfjord and Ullsfjord contains a large centrally disposed gabbro mass, the southern part of which has been mapped (as previously mentioned) by Dr. B. A. O. RANDALL (Fig. 87). The gabbro is a banded hypersthene norite which is extensively saussuritized and uralitized and which displays large folds and overfolds usually related to faulting movements. Some petrographic data on the gabbroid rocks was given by BUGGE (1942) and ELDERS (1957). The Lyngen gabbro massif is inserted in the sedimentary sequence described in the stratigraphic section. Broadly the arrangement is a simple one, as the sediments on both sides of the gabbro form a monoclinal structure generally dipping to the west, with the assumed older beds on the east side of the gabbro. The structure is complicated by the attenuated limbs of small folds. The rocks display several metamorphic facies with a general east to west decrease in metamorphism from amphibolite to greenschist.

The Lyngen area as well as other parts of the Troms district contains small bodies of carbonate-bearing ultrabasic rocks, the *sagvandites*, described by BARTH (1927b), REITAN and GEUL (1959) and RANDALL (1960).

In the Birtavarre area east of Lyngsfjord (PADGET, 1955), the thick sequence of the Birtavarre group forms a synclinorium about 20 km wide with axis plunging NW. The upper part of the Birtavarre group is separated from the underlying part by a thrust plane (the *Cappis thrust*) which can be followed for 40 km. The metamorphic grade in the area is at the transition between epidote amphibolite and amphibolite facies. The lower part of the Birtavarre sequence, including the underlying "Sparagmite schists", has suffered granitization, represented by feldspathic and quartzo-feldspathic material in smaller or lesser amounts as augen, schlieren or coarse-grained pegmatite bodies. The zone of granitization is roughly but not exactly conformable to the stratigraphy, and lies higher in the sequence in the southeast part of the area than in the northwest.

The areas on the Caledonide margin on Finnish territory (Fig. 88) were described by HAUSEN (1942a, 1942b). Situated on the Norwegian–Finnish border is the phacolithic body of layered gabbro at Reisduodder Haldde. The gabbro is wholly unmetamorphosed and has a doleritic border zone; it must be younger than the deformation in the area.

The peninsula between Reisafjord and Kvænangen, mapped by Dr. HENRI BARKEY, is the most easterly part of the Ofoten–Troms district as delimited here, the areas further east being underlain by the metamorphosed psammitic rocks of the Finnmark district, which are also exposed west of

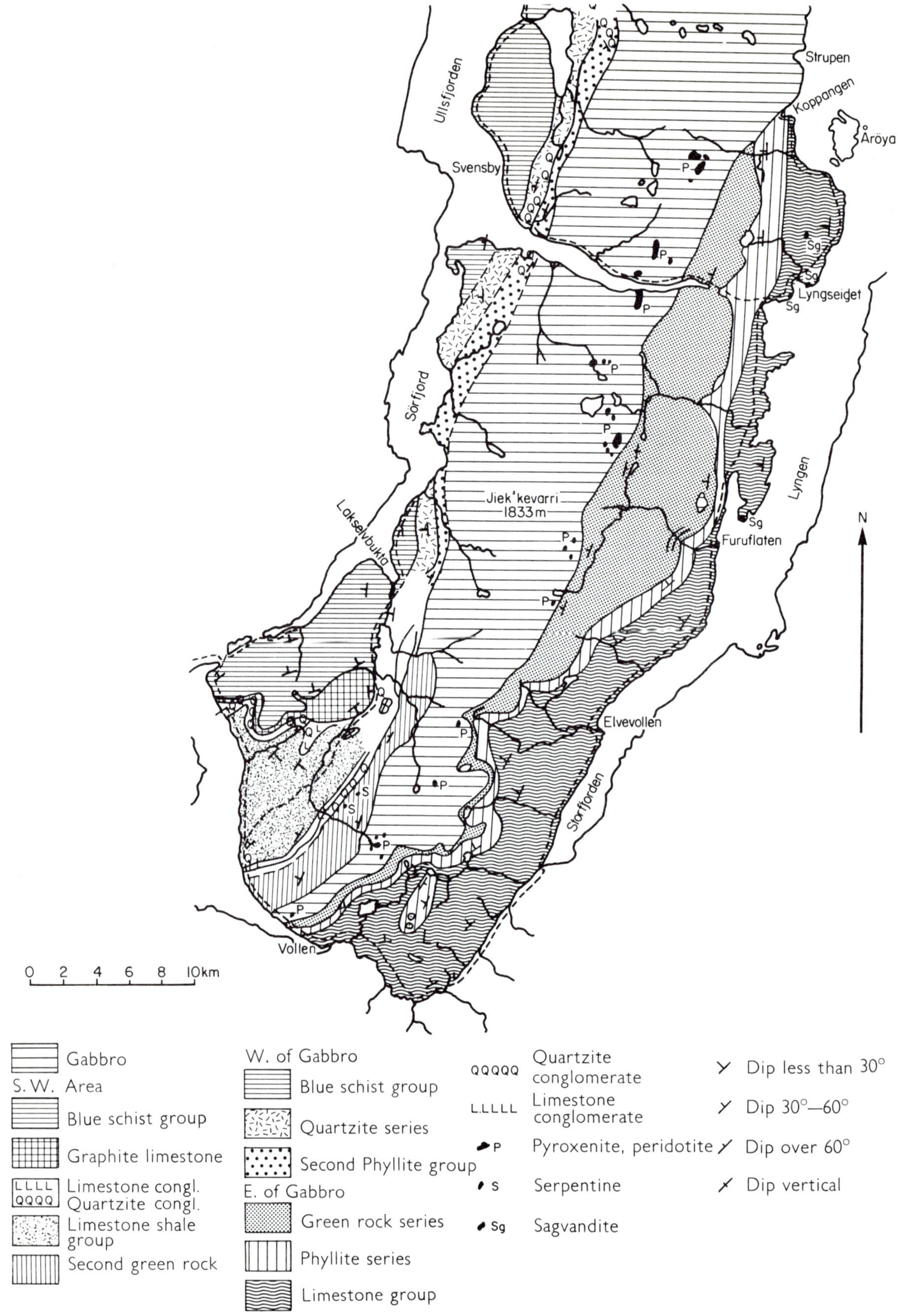

Fig. 87 *Geologic map of the southern part of the Lyngen gabbro massif and adjoining sediments.* By B. A. O. RANDALL

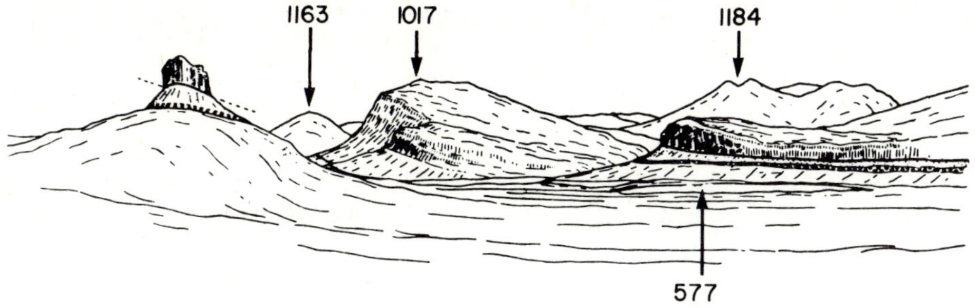

Fig. 88 *View of the Caledonide margin in Enontekiö, Finland, looking west*, from HAŪSEN (1942a). *A narrow band of the "Hyolithes zone" is between the Precambrian basement and the "hardschists" of the nappe. The outlier in Saivarri (left part of the picture) is at 69° 8′, 21° 22′*

Reisafjord. The area shows the same tectonics of basin-shaped synclines as is found in other parts of the district. In synclines of this type are found the phacolithic bodies of gabbro in Vaddasgaissa and the northern part of the peninsula. The *Vaddas gabbro* (VOGT, 1927) is situated almost exactly on the boundary between the Vaddas and the Oksfjord groups. Like the Sulitjelma gabbro the Vaddas gabbro has greenstones and amphibolites in the lower part and olivine gabbro and pyroxene gabbro in the upper parts. The sediments of the Oksfjord group at the upper side of the gabbro have suffered contact metamorphism with the gabbro, while the sediments on the lower side show the usual regional metamorphism only. A granite at Rappesvarre north of Vaddasgaisse is injected on the lower side of the Vaddas quartzite.

Addendum. KULLING's 1964 publication dealing with the Caledonide geology of the Swedish areas adjacent to the southern parts of the Ofoten–Troms district became available after completion of the manuscript. On the Swedish side, Precambrian granitic rocks and "hardschists" immediately below the Seve–Köli sediments of the great nappe are widely distributed. In many parts of the area the "hardschists", being in the main strongly mylonitized sediments, have been mapped up to the international border. These "hardschists" are not recognized as a separate unit by geologists working on the Norwegian side, and thus nothing is known of their extension on Norwegian ground, except that KULLING found them in the basal part of the Rombak group in the Narvik area. The deposits of the Rombak and Narvik groups have their equivalents on the Swedish side, where a limestone of the Rombak group is known as the *Nuolja limestone.* The Reppi calcareous schist in the Tysfjord area extends on the Swedish side south of Lake Sitasjaure (68° N) and is overlain by the *Patta amphibolite* or *greenstone.* Further east, where the Reppi formation passes into amphibole porphyroblast mica schists, the overlying Patta amphibolite shows preserved volcanic structures. The assumed volcanic rocks above the Reppi schist thus have equivalents of an undoubted volcanic nature on the Swedish side.

On the geologic maps a discrepancy is apparent between the Swedish and Norwegian sides in the Pältsa area (the northernmost part of Sweden at 69° N) where KULLING's map shows granitic gneisses, while PADGET's map (1955) shows sparagmitic schists in a corresponding position in an adjacent area north of Lake Kilpisjärvi.

The Stipok outlier north of Lake Vastenjaure (Chapter XXIII) is, according to KULLING, part of the Rødingsfjell nappe (uppermost thrust unit). It has an extension on the Norwegian side in the Linnajavrre area (67° 35′), mapped by FOSLIE (1942).

CHAPTER XIII

THE FINNMARK DISTRICT

Finnmark is the most northerly and easterly part of Norway. The *Finnmark district*, in a geological sense, is delimited to the west by the 22° meridian. The metamorphic psammitic rocks of assumed late Precambrian and Eocambrian age, that are found in the lower part of the sequence in the northern parts of the Troms district, are widely distributed to the east in Finnmark and together with meta-pelites and gneisses they underlie the northwestern parts of the district. An anticlinal zone exposing the old Precambrian (Karelian) rocks of the *Raipas suite* is almost continuous from Kvænangen in the southwest to Repparfjord in the north. In a synclinal depression northwest of the windows is a large massif of igneous and metamorphic rocks of distinctive types. The southeastern and eastern parts of the district are underlain by the unmetamorphosed late Precambrian and Eocambrian to Cambrian deposits of the *Finnmark suite*.

STRATIGRAPHY

In Finnmark (see Fig. 102, p. 131), large areas are underlain by unmetamorphic sedimentary rocks—sandstones, siltstones, shales and dolomites. Glacial deposits (tillites) are conspicuous among these sedimentary rocks. The clearing up of the stratigraphic relationships of the Finnmark deposits is due particularly to the work of HOLTEDAHL (1918, 1931) and of FØYN (1937). HOLTEDAHL divided the deposits into two main parts: an older division including thick formations of dolomite, and a younger division with the tillites at the base, separated from the older division by a small-angle erosional unconformity. A more detailed stratigraphical column of the younger division was made by FØYN, who found that the tillite-bearing sediments passed upwards into thick series of sediments, the upper parts of which contained Cambrian fossils.

It was thus definitely proved that the Finnmark deposits were of the same age as the Sparagmite group in southern Norway, as had been maintained by O. HOLTEDAHL. In a synopsis on the stratigraphy of these deposits HOLTEDAHL (1961a) classed the basal part of the younger division of the Finnmark suite, from the lower tillite to the oldest Lower Cambrian fossiliferous layer, as Eocambrian. In southern Norway, as mentioned previously, the younger part of the Sparagmite group, from the tillite-bearing Moelv formation, is classed as Eocambrian.

The older, late Precambrian part of the Finnmark suite is represented by the *Porsanger group* at Porsanger and Tana and by the *Bossekop group* at Alta. READING (1965) divided the younger parts of the sequence at Tana into the *Vestertana group* (Eocambrian and Lower Cambrian(?)) and the *Digermul group* (Cambrian and earliest Ordovician). The Eocambrian and Lower Cambrian deposits which extend to the southwest along the Caledonide margin have been called the "Hyolithes zone" for many years.

Deposits of the older, Precambrian part of the Finnmark sequence occur at Porsangerfjord as a nappe above rocks of the "Hyolithes zone" (HOLTEDAHL, 1918). Here the Porsanger group contains 500 m of light-coloured sandstone overlain by a massive and pure dolomite with beautiful examples of the concentric structures called *stromatolites* by HOLTEDAHL, who objected to considering these features as real fossils and to giving them generic names. In Lakselvdal, on the west side of the valley south of Porsangerfjord, BRIAN WHITE (written communication) found a sequence with 200 m of orthoquartzite (the *Skallenes formation*) and an overlying formation (the *Baukenes formation*, 300 m) of alternating sandstones and siltstones passing upwards into red and green shales. The *Skjånes dolomite formation* (200 m+) rests upon an eroded surface in the underlying formation and corresponds to HOLTEDAHL's Porsanger dolomite. Ripple marks and mud cracks are common in the deposits, as is generally the case in the older parts of the Finnmark sequence.

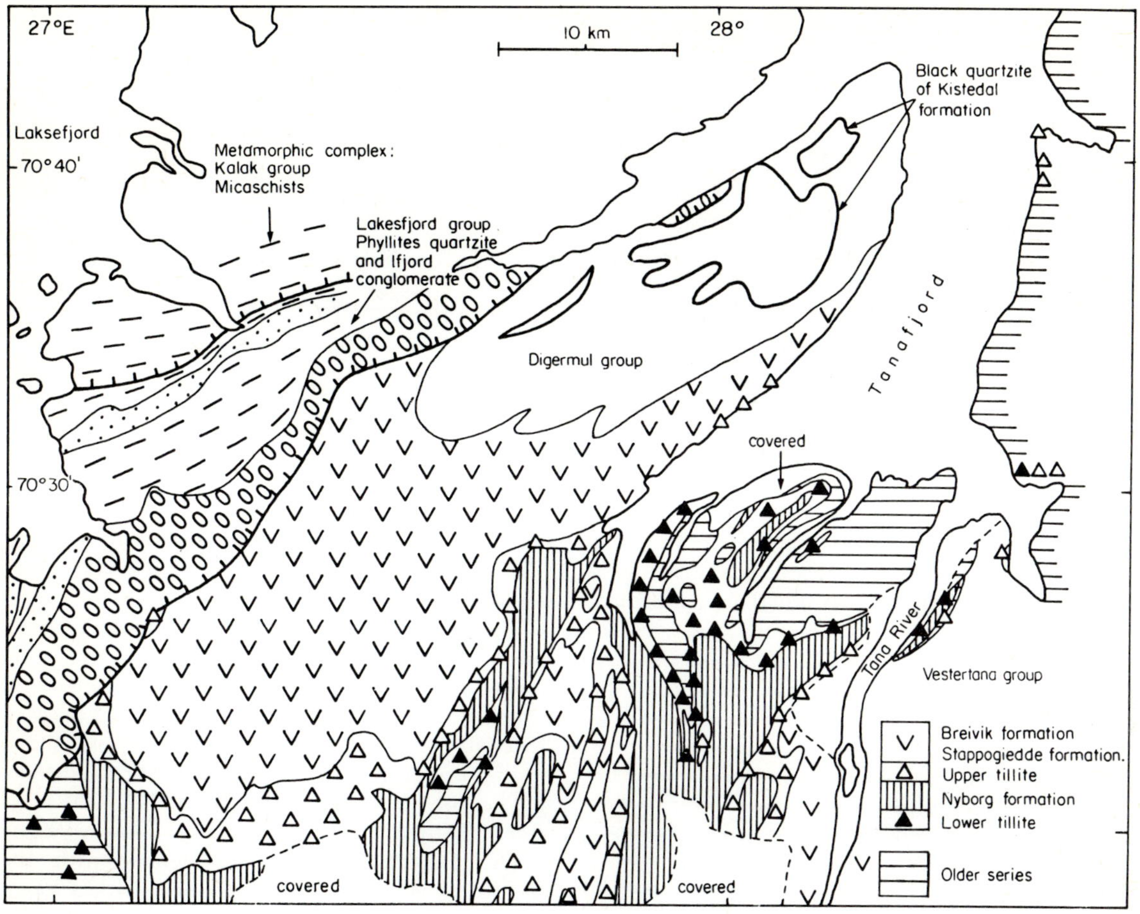

Fig. 89 *Geologic map of the Tana–Laksefjord district*, based on maps by Føyn (in Holtedahl and others, 1960) and by Reading (1965)

In the Tana district (Føyn, 1937) the Porsanger group is represented by about 1200 m of light-coloured sandstones and dark-coloured siltstones and shales. In the topmost preserved part are thin layers of dolomite, which may be equivalent to the Skjånes dolomite in the Porsanger district. The erosional unconformity with the overlying Eocambrian tillite is shown by the fact that the Porsanger group sequence becomes thicker and contains higher formations when the boundary is traced from south to north. Sediments of the older division extend along the north coast of the Varanger peninsula east of Tanafjord. Holtedahl (1918) gives a section of sandstones, shales and dolomites of total thickness about 4000 m.

Deposits of the older division are found in the *Alta window* southwest of the Porsanger district, where they rest unconformably upon the old Precambrian rocks of the Raipas suite (Føyn, 1964). The Bossekop group at Alta includes about 100 m of massive reddish quartzite with subordinate intercalations of shale. As in the Tana district, younger members of the group wedge in below the overlying tillite in a northerly direction. At Kvænangen southeast of Alta, Milnes and Ritchie (1962) reported 600 m of quartzites with abundant layers of black shale (their "Quartzite group") below the overlying tillite.

The rocks of the older part of the Finnmark deposits are widely distributed among both the autochthonous and allochthonous rocks, but little is known of their stratigraphy.

Reitan (1963) described the *Doggelv formation* (almost pure, light orthoquartzites) and the *Lomvann formation* (dark sandstones and shales) with a total thickness of 2500 to 4000 m, as resting uncomformably upon greenstones of the old Precambrian (Karelian) Raipas suite in the *Komagfjord window*—the northeasterly part of the Precambrian windows of Finnmark. In his paper of 1965 Reitan indicated the possibility of these deposits belonging to the older part of the Finnmark sequence, and being correlates of the Bossekop group at Alta.

The younger part of the Finnmark suite is of maximum thickness in the *Tana district* (Fig. 89).

Fig. 90 *Tillite at Bigganjargga resting on striated quartzite.* Photo KNUT BJØRLYKKE

Here, as stated previously, the Eocambrian and younger deposits rest upon the Precambrian part of the sequence with an erosional unconformity. About 30 km south of the head of Tanafjord the Eocambrian deposits rest directly upon the Precambrian substratum. The unconformity was caused by an uplift in the south with a tilt of 1–2° in a direction N 30° E.

In the basal part of the Vestertana group are two horizons of tillite, each about 50 m thick, separated by 400 m of reddish brown sandstone—the *Nyborg formation.* The tillites are brown coloured with a matrix of sand and silt grade and parts of them show detectable sorting and stratification. The erratics are of irregular shape, mostly with rounded edges, of all sizes up to 2 m, but seldom larger than 0.5 m. Striated boulders have not been found at Tanafjord. The erratics are of dolomite and crystalline rocks, the latter of the same general types as those found in the Precambrian complexes south of Tanafjord. Boulders of dolomite predominate in the lower tillite, while in the upper tillite most of the boulders are of crystalline rocks. Evidently the tillites in the Tanafjord district were deposited as glacial marine sediments, the boulders being dropped from icebergs which came from a glaciated area in the south. The sandstone of the Nyborg formation, intervening between the tillites, also becomes coarser when followed from north to south.

Above the upper tillite is a sequence measuring 2500–3000 m, divided into five formations by READING (1965). The two lower formations without fossils, together with the underlying tillites and the Nyborg formation, belong to the Vestertana group; the three upper fossiliferous formations form the Digermul group. Most of the sequence consists of cross-bedded and flat-bedded orthoquartzites and lithic sandstones with thin-bedded greywacke sandstones and shales. The former were evidently deposited in shallow water, probably marine, while some of the greywackes and shales were deposited in deeper water.

The sequence above the upper tillite is given below in ascending order:

The *Stappogiedde formation*, 400 m, consists of some 40 m of quartzitic sandstone at the base, a lower member of blue-green and red-violet shale, and an upper member of red quartzitic sandstone.

The *Breivik formation* consists of 500–600 m of green siltstone with quartzites in the lower half.

The Digermul group commences with the *Duolbasgaissa formation*, 500 m thick, with a lower member of thin-bedded quartzite and an upper member of massive bedded quartzite. A quartzite in the middle part of the upper member has yielded Holmia sp. There are further trail fossils present—Syringomorpha, Monocraterion and Baueria.

The *Kistedal formation*, with an estimated thickness of 700 m, has two lower members of sandstone and shale with Middle Cambrian Paradoxides fauna. The richest fauna in the upper member contains Paradoxides paradoxissimus, P. cf. davidis, Doryagnostus incertus and other agnostids and Billingsella sp. This is the fossil horizon first detected by FØYN. Separated by about 20 m of black quartzite is a member of black shale with fossils of latest Cambrian or perhaps earliest Ordovician (Tremadoc) age: Boeckaspis mobergi, Beltella n.sp. and Peltura sp. The uppermost member in the Kistedal formation is 300 m of grey quartzite.

The *Berlogaissa formation* tops the preserved sequence in the core of a syncline with an estimated thickness of 300 m of dark grey to black sandstone. Occasional layers of black shale contain beautifully preserved fossils of earliest Ordovician (Tremadoc) age: Dictyonema flabelliforme, Peltocare sp., Parabolina sp. and Saltaspis sp.

Fig. 91 *Striated surface of the quartzite beneath the Bigganjargga tillite.* Photo KNUT BJØRLYKKE

In the Tana section there is no lithologic break of any importance between the tillites and the overlying fossiliferous Cambrian deposits. FØYN (1964), on lithologic criteria, correlates the Stappogiedde formation with the lower part of the "Hyolithes zone", and the Breivik formation with the overlying part of the "Hyolithes zone", including the Platysolenites horizon. According to this correlation, the Breivik formation should correspond to the oldest part of the fossiliferous Lower Cambrian in Scandinavia and the Stappogiedde formation should be the only part of the post-tillite sequence to be considered as part of the Eocambrian. The Holmia sp. in the Duolbasgaissa formation probably indicates the biozone of Holmia kjerulfi (1bα) and thus the fossil evidence does not preclude a correlation of the Breivik with the Platysolenites biozone (1aβ).

The tillite-bearing part of the Tana sequence extends to the east as far as the Varanger district. Here the older deposits beneath the tillite contain younger members towards the northeast, while west of the head of Varangerfjord the Nyborg formation rests directly upon an uneven Precambrian gneiss basement. Near the head of Varangerfjord is Bigganjargga, the famous locality in which the tillite was first recognized by HANS REUSCH (FØYN, 1960). Here the tillite (corresponding to the lower tillite horizon) has the aspect of an unsorted moraine and rests upon a striated surface of light sandstone (Figs. 90–92). KNUT BJØRLYKKE (personal communication), who has recently investigated the Bigganjargga area, found that the light sandstone, which extends to the east on the south side of Varangerfjord, lies unconformably to the overlying tillite and thus belongs to the older part of the Finnmark sequence. He found striated facets on the boulders in the tillite. As the tillite is of small thickness and very restricted distribution, it is interpreted as the erosional remnant of a cover of ground moraine.[1] Badly sorted and stratified conglomerates and a well-sorted conglomerate with dolomite boulders in the surrounding areas are interpreted as fluvioglacial deposits corresponding to the lower tillite. The upper tillite has a development similar to that in the Tana district.

The tillites and the Nyborg formation wedge out to the west and southwest of the Tana district, but in the Alta district a 10 m thick tillite, correlated with the upper tillite at Tana, rests unconformably upon the deposits of the Bossekop group, as noted previously. The tillite is overlain by a thin coarse-grained sandstone with conglomerates and further by red and green shales with sandstone—the Borras group, corresponding to the lower part of the "Hyolithes zone" (FØYN, 1964). The tillite is also present in the *Kvænangen window* southwest of Alta. Tillites of terrestric type resting directly upon the Precambrian basement beneath the deposits of the "Hyolithes zone" are present at the Caledonide margin south of Alta (between 22° 30′ and 23° 30′ E) (HOLMSEN, 1956, 1957).

The Eocambrian and Lower Cambrian deposits of the "Hyolithes zone" extend along the Caledonide margin from Lake Torneträsk in Sweden to

Fig. 92 *Close-up of weathered surface of the Bigganjargga tillite.* Photo KNUT BJØRLYKKE

[1] v. GAERTNER (1943) gave an alternative interpretation of the deposition of the tillite.

west of the Porsanger district, where they can be traced in connection with the Tana deposits, though in parts of the area between Porsanger and Tana most of the sequence has been cut out under the overlying nappe.

VOGT (1918) investigated the "Hyolithes zone" along its extension on Swedish, Norwegian and Finnish territory from Lake Torneträsk in the south to the Reisa Valley (69° 20′ N) and found the following sequence, in ascending order:

1. basal conglomerate 2. thin green shale 3. sandstone with layers of shale 4. red and green variegated shale with Platysolenites antiquissimus 5. green shale with layers of limestone in both the lower and upper parts. The limestone contains Strenuella aff. primæva, probably indicating the Holmia kjerulfi zone. The uppermost layer preserved beneath the nappes is of black alum shales. The different members of this sequence could be traced throughout the area, the whole thickness increasing from 113 m in the south to 170 m in the north, within a distance of 170 km.

The deposits of the "Hyolithes zone" become still thicker further to the northeast. At Halkkavarre, on the east side of the valley inside Porsangerfjord, there are some 300 m of these deposits below the overlying nappe. From the base there is, according to FØYN (personal communication), a basal sandstone with conglomerates, 20 m thick, a further 80 m of bluish green mudstone with red-coloured bands, and 20 m of grey sandstone; this sequence can be correlated lithologically with the Stappogiedde formation at Tana. The overlying 130 m of alternating sandstone, siltstone and mudstone of greyish green, partly reddish colour contains a Platysolenites horizon in the middle part and can be correlated with the Breivik formation at Tana.

Addendum. In the summer of 1966 FØYN found Platysolenites in the Breivik formation, thus proving the correctness of the correlations stated previously. The paper of READING and WALKER (1966) gives a detailed account of palaeogeography and sedimentology of the tillites and other sediments of the Tana sequence.

On Sørøy island, recently investigated by RAMSAY and STURT (1963 and written communication), the following stratigraphic succession has been established with the aid of well-preserved sedimentary structures in the basic *Klubben quartzite formation*:

Hellefjord schist group: Monotonous, flaggy, fine-grained garnet–amphibole schist with psammitic laminae, sometimes containing diopside.

Breivik group: Upper staurolite mica schist
White quartzite
Lower graphite schist

Falkenes limestone: Limestone
Calcareous silicate schist

Storelv group: Upper garnet–mica schist
Mixed schists and psammites

Klubben quartzite

Limestones containing archæocyathids have been discovered, and preliminary investigations of these fossils indicate Lower Cambrian age for at least part of the succession. A correlation based on lithology of the Sørøy sequence with that of any other district is not available. Lower Cambrian age ought to correspond to parts situated low down in the Cambro–Silurian sequences of neighbouring districts.

On Magerøy island a sequence of slightly metamorphic dark-coloured interchanging arenaceous and argillaceous rocks is exposed east of Honningsvåg (71°, 26°) (REITAN, in HOLTEDAHL and others, 1960). These are overlain, apparently with an uncomformable contact, by beds of limestone and limestone conglomerate. On an excursion during the 1960 Geological Congress, fossils indicative of Lower Silurian age (pentamerids, favositids) were discovered in the limestones.

GEOLOGIC SETTING

The Caledonide margin, marked by a thin strip of autochthonous sediments beneath the nappes, makes a rather evenly curved line parallel to the isophypses of the peneplained Precambrian basement, which slopes down to the north and northwest. The slope is about 1° 20′ inside Porsangerfjord and Tanafjord, but is about thrice as large inside Altafjord (HOLTEDAHL, 1918). In more southerly parts of the district VOGT (1918) found gradients between 1° 45′ and 3° 30′ with a mean of 2° 30′.

In the Reisa area (about 69° 20′, 22°) SKJERLIE and TAN (1961) found a nappe of unmetamorphosed dark sandstone (the *Jerta sandstone*) above the "Hyolithes zone". Dolomite, conglomerates and a tillite were also found in the nappe.

In the *Porsanger district* there is a nappe of sandstone and dolomite of the Porsanger group underlain by as much as 300 m of "Hyolithes zone" sediments. There is no sharp boundary with the overlying nappe of Porsanger sandstone, as rocks of the nappe are mixed and kneaded together with the underlying rocks in a thick zone. Along the valley south of the head of Porsangerfjord the Precambrian basement is buckled in low and broad folds, which do not affect the overlying nappe. The sediments below the nappe thin out over the anticlines, and thus the emplacement of the nappe cannot be younger than the buckling of the basement. South of Porsangerfjord the allochthonous Porsanger sandstone can be followed eastwards to Antebaktoaivve (70° 10′, 27° 10′). The thrust plane passes down gradually to lower horizons in the underlying sequence, so that at Antebaktoaivve only the basal conglomerate of the "Hyolithes zone" is present beneath the nappe. Further east, deposits older than the basal part of the "Hyolithes zone" rest directly upon the Precambrian basement, and here the autochthonous sequence is not sheared off by a thrust. Folding and gliding within the sequence may, however, compensate for movement along the thrust. At Tanafjord the autochthonous sequence, containing thick competent formations, lies in broad folds as much as 3 km wide, with axial trend NE to NNE. Most of the folds are overturned to the SE, and the less competent rocks have developed an axial plane cleavage dipping NW. In the southern part of the area (not covered by the map in Fig. 89) the rocks are folded on E–W axes against a wall in the Precambrian basement. The nappe of metamorphic rocks in the northern part of the Tana district was moved in a direction to SSE above the autochthonous sediments, i.e. in a direction oblique to the axes of the earlier folds, which were deformed and disturbed during the thrusting.

The large Varanger peninsula, west of Tanafjord and north of Varangerfjord, is underlain by unmetamorphic sediments of the Finnmark sequence, most of them of the older division. The interior parts of the peninsula are largely unknown. Along the north coast thick formations of sandstones and dolomites form large folds with axial trend between NE and NNE, while in the south, on the north side of Varangerfjord, the folding is very faint; some thrusts have, however, caused repetition of the tillite-bearing deposits.

The areas with metamorphic rocks which are always in allochthonous positions above the unmetamorphosed Finnmark deposits will now be described.

In the Reisa valley in the southeast, the metamorphic nappe rests above the Jerta sandstone, previously mentioned (Skjerlie and Tan, 1961). The rocks are described as sparagmitic schists, commonly with a granoblastic structure and usually with 30 to 40% feldspar, predominantly potash feldspar. The sparagmitic schists have been granitized over a small area. Small (10–15 m) lenses of amphibolite occur as intrusions in the sparagmitic schists. A large body of dunite (about 500 m) did not appear to be in contact with them. In parts of the area the sparagmitic schists are overlain on a tectonic boundary—shown by a 1–5 m thick crush zone—by a complex of greenschists with some layers of mica schists, quartzites and limestones. The sparagmitic schists have lineations trending NW or WNW, while dragfolds indicate movement to the SE or ESE, parallel to the direction of lineation.

In its extension further to the northeast in the Alta district the metamorphic nappe was mapped and described by Holtedahl (1918) (Fig. 93). Here the nappe forms a wide and flat syncline and rests upon autochthonous sediments of the "Hyolithes zone". The nappe has a lower division of strongly schistose feldspathic meta-psammites with some layers of phyllite and an upper division (about 500 m) of flagstones (quarried in the Alta district). Interlayered between the two divisions is a 400 to 600 m thick zone of granitic gneisses. There is a gradual transition between sedimentary schists and granitic rock, but the transition may take place within a few metres. The granitic gneisses contain dykes of massive aplitic granite and pegmatite; the pegmatite veins are often folded. The granitic part of the complex disappears towards the southwest; intrusive gabbros and amphibolites also occur in this district.

In the northern part of the Alta district (about 70° 3′, 22° 40′) west of Talvik, Geukens and Moreau (1960) mapped and described two nappes resting upon the autochthonous or parautochthonous deposits of the Bossekop group. The lower nappe contains not very strongly metamorphosed quartzites, coarse, partly conglomeratic meta-arkoses, dolomites and also greenstones with pillow structure. The upper nappe unit contains garnet gneisses, pyroxene gneisses and dark amphibolitic rocks. The structure is more

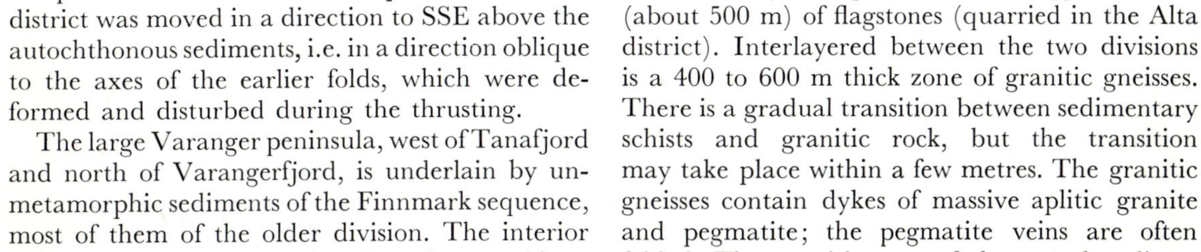

Fig. 93 *Section from the Caledonide margin (in the south) to the Alta window.* From O. Holtedahl (in Holtedahl and others, 1960)

complicated than in the lower nappe; there is intense folding on N–S axes, and the whole complex is transected by basic dykes and sills. The nappe dips rather strongly towards the large massif of igneous rocks to the northwest, and the high grade of metamorphism may be due to the proximity to the igneous rocks.

In the area between Repparfjord and Porsangerfjord (STRAND, 1951) lies 200 m of light-coloured quartz schists and quartzites with subordinate mica schist and dolomite, overlain by a thick oligoclase-bearing mica schist. In the Repparfjord window the metamorphic complex rests directly upon the Precambrian basement; nothing is left of the autochthonous sequence and the basement rocks are to some extent influenced by Caledonian tectonics (REITAN, 1963). T. LYNNEBERG, an Oslo graduate student who mapped the area north of Repparfjord, found a persistent horizon with quartzites and numerous intrusions of amphibolite, 350 m above the base of the mica schist. The area on both sides of Porsangerfjord (except for the inner parts with unmetamorphic rocks) and also the areas at Laksefjord and Tanafjord in the east are underlain by metamorphosed psammitic and pelitic rocks. Thus in the Nordkyn peninsula, between Laksefjord and Tanafjord, alternating quartzites, quartz schists and phyllites form beautiful folds trending NNE.

More detailed information on metamorphic allochthonous rocks was given by FØYN (1937, 1960) for the areas at Tanafjord and Laksefjord (see Fig. 89). There is a lower nappe of low grade rocks of the *Laksefjord group* with the coarse *Ifjord conglomerate*, *Landersfjord quartzite* and *Friarfjord phyllite*. Some of the rocks are meta-arkoses resembling the south Norwegian sparagmites. The Laksefjord group nappe is thick in the west (the Landersfjord quartzite is at least 1000 m thick), but thin in the east, where a window displays underlying dolomite of the Porsanger group. An upper nappe with highly metamorphic rocks of the *Kalak group*—alternating quartzites and mica schists—is probably an integral part of the metamorphic complex of Finnmark. At the head of Laksefjord, gneisses of the Precambrian basement occur as a number of upthrusted plates dipping 20° to 30° NNW. In some cases parts of the overlying dolomites occur between these plates. The relationship between folding and thrusting is the same as with the unmetamorphosed sediments in the south, previously mentioned.

If the thrust below the metamorphic rocks is followed northeast to the *Berlevåg area* east of the outer parts of Tanafjord, no sharp thrust line is present, but there is an increase in metamorphism and intensity of folding towards the northwest. The thrust at this point thus seems to die out at a hinge, as is also the case with the nappe of unmetamorphic rocks in the area inside Laksefjord.

In the Finnmark metamorphic complex, psammitic rocks seem to predominate in the southwestern parts, while pelitic rocks are of equally great importance in the northeastern parts. Even though late Precambrian to Eocambrian age is probable for a large proportion of the psammitic rocks, parts of the deposits may well be younger.

The northwestern part of the Porsanger peninsula (west of Porsangerfjord) with adjacent islands is underlain by gneisses. On Kvaløy (northwest of the Repparfjord window) KNUT HEGGELUND, an Oslo graduate student, mapped a complex of gneisses coming above quartzites and mica schists in the southernmost part of the island. The gneisses, with layers of quartzite and concordant bands of amphibolite, are of supracrustal origin. Charnockitic rocks with orthoclase and hypersthene occur as small patches (remnants of a nappe?) with tectonic contacts to the underlying gneisses. On Magerøy island massive gneisses and granites have been reported in the western and eastern parts of the island, while North Cape consists of plagioclase gneisses with bands of amphibolite. HOLTEDAHL (1944) found a tillite-like conglomerate together with quartzites and dolomites at Duksfjord in the northern part of the island. The sediments with Silurian fossils previously reported in the area around Honningsvåg (71°, 26°) probably belong to a tectonic unit below a nappe of the granites and gneisses. According to investigations by J. J. C. GEUL (HOLTEDAHL and others, 1960) the sediments contain a massif of gabbros and ultrabasic rocks with fine-grained congealing contacts and contact aureoles. To the west (away from the gabbros) the pelitic parts of the sediments pass into garnet mica schists by increasing metamorphic grade.

LAYERED GABBROS AND ASSOCIATED ROCKS

The Finnmark district contains a massif consisting to a large extent of basic and ultrabasic igneous rocks, but also including meta-sediments, particularly quartzitic, carbonatites and alkaline rocks. It consists of the islands of Stjernøy and Seiland, part of Sørøy, and the peninsula between Kvænangen and Altafjord, and is delimited to the southeast by the narrow Langfjord. The first modern description of the geology and petrography of part of the massif was given by BARTH (1927). After 1950 more of the area was investigated in detail under the auspices of the Norwegian Geological Survey (BARTH, 1953, 1961; KRAUSKOPF, 1954; HEIER, 1961; OOSTEROM, 1956, 1963).

JENS HYSINGJORD described the geology and petrography of the western part of Stjernøy in an unpublished university thesis (Oslo, 1960).

According to OOSTEROM (1963) the Finnmark complex can be divided into three separate parts, listed below from oldest to youngest:

A metamorphic complex of gneissic gabbroid and syenitic rocks; the inclusion of concordant bands of quartz-rich granulites and metalimestones prove it to be largely of supracrustal origin.

A sequence of mafic and ultramafic rocks, including large masses of periodotites and dunites.

A suite of carbonatites and alkaline rocks.

Events of metamorphism occurred between the formation of the first and second, and between the formation of the second and third, parts of the complex.

To the southeast and east the massif is situated above the metamorphic psammitic sequence of Finnmark, but little is known of the relationship of these rocks to the overlying rocks of the massif. On Sørøy the investigations of RAMSAY and STURT have shown that rocks of the second and third parts of the massif have intruded the island's Cambrian age sediments; time relationships have been established between the intrusions and the tectonic history of the area.

The dominant rock in the old metamorphic complex has been described as gabbro gneiss; it is a fine- to medium-grained gneissic rock consisting mainly of plagioclase and pyroxene. Olivine, common in the rocks of the gabbro sequence, has not been found in the gabbro gneisses, which on the other hand contain biotite not found in the first group of rocks. In the northern parts of Stjernøy medium- to coarse-grained amphibolites occur; these must be considered as more low grade equivalents of the gabbro gneiss. Intercalated in the gabbro gneiss and the amphibolite are bands of syenite gneiss, which occurs in the northern part of Stjernøy as bands 50 to several hundred metres thick, continuous for several kilometres. It would be very difficult to distinguish the gabbro gneiss from the banded gabbros in the intrusive part of the complex, were it not for the occurrence of granulites rich in quartz, metalimestones and lime silicate rocks as concordant bands in the gabbro gneiss, and amphibolite. The quartz granulites are fine- to medium-grained rocks of gneissic foliation marked by streaks of bluish quartz with the percentage of quartz varying between 45 and 80, other minerals present being feldspars, garnet and hypersthene. The granulites occur in a wide continuous zone in the southwest part of Seiland, where BARTH (1927) found rounded zircons which proved the sedimentary origin of the rocks.

Except for the amphibolites previously mentioned, the rocks have been metamorphosed in granulite facies with pyroxenes, garnet and mesoperthitic feldspars as characteristic minerals. The granoblastic structure of the rocks also shows this metamorphic recrystallization.

The ultramafic rocks present in the intrusive complex are comprised of dunites, several types of peridotite and olivine metagabbro. In the eastern part of Stjernøy occurs a body covering 30 km^2; it is very inhomogeneous and has inclusions of various types of ultrabasic rocks, which give the impression that pyroxene peridotite has almost completely digested a pre-existing dunite in a

Fig. 94 *Alternating layers of anorthositic gabbro and metagabbro, Rottenhallerfjell, Stjernøy (70° 19′, 22° 44′).* From OOSTEROM (1963)

Fig. 95 *Limestone with archaeocyathids*. Sørøy, Finnmark. Photograph kindly furnished by DR. BRIAN STURT. The fossil will be described by PROFESSOR C. H. HOLLAND in a forthcoming publication

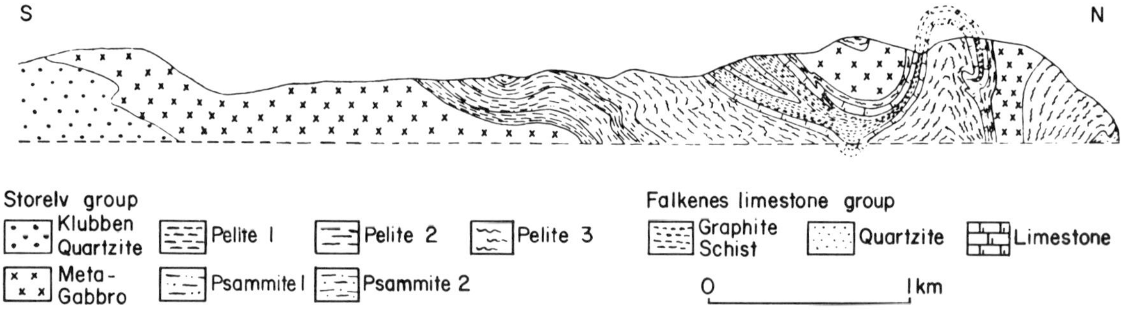

Fig. 96 *Section along the west side of the inner part of Sandøyfjord, Sørøy (70° 40′, 22° 58′)*. From RAMSAY and STURT (1963)

Fig. 97 *Banded olivine gabbro with an intruded ultrabasic lens. Reinsfjord, northeast side of Kvænangen (70° 6′, 21° 35′)*. Photo P. R. HOOPER

penecontemporaneous process of formation. A large body on Seiland consists partly of dunite, partly of pyroxene peridotite. Troctolites have also been found on Seiland, as well as fine-grained mafic dykes (lamporphyres) related to the ultramafic rocks.

The layered gabbros show concentrations of dark minerals in streaks and bands, but banding also occurs on a much larger scale. On Stjernøy there are layered sequences of metagabbro (20–25% plagioclase) with anorthosite (80–90% plagioclase); the alternating bands measure from a few decimetres to some metres, the maximum being 10–20 m thick layers of anorthosite (Fig. 94). In the southwest part of Seiland, OOSTEROM (1956) described a 500 m thick sequence of gabbro with four main layers of peridotite, each several tens of metres thick. Other sequences contain layers of syenitic rocks, diopside–garnet-bearing marbles, spinel-rich bands and garnet–sillimanite rocks (BARTH, 1961). The layered gabbros have concordant contacts with the rocks of the supracrustal complex. Neither apparent contact metamorphism nor igneous brecciation has been observed, and the contacts are gradational.

The carbonatites are restricted in their main occurrences to Stjernøy and the mainland south of the island. They contain calcite and biotite and less frequently hornblende as main constituents and are rich in apatite, which occurs in varying amounts with an average of 4–5%. The carbonatites also occur as dykes, and generally have gradational borders with the country rock. Connected with the carbonatites are hornblendites—coarse-grained black rocks consisting almost exclusively of hornblende crystals as large as 20 cm. The hornblendites have gradational borders against ultramafic rocks and gabbro gneisses, and were evidently formed from these rocks by a process of metasomatism. HEIER (1961) suggests that the carbonatites were formed by the intrusion of a potassic carbonate magma rich in K_2CO_3 and H_2O, which transformed original basic rocks into biotite-rich carbonatites, and at the same time caused the formation of hornblendites and hydration of the granulite facies minerals of the old complex.

The alkaline rocks present are nepheline syenites and nepheline–albite (canadite) pegmatites. While the former rocks are rather rich in potassium (with perthite feldspar), the latter are soda dominant with no potassium mineral other than biotite. The pegmatites occur in numerous dykes, some of which are thick. On Seiland a varied assemblage of dykes, including granitic ones, was described by BARTH (1927, 1953).

On Sørøy, igneous rocks evidently belonging to the Finnmark complex occur as intrusions in the sedimentary sequence of the island which contains Lower Cambrian fossils, as noted in the stratigraphic section (STURT, 1961; RAMSAY and STURT, 1963 and written communication; STUMPFL and STURT, 1965) (see Fig. 95). The sequence was folded in at least two phases. The first phase involved the formation of folds varying in style from overturned asymmetric to isoclinal. The structural investigations demonstrate that the large-scale recumbent folds overturned towards the east are anticlines. Axial trends of the early (F_1) folds are extremely variable and display maxima between NNW and NE. These early structures were refolded by a second generation of concentric folds whose axes vary from a N–S trend in the west and east of the island to an E–W trend in the central portion (Fig. 96). The meta-sedimentary rocks are in almandine amphibolite facies, although several phases of differing grade can be distinguished. The emplacement of the major

Fig. 98 *Density stratified layers in norite, between Olderfjord and Reinsfjord*, northeast side of Kvænangen. Photo P. R. HOOPER

basic and ultrabasic rocks took place during the later stages of the development of the older (F_1) structures, during the main regional metamorphism and during the second folding (F_2). A highly varying suite of alkaline rocks was emplaced broadly syntectonically with the second folding (F_2). Fenitization occurred during the alkaline intrusions.

A paper in press by B. A. Sturt, J. A. Miller and F. J. Finch records a number of K–A determinations on the alkaline rocks of Sørøy, giving ages of 480 to 490 m. y. An early Caledonian phase of folding and metamorphism is thus evident, probably corresponding to the Trondheim phase in south Norway (early Ordovician).

The peninsula around Langfjord (70° 10′, 21° 45′) and the islands of Silda and Loppa have been investigated by a research group led by Dr. P. R. Hooper (Ball and others, 1963; Hooper, written communication). A sequence of metasediments (quartzites, limestones and pelitic schists) was in general found to be the same as on Sørøy. It is folded into major anticlines and synclines with steep axial planes and axial trend swinging from NNE to NW. There is also folding of other (partly earlier) phases in other axial directions. Parts of the area are occupied by granulite facies gneisses and two-pyroxene granulites (the fine-grained gabbros of the 1963 paper). These rocks occur in the low ground around the fjords in the eastern parts of the Langfjord peninsula and are separated by undulating, faulted, but essentially flat thrust planes, from the large massifs of coarse-grained banded gabbros which form the high interior parts of the peninsula (Figs. 97, 98).

CHAPTER XIV

OLD RED DEPOSITS OF LATE SILURIAN TO MIDDLE DEVONIAN AGE

In those parts of the North Atlantic region where important Caledonian crustal movements took place, sediments of latest Silurian and Devonian age are present in continental Old Red facies. As described in a previous section, the Ringerike sandstone in the Oslo region is a late Silurian Old Red deposit with a gradual transition from underlying marine deposits. In other parts of Scandinavia the Old Red deposits rest with a profound angular unconformity upon metamorphosed Caledonian supracrustal and intrusive rocks (Fig. 99). The Scandinavian Old Red is not widely distributed, being restricted to three main districts (apart from the Oslo region), all of them in southern Norway. One is on the west coast between Sognefjord and Nordfjord (the *Vestlandet region*), the second is in the coastal area outside the mouth of Trondheimsfjord (the *Trondheimsleden district*), and the third is a small area at Lake Røragen east of Røros. To some extent the limited distribution of the Old Red deposits is due to subsequent removal of rocks, but it may also be due to the fact that deposition took place in a number of isolated internal basins under high local relief. Much of the sediments are more coarse grained than is usual for Old Red sediments in other regions, and breccias and conglomerates attain great thicknesses. In contrast to the underlying rocks the Old Red sediments are unmetamorphic (except for some low grade recrystallization in zones of tectonic movement), but the rocks are hard and well lithified.

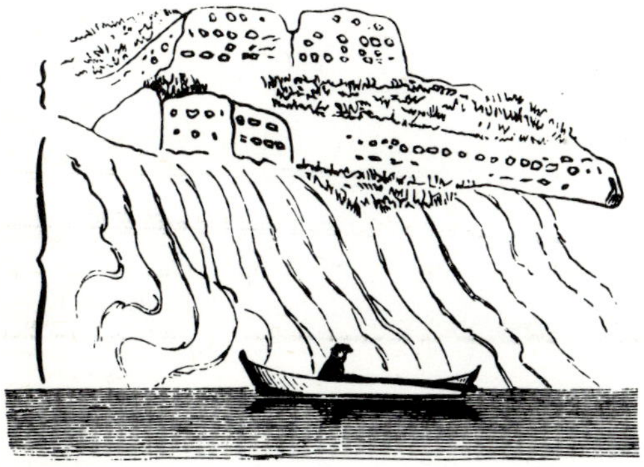

Fig. 99 *Unconformity between Cambro–Silurian schists and overlying Devonian basal breccia. South part of Bremangerland (61° 47′, 5°).* From H. Reusch

The Old Red deposits of the Vestlandet region (C. F. Kolderup, 1928) occur in five separate areas between 60° 50′ and 61° 55′ N. Three of the areas (those of *Solund*, *Kvamshesten* and *Håsteinen*) are situated in synclinal depressions in the basal gneiss, and are thus directly underlain by Caledonian supracrustal or intrusive rocks. In all the areas coarse breccias and conglomerates lie at the base of the series. The lowest parts of these contain large angular blocks from the fractured substratum *in situ*, the fissures being filled with red sandstone. In the upper parts of the sequence greyish green or red sandstones with subangular, mostly coarse grains occur. The sandstones are arkoses; chemical analyses give 58–77% SiO_2, and show that the material was derived from nearby rocks. Thus the sandstone in the Kvamshesten area, resting upon a mangerite syenite, is extremely rich in feldspar from the syenite (58% SiO_2 in the sandstone). Interlayered with the sandstones are conglomerates, less coarse than the basal breccias and conglomerates.

The southernmost area is on the islands of Byrknesøyene and Holmengrå south of the mouth of Sognefjord, where the Old Red rests upon anorthositic rocks of the Bergen arcs. Here only breccias and conglomerates from the lower part of the sequence are preserved.

Most of the Solund area lies in the Solund islands outside the mouth of Sognefjord. The deposits on Bulandet and Værlandet islands may be regarded as the northern part of the same area separated by the sea. The whole area is roughly in the form of a triangle with apex pointing east. Most of the deposits are very coarse breccias and conglomerates, but coarse sandstones occur in the central parts; the total thickness is estimated to be 2500–3000 m. Tectonically the area forms a basin with the sediments dipping towards the interior; there are steep dips in the north and more moderate dips (20–40°) on the south side. Over most of the area the lower boundary of the deposits is undisturbed, but strong tectonic movements after deposition are shown by sheets of gabbro overthrust above the Devonian sediments on planes dipping east. Both the gabbro and the underlying sediments are strongly deformed and mylonitized at the contact.

The Solund area contains the only igneous rocks found in the Norwegian Old Red. Five irregular sills (or volcanic layers?) of keratophyre occur in the sediments near the basement of underlying rocks. There are also dykes of granite and quartz, and small lenses of spherulitic rock.

The deposits of the Kvamshesten area (15 by 5 km) rest in the west upon a massif of mangerite syenite. All the deposits dip east. The lower division of basal breccias and conglomerates occurring in the west measures a maximum of 600 m, and is followed by nearly 4000 m of red and green sandstones with conglomerates. The uppermost division (about 700 m) consists of alternating sandstones and conglomerates, the latter generally the thicker of the two. In a measured section there were 80 alternations of these beds in a thickness of 350 m.

The Håsteinen area north of Kvamshesten extends for 10 km in an E–W direction and contains a 1000 m sequence, chiefly of coarse breccias and conglomerates. In both the Kvamshesten and Håsteinen areas rocks of the substratum are found as thrust masses above the Devonian deposits. There are in addition post-Devonian tectonics of faulting.

The *Hornelen area* just south of Nordfjord is the largest of the Norwegian Old Red areas and measures 55 by 20 km, with an ENE elongation.

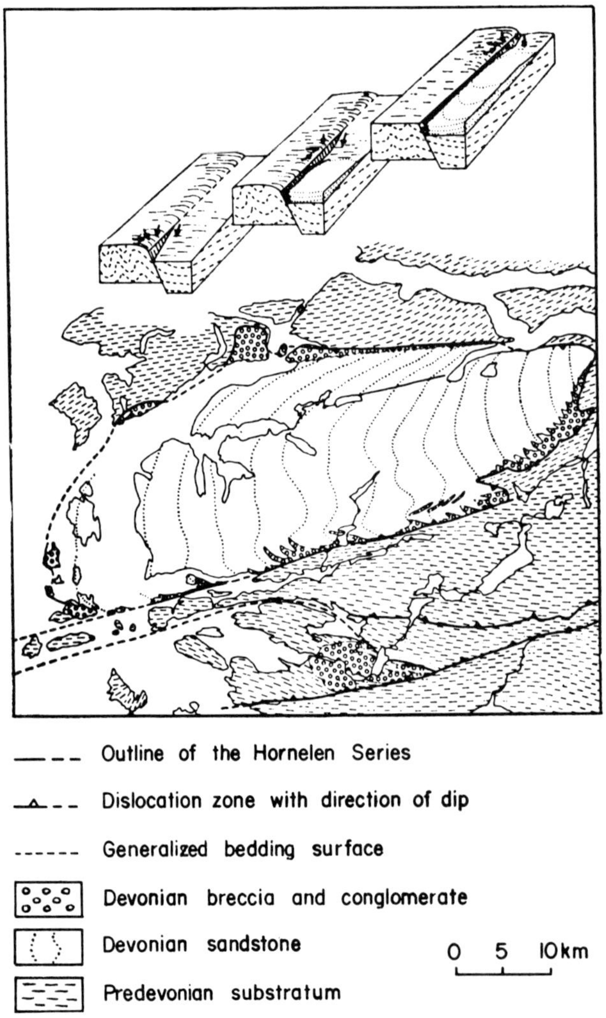

Fig. 100 *Geologic map of the Hornelen Old Red basin.* From Bryhni (1964). The upper picture shows how the basin is assumed to have been filled during downfaulting progressing from west to east

Fig. 101 *View along the north margin of the west part of the Hornelen Old Red basin, looking east.* Nordfjord is seen in the left distance. Ålfoten, a branch of Nordfjord, in the middle distance, is 15 km distant from the foreground. The marginal breccia forms a wall at the north side of the basin. Copyright Widerøes Flyveselskap A/S, Oslo

Very interesting relationships in the area have recently been cleared up by the investigations of BRYHNI (1963, 1964) (Figs. 100, 101). On the west side, the basal breccias normally rest upon Caledonian rocks, and on the other sides the area is limited by faults. The disposition of the bedding simulates an amphitheatre with a regular eastwards dip in the middle part. This gives an estimated stratigraphic thickness of 20 000 to 25 000 m. The major part of the sediments is of green sandstone, but at the faulted contacts to the underlying rocks there are coarse breccias with boulders up to 1 m^3. Coarse bedding in the breccias is indicated by occasional graded bedding, by intercalated sandstones and by imbricated fragments. In spite of their great thickness (several kilometres) the breccias intertongue basinwards into sandstone within a distance of 1,5 km. It is impossible to ascribe this great thickness to tectonic repetitions, as no thrusts can be seen in spite of the splendid exposures. According to BRYHNI's interpretation, deposition in the Hornelen area took place in a basin, down-faulted between crystalline blocks in the north and south. As deposition proceeded, the faulting of the basin moved eastwards and the earlier deposited sediments were tilted. It is implied in this interpretation that the Old Red deposits of Hornelen have never been much more widely distributed than they are today.

A Psilophyton was found in the Solund area, while a richer flora and fauna was found to occur in the Hornelen area. According to HØEG (1931, 1935), the presence of the plant genera Asteroxylon, Barrandeina, Hyenia (named after Hyen in Nordfjord), Dawsonites and Pectinophyllum (Barinophyllum) indicate Middle Devonian age. JARVIK (1948) described fish remains, Gyroptychius (?) kiæri and fragments of Porolepiformes type, which suggest a correlation with the lower part of the Scottish upper Middle Devonian.

In the Trondheimsleden region we shall deal firstly with the Old Red deposits occurring in a strip along the southeast coast of Hitra (in the northeast part of the island), and on the islands southeast of Smøla, separated from the Hitra outcrops by 30 km of intervening sea (REUSCH, 1914). The deposits can be traced some 25 km further southwest over a few small islands to Inngrip off Kristiansund. The Old Red deposits overlie the Caledonide rocks of Hitra and Smøla on a northeast striking plane with dip varying from about 45° southeast to steep and overturned in some places; in one locality the boundary plane is mylonitized. To the southeast, horizontal layers have been observed, indicating the centre of a syncline, but the southeast wing of the syncline has not been exposed. The deposits commence with a very coarse conglomerate, the lower part of which is a breccia *in situ* with angular blocks several metres across from the underlying complex of igneous rocks; conglomerates with rounded boulders may be found filling the fissures. An important locality is that of Balsnesaune on Hitra, where the basal conglomerate measures 20 to 30 m and is overlain by 500 m of very coarse-grained sandstone or grit with some thin layers of dark, fine-grained sandstone and silty shale. In these fine-grained sediments REUSCH found the phyllocarid Dictyocaris—determined as D. slimoni by STØRMER (1935)—and eurypterid remains. The fossil indicates late Silurian (Downtonian) age.

Possibly as a continuation to the northeast of the narrow syncline southeast of Hitra and Smøla, the Old Red deposits form a wide syncline on Storfosen and adjacent islands which lie between Hitra and the Ørlandet peninsula (RICHTER, 1947). The syncline has an axial pitch of 20° east and is closed to the west. The south wing of the syncline consists of coarse conglomerates with boulders of maximum size 0,5 m. The imbrication of the boulders indicates that they were transported from the south. RICHTER mentions the great likeness of the conglomerate to the Nagelfluh of the Alpine molasse. In the central and northern part of the syncline there are greyish green micaceous sandstones and partly coarse-grained arkoses. RICHTER considers the conglomerate in the south wing of the syncline (thickness 1800 m) to be equivalent to the sandstones in the north wing (thickness 700 m). The Old Red deposits, separated from Storfosen by an intervening area with older rocks (due to faulting), appear again in the Ørlandet peninsula, where they form a flat syncline. Here the main part of the rocks are conglomerates. VOGT (1929) found an area with similar Old Red deposits exposed in islands and skerries off the coast, north of Ørlandet. One area measuring 15 by 7 km has its centre about 63° 55′, 9° 40′; there is a smaller area southwest of this.

More or less well-preserved plant fossils have been found on Storfosen and Ørlandet and on the islands at the coast. HØEG (1945) referred the plants to Psilophyton, Hostimella and Hyenia. The presence of the genus Hyenia is indicative of Middle Devonian age—perhaps a low horizon in the Middle Devonian.

In the *Røragen district* (62° 35′, 11° 50′) the Old Red deposits occupy an area no more than 6 by 2 km; the rocks strike northeast and generally dip about 45° to the southeast (GOLDSCHMIDT, 1913). They rest upon mica schists of the Røros district, and partly upon augen gneisses. The lower part of the sequence consists of relatively fine-grained sediments—a basal conglomerate with boulders of moderate size, up to 80 m thick, overlain by 130 m of grey sandstone and shale with plant fossils. A thin red sandstone lies above the grey sandstone, but in parts of the area it rests directly upon the Caledonide rocks. The upper part of the sequence consists of conglomerates and breccias. In the southwestern, smaller part of the area, adjacent to a large serpentinite body at Feragen, is a 200 m thick serpentine conglomerate. In the northeast is a 700 m thick breccia with large fragments, up to 1 m across, of mica schists and other rocks. The plant fossils in the dark sandstone and shale are Psilophyton goldschmidti and Drepanocladus, indicating Lower Devonian age, according to HALLE and HØEG (1935).

GOLDSCHMIDT assumed that the Devonian had sunk down and had been tilted by faulting, but that no folding took place after deposition of the sediments. HOLMSEN (1963) reported observations which contradict this assumption. Parts of the sediments show folds of small dimensions that cannot have had their origin in faulting. He also reported recrystallization of the rocks on planes of tectonic movement; some shales thus became lustrous.

We have seen that the deposition of the Norwegian Old Red deposits took place over a comparatively long period of time—from the late Silurian to the Middle Devonian—and that the deposits are not of the same age in all areas, being older in Trondheimsleden and Røragen than in Vestlandet. Rather strong tectonic movements took place during deposition (as is evident in the case of the Hornelen area) and afterwards. For the post-Middle Devonian earth movements VOGT (1928) coined the term *Svalbardian* as a separate tectonic phase. As emphasized by VOGT, it is probable that this late or posthumous phase also had its effects outside the areas with Devonian sediments.

CHAPTER XV

ECONOMIC GEOLOGY

Before entering upon a description of the mineral resources of the Norwegian Caledonides it may be appropriate to mention that Norway has important ore deposits in production in the Precambrian parts of the country also. These deposits consist of ilmenite and molybdenite in the southwest part of the country and sedimentary iron ore in the northeast part, south of Kirkenes in Finnmark.

Among the ore deposits of the Caledonides, those of cupriferous pyrite have been of great economic importance for more than three centuries. They are found in sedimentary and volcanic rocks of eugeosynclinal facies; a genetic connection with gabbroid intrusive rocks has been generally assumed. A large part of the deposits occurs in the Trondheim region, but they are found in beds of similar facies along the west coast of southern Norway, in the Grong area and in Rana and Salta in northern Norway (Sulitjelma); they also occur in Sweden. The deposits contain pyrite as the main mineral, in some cases substituted by pyrrhotite. Chalcopyrite and sphalerite are present in almost all the deposits, and galena commonly occurs in small amounts. The composition of the ores is 42–50% S, 36–45% Fe, 0,5–3,0% Cu and an amount of Zn more or less equal to the amount of Cu. The composition varies greatly, even between different parts of the same ore body.

The deposits occur as bodies of compact ore of varying shape and dimensions; often the bodies are lens-shaped, or of elongated fish-like or ruler-like shape of length greatly exceeding the other dimensions. An example often cited is the ore body in *Kongens grube* (the King's mine) in the Røros district, of length about 3 km, width between 50 and 120 m and thickness between 1 and 4 m with a maximum of 8 m. The direction of elongation of the ore bodies coincides with the direction of tectonic axes in the country rocks (Vogt, 1952). Other ore bodies are plate-like or layer-like, as is the *Folldal deposit*, recently described by Geis (1961a) and by Page (1964). The larger of the ore bodies, which possibly consists of coalesced lenses, contains as much as 5 to 15 million tons. A publication of Foslie (1926) contains data on the shape and ore reserves (at that time) of the deposits. The students of the Norwegian pyrite deposits, J. H. L. Vogt, Foslie (1926, 1927), Carstens (1935, 1936) and Vogt (1935) assumed a genetic connection between magmatic activity and ore deposition. But no proof has ever been given of a genetic connection between an intrusive body and an adjacent ore body; the theory rests on the general geological association of the two sorts of phenomena. While J. H. L. Vogt and Th. Vogt regarded the ores as in the main of liquid magmatic origin, C. W. Carstens advocated formation by hydrothermal solutions, leading to a replacement of the original rocks. In more recent times, the epigenetic nature of the deposits was not in doubt until Oftedahl (1958a, 1958b) advanced a theory of exhalative sedimentary ore formation, the sulphides being precipitated from volcanic exhalations. Sedimentary, syngenetic ore formation seems improbable, particularly in the case of sharply delimited large and elongated bodies. Also inconsistent with the syngenetic theory are the metasomatic changes in the wall rocks of the ores, reported by Foslie (1940) and Gjelsvik (1965) in the *Skorovatn deposit* of the Grong district and by Carstens (1944) in the *Løkken deposit* of the Trondheim region.

Some of the cupriferous pyrite deposits occur together with pyrite deposits of undoubted sedimentary origin. According to Carstens (1924) they occur as 0,5 to 3 m thick layers interbedded with meta-basaltic greenstones, accompanied by layers of black carbonaceous schists, chert (jasper) and layers with stilpnomelane (iron-rich chlorite) (see Chapter V). Only one of the copper-free sedimentary pyrite deposits has been of economic value, namely that worked on Stord island off the west coast of southern Norway.

A special type of sulphide deposit is the zinc-lead deposit, possibly connected with the cupriferous pyrite deposits by intermediate links. A typical example is that mined at Bleikvassli in northern Norway (VOKES, 1963). Non-ferrous sulphides account for some 25% of the total sulphides, with sphalerite and, to a lesser degree, galena predominating.

Lead deposits discovered more recently consist of disseminations of galena and sphalerite in autochthonous and allochthonous Eocambrian and Cambrian sandstones along the Caledonide margin, the more important of them occurring in Sweden. In Norway these deposits occur in the Drevsjø area on the international border (61° 50′ N), and deposits north of Gjøvik on the west side of Lake Mjøsa are promising.

Occurrences of sulphide ores of magmatic origin are the nickeliferous pyrite deposits in basic and ultrabasic igneous rocks. Some Caledonide deposits of this type have been worked; one of them is the *Hosanger deposit* in the Bergen district (BJØRLYKKE, 1949). The Råna norite body in Ofoten contains large deposits of low grade nickel ore. A few molybdenite deposits in granitic rocks have been worked.

There are two types of sedimentary oxidic iron ores. One type is represented by the iron ores of Fosdalen in the Snåsa syncline. It occurs in a volcanic succession of greenstones and keratophyres in the lower part of the Hovin group and has been traced more or less continuously over a distance of 130–140 km. The ore minerals present are magnetite, pyrite and a little chalcopyrite; the ores contain 40–50% Fe and 2–3% S (in the pyrite). CARSTENS (1955) interpreted the ores as being of sedimentary origin, the iron having been introduced by volcanic exhalations. They should thus have an origin similar to that of the sedimentary sulphide deposits, formed, however, under oxidizing conditions.

The second type of sedimentary iron ore is the *Nordland type*. The richest deposits of this type occur in Dunderlandsdalen in the Rana district and in Ofoten. The deposits in Dunderlandsdalen occur in a sequence of mica schists, crystalline limestones and dolomites (J. A. W. BUGGE, 1948). The ores contain 30–35% Fe with hematite or magnetite or both. In the Ofoten districts some of the ores are rich in manganese, but in most of the ores the manganese is present in silicates. According to BUGGE the ore deposition was wholly a sedimentary process; there are no volcanic rocks in the ore-bearing sequence, which must have been deposited under conditions of strong weathering and geochemical differentiation of the elements.

The Gneiss region contains iron ores of *Raudsand type*, mined at Raudsand on Sunndalsfjord (GEIS, 1965). The ores occur as sills and schlieren in amphibolite, and contain about 19% Fe, 5,5% TiO_2 and 0,2% V. Possibly the ore-bearing rock is related to the gabbros of the same district—the dolerites described by GJELSVIK (1952) and CARSTENS (1957a), which have iron ore concentrations in some cases.

The iron ores of the Lofoten district in northern Norway (not worked) have been mentioned previously.

The numerous bodies of dunite and serpentinite contain segregations of chromite. A few of the larger of these deposits have been worked.

Gold has been mined from quartz-bearing dykes on Bømlo island off the west coast of southern Norway; deposits in Bindalen have been under investigation.

Norway's total production of pyrite and copper ores amounted to about 40 million tons, according to POULSEN (1960) and only a small part of this has not come from Caledonide deposits. The total amount of copper metal extracted from the ores produced was about 900 000 tons.

The following figures give average annual production for the period 1961–1963, the last years for which official statistics are available. Twelve mining companies working sulphide deposits produced (annually) about 750 000 tons of pyrite, of which the *Lökken mine* accounted for 260 000 tons with 2% Cu and the *Skorovatn mine* 150 000 tons with 1,4% Cu. To this must be added 28 000 tons of copper concentrate (20–25% Cu), of which 18 000 tons came from the *Sulitjelma mines*. In addition there were 22 000 tons (about 50% Zn) of zinc concentrate and 5000 tons of lead concentrate (about 60% Pb) produced.

The *Raudsand mine* produced 110 000 tons of iron ore (63% Fe) and 3300 tons of ilmenite concentrate (30% TiO_2). The *Fosdalen mine* produced about 240 000 tons of iron ore (65% Fe). The deposits in Dunderlandsdalen came into production in 1965 with a planned annual output of 400 000–500 000 tons of iron ore concentrate.

Lists and maps of the mines and claims of Norway are contained in the publications of FOSLIE (1925) (southern Norway) and POULSEN (1964) (northern Norway).

Limestone and dolomites are worked in many districts, and almost all of them come from Caledonide districts (including the southern unfolded part of the Oslo region). Fauske in the Salta district is renowned for its many fine varieties of ornamental marble.

Trondhjemites ("white granites") are utilized in the granite industry.

Roofing-slates and flags have been produced in a great number of places—Voss, Valdres, Otta, Oppdal and Alta are particularly well known.

Graphite, formed by recrystallization of carbonaceous sediments, is produced at Skaland on Senja island in northern Norway.

Many of the dunites and peridotites have been metamorphosed into talc rocks, utilized as soap stones or as talc flour.

Quartz and quartzite are utilized in a number of places. Feldspar is produced at Hundholmen in the Tysfjord district. Production of albite and nepheline from nepheline syenite at Stjernøy, Finnmark commenced a few years ago.

		MJØSA DISTRICT (South Norway)		FINNMARK REGION (North Norway) TANA DISTRICT
MIDDLE CAMBRIAN TO EARLIEST ORDOVICIAN		*Black alum shales*	Digermul group	*Berlogaissa formation* *Kistedal formation*
LOWER CAMBRIAN		*Grey shales and siltstones with* Holmia, Volborthella *and* Platysolenites	Digermul group	*Duolbasgaissa formation*
			Vesterana group	*Breivik formation with* Platysolenites
		← Disconformity →		
EOCAMBRIAN	Sparagmite group	*Vemdal sandstone* *Ekre shale* *Moelv tillite* *Moelv sparagmite*	Vesterana group	*Stappogiedde formation* *Upper tillite* *Nyborg formation* *Lower tillite*
				← Low angle unconformity →
LATEST PRECAMBRIAN	Sparagmite group	*Biri limestone* *Biri conglomerate* *Brøttum formation* *Elstad sandstone* (in Fåvang)	Porsanger group	*Thick sandstones, shales and dolomites*

Fig. 102 Latest Precambrian and Eocambrian stratigraphy of Norway (also including the nearest overlying fossiliferous deposits)

	BRITISH CHRONOSTRATIGRAPHY	OSLO STRATIGRAPHY	OSLO REGION: Oslo district	OSLO REGION: Mjøsa district	HARDANGERVIDDA	BERGEN DISTRICT	TRONDHEIM REGION: Western part: HORG SYNCLINE	TRONDHEIM REGION: Eastern part MERAKER
	Downton	ét. 10	Ringerike sandstone					
SILURIAN	Ludlow	Upper Spiriferid series ét. 9	Shales Limestone	Bruflat sandstone (Ludlow–Wenlock)				
SILURIAN	Wenlock	Lower Spiriferid series ét. 8	Limestone Mainly shales					
SILURIAN	Llandovery	Pentamerus ser. ét. 7	Marls ←— Pentamerus limestone	Graptolite shales Pentamerus limestone —→		Schists with monograptids		Black schists with Rastrites
SILURIAN	Llandovery	Stricklandia series ét. 6	Marly shales	Sandstone		Schists with Stricklandia	Horg group: Sandå formation ←Lyngestein conglomerate→	←Thin quartz conglomerate→
ORDOVICIAN	Ashgill	Tretaspis series ét. 4c–5b	Calcareous sandstone Gastropod limestone Tretaspis shale			Gastropod limestone ←Moberg conglomerate→	Upper Hovin group: Hovin sandstone ←Volla conglomerate→	
ORDOVICIAN	Caradoc	Chasmops series ét. 4aβ–4b	Nodular marlstones and impure limestones	Mjøsa limestone Siltstones and sandstones		Micaschists limestones, greenstones and other volcanics (Caradoc–Middle Cambrian)	Lower Hovin group: Hareklett formation (Cryptolithus) Dicranograptus shale Hølonda formation Bogo shale Bjørkum shale (Trinucleus)	
ORDOVICIAN	Llandeilo-Llanvirn	Ogygiocatris caris series ét 4aα	Mainly black shales		Grey phyllites			Phyllites and greywackes Greenstone Limestone Phyllites
ORDOVICIAN	Arenig	Asaphus series ét. 3b–c	Orthoceras limestone (Stein limestone) Graptolite shales		Limestone Quartzite		←Stokkvola conglomerate→	
ORDOVICIAN	Tremadoc	Ceratopyge series ét. 2e–3a	Ceratopyge limestone Black shales		Black alum shale phyllites (Tremadoc–Middle Cambrian)		Støren group: Thick meta-basalts	Schist with Dictyonema Alternating sediments and volcanics ←Gudå conglomerate→
CAMBRIAN	Upper Cambrian	Olenid series ét. 2a–d	Alum shales (Upper–Middle Cambrian)					
CAMBRIAN	Middle Cambrian	Paradoxides series ét. 1c–d					Røros group: Mainly micaschists	Drøia schists (Gula schists)
CAMBRIAN	Lower Cambrian	Holmia series ét. 1a–b	Grey shales and siltstones		Thin conglomerate			

Fig. 103 Cambro–Silurian stratigraphy of some important districts in Norway

SELECTED BIBLIOGRAPHY

ABBREVIATIONS USED IN SELECTED BIBLIOGRAPHY

Arch. Math. Naturv.	= Archiv for Mathematik og Naturvidenskab, Kristiania
Årbok Univ. Bergen, Mat.-Nat. Ser:	= Universitetet i Bergen Årbok. Mat.-naturv. rekke Bergens Museums Årbok Mat-naturv. rekke Bergen (from 1948)
Geol. Fören. Stockh. Förh.	= Geologiska Föreningens i Stockholm Förhandlingar
Kgl. Norske Vidensk. Selsk. Forh; Kgl. Norske Vidensk. Selsk. Skr.	= Det Kongelige Norske Videnskabers Selskab, Trondheim. Forhandlinger, Skrifter
Norsk Geol. Tidssker.	= Norsk Geologisk Tidsskrift, Oslo (Kristiania)
Norg. Geol. Undersøkelse.	= Norges Geologiske Undersøkelse (Geological Survey of Norway, publications of) Oslo (Kristiania)
Nyt Mag. Naturv.	= Nyt Magasin for Naturvidenskaberne, Kristiania
Sveriges Geol. Undersökn. Årsbok, Ser. Ca: Avhandl. Uppsat.; Ser C: Avhandl. och Uppsat.	= Sveriges Geologiska Undersøkning, Stockholm

The following abbreviations have been taken from *World List of Scientific Periodicals*, Butterworth's, London, 1966.

Skr. VidenskSelsk. Christiania Mat.-naturv. Kl. (to be followed by number and year)	= Skrifter udg.af Videnskabsselskabet i Christiania: Matematisk-naturvitenskapelige Klasse
Skr. norske Vidensk-Akad. Mat.-naturv. Kl. (to be followed by number and year)	= Skrifter utgitt av det Norske Videnskapsakademi i Oslo. Matematisk-naturvitenskapelige Klasse
Avh. norske VidenskAkad. Oslo Mat.-naturv. Kl. (to be followed by number and year)	= Avhandlinger ugitt av det Norske videnskaps-akademi i Oslo

Ackermann, K. J., R. Nicholson and B. J. Walton (1960). Mineral development and deformation in metasedimentary rocks in the Glomfjord region, north Norway. *Intern. Geol. Congr. 21st Copenhagen, Rep. Session, Norden*, **XIX**, 54–63.

Appleyard, E. C. (1965). Preliminary description of the Dönnesfjord area, Söröy. *Norg. Geol. Undersøkelse.*, **231**, 143–164.

Asklund, B. (1961). The extension of the Serv nappe in the Scandinavian mountain chain. *Sveriges Geol. Undersökn. Arsbok. Ser. Ca: Avhandl. Uppsat.*, **584**, 28 pp.

Bailey, E. B. and O. Holtedahl (1938). Northwestern Europe Caledonides, *Reg. Geologie der Erde*, Band 2, Abschn. 2. Leipzig, 76 pp.

Ball, T. K., C. B. Gunn, P. R. Hooper and D. Lewis (1963). Preliminary geological survey of the Loppen district, west Finnmark. *Norsk Geol. Tidsskr.*, **43**, 215–246.

Banham, P. H. (1966). Fault vein mineralization as a result of shearing in Norwegian basement rocks. *Norsk Geol. Tidsskr.*, **46**, 181–192.

Banham, P. H. and R. E. Elliot (1965). Geology of the Hestbrepiggen area. *Norsk Geol. Tidsskr.*, **45**, 189–198.

Barth, T. F. W. (1927a). Die Pegmatitgänge der kaledonischen Eruptivgesteine im Seiland- Gebiete. *Skr. norske VidenskAkad.*, **1927,** No. 8, 123 pp.

– (1927b). Sagvandite, a magnesite-bearing igneous rock. *Norsk Geol. Tidsskr.*, **9**, 271–303.

– (1938). Progressive metamorphism of sparagmite rocks. *Norsk Geol. Tidsskr.*, **18**, 54–65.

– (1953). Layered gabbro series at Seiland. *Norg. Geol. Undersøkelse.*, **184**, 191–200.

– (1961). Garnet–sillimanite and garnet–spinel bands in the layered gabbro series at Seiland. *Bull. Geol. Inst. Univ. Uppsala*, **40**, 17–24.

Battey, M. H. (1960). Observations on the peridotites and pyroxenites of the Jotunheim complex of Norway. *Intern. Geol. Congr. 21st Copenhagen, Rep. Session, Norden*, **XIII**, 198–207.

– (1965). Layered structure in rocks of the Jotunheim complex, Norway. *Mineral. Mag. (London)*, **34**, 35–51.

Beswick, A. E. (1963). *See* Chadwick and others.

Birkeland, T. (1958). Geological and petrological investigations in northern Trøndelag. *Norsk Geol. Tidsskr.*, **38**, 327–420.

Björlykke, H. (1949). Hosanger nikkelgruve. *Norg. Geol. Undersøkelse.*, **172**, 38 pp.

Björlykke, K. (1965a). The geochemistry and mineralogy of some shales from the Oslo region. *Norsk Geol. Tidsskr.*, **45**, 435–456.

– (1965b). Bigganjargga-tilliten. *Geologi (Helsinki)*, **129**.

– (1965c). The Eocambrian stratigraphy of the Björånes window. *Norg. Geol. Undersøkelse.*, **234**, 5–14.

– (1966). Sedimentary petrology of the Sparagmites of the Rena district. *Norg. Geol. Undersøkelse.*, **238**, 5–53.

Björlykke, K. O. (1898). Geologisk kart med beskrivelse over Kristiania* by. *Norg. Geol. Undersøkelse.*, **25**, 86 pp.

– (1902). Fra Hardangervidda. *Norg. Geol. Undersøkelse.*, **34**, No. 2, 51–76.

– (1905). Det central Norges fjeldbygning. *Norg. Geol. Undersøkelse.*, **39**, 595 pp.

– (1908). Jæderens geologi. *Norg. Geol. Undersøkelse.*, **48**, 160 pp.

Björlykke, U. (1963). Foreløbige meddelelser fra kartbladet Essandsjø. *Norg. Geol. Undersøkelse.*, **223**, 5–9.

Blake, D. H. (1962). A new Lower Ordovician graptolite fauna from the Trondheim region. *Norsk Geol. Tidsskr.*, **42**, 223–238.

– (1963). *See* Chadwick and others.

* The name of the city of Oslo was Kristiania (Christiania) from 1624 until 1925, when the old name was taken up again.

BRADSHAW, R. (1960). *See* HOLLINGWORTH and others.

BROCH, O. A. (1964). Age determinations of Norwegian minerals up to March 1964. *Norg. Geol. Undersøkelse.*, **228**, 84–113.

BROCH, O. A., F. ISACHSEN, O. ISBERG and T. STRAND (1940). Bidrag til Skudenes-sedimentenes geologi. *Norg. Geol. Undersøkelse.*, **155**, 41 pp.

BRÖGGER, W. C. (1882). Die silurischen Etagen 2 und 3. *Universitets-program* 1882, Kristiania, Oslo.

– (1887). Geologisk kart over øerne ved Kristiania. *Nyt Mag. Naturv.*, **31**, 162–195.

– (1893). Lagfölgen på Hardangervidda og den såkaldte "Höifjeldskvarts". *Norg. Geol. Undersøkelse.*, **11**, 142 pp.

BRYHNI, I. (1962). Structural analysis of the Gröneheia area. *Norsk Geol. Tidsskr.*, **42**, 331–369.

– (1963). Relasjonen mellom senkaledonisk tektonik og sedimentasjon ved Hornelens og Håsteinens devon. *Norg. Geol. Undersøkelse.*, **223**, 10–25.

– (1964). Migrating basins on the Old Red continent. *Nature*, **202**, 384–385.

– (1967). Reconnaissance studies of gneisses, ultrabasites, eclogites and anorthosites in outer Nordfjord, western Norway. *Norg. Geol. Undersøkelse.*, **241**, 68 pp.

BRYN, K. Ø. (1959). Geologien på den söndre del av kartblad Essandsjö. *Norg. Geol. Undersøkelse.*, **205**, 5–16.

– (1960). Et funn av pseudotachylit i Sör-Tröndelag. *Norg. Geol. Undersøkelse.*, **211**, 8–17.

BŪBNOFF, S. VON (1930). *Geologie von Europa*, 2 Band, 1 Teil, *Kaledoniden und Varisciden*, Gebr. Borntraeger, Berlin.

BUGGE, C. (1910). Rennebu. *Norg. Geol. Undersøkelse.*, **56**, 42 pp.

– (1939). Hemsedal og Gol. *Norg. Geol. Undersøkelse.*, **153**, 83 pp.

– (1948). Kisene i fjellkjeden. *Norsk Geol. Tidsskr.*, **27**, 97–102.

– (1954). Den kaledonske fjellkjede i Norge. *Norg. Geol. Undersøkelse.*, **189**, 79 pp.

BUGGE, C. and S. FOSLIE (1922). Norsk arsenmalm og arsenikfremstilling. *Norg. Geol. Undersøkelse.*, **106**, 33 pp.

BUGGE, J. A. W. (1942). Gabbrofeltene i Troms. *Norg. Geol. Undersøkelse.*, **157**, 82–89.

– (1948). Rana gruber. *Norg. Geol. Undersøkelse.*, **171**, 149 pp.

CARSTENS, C. W. (1918). Norske peridotitter. *Norsk Geol. Tidsskr.*, **5**, 1–73.

– (1920). Oversigt over Trondhjemsfeltets bergbygning. *Kgl. Norske Vidensk. Selsk. Skr.*, **1919**, No. 1, 152 pp.

– (1924a). Der unterordovicische Vulkanhorizont in dem Trondhjemgebiet. *Norsk Geol. Tidsskr.*, **7**, 185–270.

– (1924b). Der unterordovicische Vulkanismus auf Smølen. *Vid. Selsk.*, **1924**, No. 19, 28 pp.

– (1944). Om dannelsen av de norske svovelkisforekomster. *Kgl. Norske Vidensk. Selsk. Forh.*, **17**, 27–54.

– (1928). Petrologische Studien im Trondhjemgebiet. *Kgl. Norske Vidensk. Selsk. Skr.*, **1928**, No. 1, 96 pp.

– (1935). Zur Genesisder Kiesvorkommen des Trondjemgebietes. *Kgl. Norske Vidensk. Selsk. Skr.*, **1935**, No. 11, 38 pp.

– (1951). Lökkenfeltets geologi. *Norsk Geol. Tidsskr.*, **29**, 9–25.

CARSTENS, H. (1955). Jernmalmene i det vestlige Trondheimsfelt og forholdet til kisforekomstene. *Norsk Geol. Tidsskr.*, **35**, 211–220.

– (1957a). Gabbros and associated titaniferous ores in the west Norwegian gneisses. *Kgl. Norske Vidensk. Selsk. Skr.*, **1957**, No. 3, 67 pp.

– (1957b). Gabbro, anorthosite and titaniferous iron ore in Flakstadøy, Lofoten. *Kgl. Norske Vidensk. Selsk. Skr.*, **1957**, No. 4, 24 pp.

– (1960). Stratigraphy and volcanism in the Trondheimsfjord area. Excursion guide. *Norg. Geol. Undersøkelse.*, **212b**, 23 pp.

CHADWICK, B., H. D. BLAKE, A. E. BESWICK and J. W. ROWLING (1963). The geology of the Fjeldheim–Gåsbakken area. *Norg. Geol. Undersøkelse.*, **223**, 43–60.

CHALOUPSKÝ, J. (1963). Notes on the geology of an area west of Stören. *Norg. Geol. Undersøkelse.*, **223**, 61–66.

COWAN, D. R. *See* ELLIOT and COWAN (1966).

DIETRICHSON, B. (1945a). Geologiske undersökelser i Espedalen. *Norg. Geol. Undersøkelse.*, **163**, 45 pp.

– (1945b). Sedimentene på det undre skyvedekke i Gudbrandsdalen. *Norsk Geol. Tidsskr.*, **25**, 53–69.

– (1950). Det kaledonske knuteområde i Gudbrandsdalen. *Norsk Geol. Tidsskr.*, **28**, 65–143.

– (1953). Pseudotachylit fra kaledonske skyvesoner. *Norg. Geol. Undersøkelse.*, **184**, 23–70.

– (1960). Intact recrystallized rocks along the movement zone of the Jotun norite nappe. *Intern. Geol. Congr. 21st Copenhagen, Rep. Session, Norden*, **XIX**, 64–88.

DONS, J. A. (1956). Kartkatalog. Norges berggrunn. (Catalogue of maps. Bed-rock geology of Norway). *Norg. Geol. Undersøkelse.*, **193**, 146 pp.

– Editor (1960). Excursions in Norway. Seventeen guide-books prepared for the XXI International Geological Congress 1960. *Norg. Geol. Undersøkelse.*, **212a–r**.

– *See* HOLTEDAHL and DONS (1957).

DU DRESNAY, R. (1950). Quelques observations dans le district Kongsvoll–Snöhetta. *Norsk Geol. Tidsskr.*, **28**, 157–170.

DU RIETZ, T. (1949). The Nasafjäll region. *Geol. Fören. Stockh. Förh.*, **58**, 425–438.

ELDERS, W. A. (1957). A preliminary note on a xenolith-rich dyke from Lyngen. *Acta Borealia, A. Sci.*, **12**, 13 pp.

ELLIOT, R. B. and D. R. COWAN (1966). The petrochemistry of the amphibolites of the Holleindalen greenstone group. *Norsk Geol. Tidsskr.*, **46**, 309–326.

– *See* BANHAM and ELLIOT (1965).

ENGLUND, J.-O. (1966). Sparagmittgruppens bergarter ved Fåvang, Gudbrandsdalen. *Norg. Geol. Undersøkelse.*, **238**, 55–103.

ESKOLA, P. (1921). The eclogites of Norway. *Skr. VidenskSelsk., Christ.*, **1921**, No. 8, 118 pp.

FEYLING-HANSSEN, R. (1948). Basalkonglomerat ved Stranddal i Ryfylke. *Norsk Geol. Tidsskr.*, **27**, 104–105.

FOERSTE, A. F. *See* KIAER (1932).

FLINN, D. (1961). On deformation of thrust planes in Shetland and the Jotunheimen area in Norway. *Geol. Mag.*, **98**, 245–256.

FOSLIE, S. (1925). Syd-Norge gruber og malmforekomster. *Norg. Geol. Undersøkelse.*, **126**, 89 pp.

– (1926). Norges svovelkisforekomster. *Norg. Geol. Undersøkelse.*, **127**, 122 pp.

– (1940). Skorovas kisfelt i Grong. *Norsk Geol. Tidsskr.*, **19**, 115–116.

– (1941). Tysfjords geologi. *Norg. Geol. Undersøkelse.*, **149**, 298 pp.

– (1942). Hellemobotn og Linnajavrre. *Norg. Geol. Undersøkelse.*, **150**, 119 pp.

– (1949). Håfjellsmulden i Ofoten. *Norg. Geol. Undersøkelse.*, **174**, 129 pp.

– (1955). Kisdistriktet Varaldsøy–Ølve i Hardanger. *Norg. Geol. Undersøkelse.*, **147**, 106 pp.

– *See* BUGGE and FOSLIE (1922).

FOSLIE, S. and T. STRAND (1956). Namsvatnet med en del av Fröyningsfjell. *Norg. Geol. Undersøkelse.*, **196**, 82 pp.

FÖYN, S. (1937). The Eocambrian series of the Tana district. *Norsk Geol. Tidsskr.*, **17**, 65–164.

– (1964). Den tillitførende formasjonsgruppe i Alta. *Norg. Geol. Undersøkelse.*, **228**, 139–150.

– *See* HOLTEDAHL and others (1960).

GAERTNER, H. R. VON (1943). Bemerkungen über den Tillit von Bigganjargga. *Geol. Rundschau*, **34**, 226–231.

GEIS, H.-P. (1961a). Strukturelle iakttagelser ved noen norske kisforekomster. *Norsk Geol. Tidsskr.*, **41**, 173–196.

– (1961b). Zur Spilitbildung. *Geol. Rundschau*, **51**, 375–384.

– (1962). Aus dem Übergang vom Oberbau in den Unterbau. Ein Beispiel aus West-Norwegen. *Geol. Rundschau*, **52**, 184–188.

– (1965). Eisen- Titanlagerstätten bei Raudsand in West-Norwegen. *Norg. Geol. Undersøkelse.*, **234**, 15–52.

GETZ, A. (1890). Graptolitførende skiferzoner i det trondhjemske. *Nyt Mag. Naturv.*, **31**, 31–42.

GEUKENS, F. and J. MOREAU (1960). Contribution a la géologie de la région de Talvik (Finnmark). *Intern. Geol. Congr. 21st Copenhagen, Rep. Session, Norden*, **IX**, 192–200.

GEUL, J. J. C. *See* REITAN and GEUL (1959).

GJELSVIK, T. (1946). Anorthositkomplekset i Heidal. *Norsk Geol. Tidsskr.*, **26**, 1–58.

– (1951). Oversikt over bergartene i Sunnmöre og tilgrensende deler av Nordfjord. *Norg. Geol. Undersøkelse.*, **179**, 45 pp.

– (1952). Metamorphosed dolerites in the gneiss area of Sunnmöre. *Norsk Geol. Tidsskr.*, **30**, 33–134.

– (1953). Det nordvestlige gneisområde i det sydlige Norge. *Norg. Geol. Undersøkelse.*, **184**, 71–94.

– (1965). Geokjemiske undersøkelser av sidebergarten—Skorvatn. *Geologi (Helsinki)*, **9–10**, 133.

GOLDSCHMIDT, V. M. (1912a). Ein kambrisches Konglomerat von Finse und dessen Metamorphose. *Skr. VidenskSelsk., Christ.*, **1912**, No. 18, 18 pp.

– (1912b). Die kaledonische Deformation der südnorwegischen Urgebirgstafel. *Skr. VidenskSelsk., Christ.*, **1912**, No. 19, 11 pp.

– (1913). Das Devongebiet am Röragen bei Röros. Mit einem paläontologischen Beitrag von A. G. Nathorst. *Skr. VidenskSelsk., Christ.*, **1913**, No. 9, 27 pp.

– (1915). Die Kalksilikatgneise und Kalksilikatglimmerschiefer des Trondhjem-Gebietes. *Skr. VidenskSelsk., Christ.*, **1915**, No. 10, 37 pp.

– (1916a). Konglomeraterne inden Høifjeldskvartsen. *Norg. Geol. Undersøkelse.*, **77**, 61 pp.

– (1916b). Übersicht der Eruptivgesteine im kaledonischen Gebirge zwischen Stavanger und Trondhjem. *Skr. VidenskSelsk., Christ.*, **1916**, No. 2, 140 pp.

– (1921). Die Injektionsmetamorphose im Stavanger-Gebiete. *Skr. VidenskSelsk., Christ.*, **1920**, No. 10, 142 pp.

– (1925). Über fossilführende unterkambrische Basalablagerungen bei Ustaoset. *Fennia*, **45**, No. 1, 1–11.

– (1943). Om friksjonsglas (pseudotachylit) i fjellkjeden. *Geol. Fören. Stockh. Förh.*, **65**, 83–84.

GREEN, D. H. *See* MCDOUGALL and GREEN (1964).

GRENDER, G. C. (1964). A petrographic study of some Eocambrian sedimentary rocks from Mjøsa and Tanafjord. *Norsk Geol. Tidsskr.*, **42**, 103–142.

GRÖNHAUG, A. *See* GUSTAVSON and GRÖNHAUG (1960).

Gunn, C. B. *See* Ball and others (1963).

Gustavson, M. (1963). Grunnfjellsvinduer i Dividalen. *Norg. Geol. Undersøkelse.*, **223**, 92–105.

– (1967). The Caledonian mountain chain in the south Troms and Ofoten areas. I. Basement rocks and Caledonian metasediments. *Norg. Geol. Undersøkelse.*, **239**, 162 pp.

Gustavson, M. and A. Grönhaug (1960). En geologisk undersøkelse av den nordvestlige del av kartblad Börgefjell. *Norg. Geol. Undersøkelse.*, **211**, 26–74.

Hadding, A. *See* Kiar (1932).

Hagemann, F. (1966). Silurian bentonites in the Oslo region. *Norg. Geol. Undersøkelse.*, **242**, 44–61.

Hagemann, F. and N. Spjeldnaes (1955). Notes on bentonites (K-bentonites) from the Oslo–Asker district. *Norsk Geol. Tidsskr.*, **35**, 29–51.

Hagemann, F. *See* Spjeldnaes and Hagemann (1955); Skjeseth and Hagemann (1960).

Hausen, H. (1942a). Der Abschmitt Finnlands der kaledonischen Überschiebungszone. *Soc. Sci. Fennica, Commentationes Phy.-Math.*, **9**, 107 pp.

– (1942b). Das Halditjokkomassiv. *Acta Acad. Aboensis, Math. Phys.*, **13**, No. 7, 83 pp.

Hedström, H. (1930). Om ordoviciska fossil från Ottadalen. *Avhandl. Norske Videnskaps-Akad, Oslo I, Mat.-Naturv. Kl.*, **1930**, No. 10, 10 pp.

Heier, K. S. (1960). Petrology and geochemistry of high grade metamorphic and igneous rocks on Langøy. *Norg. Geol. Undersøkelse.*, **207**, 246 pp.

– (1961). Layered gabbro, hornblendite, carbonatite and nepheline syenite on Stjernøy. *Norsk Geol. Tidsskr.*, **41**, 109–155.

Henningsmoen, G. (1952). Early Middle Cambrian fauna from Rogaland. *Norsk Geol. Tidsskr.*, **30**, 13–31.

– (1960). Cambro-Silurian deposits of the Oslo region. "Geology of Norway." *Norg. Geol. Undersøkelse*, **208**, 130–150.

Henningsmoen, G. and N. Spjeldnaes (1960). Paleozoic stratigraphy of the Oslo region, Eocambrian stratigraphy of the Sparagmite region. Excursion guide. *Norg. Geol. Undersøkelse.*, **212n**, 30 pp.

Hernes, I. (1954). Eclogite amphibolites in the Molde peninsula. *Norsk Geol. Tidsskr.*, **33**, 163–184.

– (1956a). Surnadalssynklinalen. *Norsk Geol. Tidsskr.*, **36**, 26–39.

– (1956b). Geologisk oversikt over Molde–Kristiansundområdet. *Kgl. Norske Vidensk. Selsk. Skr.*, **1955**, No. 5, 17 pp.

– (1963). Norwegian minerals and rocks of possible Caledonian origin. *Nature*, **199**, 686.

– (1964). Anzeichen einer frühkaledonischen Metamorphose. *Neues Jahrb. Mineral. Monatsh.*, **4**, 110–116.

– (1965a). Die kaledonische Schichtenfolge in Mittelnorwegen. *Neues Jahrb. Geol. Palaeontol. Monatsh.*, **2**, 69–84.

– (1965b). Geologisk oversikt over Tingvollområdet. Tingvoll bygdebok, 29 pp.

Hödal, J. (1945). Rocks of the anorthosite kindred in Vossestrand. *Norsk Geol. Tidsskr.*, **24**, 129–243.

Höeg, O. A. (1931). Notes on the Devonian flora of western Norway. *Kgl. Norske Vidensk. Selsk. Skr.*, **1931**, No. 6, 33 pp.

– (1935). Further contributions to the Middle Devonian flora of western Norway. *Norsk Geol. Tidsskr.*, **15**, 1–17.

– (1945). Contributions to the Devonian flora of western Norway. III. *Norsk Geol. Tidsskr.*, **25**, 183–192.

– *See* Kiaer (1932).

Högbom, A. G. (1913). Fennoskandia. *Handbuch der Reg. Geologie*, IV Band, 3 Abt., Heidelberg, 197 pp.

Hollingworth, S. E., M. K. Wells and R. Bradshaw (1960). Geology and structure of the Glomfjord region. *Intern. Geol. Congr. 21st Copenhagen, Rep. Session, Norden*, **XIX**, 33–42.

Holmes, M. (1966). Structure of the area north of Örnes, Nordland, Norway. *Norg. Geol. Undersøkelse.*, **242**, 62–93.

– *See* Rutland and others (1960).

Holmsen, G. (1932). Rana. *Norg. Geol. Undersøkelse.*, **136**, 107 pp.

– (1935). Nordre Femund. *Norg. Geol. Undersøkelse.*, **144**, 55 pp.

– (1937). Søndre Femund. *Norg. Geol. Undersøkelse.*, **148**, 42 pp.

Holmsen, P. (1950). Tynset. *Norg. Geol. Undersøkelse.*, **175**, 7–42.

– (1955). Trekk av Oppdalsfeltets geologi. *Norsk Geol. Tidsskr.*, **35**, 135–150.

– (1956). Hyolithes-zonens basale lag i Vest-Finnmark. *Norg. Geol. Undersøkelse.*, **195**, 65–72.

– (1957). De eokambriske lag under Hyolithes-zonen mellem Carravarre og Caskias, Vest-Finnmark. *Norg. Geol. Undersøkelse.*, **200**, 47–50.

– (1963). On the tectonic relations of the Devonian complex of the Røragen area. *Norg. Geol. Undersøkelse.*, **223**, 127–138.

– (1965). On the nature of the basement contact. *Norg. Geol. Undersøkelse.*, **234**, 64–70.

– *See* Strand and Holmsen (1960).

Holmsen, P. and Chr. Oftedahl (1956). Ytre Rendal og Storelvdal. *Norg. Geol. Undersøkelse.*, **194**, 173 pp.

Holmsen, P. and S. Skjeseth (1966). Trysil-hevningen mellom Osensjöen og Jordet i Trysil. Med beskrivelse av diamantborkjerne ved J. P. Nystuen. *Norg. Geol. Undersøkelse.*, **242**, 119–129.

Holtedahl, H. (1950). Geological and petrographical investigations in the northwestern part of the Oppdal quadrangle. *Arbok Univ. Bergen, Mat.-Nat. Ser.*, **1949**, No. 7, 60 pp.

HOLTEDAHL, O. (1915). Iagttagelser over fjeldbygningen omkring Randsfjordens nordende. *Norg. Geol. Undersøkelse.*, **75** [1], 32 pp.

– (1918). Bidrag til Finmarkens geologi. *Norg. Geol. Undersøkelse.*, **84**, 314 pp.

– (1920). Paleogeography and diastrophism in the Atlantic–Arctic regions during Paleozoic time. *Am. J. Sci.*, **49**, 1–25.

– (1921). Engerdalen. *Norg. Geol. Undersøkelse.*, **89**, 74 pp.

– (1931). Additional observations on the rock formations of Finnmark. *Norsk Geol. Tidsskr.*, **11**, 241–278.

– (1936). Trekk av det skandinaviske fjellkjedestrøks historie. Nordiska (19. skand.) Naturforskaremötet i Helsingfors 1936, 129–145.

– (1938). Geological observations in the Oppdal–Sunndal–Trollheimen district. *Norsk Geol. Tidsskr.*, **18**, 29–53.

– (1941). Om fjellbygningen i Oppdal–Sunndal-strøket. *Norsk Geol. Tidsskr.*, **20**, 274–276.

– (1944). On the Caledonides of Norway. *Skr. norske Vidensk Akad.*, **1944**, No. 4, 31 pp.

– (1953). Norges geologi. *Norg. Geol. Undersøkelse.*, **164**, 2 vols., 1118 pp.

– (1959). Noen iakttagelser fra Grönnsennknipa. *Norg. Geol. Undersøkelse.*, **205**, 90–106.

– (1960a). Editor: Geology of Norway. *Norg. Geol. Undersøkelse.*, **208**, 540 pp.

– (1960b). Stratigraphy of the Sparagmite group including "the sandstone divisions of Finnmark", "Geology of Norway". *Norg. Geol. Undersøkelse.*, **208**, 111–127.

– (1960c). Devonian. Including Downtonian in the Hitra district. "Geology of Norway". *Norg. Geol. Undersøkelse.*, **208**, 285–297.

– (1960d). On supposed marginal faults and the oblique uplift of the landmass in Cenozoic time. "Geology of Norway". *Norg. Geol. Undersøkelse.*, **208**, 351–357.

– (1960e). Features of the geomorphology. "Geology of Norway". *Norg. Geol. Undersøkelse.*, **208**, 507–531.

– (1961a). The "Sparagmite Formation" (Kjerulf) and "Eocambrian" (Brögger) of the Scandinavian peninsula. *Congr. Geol. Intern. Compt. Rend. 20th, Mexico.* El Sistema Cambrico, Symposium, *Moscow, Part III*, 9–43.

– (1961b). Grensen fyllit—Valdressparagmit i strøket sydøst for Grönnsennknipa. *Norg. Geol. Undersøkelse.*, **213**, 96–99.

– (1964). The Caledonides of Norway. *Tectonics of Europe. Intern. Geol. Congr. Moscow*, 74–83.

– *See* BAILEY and HOLTEDAHL (1938).

HOLTEDAHL, O. and J. A. DONS (1957). Geological guide to Oslo and district. *Skr. norske Vidensk Akad.*, **1957**, No. 3, 86 pp. (out of print, a new edition has appeared, Universitetsforlaget, Oslo).

HOLTEDAHL, O., S. FÖYN and P. H. REITAN (1960). Aspects of the geology of northern Norway. Excursion guide. *Norg. Geol. Undersøkelse.*, **212a**, 66 pp.

HOOPER, P. R. *See* BALL and others (1963).

ISACHSEN, F. *See* BROCH and others (1940).

ISBERG, O. *See* BROCH and others (1940).

JARVIK, E. (1948). On the Middle Devonian crossopterygians from the Hornelen field. *Årbok. Univ. Bergen, Mat.-Nat. Ser.*, **1948**, No. 8, 48 pp.

JONES, M. A. *See* RUTLAND and others (1960).

JÖRGENSEN, P. (1964). Mineralogical composition of two Silurian bentonite beds from Sundvollen. *Norsk Geol. Tidsskr.*, **44**, 227–234.

KALDHOL, H. (1909). Fjeldbygningen i den nordöstlige del av Ryfylke. *Norg. Geol. Undersøkelse.*, **49**, 59 pp.

KALSBEEK, F. (1964). Zircons from some metamorphic rocks in the Stavanger area. *Norsk Geol. Tidsskr.*, **44**, 11–18.

KAUTSKY, G. (1946). Neue Gesichtspunkte zu einigen nordskandinavischen Gebirgsproblemen. *Geol. Fören. Stockh. Förh.*, **68**, 589–602.

– (1953). Der geologische Bau des Sulitelma–Salojauregebietes. *Svenges Geol. Undersokn. Arsbok, Ser. C: Avhandl. och Uppsat.*, **528**, 228 pp.

KIAER, J. (1902). Etage 5 i Asker. *Norg. Geol. Undersøkelse.*, **34**, No. 1, 111 pp.

– (1908). Da Obersilur im Kristianiagebiete. *Skr. VidenskSelsk., Christ.*, **1906**, II, 595 pp.

– (1924). The Downtonian fauna of Norway. I. Anaspida. *Skr. VidenskSelsk., Christ.*, **1924**, No. 6, 139 pp.

– (1930). Den fossilførende ordovicisk-siluriske lagrekke på Stord. *Arbok Bergens Museum, Mat.-Nat. Ser.*, **1929**, No. 11, 92 pp.

– (1932). The Hovin group in the Trondheim area. With paleontological contributions by O. A. Höeg, A. Hadding, F. R. C. Reed, T. Strand, Leif Störmer and the author. *Skr. norske Vidensk Akad.*, **1932**, No. 4, 176 pp.

KILDAL, E. S. (1966). Note on the geology of the archipelago NW of Bergen. *Norg. Geol. Undersøkelse.*, **242**, 130–134.

KISCH, H. J. (1962). Petrographical and geological investigations in the south-western Tydal region. Thesis, Amsterdam, 136 pp.

KOLDERUP, C. F. (1899). Lofotens og Vesteraalens gabbrobergarter. *Arbok Bergens Museum, Mat.-Nat. Ser.*, **1898**, VII, 54 pp.

Kolderup, C. F. (1928). Das Old Red im westlichen Norwegen. *Fennia*, **50**, No. 23, 17 pp.

Kolderup, C. F. and N.-H. Kolderup (1940). Geology of the Bergen arc system. *Bergens Museum Skr.* No. **20**, 137 pp.

Kolderup, N.-H. (1921). Der Mangeritsyenit und umgebende Gesteine in Söndfjord. *Arbok Bergens Museum, Mat.-Nat. Ser.*, **1920–1921**, No. 5, 71 pp.

– (1924). Die Injektionsmetamorphose des Dolomitmarmors im Mögstergebiet. *Arbok Bergens Museum, Mat.-Nat. Ser.*, **1923–1924**, No. 1, 64 pp.

– (1928). Fjeldbygningen i ströket mellem Nordfjord og Sognefjord. *Arbok Bergens Museum, Mat.-Nat. Ser.*, **1928**, No. 1, 228 pp.

– (1929). En vestnorsk kisförende kvartskeratofyr. *Arbok Bergens Museum, Mat.-Nat. Ser.*, **1929**, No. 4, 22 pp.

– (1931). Oversikt over den kaledoniske fjellkjede på Vestlandet. *Arbok Bergens Museum, Mat.-Nat. Ser.*, **1931**, No. 1, 43 pp.

– (1941). Trekk av Sunnhordlands geologi. *Norsk Geol. Tidsskr.*, **21**, 183–185.

– (1950). Gloppen-antiklinalen. *Arbok Univ. Bergen, Mat.-Nat. Ser.*, **1950**, No. 2, 9 pp.

– (1955). Sammenhengen mellom skifre, gneiser og amfiboliter i "Nordvest-tavlen". *Geol. Fören. Stockh. Förh.*, **77**, 257–264.

– (1960a). Origin of Norwegian eclogites in gneisses. *Norsk Geol. Tidsskr.*, **40**, 73–76.

– (1960b). The relationship between Cambro-Silurian schists and the Gneiss complex in the Deep-Caledonides of Sogn and Fjordane. Excursion guide. *Norg. Geol. Undersøkelse.*, **212c**, 31 pp.

Kolderup, N.-H. and I. Th. Rosenqvist (1951). Giant garnet crystals from Gjölanger. *Arbok Univ. Bergen, Mat.-Nat. Ser.*, **1950**, No. 6, 9 pp.

Kollung, S. (1963). Metamorfe og eruptive bergarter på Hitra. *Norg. Geol. Undersøkelse.*, **223**, 161–222.

Krauskopf, K. B. (1954). Igneous and metamorphic rocks of the Öksfjord peninsula. *Norg. Geol. Undersøkelse.*, **188**, 29–50.

Kulling, O. (1961). On the age and tectonic position of the Valdres sparagmite. *Geol. Fören. Stockh. Förh.*, **83**, 210–214.

– (1962). Berggrunnen inom Lapplandsfjällen. *Sveriges Geol. Undersokn. Ser. Ba:* **16**, 225–270.

– (1964). Översikt över norra Norrbottenfjällens kaledonberggrund. *Sveriges Geol. Undersokn, Ser. Ba*: **19**, 166 pp.

Kulling, O. and P. Geijer (1960). The Caledonian mountain chain in the Torneträsk–Ofoten area. The Kiruna iron ore field, Swedish Lappland. *Intern. Geol. Congr. Excursion Guides Nos. A25 and C20, Geological Survey of Sweden*, 76 pp.

Kvale, A. (1937). Et kaledonisk intrusiv- og effusivfelt på Stord. *Arbok Bergens Museum, Mat.-Nat. Ser.*, **1937**, No. 1, 138 pp.

– (1946a). Petrologic and structural studies in the Bergsdalen quadrangle. Part I. Petrography. *Arbok Bergens Museum, Mat.-Nat. Ser.*, **1945**, No. 1, 201 pp.

– (1946b). Noen bemerkninger om Telemarkformasjonen på Vestlandet. *Arbok Bergens Museum, Mat.-Nat. Ser.*, **1945**, No. 5, 18pp.

– (1948). Petrologic and structural studies in the Bergsdalen quadrangle. Part II. Structural geology. *Arbok Bergens Museum, Mat.-Nat. Ser.*, **1946–1947**, No. 1, 255 pp.

– (1953). Linear structures and their relation to movement in the Caledonides of Scandinavia and Scotland. *Quart. J. Geol. Soc.*, **109**, 51–73.

– (1955). Dekkestrukturer mellom Bergensbuene og Sogn. *Norsk Geol. Tidsskr.*, **35**, p. 221.

– (1960a). The nappe area of the Caldeonides in western Norway. Excursion guide. *Norg. Geol. Undersøkelse.*, **212e**, 43 pp.

– (1960b). Earthquakes. "Geology of Norway". *Norg. Geol. Undersøkelse.*, **208**, 490–506.

Landmark, K. (1949). Geologiske undersökelser Luster–Böverdalen. *Arbok Univ. Bergen, Mat.-Nat. Ser.*, **1948**, No. 1, 57 pp.

– (1951a). Tverfoldning i den kaledonske fjellkjede. *Norsk Geol. Tidsskr.*, **29**, 241–247.

– (1951b). Dykes of oligoclasite in amphibolite near Tromsö. *Acta Borealia, A. Sci.*, **1**, 23 pp.

Lappin, M. A. (1960). On the occurrence of kyanite in the eclogites of the Selje and Åheim districts. Nordfjord. *Norsk Geol. Tidsskr.*, **40**, 289–296.

– (1966). The field relationships of basic and ultrabasic masses in the Basal gneiss complex of Stadtlandet and Almklovdalen, southwest Norway. *Norsk Geol. Tidsskr.*, **46**, 439–496.

Lapworth, Ch. (1905). Notes on the graptolites from Bratland, Gausdal. *Norg. Geol. Undersøkelse.*, **39**, appendix, 15 pp.

Lewis, D. *See* Ball and others (1963).

Lindström, M. (1958). Towards a further hypothesis of the Scandinavian Caledonides. *Geol. Fören. Stockh. Förh.*, **80**, 363–380.

Löberg, B. E. (1965). Undersökelser av Sparagmitt-gruppen—Gausdal. *Geologi* (*Helsinki*), **9–10**, 136.

LOESCHKE, J. (1967a). Zur Stratigraphie und Petrographie des Valdres-Sparagmites und der Mellsenn-Gruppe bei Mellane, Valdres. *Norg. Geol. Undersøkelse.*, **243** (in press).

– (1967b). Zur Petrographie des Valdres-Sparagmites zwischen Bitihorn und Langsuen, Valdres. *Norg. Geol. Undersøkelse.*, **243** (in press).

LUND, P. R. (1965). En geologisk undersökelse på den söröstre del av kartbladet Salangen. *Norg. Geol. Undersøkelse.*, **234**, 74–102.

MAJOR, H. (1946). Noen profiler fra eldste silur i Oslofeltets nordlige deler. *Norsk Geol. Tidsskr.*, **26**, 59–142.

MARKLUND, N. (1952). A Cambro-Silurian type-section in the Sarvas region. *Geol. Fören. Stockh. Förh.*, **74**, 353–384.

MARLOW, W. (1935). Folldal. *Norg. Geol. Undersøkelse.*, **145**, 5–23.

McDOUGALL, I. and D. H. GREEN (1964). Excess radiogenic argon in pyroxenes and isotopic ages on minerals from Norwegian eclogites. *Norsk Geol. Tidsskr.*, **44**, 183–196.

MERCY, E. L. P. and M. J. O'HARA (1965). Chemistry of some garnet-bearing rocks from south Norwegian peridotites. *Norsk Geol. Tidsskr.*, **45**, 323–332.

– *See* O'HARA and MERCY (1963).

MILNES, A. G. and A. RITCHIE (1962). A contribution to the geology of the Kvaenangen window. *Norsk Geol. Tidsskr.*, **42**, 77–102.

MOREAU, J. *See* GEUKENS and MOREAU (1960).

MÜNSTER, TH. (1900). Kartbladet Lillehammer. *Norg. Geol. Undersøkelse.*, **30**, 49 pp.

MURET, G. (1960). Partie S.E. de la culmination de Romsdal. *Intern. Geol. Congr. 21st Copenhagen, Rep. Session, Norden*, **XIX**, 28–32.

NATERSTAD, J. (1965). Cambro–Silurian stratigraphy on the northwest side of the Hardangerfjord. *Norsk Geol. Tidsskr.*, **45**, p. 365.

NATHORST, A. G. *See* GOLDSCHMIDT (1913).

NICHOLSON, R. (1966). On the relations between volcanic and other rocks in the fossiliferous east Lommivann area, Norwegian Sulitjelma. *Norg. Geol. Undersøkelse.*, **242**, 143–156.

NICHOLSON, R. and B. J. WALTON (1963). The structural geology of the Naevervatn–Storglomvatn area. *Norsk Geol. Tidsskr.*, **43**, 1–58.

– *See* ACKERMANN and others (1960); RUTLAND and NICHOLSON (1965).

NICKELSEN, R. P. (1967). The structure of Mellene and Heggeberg. *Norg. Geol. Undersøkelse.*, **243** (in press).

NILSSON, G. (1964). Berggrunnen inom Blåsjöområdet. *Sveriges Geol. Undersokn. Arsbok, Ser. C: Avhandl. och Uppsat.*, **595**, 70 pp.

NISSEN, A. (1965). En petrografisk-mineralogisk undersökelse i området syd for Majavann. *Norg. Geol. Undersøkelse.*, **234**, 103–159.

NYSTUEN, J. P. (1965). Nye observasjoner til belysning av Trysilhevningen. *Geologi (Helsinki)*, **9–10**, 139.

– *See* HOLMSEN and SKJESETH (1966).

OFTEDAHL, CHR. (1945). Om tillitene i det central-norske sparagmitområde. *Norsk Geol. Tidsskr.*, **25**, 285–294.

– (1948). Deformation of quartz conglomerates in central Norway. *J. Geol.*, **56**, 476–487.

– (1950a). Petrology and geology of the Rondane area. *Norsk Geol. Tidsskr.*, **28**, 199–225.

– (1950b). Övre Rendal. *Norg. Geol. Undersøkelse.*, **177**, 7–27.

– (1954). Dekketektonikken i den nordlige del av det östlandske sparagmitområde. *Norg. Geol. Undersøkelse.*, **188**, 5–20.

– (1956). Om Grongkulminasjonen og Grongfeltets skyvedekker. *Norg. Geol. Undersøkelse.*, **195**, 57–64.

– (1958a). Oversikt over Grongfelets skjerp og malmforekomster. *Norg. Geol. Undersøkelse.*, **202**, 67 pp.

– (1958b). A theory of exhalative-sedimentary ores. *Geol. Fören. Stockh. Förh.*, **80**, 1–19.

– (1959). Om vulkanittene i den kaledonske fjellkjede i Norge. *Norsk Geol. Tidsskr.*, **39**, 263–265.

– (1961). On the genesis of the gabbroid rock bodies in the Norwegian Caledonides. *Bull. Geol. Inst. Univ. Uppsala*, **40**, 87–94.

– (1964). The nature of the basement contact. *Norg. Geol. Undersøkelse.*, **227**, 5–12.

– (1965). Kaledonske fjellkjede-faser i det eugeosynklinale Trondheimsfelt. *Geologi (Helsinki)*, **9–10**, 141.

– (1966). Note on the main Caledonian thrusting in northern Scandinavia. *Norsk Geol. Tidsskr.*, **46**, 237–244.

– *See* HOLMSEN and OFTEDAHL (1956).

O'HARA, M. J. *See* MERCY and O'HARA (1965).

O'HARA, M. J. and E. L. P. MERCY (1963). Petrology and petrogenesis of some garnetiferous peridotites. *Trans. Roy. Soc. Edinburgh*, **65**, No. 12, 251–314.

OOSTEROM, M. G. (1956). Some notes on the Lille Kufjord layered gabbro. *Norg. Geol. Undersøkelse.*, **195**, 73–88.

– (1963). The ultramafites and layered gabbro sequences in the granulite facies rocks on Stjernöy. *Leidse Geol. Mededel.*, **28**, 179–293.

OXAAL, J. (1919). Dunderlandsdalen. *Norg. Geol. Undersøkelse.*, **86**, 82 pp.

PADGET, P. (1955). The geology of the Caledonides of the Birtavarre region. *Norg. Geol. Undersøkelse.*. **192**, 107 pp.

PAGE, N. J. (1964). The sulfide deposits of Nordre Gjetryggen gruve, Folldal. *Norg. Geol. Undersøkelse.*, **228**, 217–269.

PEACEY, J. S. (1963). Deformation in the Gangåsvann area. *Norg. Geol. Undersøkelse.*, **223**, 275–293.

– (1964). Reconnaissance of the Tömmerås anticline. *Norg. Geol. Undersøkelse.*, **227**, 13–84.

POULSEN, O. A. (1960). Exploitation of mineral raw materials in Norway. "Geology of Norway". *Norg. Geol. Undersøkelse.*, **208**, 532–540.

– (1964). Norges gruver og malmforekomster. II. Nord-Norge. *Norg. Geol. Undersøkelse.*, **204**, 101 pp.

PROST, A. (1964). Observations sur la fenêtre antécambrienne de la Snödöla et sur la région septentrionale du "Fjell" de Ringebu. *Bull. Soc. Geol. France* [7], **3**, 351–357. Paris.

RAMBERG, H. (1944). En undersökelse av Vestrandens regionalmetamorfe bergarter. *Norsk Geol. Tidsskr.*, **23**, 1–174.

– (1966). The Scandinavian Caledonides as studied by centrifuged dynamic models. *Bull. Geol. Inst. Univ. Uppsala*, **43**, 1–73.

RAMBERG, I. (1965b). Regionalgeologiske undersökelser i Rösvatnområdet. *Geologi (Helsinki)*, **9–10**, p. 141.

– (1967). Kongsfjell-områdets geologi, Helgeland, Nord-Norge. *Norg. Geol. Undersøkelse.*, **240**, 152 pp.

RAMSAY, D. M. and B. A. STURT (1963). A study of fold styles, their associations and symmetry relationships from Söröy. *Norsk Geol. Tidsskr.*, **43**, 411–430.

– *See* STURT and RAMSAY (1965).

RANDALL, B. A. O. (1960). Sagvandites of Lyngen, Troms. *Intern. Geol. Congr. 21st Copenhagen, Rep. Session, Norden*, **XIII**, 443–451.

READING, H. G. (1965). Eocambrian and Lower Paleozoic geology of the Digermul peninsula. *Norg. Geol. Undersøkelse.*, **234**, 167–191.

READING, H. G. and R. G. WALKER (1966). Sedimentation of Eocambrian tillites and associated sediments in Finnmark, Northern Norway. *Palaeogeography, Palaeoclimatology, Palaeoecology*, Vol. 2, Elsevier, Amsterdam, pp. 177–212.

REED, F. R. C. *See* KIAER (1932).

REITAN, P. H. (1963). The geology of the Komagfjord tectonic window. *Norg. Geol. Undersøkelse.*, **221**, 71 pp.

– (1965). Correlations of the Doggelv and Lomvann formations, Finnmark. *Norg. Geol. Undersøkelse.*, **234**, 192–195.

– *See* HOLTEDAHL and others (1960).

REITAN, P. H. and J. J. C. GEUL (1959). On the formation of carbonate-bearing basic rocks at Kviteberg, Lyngen. *Norg. Geol. Undersøkelse.*, **205**, 111–127.

REKSTAD, J. (1902). Fra Hardangervidden. *Norg. Geol. Undersøkelse.*, **34**, No. 2, 31–50.

– (1903). Fra höifjeldsströget mellem Haukeli og Hemsedalsfjeldene. *Norg. Geol. Undersøkelse.*, **36**, No. 4, 54 pp.

– (1905). Fra Indre Sogn. *Norg. Geol. Undersøkelse.*, **43**, IV, 43 pp.

– (1907). Folgefonnshalvöens geologi. *Norg. Geol. Undersøkelse.*, **45**, I, 47 pp.

– (1909). Geologiske iagttagelser fra Söndhordland. *Norg. Geol. Undersøkelse.*, **49**, IV, 26 pp.

– (1917a). Vega. *Norg. Geol. Undersøkelse.*, **80**, 83 pp.

– (1917b). Fjeldströket Fauske–Junkerdalen. *Norg. Geol. Undersøkelse.*, **81**, IV, 70 pp.

– (1919). Geologiske iakktagelser på strekningen Folla–Tysfjord. *Norg. Geol. Undersøkelse.*, **83**, I, 60 pp.

– (1925a). Hatfjelddalen. *Norg. Geol. Undersøkelse.*, **124**, 36 pp.

– (1925b). Traena. *Norg. Geol. Undersøkelse.*, **125**, 36 pp.

– (1929). Salta. *Norg. Geol. Undersøkelse.*, **134**, 73 pp.

REUSCH, H. (1884). Geologiske notiser fra Kristianiaegnen. *Nyt Mag. Naturv.*, **28**, 105–152.

– (1888). Bömmelöen og Karmöen med omgivelser. Kristiania (Oslo), 422 pp.

– (1905). Voss. *Norg. Geol. Undersøkelse.*, **40**, 76 pp.

– (1908). Tekst til geologisk kart over fjeldströkene mellem Jostedalsbreen og Ringerike. *Norg. Geol. Undersøkelse.*, **47**, 40 pp.

– (1913). Tekst til geologisk oversiktkart over Söndhordland og Ryfylke. *Norg. Geol. Undersøkelse.*, **64**, 83 pp.

– (1914). Nogen bidrag til Hitterens og Smölens geologi. *Norg. Geol. Undersøkelse.*, **69**, IV, 50 pp.

RICHTER, M. (1943). Das Alter des westnorwegischen Grundgebirges. *Geol. Rundschau*, **34**, 154–161.

– (1947). Die Devonmolasse am Ausgang des Trondheimfjords. *Z. Deut. Geol. Ges.*, **99**, 1–7.

– (1958). Die Halbinsel Örlandet am Ausgang des Trondheimfjordes. *Geologie (Berlin)*, **7**, 757–778.

RITCHIE, A. *See* MILNES and RITCHIE (1962).

ROSENDAHL, H. (1932). Bidrag til Varangernesets geologi. *Norsk Geol. Tidsskr.*, **12**, 487–506.

– (1933). Littérature géologique concernant la Norvège avec Svalbarde etc.—1920–1932. Avec des suppléments 1916–1919. *Norsk Geol. Tidsskr.*, **13**, 129–301.

ROSENQVIST, I. TH. (1941). The Lönset anticline in the Oppdal area. *Norsk Geol. Tidsskr.*, **21**, 25–48.

– (1942). Om öiegneisdannelse i fjellkjeder. *Norsk Geol. Tidsskr.*, **21**, 165–180.

– (1943). Metamorphism and metasomatism in the Oppdal area. *Norsk Geol. Tidsskr.*, **22**, 106–202.

ROSENQVIST, I. TH. (1956). Granulitt-mangeritt-eklogitt. *Norsk Geol. Tidsskr.*, **36**, 41–47.

– *See* KOLDERUP and ROSENQVIST (1951).

ROWLING, J. W. *See* CHADWICK and others (1963).

RUTLAND, R. W. R. (1959). Structural geology of the Sokumvatn area. *Norsk Geol. Tidsskr.*, **39**, 287–337.

RUTLAND, R. W. R., M. HOLMES and M. A. JONES (1960). Granites of the Glomfjord area. *Intern. Geol. Congr. 21st Copenhagen, Rep. Session, Norden*, **XIX**, 43–53.

RUTLAND, R. W. R. and R. NICHOLSON (1965). Tectonics of the Caledonides of part of Nordland, Norway. *Quart. J. Geol. Soc.*, **121**, 73–109.

SCHAAR, G. (1962). Petrological and geological investigations in the Sylene–Skardörfjell region. Thesis, Amsterdam, 74 pp.

SCHETELIG, J. (1913). Hitteren og Smölen. *Norsk Geol. Tidsskr.*, **2**, No. 10, 27 pp.

SCHIÖTZ, O. E. (1902). Den sydöstlige del av Sparagmit–Kvartsfjeldet i Norge. *Norg. Geol. Undersøkelse.*, **35**, 131 pp.

SJÖGREN, HJ. (1900). Öfversigt af Sulitelma-områdets geologi. *Geol. Fören. Stockh. Förh.*, **22**, 437–462.

SKJERLIE, F. J. (1957). Geological investigations between Fjaerlandsfjorden and Sogndalsdalen. *Arbok Univ. Bergen, Mat.-Nat. Ser.*, **1957**, No. 10, 67 pp.

SKJERLIE, F. J. and T. H. TAN (1961). The geology of the Caledonides in the Reisa valley area. *Norg. Geol. Undersøkelse.*, **213**, 175–196.

SKJESETH, S. (1958). Uran i kambrisk alunskifer i Oslofeltet og tilgrensende områder. *Norg. Geol. Undersøkelse.*, **203**, 100–111.

– (1962). "Trysilhevningen". Kambro-ordovicisk stratigrafi i Femundtrakten. *Norg. Geol. Undersøkelse.*, **215**, 102–113.

– (1963). Contributions to the geology of the Mjösa district and the classical Sparagmite area in southern Norway. *Norg. Geol. Undersøkelse.*, **220**, 126 pp.

– *See* HOLMSEN and SKJESETH (1966).

SKJESETH, S. and H. SÖRENSEN (1953). An example of granitization in the central zone of the Caledonides in northern Norway. *Norg. Geol. Undersøkelse.*, **184**, 154–180.

SKJESETH, S. and F. HAGEMANN (1960). Aspects of the geology of the central parts of the Oslo region. Excursion guide. *Norg. Geol. Undersøkelse.*, **212r**, 12 pp.

SKORDAL, A. J. (1948). Vulkaniter og sedimenter på den söröstre del av Stord. *Arbok Univ. Bergen, Mat.-Nat. Ser.*, **1948**, No. 2, 58 pp.

SMITHSON, S. B. (1964). Forelöpige resultater av tyngdemålinger over Jotun-dekkene. *Tidsskr. Kjemi, Bergvesen Met.*, **24**, No. 5, 97–101.

SÖRBYE, R. C. (1949). Geological studies in the northwestern part of the Haugesund peninsula. *Arbok Univ. Bergen, Mat.-Nat. Ser.*, **1948**, No. 6, 79 pp.

– (1951). Bergen–Jotunbergarter i fjellet Napen, Suldalsheim. *Arbok Univ. Bergen, Mat.-Nat. Ser.*, **1950**, No. 4, 15 pp.

– (1954). Kaledonidene i nordöstre Ryfylke og på Haugesundshalvöya. *Norsk Geol. Tidsskr.*, **33**, 235–238.

– (1965). Noen "höifjellsgneis"—problemer vest for Saudafjorden. *Norsk Geol. Tidsskr.*, **45**, 162.

SÖRENSEN, H. (1955). A petrographical and structural study of the rocks at Engenbreen, Holandsfjord. *Norsk Geol. Tidsskr.*, **191**, 71–102.

– *See* SKJESETH and SÖRENSEN (1953).

SPJELDNAES, N. (1957). The Silurian–Ordovician border in the Oslo district. *Norsk Geol. Tidsskr.*, **37**, 355–371.

– (1959). Traces of Eocambrian orogeny in southern Norway. *Norsk Geol. Tidsskr.*, **39**, 83–86.

– (1963). A new fossil (Papillomembrana) from the Upper Precambrian of Norway. *Nature*, **200**, 63–64.

– (1964). The Eocambrian glaciation in Norway. *Geol. Rundschau*, **54**, 24–45.

– (1966). Silurian tidal sediments from the base of the Ringerike formation, Oslo region, Norway. *Norsk Geol. Tidsskr.*, **46**, 497–510.

– *See* HENNINGSMOEN and SPJELDNAES (1960).

SPJELDNAES, N. and F. HAGEMANN (1955). Ordoviciske og siluriske bentonitlag i Oslofeltet. *Norsk Geol. Tidsskr.*, **35**, p. 207.

STEENKEN, W. F. (1957). Geology and Petrology of the region south of Russånes, Thesis, 's-Gravenhage, 120 pp.

STÖRMER, L. (1925). On a Lower Cambrian fauna at Ustaoset. *Fennia*, **45**, No. 1, 12–22.

– (1934). Spor av kaledonisk overskyvning i Nordmarka. *Norsk Geol. Tidsskr.*, **13**, 116–128.

– (1935). Dictyocaris, Salter, a large crustacean from the Upper Silurian and Downtonian. *Norsk Geol. Tidsskr.*, **15**, 265–298.

– (1938). Tektoniske iakttagelser på Bygdöy. *Norsk Geol. Tidsskr.*, **18**, 66–69.

– (1941). Dictyonema shales outside the Oslo region. *Norsk Geol. Tidsskr.*, **20**, 161–167.

– (1944). Kaledonisk dekke-tektonik på Hadeland? *Norsk Geol. Tidsskr.*, **22**, 74–91.

STÖRMER, L. (1953). The Middle Ordovician of the Oslo region. Introduction to stratigraphy. *Norsk Geol. Tidsskr.*, **31**, 37–141.

– (1964). Noen tektoniske iakttagelser på Ustaoset. *Norsk Geol. Tidsskr.*, **44**, 171–175.

– (1965). Kambro-silursedimentene i Syd-Norge og kaledoniske jordskorpebevegelser. *Norsk Geol. Tidsskr.*, **45**, 161–162.

– *See* KIAER (1932).

STRAND, T. (1932). A Lower Ordovician fauna from the Smöla island. *Norsk Geol. Tidsskr.*, **11**, 356–365.

– (1938). Nordre Etnedal. *Norg. Geol. Undersøkelse.*, **152**, 71 pp.

– (1945). Structural petrology of the Bygdin conglomerate. *Norsk Geol. Tidsskr.*, **24**, 14–31.

– (1948). On the Middle Cambrian of Hadeland. *Norsk Geol. Tidsskr.*, **27**, 89–96.

– (1949). On the gneisses from a part of the northwestern Gneiss area in southern Norway. *Norg. Geol. Undersøkelse.*, **173**, 45 pp.

– (1951a). The Sel and Vågå map areas. *Norg. Geol. Undersøkelse.*, **178**, 116 pp.

– (1951b). Slidre. *Norg. Geol. Undersøkelse.*, **180**, 54 pp.

– (1952). Raipas og kaledon i ströket omkring Repparfjord. *Norg. Geol. Undersøkelse.*, **183**, 22–31.

– (1953). The relation between the basal gneiss and the overlying meta-sediments in the Surnadal district. *Norg. Geol. Undersøkelse.*, **184**, 100–123.

– (1954). Aurdal. *Norg. Geol. Undersøkelse.*, **185**, 71 pp.

– (1955). Sydöstligste Helgelands geologi. *Norg. Geol. Undersøkelse.*, **191**, 56–70.

– (1956). Gabbrokonglomeratet i Sjodalen. *Norg. Geol. Undersøkelse.*, **195**, 43–56.

– (1958). Greenschists from the south-eastern part of Helgeland. *Norg. Geol. Undersøkelse.*, **203**, 112–129.

– (1959). Valdres-sparagmittens stratigrafiske stilling. *Norg. Geol. Undersøkelse.*, **205**, 184–198.

– (1960a). Cambro-Silurian deposits outside the Oslo region. "Geology of Norway". *Norg. Geol. Undersøkelse.*, **208**, 151–165.

– (1960b). The pre-Devonian rocks and structures in the region of Caledonian deformation. "Geology of Norway". *Norg. Geol. Undersøkelse.*, **208**, 170–284.

– (1961). The Scandinavian Caledonides. A review. *Am. J. Sci.*, **259**, 161–172.

– (1962). On the age and tectonic position of the Valdres sparagmite. *Geol. Fören. Stockh. Förh.*, **84**, 230–231.

– (1963). Noen stratigrafiske aldersspørsmål i Grong-feltet og den sydöstligste del av Helgeland. *Norg. Geol. Undersøkelse.*, **223**, 294–297.

– (1964a). Otta-dekket og Valdres-gruppen i strøkene langs Böverdalen og Leirdalen. *Norg. Geol. Undersøkelse.*, **228**, 280–288.

– (1964b). Geology and structure of the Prestberget area. *Norg. Geol. Undersøkelse.*, **228**, 289–310.

– *See* KIAER (1932); BROCH and others (1940); FOSLIE and STRAND (1956).

STRAND, T. and P. HOLMSEN (1960). Stratigraphy, petrology and Caledonian nappe tectonics of central southern Norway. Caledonized basal gneisses in a northwestern area (Oppdal–Sunndal). Excursion guide. *Norg. Geol. Undersøkelse.*, **2121**, 31 pp.

STUMPFL, E. F. and B. A. STURT (1965). A preliminary account of the geochemistry and ore mineral parageneses of some Caledonian basic igneous rocks from Söröy. *Norg. Geol. Undersøkelse.*, **234**, 196–230.

STURT, B. A. *See* RAMSAY and STURT (1963); STUMPFL and STURT (1965).

STURT, B. A. and D. M. RAMSAY (1965). The alkaline complex of the Breivikbotn area, Söröy, northern Norway. *Norg. Geol. Undersøkelse.*, **231**, 1–142.

TAN, T. H. *See* SKJERLIE and TAN (1961).

TIMOFEV, B. V. (1963). On some organic remains from the Eocambrian of Norway. *Norsk Geol. Tidsskr.*, **43**, 473–476.

TÖRNEBOHM, A. E. (1896). Grunddragen af det centrala Skandinaviens bergbygnad. *Kgl. Svenska Vetenskapsacad. Handl.*, **28**, No. 5, 212 pp.

TORSKE, T. (1965). Geology of the Mostadmarka and Selbustrand area, Tröndelag. *Norg. Geol. Undersøkelse.*, **232**, 83 pp.

VOGT, J. H. L. (1890). Norske ertsforekomster. VII. Foldalens kisfelt. *Arch. Math. Naturv.*, **13**, 202–270.

– (1897). Norsk marmor. *Norg. Geol. Undersøkelse.*, **22**, 364 pp.

– (1907). Über die schräge Senkung and spätere schräge Hebung des Landes im nördlichen Norwegen. *Norsk Geol. Tidsskr.*, **1**, No. 6, 47 pp.

VOGT, TH. (1910). Om eruptivbergarterne paa Langöen i Vesteraalen. *Norg. Geol. Undersøkelse.*, **53**, VI, 39 pp.

– (1916). Petrographisch—chemische Studien an einigen Assimilations-Gesteinen der nordnorwegischen Gebirgskette. *Skr. VidenskSelsk., Christ.*, **1915**, No. 8, 34 pp.

– (1918). Geologiske studier langs den östlige del av fjeldkjeden i Tromsö amt. *Norsk Geol. Tidsskr.*, **4**, 260–266.

– (1922). Bidrag til fjeldkjedens stratigrafi og tektonik. *Geol. Fören. Stockh. Förh.*, **44**, 714–739.

VOGT, TH. (1924). Forholdet mellem Sparagmitsystemet og det marine underkambrium ved Mjøsen. *Norsk Geol. Tidsskr.*, **7**, 281–385.

– (1927). Sulitjelmafeltets geologi og petrografi. *Norg. Geol. Undersøkelse.*, **121**, 560 pp.

– (1928). Den norske fjellkjedes revolusjonshistorie. *Norsk Geol. Tidsskr.*, **10**, 97–115.

– (1929a). Et fund av nye fiskefossiler fra sandstenerne på Jelöen. *Norske Vidensk. Akad. Årbok*, **1928**, 21–22.

– (1929b). Undersökelser over den underdevonske konglomerat-sandstenserie i Ytre Tröndelagen. *Norg. Geol. Undersøkelse.*, **133**, 59–61.

– (1935). Origin of the injected pyrite deposits. *Kgl. Norske Vidensk. Selsk. Skr.*, **1935**, No. 20, 17 pp.

– (1936). Orogenesis in the region of Paleozoic folding of Scandinavia and Spitsbergen. *Intern. Geol. Congr. Rep. 16th Session*, 933–955. Washington.

– (1940). Geological notes on the Dictyonema locality and the upper Gauldal valley. *Norsk Geol. Tidsskr.*, **20**, 171–191.

– (1942). Trekk av Narvik–Ofoten-traktens geologi. *Norsk Geol. Tidsskr.*, **21**, 198–209.

– (1945). The geology of part of the Hölonda–Horg district. *Norsk Geol. Tidsskr.*, **25**, 449–528.

– (1952). Flowage structures and ore deposits in the Caledonides of Norway. *Intern. Geol. Congr., 18th, London, Rep. Session, Norden, Part XIII*, 240–244.

– (1953). Biri-dekket og Moelv-vinduet ved Mjösa. *Kgl. Norske Vidensk. Selsk. Forh.*, **25**, No. 27, 131–138.

VOKES, F. M. (1956). The copper deposits of the Birtavarre district. *Norg. Geol. Undersøkelse.*, **199**, 239 pp.

– (1960). Mines in south and central Norway. Guide to excursion No. C10. *Norg. Geol. Undersøkelse.*, **212m**, 73 pp.

– (1963). Geological studies in the Caledonian zinc-lead orebody at Bleikvassli. *Norg. Geol. Undersøkelse.*, **222**, 126 pp.

WALKER, R. G. *See* READING and WALKER (1966).

WALTON, B. J. *See* ACKERMANN and others (1960); NICHOLSON and WALTON (1963).

WEGMANN, C. E. (1925a). Sur le lambeau de recouvrement de la Gula. *Compt. Rend. Soc. Geol. France*, **1925**, 158–159.

– (1925b). Sur une déformation tardive de la chaîne calédonienne scandinave. *Compt. Rend. Soc. Geol. France*, **1925**, p. 213.

– (1926). Sur le rôle tectonique de quelques gneiss oeillés de la chaîne calédonienne scandinave. *Eclogae Geol. Hevl.*, **19**, 641–643.

– (1935). Zur Deutung der Migmatite. *Geol. Rundschau*, **26**, 305–350.

– (1959). La flexure axiale de la Driva et quelques problèmes structuraux des Calédonides scandinaves. *Norsk Geol. Tidsskr.*, **39**, 25–74.

WELLS, M. K. *See* HOLLINGWORTH and others (1960).

WERENSKIOLD, W. (1911). Söndre Fron. *Norg. Geol. Undersøkelse.*, **60**, 107 pp.

WOLFF, F. CHR. (1960). Forelöpige meddelelser fra kartbladet Verdal. *Norg. Geol. Undersøkelse.*, **211**, 212–230.

– (1964). Stratigraphical position of the Gudå conglomerate zone. *Norg. Geol. Undersøkelse.*, **227**, 85–91.

– (1965). Merakerfeltets inverterte kambro-silur. *Geologi (Helsinki)*, **9–10**, p. 144.

YOCHELSON, E. L. (1963). Gastropods from the Otta conglomerate. *Norsk Geol. Tidsskr.*, **43**, 75–82.

ZACHRISSON, E. (1964). The Remdalen syncline. *Sveriges Geol. Undersokn. Arsbok, Ser. C: Avhandl. och Uppsat.*, **596**, 53 pp.

Part II

THE SWEDISH CALEDONIDES

by

OSKAR KULLING

PREFACE

Professor L. U. de Sitter first suggested to me that I prepare a Geology of the Scandinavian Caledonides in cooperation with T. Strand of Oslo.

The Swedish Caledonides are known for the most part as a result of reconnaissance mapping. Some of the material for this book was gathered by the writer in investigations which began in 1924 in the Southern Lapland mountains, and are still in progress in the Northern Lapland mountains. Information obtained from these, the Norrbotten mountains, has been only partly developed (Kulling, 1964). The same can be said of the geological evidence gathered in the southern Swedish Caledonian mountains, mainly by Asklund, Thorslund, Strömberg and cooperators. This description of the Swedish Caledonides is therefore of a rather preliminary nature, but the writer naturally speaks with some authority after forty years' field work in the mountains.

Acknowledgements

I am indebted to the Director of the Geological Survey of Sweden, K. A. Lindbergson, for permission to write my part of the book whilst carrying out my survey duties; also to Professor L. U. de Sitter for suggesting the task to me. Professor I. Hessland has kindly assisted with some linguistic improvements.

O. Kulling

CHAPTER XVI

INTRODUCTORY NOTES ON GEOGRAPHY AND GEOLOGY

GEOGRAPHICAL RANGE AND DIVISION

The Caledonides of Sweden cover the northwestern part of the country only, while the greater part of Norway is composed of Caledonide rocks. The Swedish part of the old mountain range is about 900 km in length in the international border zone, and its broadest part is in Central Jämtland, where its width is 150 km. From north to south the geographical divisions used in this book are as follows (see Fig. 104):

THE LAPLAND MOUNTAINS — THE NORTHERN MOUNTAINS

1. The northern Norrbotten mountains, including the isolated Pältsa mountain area in the North
2. The southern Norrbotten mountains
3. The northern Västerbotten mountains
4. The southern Västerbotten mountains

THE SOUTHERN MOUNTAINS

5. The northern Jämtland mountains and the Tåsjö area of northwestern Ångermanland
6. The Storsjö–Föllinge lowland
7. The middle-west Jämtland mountains
8. The southern Jämtland mountains
9. The Härjedalen mountains
10. The northern Dalarne mountains

The Swedish Caledonide area consists of a large number of morphological features. The greater part of it is composed of high and low mountains. East of the mountainous part a premontan region of Caledonide bedrock extends in places between the mountains proper and the Precambrian terrain east of the Caledonian range; this terrain consists mainly of hilly areas and monadnock plains. Within the high and low mountains plateaux and depressions can be distinguished; the latter are particularly low-lying parts of the highland region. The premontan Caledonide areas consist of plains and dissected plateaux of moderate altitude.

The above-named geographical districts of the Caledonides differ from each other in several respects such as morphology, communication and accessibility, density of population, etc. Some comments will be made on the above divisions.

The Norrbotten mountains are for the most part high mountains. In their northern and middle parts this high mountainous character predominates right to the eastern mountain border zone. In the southeastern part, low mountains and dissected plateaux predominate. No roads cross the Norrbotten mountains; some roads, however, reach the eastern frontal zone. In the Torneträsk area the Kiruna–Narvik railway crosses the northern part of the mountains. The Norrbotten mountains are largely unpopulated.

The Västerbotten Caledonian area has a premontan region east of the mountain region. The high mountains, together with adjacent Norwegian high mountains, are arranged in an eastern and a western zone. The main part of the western zone is on the Norwegian side of the international border. Low mountains and plains or depressions extend between the high mountains. The premontan region is dominated by plateaux, but areas of more varying relief also occur. The plateaux are often dissected, and steps divide the plateau areas into several levels.

There is no railway through the Västerbotten mountains. In several parts, roads, some of which cross the international border into Norway, enter the Caledonian area and branch off. Forty years ago, when starting bedrock-mapping in the Västerbotten Caledonides the writer had to travel by foot and by boat. However, roads were under construction in several parts of the mountains. Small settlements and solitary villages (a lot of them now abandoned) occur in most low-lying districts of the mountainous area. The high mountain areas are barren country, apart from in the vicinity of the roads through the areas mentioned.

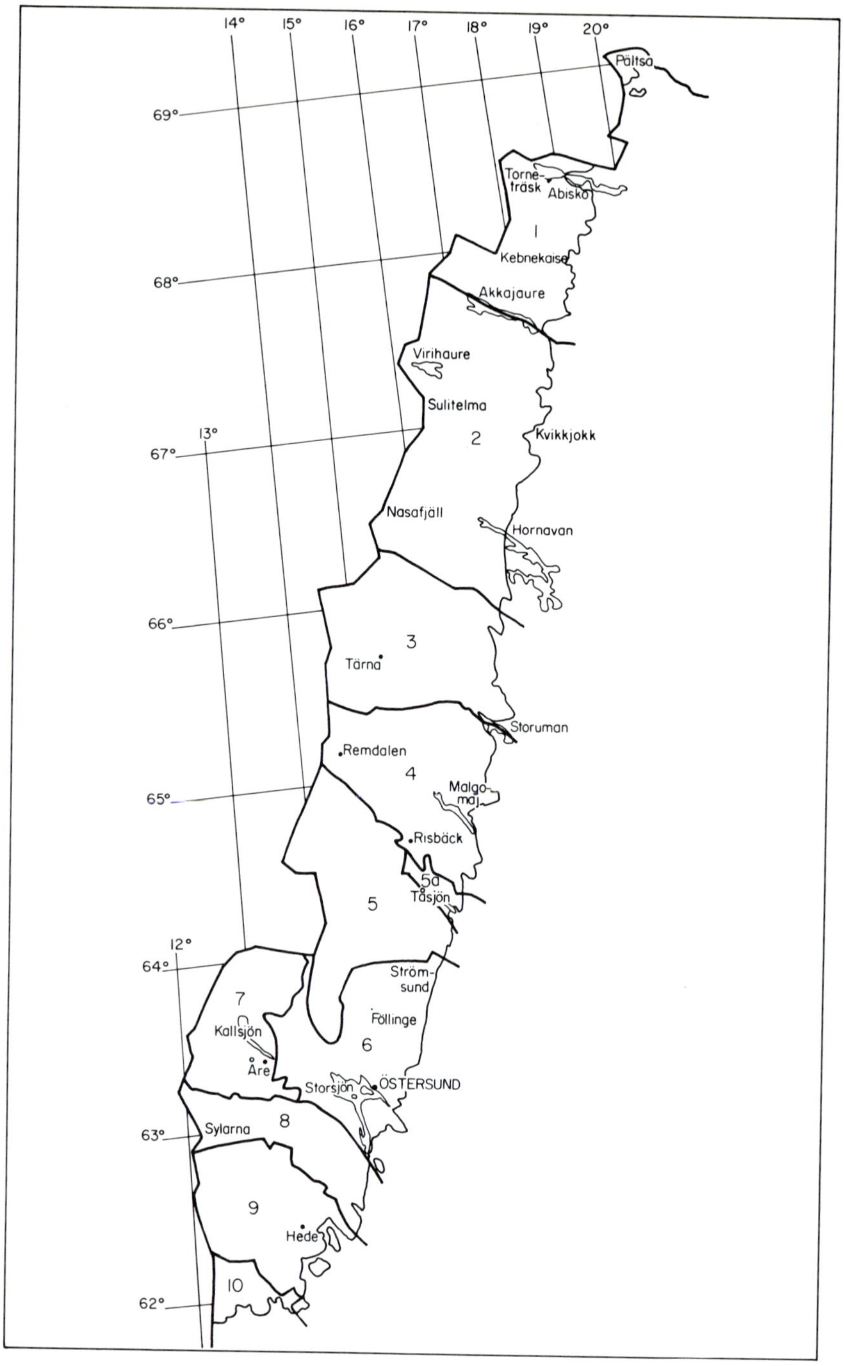

Fig. 104 *Geographical division of the Swedish Caledonides*

1 The northern Norrbotten mountains, including the isolated Pältsa area in the north
2 The southern Norrbotten mountains
3 The northern Västerbotten mountains
4 The southern Västerbotten mountains
5 The northern Jämtland mountains and the Tåsjö area of northwestern Ångermanland
6 The Storsjö–Föllinge lowland
7 The middle–west Jämtland mountains
8 The southern Jämtland mountains
9 The Härjedalen mountains
10 The northern Dalarne mountains

The mountains of Jämtland cover the northwestern, western and southern parts of the province; mostly low mountains, with some scattered high mountains, occur in the area. *The Storsjö–Föllinge lowland* in the eastern part of the Jämtland Caledonides is a flat depression in its southern part—the Storsjö area proper. Its northern part is characterized by plains with monadnocks. Farther to the north and northwest it is transformed into a dissected plateau province which has its northern continuation in the southeastern Västerbotten border area.

Within the Caledonides of Jämtland, the Storsjö lowland is an old, settled region and has for long been the main passage into Norway. The centre of this settled country, the town of Östersund, is the only one in the entire Swedish Caledonian mountain range. It is a centre for railways going in all directions. A system of roads connects Östersund with most other parts of Jämtland.

The *Härjedalen mountains* and the *northernmost Dalarne mountains* with Caledonide bedrock consist of a southeastern dissected plateau province and a western, low mountains province. Some roads enter the mountain area, which has scattered settlements in the valleys, and some isolated villages.

CALEDONIAN BEDROCK

The Swedish part of the Caledonides strikes SSW–NNE. On the eastern border of the range, unmetamorphosed or little-altered sedimentary rocks rest on old Precambrian rocks. Above and to the west of this narrow zone of autochthonous sedimentary rocks follow the Caledonian thrust rocks of various types. Low grade metamorphic sediments usually rest on the autochthonous beds. However, higher grade metamorphic sediments or cataclastic to mylonitic igneous rocks often lie directly on the autochthonous sediments or on the low grade metamorphic ones. The whole of the Swedish Caledonides, disregarding the small amounts of exposed autochthonous sediments in the eastern rim and along the boundaries of some windows of old Precambrian rocks, consist of thrust rocks.

The geographical division of the Caledonian mountain range runs transversely or in a NW–SE direction. The course of the Caledonian structures and geological–tectonic units is usually conformable with the main trend of the range, or NNE–SSW directed. Post-orogenic denudation has carried away a major portion of the bedrock of the old mountain chain. After the primary alpine topography was destroyed and the area of extent of the range was to a certain degree basally levelled, the late Tertiary (probably Pliocene) uplift of the old orogenic belt created the existing topographical chain, the *Scandes* (LJUNGNER, 1958). Thus, the Swedish mountains are not the original remains of an orogenic belt. Nor do they represent roots of the Caledonides, as their main parts are composed of flat-lying nappes with their roots to the west of the international border with Norway. There is a close connection between the existing topography of the Swedish Caledonides and their geology. The high mountains are in the main composed of high grade metamorphic rocks such as amphibolites, garnet–mica schists, etc., whereas in the low mountains a great many kinds of lower grade metamorphic rocks predominate. In the lowland areas, especially in the Storsjö–Föllinge area, the bedrock is composed of little-altered or unaltered sedimentary rocks. The low-lying plateaux in the eastern part of the Caledonides, for instance in Härjedalen and northern Jämtland, are for the main part composed of hard quartzitic rocks.

The eastern front of the Caledonides generally has the characteristic appearance of a steep fortress wall with projections towards the east, see Figs. 105–107. The mountain walls often run from the eastern frontier deep into the mountainous area and comprise the walls of the fjord-like valleys containing rivers and long, narrow lakes. The accentuated wall-like appearance of the Caledonides at the eastern border is caused by the geotectonics of the front range. The upper portion of the bedrock is composed of hard thrust rocks such as quartzites, arkoses, schists and igneous rocks, often more or less cataclastic, lying on soft rocks such as shales and sandstones. How far east the thrusts once advanced is unknown—one can only fix their actual erosional limit.

In east Jämtland, particularly in the eastern part of the Storsjö–Föllinge area, a mountain wall is not present, because the thrust rocks of that area are composed of little-altered sediments such as limestones, shales and sandstones, i.e. rocks of about the same habit as the underlying autochthonous rocks.

The narrow eastern zone of autochthonous sedimentary rocks is to a great extent fossiliferous, with Cambrian and Ordovician fossils. The formations within the thrusted sedimentary rocks in the Storsjö–Föllinge area contain not only Cambrian and Ordovician but also Silurian fossils. The autochthonous sedimentary rocks and the eastern part of the thrusted Cambro–Silurian is of so-called Eastern facies, i.e. of Baltic facies. Towards the west the facies conditions change, and in the western part of the Storsjö–Föllinge area the sedimentary rocks are of miogeosynclinal type, with thick greywacke beds, and exceptionally

Fig. 105 *The Mt. Tjeggelvaspakte portion of the Caledonian thrust front on the eastern border of the southern Norrbotten mountains.* Mt. Tjeggelvaspakte lies on the northern shore of Lake Tjeggelvas. The mountain wall is composed of mylonitic gneissose rocks of Middle thrust rock position. The scree- and vegetation-covered lower part of the mountain consists of autochthonous shales and sandstones of the Hyolithus group. Immediately to the east of the mountain the old Precambrian crystalline basement outcrops. There are no Lower thrust rocks in this part of the Caledonides. Looking northwest from a cottage on the shore. Photo O. KULLING (1928)

with some thin volcanic ash beds. From the Föllinge area a zone of sedimentary rocks of varying width, of western facies, runs northward into the southeastern part of the Västerbotten mountain border area. The last-named zone of fossiliferous sediments is often bounded to the east and west by a hard quartzite—the unfossiliferous *Vemdal quartzite.* According to the writer's interpretation of the tectonics of the area, the fossiliferous sediments form a synclinorium overfolded to the ESE, and the entire sedimentary sequence belongs to the Lower thrust rocks. Vemdal quartzite areas—

Fig. 106 *The upper part—the mountain wall and crest area of Mt. Puoivitsaråtsasvare on the northeastern shore of eastern Paijeb Kaitumjaure,* is an outlying part of the Middle thrust rocks, and is composed of mylonitic granite—syenite. The scree- and vegetation-covered lower part of the mountain (according to surveys in the surrounding areas) consists of autochthonous shales and sandstones of the Hyolithus group, lying on an old Precambrian crystalline basement. North of Mt. Puoivitsaråtsasvare, on Mt. Kårjåtjåkko, a sheet of Lower thrust rocks occurs between the two mentioned above rock complexes. This thrust sheet seems to have thinned out to the south, on Mt. Puoivitsaråtsasvare. Looking northwest from the east slope of Mt. Tjikkopakte. Paijeb Kaitumjaure lies in the southeastern part of the northern Norrbotten mountains, on the eastern edge of the Caledonides. Photo J. VON FEILITZEN (1949)

Fig. 107 *Looking eastward along the Ladtjovagge U-shaped valley across the eastern front of the Caledonides south of Kebnekaise.* The bedrock of the valley is composed of the old Precambrian crystalline basement of the chain and the mountain slopes of rather flat-lying overthrust rocks. The autochthonous sediments of the Hyolithus group are mostly wholly covered by debris. Photo V. Schytt

a group of quartzite formations with intercalated slate formations—occur also in several other parts of the eastern Caledonide district such as the eastern border areas of the Västerbotten and the Norrbotten mountains, and occupy vast expanses in southern parts of the Caledonides (i.e. in the southwestern part of southern Jämtland, in the southeastern part of Härjedalen, and in northern Dalarna). Stratigraphically, the Vemdal quartzite lies directly below the fossiliferous Cambro–Silurian. In northern Jämtland, in Ångermanland and in southern Västerbotten, the Lower thrust rocks consist of fossiliferous Cambro–Silurian, Vemdal quartzite, real tillite rocks, a series of late-Precambrian arkoses (the Sparagmite group), and early Precambrian igneous rocks of low grade metamorphism. The *Tillite formation* is intercalated between the Vemdal quartzite and the Sparagmite. Another large area of thrusted, little-altered sparagmite and Vemdal quartzite is situated in the southern Härjedal–northern Dalarne mountains; scattered erratics of tillite have been reported in this area. Arkoses of middle grade belonging to the Sparagmite group occur in the eastern part of northern Västerbotten and in the southern Norrbotten mountains, and belong to a tectonic complex, the *Stalon complex*, which is thrusted above the low grade complex of which the Vemdal quartzite and sparagmite are important members. Arkosic rocks of middle grade and of an analogous tectonic position are widely distributed in the middle zone of the Härjedalen mountains, but small areas of such rocks also occur in other parts of the southern mountains. These arkoses are assumed to belong to the Sparagmite group. Since 1955 (due to Strömberg) this sparagmite complex has been designated the *Serv complex.*

In the Swedish Caledonides areas of various sizes containing old Precambrian igneous rocks occur. In some areas they are of primary habit and quite well preserved, like the igneous rocks of the basement to the east of the mountains. Wide expanses of the rocks are, however, strongly cataclastic to mylonitic and so altered that their primary composition is hard to decipher.

The best-preserved Precambrian igneous rocks are found in the *Sjangeli–Rombak* and *Kuokel windows*, situated in the northwestern part of the Norrbotten mountains. Here the basement rocks are uncovered by erosion. At the Nasafjäll window in the international border zone of the southwestern Norrbotten mountains, and in several basement rock occurrences in the Lower thrust rocks (the biggest one being the Olden bedrock area in northwestern Jämtland) the igneous rocks are in the main only slightly metamorphosed. The high grade transformed basement rocks—the mylonitic to gneissic igneous rocks—belong to the Middle thrust rocks, to which belong also the above-named Stalon and Serv complexes. The biggest areas of highly transformed Precambrian igneous rocks are as follows: the *Akkajaure area* in the middle Norrbotten mountains; the "granite–mylonite" complex in the Offerdal–Hotagen part of northern Jämtland; the *Fuda complex* in the eastern part of the southern Jämtland mountains; and the *Tännäs augen gneiss* in southern Härjedalen.

However, all these sedimentary and igneous rocks together cover only the smaller part of the Swedish Caledonides. The greater part of the mountain range is composed of "mountain schists" of middle to high metamorphic grade. In some middle grade sedimentary formations in the western parts of the Swedish Caledonides determinable fossils of Upper Ordovician and Lower Silurian age have been discovered. Scattered, poorly preserved fossils have been found in formations older than the above-mentioned deposits. Because of these fossil discoveries, the entire sequence of middle to high grade schists has been called the *western Cambro–Silurian rocks*, belonging to eugeosynclinal facies. However, no traces of Cambrian fossils have been detected in the mountain schists. Basic igneous rocks, often in the form of amphibolite, are common in most mountain schist areas. The biggest complex of amphibolite rocks is called the *Eastern amphibolite.* Large areas of amphibolite occur in the northern and middle parts of the Norrbotten mountains. In the southern Västerbotten mountains and in northern Jämtland the southern continuation of the same amphibolite is present as a large number of lobate amphibolite occurrences. In Jämtland and Härjedalen, the amphibolite complex lies in the western part of the mountains. The Sylarne mountains amphibolite lying on the international border zone is the southernmost Swedish portion of this basic complex. The amphibolite complex may be looked upon as a marker zone—the backbone of the Swedish Caledonides. The most altered rocks of the high grade mountain schists are the migmatite rocks. They are distributed mainly in the Eastern amphibolite zone, generally immediately below the amphibolite. In the same NNE–SSW trending zone, where no amphibolite, or only scattered lenses of amphibolite occur, migmatite to feldspar–porphyroblastic schists can also be found. A western zone of migmatization occurs in certain parts of the international border zone also. These highly transformed schists occur mainly on the Norwegian side of the boundary and compose the lower part of probably the uppermost tectonic unit of the entire Scandinavian Caledonides.

In the foregoing, the writer has presented a short synopsis of the Swedish Caledonian rocks without laying stress on any tectonic subdivisions. The distribution of rocks in the different tectonic units of the chain is given subsequently.

CHAPTER XVII

HISTORICAL REVIEW AND MAIN PRINCIPLES OF THE TECTONIC DIVISION

Forty years ago when the author first encountered Swedish Caledonian problems, the prevailing opinion among Scandinavian geologists was that Törnebohm's theory of large Caledonian thrustings was wrong. The geologists studying the mountains considered the observed thrusts to be short transported units, mostly of parautochthonous origin. As for the large bodies of amphibolite and granite–syenite in the mountain rock sequence, the main opinion was that the igneous rocks represented magmas from a western, highly activized zone of the Caledonian geosyncline, from which the magmas were transported at rather high stratigraphic levels as conformably running intrusions for a considerable distance to the east, and then solidified. The regional metamorphism of the mountain schists was thought to be partly caused by igneous activity, the heat of the intruded rocks. No fossils of a guiding character have so far been discovered in the Swedish middle to high grade mountain schists of supposed eugeosynclinal origin. A few badly preserved fossils, mainly stem segments of *Pelmatozoa* observed in limestones, indicate the presence of Ordovician or Silurian rocks in the so-called western Cambro–Silurian rocks of the Swedish mountain schists. However, on the Norwegian side of the international border well-preserved fossils of Ordovician and Silurian age have been known for a long time in several Caledonian slates and limestones. In Jämtland and in the eastern border zone of the range, little-altered sediments below the mountain schists have yielded a great many fossils of Cambrian, Ordovician and Silurian age for many years and are designated the *eastern Cambro–Silurian deposits*.

In 1925, while surveying an area in the central Västerbotten mountains, the author discovered determinable fossils in two formations: one, the coral rich limestone of uppermost Ordovician age, the *Slätdal limestone*; the other, the graptolite holding slate, the *Broken slate*, of Lower Silurian age. The two formations lie in the middle part of a stratigraphical sequence with formations of characteristic qualities (Kulling, 1933).

During a reconnaissance tour of the Torneträsk area in the northern Norrbotten mountains the author came to the conclusion that the Caledonides were on the whole built up of big overthrusted masses, and that the cataclastic to mylonitic granite and syenite rocks of the Torneträsk area must be thrust basements, i.e. overthrusted Precambrian basement rocks (Kulling, 1930). The following year Asklund and Zenzén started geological investigations in the southern Swedish mountains, primarily in order to examine the tectonic nature of the unfossiliferous Vemdal quartzite, lying above the eastern Cambro–Silurian of the mountain border area. Both stated (Zenzén, 1932; Asklund, 1933) that the Vemdal quartzite was thrusted above the autochthonous eastern Cambro–Silurian sediments. They found overthrusted early Precambrian, Archean, igneous rocks in the areas examined on the boundary zone between the Vemdal quartzite and its underlying autochthonous sediments.

After these contributions, Törnebohm's pioneering ideas on Caledonian thrustings regained the leading position amongst the Scandinavian mountain geologists.

The geological map of Jämtland county by Högbom (1894) and that of Central Scandinavia by Törnebohm (1896) (both areas comprising the main part of the southern Swedish mountains) are instructive general bedrock maps. On Törnebohm's map the extent of the so-called *Big thrust units* is distinguished, including the principal nappe unit in the west and its klippen east of the continuous nappe-front. Minor thrusts below the main nappe were also marked on the map. Högbom in his description accompanying the Jämtland county map accepted Törnebohm's general views on Caledonian tectonics and also his idea on some minor thrustings below the big thrust. Thus, he noticed that in some places in the parish of Berg, southeastern Jämtland, the overthrusted Vemdal quartzite had slices of its primary Archean basement attached to the thrust unit. Asklund about forty years later gained acceptance for his overthrust theory which was based on an examination of the bedrock of the same area. However, his view is that both Högbom and Törnebohm have highly underestimated the extent of the Lower thrust rocks, i.e. the width of the Vemdal quartzite nappe; this is a significant result.

For the last thirty years Asklund has led bedrock mapping of the southern Swedish Caledonides. Thorslund has carried out the greater part of a survey of the fossiliferous Cambro–Silurian domains and examined collected fossils. Many geologists have also taken part in the geological survey of the southern mountains. However, no general geological map, corresponding to that of Högbom, has so far been worked out.

The geological mapping of the northern Swedish Caledonides—the Lapland mountains—continues also. The first general survey of the Västerbotten mountains proper was carried out in the 1920's by Quensel and Backlund. A commentary on the bedrock map has appeared (Quensel, 1960).

Since 1940 the author has been leader of the general geological survey of the Lapland mountains and has published monographs on the geology of the eastern marginal zone of the Västerbotten mountains (1942), the Caledonian range of Västerbotten county (1955) and the northern Norrbotten mountains (1964). A monograph on the southern Norrbotten mountains is in preparation.

In addition to bedrock mapping by the Geological survey in the Lapland Caledonian mountains, two other groups of geologists have carried out surveys during the last decades. These are: prospectors of the Boliden Company; and Dutch geology students, collecting thesis specimens. The geological results of the prospecting geologists and scientific research students have been of great value to the writer in his geological investigations of the Lapland Caledonides.

In most parts of the Swedish Caledonides the sequence of large rock-units is about the same from east to west or from bottom to top: 1. Autochthonous well-preserved sedimentary rocks of so-called eastern facies, with fossils. 2. Thrusted little-altered sedimentary rocks with or without fossils and with or without attached parts of their Precambrian basement rocks. 3. Distant thrusted rocks of middle to high grade, considered as coming from the eugeosynclinal area. The last-named category can mostly be subdivided into upper and lower rock complexes of different composition. In the lower complexes there are gneissic, mylonitic to cataclastic basement rocks (for the main part acid igneous rocks of Archean age) arkoses of the Sparagmite group, and so-called *hardschists*. The upper and major part of the long range thrust masses is composed of eugeosynclinal deposits: sediments, lavas, and intrusive bodies. Fossil remains are preserved in places in the metamorphic eugeosynclinal sediments.

Geological formations and large tectonic units are often named with reference to a characteristic type of rock, such as granite mylonite nappe, or to a place in which the rock assemblage or type of rock is characteristically developed, such as Vemdal Quartzite Nappe. Starting in different parts of the Swedish mountains the geologists have, for several decades, produced a great many subdivisions of the bedrock, which are difficult to compare and to correlate with each other. These systems of division are not all founded on facts, and no general system of division is in use. In this book the following principal subdivisions of the Swedish Caledonide rock complexes are used: Uppermost thrust rocks; Upper thrust rocks; Middle thrust rocks; Lower thrust rocks; Lowermost thrust rocks and the autochthonous sedimentary rocks deposited on the Precambrian basement. These subdivisions are described below. In Fig. 108 a review of the division of rocks in the large geographical segments of the Swedish mountains is given (see also Fig. 109).

Autochthonous sedimentary rocks. They occur mainly as a narrow belt in the eastern border of the Caledonides. In some window areas in the international border zone, thin autochthonous sedimentary deposits also occur. Along the eastern border of the southern mountains the sediments are composed of sandstones, shales, black shales and limestones of Cambrian and Ordovician age. In northern Västerbotten and further north only sandy and shaly beds of late Precambrian and Lower and Middle Cambrian age are met with.

Lowermost thrust rocks. These thrust units are often called the *East Jämtland nappes* and occur mainly in the eastern parts of the Storsjö and the Föllinge lowland areas. They are composed of well-preserved Cambrian, Ordovician and Silurian sediments. Exceptionally, on Frösön in the neighbourhood of Östersund, small portions of Archean basement rocks are incorporated in the thrust rocks. The East Jämtland nappes are probably of parautochthonous character. To the south of Lake Storsjön four such small nappes can be distinguished.

Lower thrust rocks. These extend for some distance in the eastern parts of the Caledonian chain in both Sweden and Norway. They are low grade metamorphic rocks and their composition differs in different segments of the mountains. Cambro–

		NORTHERN NORRBOTTEN MOUNTAINS	SOUTHERN NORRBOTTEN MOUNTAINS	NORTHERN VÄSTERBOTTEN MOUNTAINS	SOUTHERN VÄSTERBOTTEN MOUNTAINS	NORTHERN JÄMTLAND MOUNTAINS AND TÅSJÖ AREA	WESTERN AND SOUTHERN JÄMTLAND MOUNTAINS AND FÖLLINGE–STORSJÖ AREA	HÄRJEDALEN MOUNTAINS AND NORTHERNMOST DALARNE MOUNTAINS
THRUST ROCKS	UPPERMOST		*Rödinsfjäll nappe* (With *Stipok klippe* etc.)					
THRUST ROCKS	MAIN SEVE NAPPE: UPPER		*Virihaure nappe*	*Storfjäll nappe complex*				
THRUST ROCKS	MAIN SEVE NAPPE: UPPER	*Seve–Köli complex* (lower part only)	*Seve–Köli complex* (Formations with determinable fossils in the middle part of the sequence)		*Seve–Köli complex*		*Seve–Köli complex* (Lower part only)	
THRUST ROCKS	MAIN SEVE NAPPE: MIDDLE	*Pältsa gneiss complex*; *Abisko complex* (Torneträsk and adjacent areas)	*Syenite complex*	*Stalon complex*		*Serv complex* *Granite mylonite complex*	*Serv complex* *Tännäs gneiss complex*	
THRUST ROCKS	LOWER	*Rautas nappe complex*	*Vemdal nappe* (Formerly Ström quartzite nappe)		*Blaik nappe complex* {*Gittsfjäll nappe*; *Vemdal nappe* (including *Föllinge complex*) (formerly Ström quartzite nappe)	*Svaningen complex*	*Olden nappe*	*Hede–Rogen complex*
THRUST ROCKS	LOWERMOST						*East Jämtland Nappes* {*Sunne nappe*, *Ås–Husås nappe*, *Bjerme nappe*, *Skute nappe*	
AUTOCHTHONOUS SEDIMENTS		*Cambrian and Late-Precambrian sediments or Hyolithus series*		*Cambrian and Late-Precambrian sediments*	*Cambro–Ordovician sediments*			
		Precambrian basement						

Fig. 108 *Tectonic Schedule for the Swedish Caledonides* (O. KULLING, 1965)

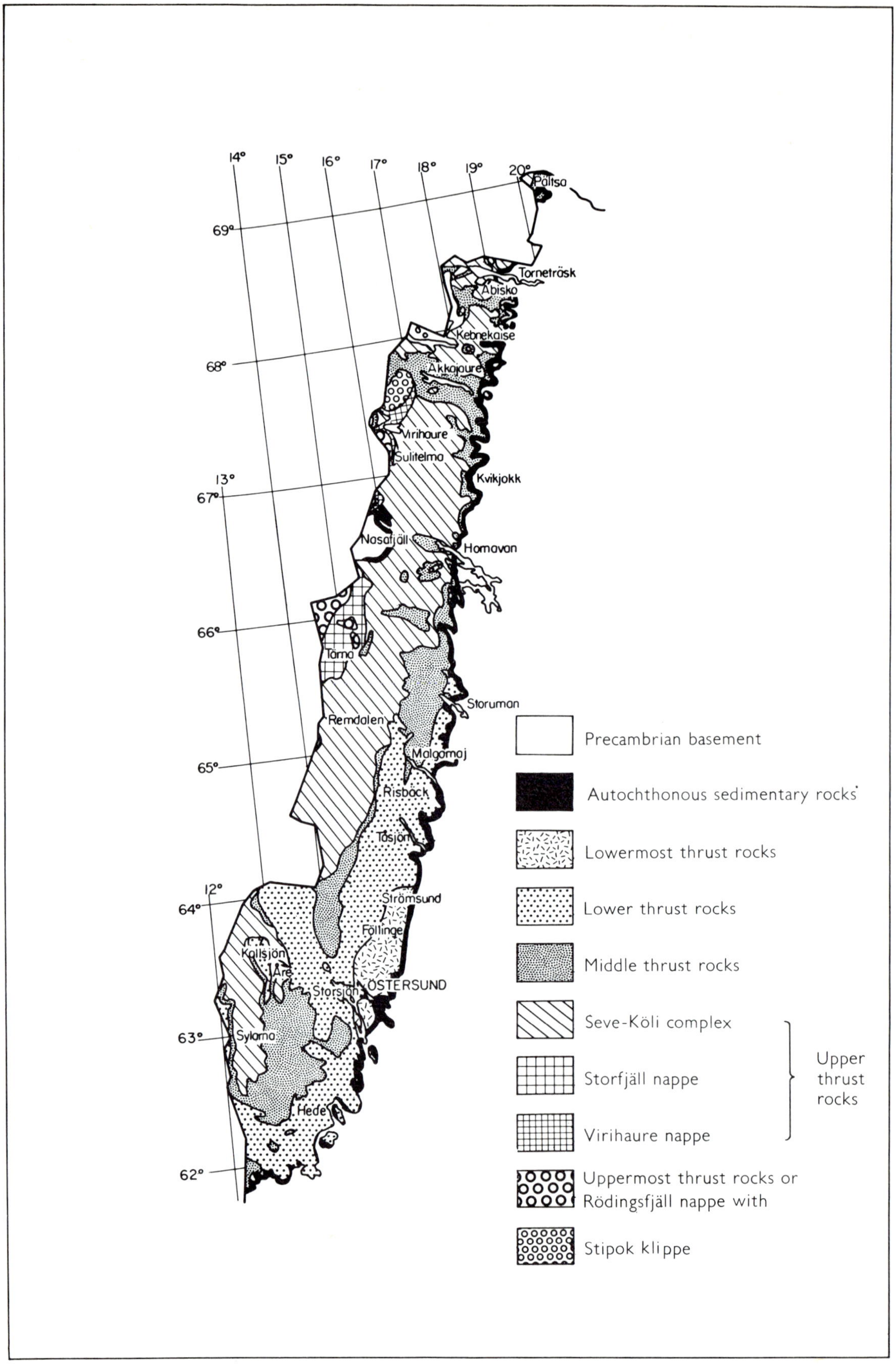

Fig. 109 *Tectonical division of the Swedish Caledonides according to* O. Kulling (1966)

Silurian and late Precambrian sediments with or without attached Archean basement rocks comprise the bedrock of the Lower thrust rocks. In the northern Swedish mountains the Lower thrust rocks are restricted to the eastern border zone. In the southern mountains the Lower thrust rocks in central Jämtland and in southern Härjedalen run from the eastern border of the chain across the international border to the west.

Middle thrust rocks. All rocks of this thrust complex are strongly tectonically altered. The sediments are mostly of biotite facies and are partly banded; the igneous rocks and coarse sediments such as arkoses are cataclastic to mylonitic. No fossils are of course found in the Middle thrust rocks, the uppermost part of which is probably of Cambrian age in some areas. The major part of the rocks mentioned is of late Precambrian and early Precambrian age. The Middle thrust rocks are widely distributed in the northern Norrbotten mountains, particularly in their southernmost part in the Akkajaure area where a segment of the rocks extends from the eastern mountain border to beyond the international border. In the southern Norrbotten mountains the rocks appear in the eastern border zone. The same can be said of their distribution in the Västerbotten mountains; a big erosional protrusion of the Middle thrust rocks is directed southwards and is surrounded by Lower thrust rocks. In the southern Swedish mountains the Middle thrust rocks extend over large areas in southern Jämtland and northern Härjedalen. Isolated portions of the thrust rocks occur in the Ansätten area to the north of Lake Storsjön, in the Fuda area to the south of the same lake, and in other scattered areas.

Upper thrust rocks. The major occurrence is called *the Seve–Köli complex*; this comprises a very large portion of the Swedish mountains and can be said to be the tectonic backbone of the entire Scandinavian Caledonides. The Seve–Köli complex is composed of medium to high grade eugeosynclinal sediments with acid to basic volcanic intercalations, but also with intrusions of mostly basic character In some relatively low grade areas of the Seve–Köli sequence, determinable fossils, mentioned in an earlier chapter, have been collected.

The names *Seve* and *Köli* have a long history in the literature of the Scandinavian Caledonides. They were introduced in 1872 by TÖRNEBOHM in his description accompanying a published geological section across the middle part of the Scandinavian Caledonides, in the area between Östersund and Levanger. TÖRNEBOHM divided the mountain schists into an upper or *Köli group* and a lower or *Seve group*. The former is characterized by low to middle metamorphic schists, often of shaly composition. The latter is of harder grit than the Köli schists, of high grade and to some extent gneissic. TÖRNEBOHM took the name Köli from an old term, the Kölin or Kiölfjället (Kjöl mountain), of the mountainous zone in the international border area between Jämtland province and Trondheim Amt. The name *Seve* goes back to the Roman author Plinius, who called the mountainous border zone between Sweden and Norway the *Sevo mons*. The name *Seveberget* (Seve mountain) was then used in the same sense as Plinius' on geographical maps of Scandinavia, chiefly during the eighteenth century.

The *Virihaure nappe* occupies a relatively narrow zone above the Seve–Köli complex in the western part of the middle Norrbotten mountains. The rocks of the nappe are of about the same composition and metamorphic grade as the underlying Seve–Köli rocks and can probably be looked upon as a westerly part of the complex, which has been folded, imbricated and overthrusted on its eastern continuation.

The *Storfjäll nappe complex* in the northwestern part of the Västerbotten mountains lies between the typical Seve–Köli rocks in the southeast to east and the Rödingsfjäll nappe or Uppermost thrust rocks in the northwest to west. The tectonics of the Storsfjäll nappe complex have only been partially clarified, and only scattered parts of its lower tectonic boundary have been verified. "Zones of strong differential movements" (ALEVA, 1950; MULDER, 1951) have been reported in the northwestern part of the nappe complex (KULLING, 1955) but a general geological–tectonic examination of this and other problems of the entire bedrock zone (somewhat provisionally designated the Storfjäll nappe complex) is highly desirable. The rocks of the nappe complex are typical eugeosynclinal types of middle to high metamorphic grade. In some parts of the bedrock zone there occur bodies of granites—the only Caledonian deposits known in the entire Swedish Caledonides.

Uppermost thrust rocks. These thrust rocks occur in the northwestern international border region of the Västerbotten mountains, and in some westerly areas of the middle Norrbotten mountains, but are most widely distributed on the Norwegian side of the international border. Some twenty years ago the writer gave the name the *Rödingsfjäll nappe* to this uppermost rock complex of eugeosynclinal deposits, of high metamorphic grade. The Norrbotten parts of the Uppermost thrust rocks appear to be klippen.

CHAPTER XVIII

THE AUTOCHTHONOUS

THE PRECAMBRIAN BASEMENT

The Basement of the Swedish Caledonides in Lapland, Ångermanland and Jämtland, is composed of early Precambrian rocks, belonging to the Karelian and Svionian cycles. Along the southernmost part of the chain, around Härjedalen and northern Dalarne, younger Precambrian rocks make up the substratum.

The *Karelian Lina granite*, *Sorsele granite*, perthite granite and syenite, together with metamorphic sedimentary rocks and volcanics of the Svionian cycle form a complicated pattern in northern Norrbotten, east of the Caledonian chain. In southern Lapland and in Jämtland the *Svionian Revsund granite* predominates. The younger Precambrian rocks of the basement along the southern portion of the Swedish part of the chain can be divided into the *Jotnian Dala sandstone* and *diabases*, the *sub-Jotnian Dala porphyries*, and the *Dala and Rätan granites*. The acid igneous rocks were separated from the overlying Dala sandstone by a period of denudation, registered by the so-called sub-Jotnian peneplane.

THE SEDIMENTS OF THE BASEMENT: INTRODUCTION

The autochthonous sediments along the eastern front of the Caledonides rest on peneplaned Precambrian rocks. When the first transgression of late Proterozoic age reached the zone of the present Caledonian border the rocks of the old land surface seemed to be partly quite fresh, partly somewhat kaolin-weathered to a depth of some metres. However, it is seldom possible to study the basal boundary zone of the sediments because of lack of exposures. Most observations on the weathered basement surface have been in the Norrbotten mountain border zone.

The autochthonous sediments contain Late Proterozoic, Cambrian and Ordovician fossils and usually cover a very narrow belt, often only a few km in width. Round the easternmost part of Lake Storsjön in central Jämtland, the belt of autochthonous sediments has a width of more than 10 km.

LATE PROTEROZOIC AND CAMBRIAN

Sediments with Cambrian fossils are found along the entire border zone. The unfossiliferous sandstones in the basal part of the sedimentary sequence were considered to be Eocambrian by some geologists. The author's earlier opinion was that these basal beds are so closely attached to the sediments lying above that they ought really to be incorporated in the Lower Cambrian sequence. However, recently discovered fossils within the lowermost sandy-shaly beds have thrown light upon their age relationships. The lowermost sediments in the northern part of the border zone will now be considered as Latest Proterozoic.

Late Proterozoic and Lower Cambrian beds

The Lower Cambrian in Sweden is stratigraphically divided into four zones, viz: 4, the zone with *Strenuella linnarssoni*, 3, the zone with *Holmia kjerulfi*, 2, the zone with *Volborthella tenuis*, and 1, the zone with *Discinella holsti*. On the Norwegian side of the border, in southern Norway, the zone with *Holmia* cf. *mickwitzi* immediately below the *Discinella holsti* zone seems to extend only locally.

Within the easterly marginal zone of the mountains the best known parts of the Late Proterozoic and Lower Cambrian beds are to be found in the Lake Torneträsk area of the northern Norrbotten mountains. The two most thoroughly examined sections are the one south of Torneträsk, above the northern slope of Luopakte, and the one to the north of the lake along Vakkejokk (see Figs. 110–112). The successions obtained show

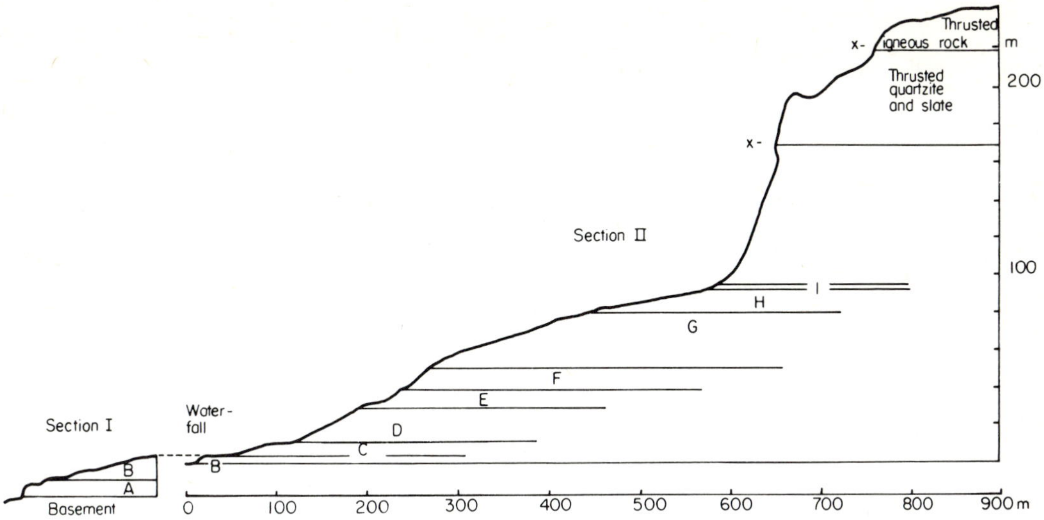

Fig. 110 *The Luopakte section across the Hyolithus Series on the northeastern slope of Mt. Luopakte to the south of Lake Torneträsk.* From O. KULLING (1960). A Lower sandstone; B Lower shale; C Middle sandstone with *Spriggia*-bearing sandy shale below the uppermost sandstone bed of the formation; D Middle shale, E–F–G Upper sandstone; H–I Upper shale with top bed of fossiliferous limestone; J Alum shale. Above the alum shale are the Lower thrust rocks of quartzite and slate with lenses of fossiliferous limestone. On the sediments lie Middle thrust rocks of old Precambrian granite–syenite.

great similarity but differ in certain respects. The Luopakte section is noted for the fossils occurring within it, the Vakkejokk section for the coarse tillite-like sedimentary breccia (Vakkejokk breccia) which is interbedded with the lower part of the sedimentary succession. After the Late Proterozoic and Lower Cambrian sediments in the Torneträsk area have been dealt with a review of the sediments in other parts of the mountain marginal zone will be given.

On the north-facing slope of Luopakte the autochthonous sediments are divided into 6 formations: the Lower sandstone formation, with an arkosic conglomerate at the base; the Lower shale formation; the Middle sandstone formation; the Middle shale formation; the Upper sandstone

Fig. 111 *The northern aspect of the northeastern corner of Mt. Luopakte.* The section along the brook in the picture—the classic Luopakte section first worked out by J. CHR. MOBERG (1908)—is to date the best-known section through the autochthonous Hyolithus group of sediments in the Norrbotten mountains. The rock wall in the background is composed of quartzite of Lower thrust rock position. The writer's Middle shale formation outcrops on either side of the lowest waterfall in the picture. The cataracts above the waterfall register the ledges of the Upper sandstone formation; cf. the diagrammatic section in Fig. 110. Photo O. KULLING (1965)

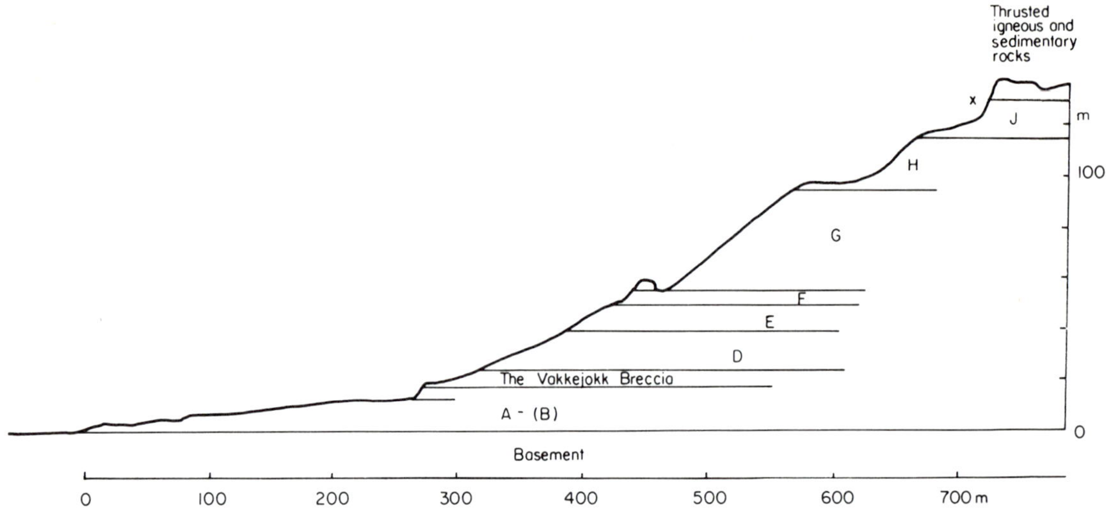

Fig. 112 *Section across the Hyolithus Series along Vakkejokk to the north of Lake Torneträsk.* The subdivision of the autochthonous sediments is about the same as in the section of Mt. Luopakte (cf. Fig. 110). From O. Kulling (1960). A–B Lower sandstone, possibly with some remnants of the Lower shale; The Middle sandstone (C) of Mt. Luopakte is not present. The Vakkejokk breccia in the section is found only as a well-defined formation to the north of Lake Torneträsk; D Middle shale; E–F–G Upper sandstone; H Upper shale; J Alum shale. Above the alum shale follow Middle thrust rocks with acid igneous rocks and metamorphosed sediments

formation; and the Upper shale formation with a thin limestone layer towards the top. Above the last-named formation follows one of schistose black shale as the uppermost formation of the autochthonous sedimentary sequence. In the Vakkejokk section and the area to the east of it, the Lower shale formation is partly absent. The above-mentioned Vakkejokk breccia lies between the Middle sandstone formation and the Middle shale formation and was discovered and named in 1930 by the author. Along the mountain front east of Vakkejokk the Lower sandstone formation sometimes has a basal conglomerate on top of the Precambrian granitic basement.

Some of the more important features of the above-mentioned formations within the autochthonous sediments around Torneträsk are given below.

The Lower sandstone formation is usually some tens of metres thick and grey in colour. Its lowermost zone is often an arkosic conglomerate. The uppermost part of the basement is brecciated *in situ* in places. Ripple-marks occur. Certain layers in the sandstone formation are current-bedded and lime-bearing. Usually the sandstone lacks traces of fossils but north of Torneträsk the author has observed fossil trails on the bedding surfaces in a shaly horizon of the formation. The Lower shale formation is made up of dark grey and sandy shales of cyclic nature. Tracks of fossils and also small thread-like features, interpreted as fossil remains, have been observed on the bedding surfaces. In the Luopakte section the writer found in summer 1965 a fragmentary fossil, measuring only a few mm in width, in the graded bedded sandy shale formation. The thickness of the Lower shale formation in Luopakte is about 12 m.

The Middle sandstone formation (Luopakte area) is 7–8 m thick and composed of grey, greenish grey and yellowish grey sandstones with interbedded shaly layers. Fossil trails and other traces of former life have been observed at several levels. Most interesting is the find in 1960 of *Spriggia annulata* Southcott in the upper part of the formation. The first specimen of this fossil species was discovered by the writer in the Middle sandstone formation some 20 km to the south of Luopakte in 1959. The first find of *Spriggia* in the Luopakte section was followed in 1965 and 1966 by finds of several new specimens of *Spriggia* fossils (see Figs. 113 and 114). Remains of another kind of fossil were discovered also. All fossil remains collected are from the upper part of a roughly 2 m thick greenish grey to light grey sequence of graded bedded sandy shales, which lies below the 2 m thick light-coloured sandstone top-bed of the formation (Fig. 115). In 1966 *Spriggia* fossils were found in the area north of Torneträsk. The fossils were collected in sandy shale with cyclic sedimentation, about 1 m below the Vakkejokk breccia (see Fig. 116).

Spriggia annulata had previously been found only in the Late Proterozoic *Pound sandstone* in the Flinders Ranges of South Australia and is looked upon as probably a primitive medusa with a pelagic mode of life. Initially the fossils were

Fig. 113 *Specimen of* Spriggia annulata *Southcott from the* Spriggia *beds at Luopakte,* ×$\frac{6}{7}$. Showing a very small fragment of another *Spriggia* fossil in the lower right edge; collected by O. KULLING, 1965. Photo INGER SIGNORELLI (1966)

Fig. 114 *Specimen of* Spriggia annulata *Southcott from the* Spriggia *beds at Luopakte,* ×$\frac{6}{7}$; collected by O. KULLING, 1960. Photo INGER SIGNORELLI (1966)

considered by the writer as belonging to the first faunal zone from the base of Lower Cambrian; he now thinks, however, that the *Spriggia* fossils are representatives of a Late Proterozoic fauna. As the Middle shale formation, lying above the Middle sandstone formation in the Luopakte area, holds fossils of the second faunal zone from the base of the Scandinavian Lower Cambrian, the first Lower Cambrian transgression is most

Fig. 115 *The locality of the* Spriggia *shale near the brook which descends northward from the northeastern steep slope of Mt. Luopakte.* The fossil-bearing shale lies below the sandstone ledge in the foreground. The (sandy) shale and sandstone compose the upper part of the writer's Middle sandstone formation, and the Middle shale formation is exposed along the lowest of the waterfalls in the background (cf. Fig. 111). Photo O. KULLING (1965)

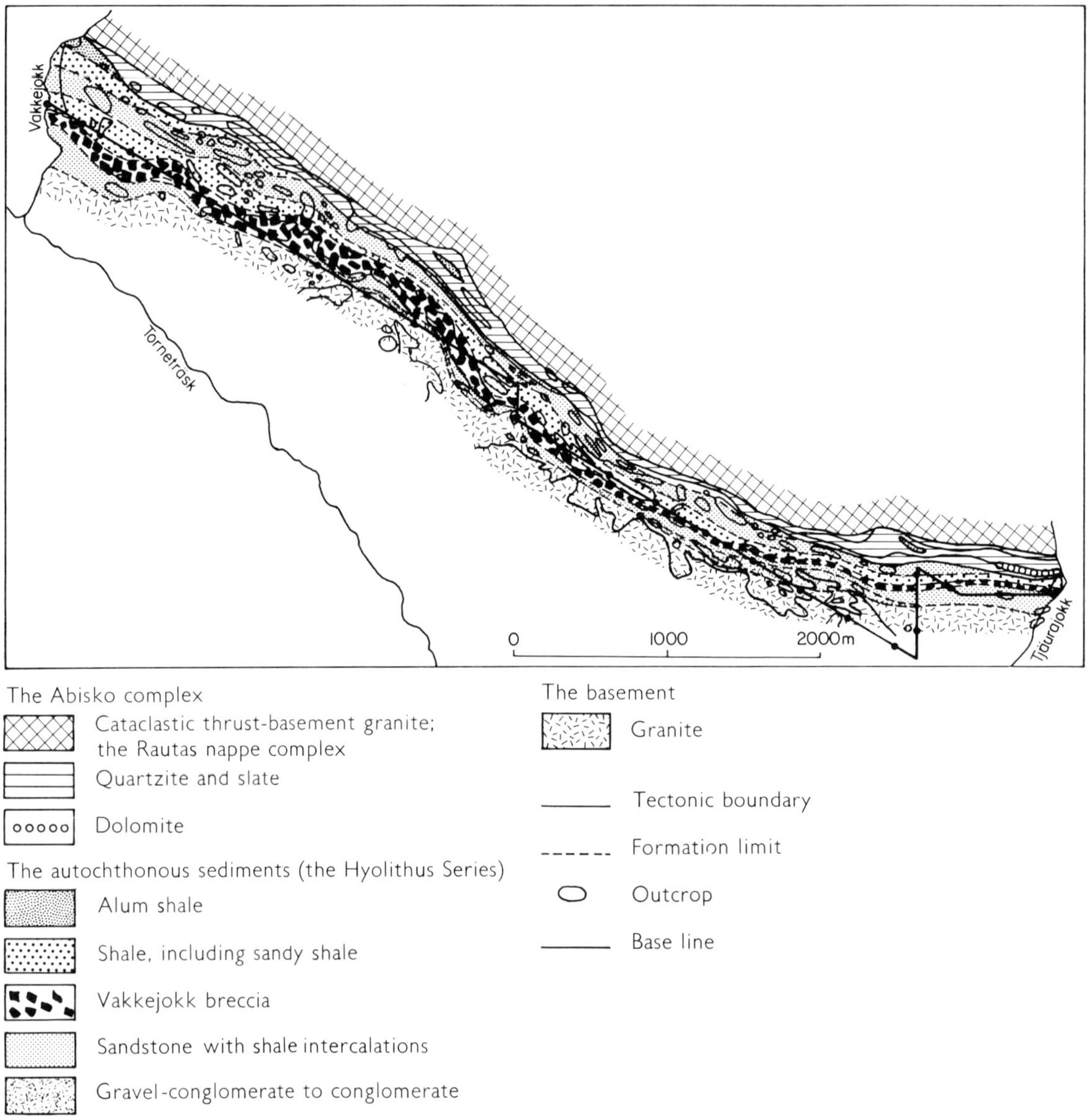

Fig. 116 *Geological map elucidating the distribution of the Vakkejokk sedimentary breccia in the area between Vakkejokk and Tjäurajokk to the north of Lake Torneträsk in the northern part of the Norrbotten mountains.* After O. Kulling (1964)

probably represented by the 2 m thick sandstone bed situated at the top of the earlier Middle sandstone formation; this formation, with regard to what has been said above, must be divided into a Lower Cambrian part and an Uppermost Proterozoic part.

As reported in the writer's earlier papers (see for instance Kulling, 1960) "a stratigraphical break" is indicated in the upper part of the Middle sandstone formation. This stratigraphical gap occurs immediately above the *Spriggia*-bearing shale. Of the 2 m top-bed sandstone mentioned, the lower 0,5 m contains scattered pebbles and angular pieces of vein-quartz, Archean igneous rocks and greenish siltstone. The siltstone fragments are usually angular and occur mostly in the lowermost bed of sandstone, sometimes in such profusion that the calcareous sandstone looks like a sedimentary breccia. The upper surface of the sandstone bed is uneven and represents a stratigraphical gap.

The Vakkejokk breccia or Vakkejokk formation lies conformably within the autochthonous sediments north of Lake Torneträsk. The breccia can be followed for about 14 km eastwards of Vakkejokk. Fragments of sandstone and shale occur in the breccia, sometimes in abundance. The sediments were probably formed from the Lower shale formation and the Lower and Middle sandstone formations by erosion, and were subsequently incorporated into the breccia whose main fragments are fine-grained Precambrian granite. The thickness of the breccia north of Torneträsk varies from about 10 to a few metres. At certain places it has a rather uneven contact with the sandstone beneath; in general, however, it lies conformably on the sandstone. The Vakkejokk breccia north of Torneträsk resembles moraine structurally but also shows features which are not easy to explain in a tillite. Sometimes the breccia proper passes upwards into an arkosic breccia-like conglomerate, which is overlain by conglomerate containing poorly rounded granite pebbles. A common type of Vakkejokk breccia contains angular fragments of granite, of variable size, embedded in a matrix of granite grit. Another type contains large and small, round to angular granite fragments enclosed in rucked-up shales. Both these varieties occur in their turn as fragments in more heterogeneous breccia—one thus speaks of *breccia-in-breccia* structure. The sandstone underlaying the breccia has been pressed together in parts and even folded into the breccia. The mode of formation of the breccia is shown in Figs. 117–122. Some remarks will be made about the probable stratigraphical relationships between the granite conglomerate north of Lake Torneträsk, and the Middle shale formation, which follows above the Vakkejokk breccia and the post-*Spriggia* sandstone of the Middle sandstone formation south of the lake. The granite conglomerate north of Torneträsk, and the post-*Spriggia* sandstone south of the lake, are both clastic rocks immediately beneath the Middle shale formation and probably bear some relationship to each other. The lowest layer of sandstone south of the lake, with abundant siltstone fragments and also fragments and pebbles of igneous rocks, could probably be a southern but not very characteristic counterpart of the Vakkejokk breccia.

The Middle shale formation rests, to the north of Lake Torneträsk, on the Vakkejokk breccia. South of the lake, where the Vakkejokk breccia has not been found in sections of autochthonous sediments studied by the author, Middle shale rocks rest on the Middle sandstone formation. The shale formation should normally correspond to the "Hyolithus Shales" of previous workers who have investigated the autochthonous sedimentary series of the eastern border area of the Norrbotten mountains. The shale formation is fossiliferous. The fossils belong to the second faunal zone from the base of the Scandinavian Lower Cambrian. In the Torneträsk area the author has found two guide fossils, *Volbortella tenuis* Schmidt and *Platysolenites antiquissimus* Eichw. in the lower part of the shales. The Middle shale formation is usually 15–20 m thick in the area under consideration. Its colour varies from red to pale grey.

The Upper sandstone formation is 50–60 m thick and consists of a large number of sandstone beds of varying type and colour. A characteristic feature of the sediments is that the bedding surfaces are criss-crossed by fossil trails and other bedding markings. The author has identified a large number of different types of trail. The Upper sandstone formation is probably chronostratigraphically equivalent to sediments with fossils of zone 3 of the Scandinavian Lower Cambrian.

The Upper shale formation, the thickness of which varies from 15–20 m, contains shale of a dark grey to greenish grey colour. The lowest bed is conglomeratic, with phosphorite grains. In Luopakte the uppermost part of the shale formation consists of fossil-rich lime-bearing shale which passes upwards into fossiliferous limestone. *Strenuella Linnarssoni* Kiaer, *Ellipsocephalus Nordenskiöldi* Linrs. and other fossils from faunal zone 4 of the Lower Cambrian have been found in the fossiliferous layers. The shale formation is the uppermost well-preserved stratigraphic unit of the autochthonous sedimentary rocks in the northern Norrbotten mountains. The schistose black-shale beds, which succeed the Upper shale formation, are supposed to be of Middle Cambrian age.

It is convenient to start a review of the autochthonous Late Proterozoic–Lower Cambrian sediments along the eastern border of the Swedish Caledonides with the type sections of the Torneträsk area described above. In the Norrbotten mountain frontier the Late Proterozoic and Lower Cambrian sediments, together with overlying black shales of probably Middle Cambrian, have been called the *Hyolithus Series* for many years. The first fossils found (by Svenonius, 1883) were specimens belonging to *Hyolithidae*, of Lower Cambrian age. At first Svenonius included not only the autochthonous sediments of Eastern facies in his Hyolithus Series but also large slices of what are now known as the thrusted rocks and the Precambrian basement. Eventually the present limits came into use.

Along the northern Norrbotten border zone, the autochthonous sediments have so far produced fossils of marker (guide) character only in the Late Proterozoic *Spriggia* zone and in Lower Cambrian faunal zones 2 and 4. It is of particular interest that shales with fossils from zone 4 in the area south of Stora Sjöfallet, in the northeastern part of the southern Norrbotten mountains, are separated from the Precambrian basement by a breccia-like conglomerate. Thus, sediments from Lower Cambrian faunal zones 1–3 are lacking. In the southernmost part of the Norrbotten mountain border zone (the Laisvall area) fossils from Lower Cambrian zone 4 have been found (Ljungner, 1939) in the uppermost part of the autochthonous sedimentary sequence, over 100 m thick, resting on a

Figs. 117–122 *Representative samples of the Vakkejokk breccia from the area north of Lake Torneträsk*

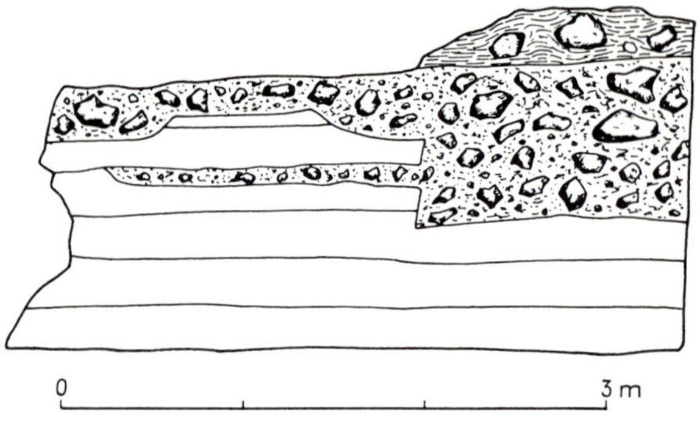

Fig. 117 *Breccia in interfingering contact with its substratum of stratified sandstone.* Shale with scattered pebbles and boulders lies on the thin breccia. From O. Kulling (1960a)

Fig. 118 *Breccia of breccia-in-breccia type.* The large block of coarse breccia with sandy matrix is enclosed by breccia with a shaly matrix. From O. Kulling (1964)

Fig. 119 *Breccia containing subrounded blocks of Precambrian granite of various sizes in a shaly matrix.* From O. Kulling (1960a)

Fig. 120 *Breccia with a shaly matrix and mainly sandstone as enclosed material.* The block in the middle of the picture is probably of granite. Photo O. KULLING (1958)

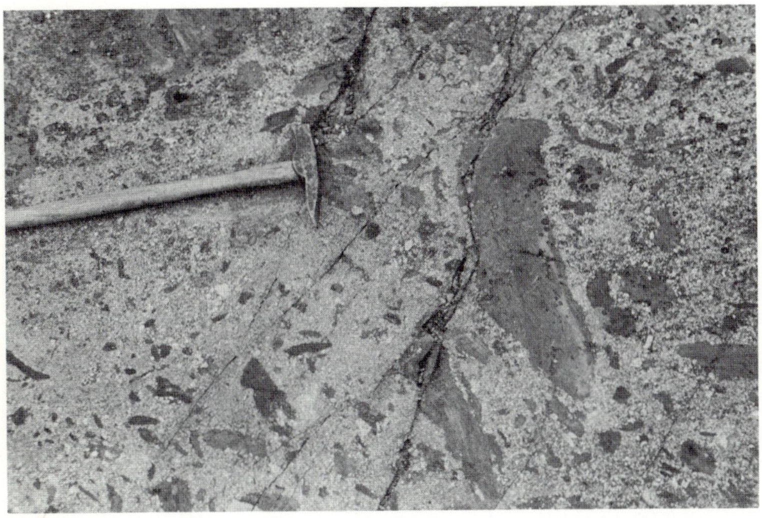

Fig. 121 *Numerous shale fragments up to 0,5 m in length enclosed in a granite–gravel conglomeratic matrix.* Photo O. KULLING and G. FREDRIKSON (1965)

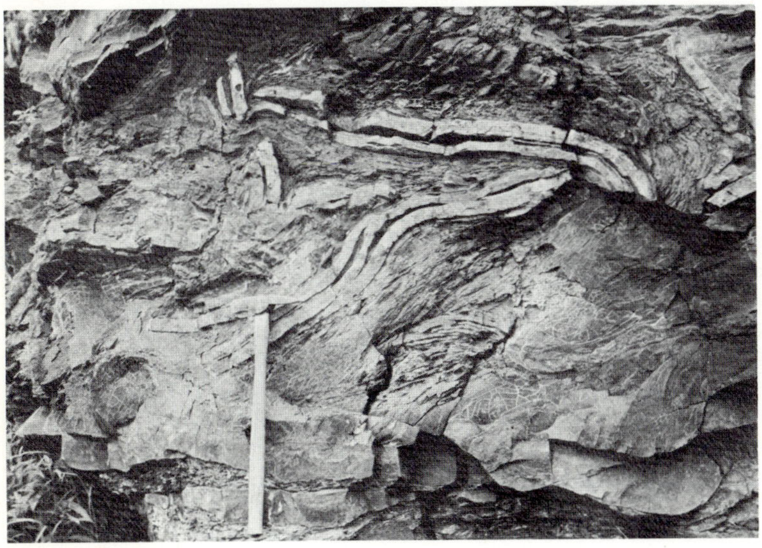

Fig. 122 *Shale with beds of sandstone lying immediately beneath the Vakkejokk breccia, thrown together during the deposition of the breccia.* Photo O. KULLING and G. FREDRIKSON (1965)

F. Kautsky (1945)		N. Marklund (1952)	
	Black shale		Black shale
43,5 m	Marly shale with scattered limestone lenses in the uppermost beds. In some of the limestone lenses fossils from the Lower Cambrian zone 4	4 m	Pyritic marlstone
		14 m	Claystone
		0,5 m	Conglomerate
		1 m	Claystone
		6 m	Fine-bedded arenaceous shale
		2 m	Dark sandstone
		5 m	Dark arenaceous shale
10,5 m	Dark grey shale	10,5 m	Claystone and siltstone
1 m	Sandstone with flakes of shale	1 m	Conglomerate with fragments of shale
16 m	Grey sandstone with a thin bed of quartz conglomerate	5 m	Sandstone
		0,1 m	Quartz–quartzite conglomerate
		8 m	Sandstone
		0,1 m	Quartz–quartzite conglomerate
20 m	White sandstone with intercalations of shale at higher levels	20 m	Sandstone
5,5 m	Dark grey sandstone with mudstone-intercalations	10 m	Mudstone
7,5 m	Dark grey mudstone		
6,5 m	Arkosic sandstone	0,5 m	Conglomerate (tillite?)
2 m	Arkose		Several m arkose
	Weathered granite		Weathered granite

Fig. 123 *Stratigraphical tables of the autochthonous sedimentary sequence of the Laisvall area, based on diamond drill core samples*

weathered granite basement. The sedimentary sequence of strata can be divided into an upper shaly part and a lower sandy part. Two sections from the Laisvall area exemplify the stratigraphy (Fig. 123). Reconnaissance work in the area some 10 km south of Laisvall by the present writer, in 1965, resulted in the discovery of a very rich trilobite fauna in a dark grey shale formation, representing Lower Cambrian faunal zone 3 (considered from below). Most probably the shale formation is a stratigraphical counterpart of the 14 m claystone formation, in the uppermost part of Marklund's stratigraphical table. The stratigraphy of the autochthonous sediments of the area is not, however, quite cleared up yet. On a visit to Laisvall mine in the summer of 1965 a fossil-bearing sample of the 1 m thick conglomerate beneath the 10,5 m shale formation was presented to the writer. The fossil was of a curious shape and rather fragmentary but inspired hopes of further fossil discoveries in the conglomerate and adjacent beds.

The direction of the transgressions of the sea in relation to the present trend of the Caledonian front of erosion to the east may be discussed with reference to the description above of the stratigraphy of the autochthonous sediments in selected sections in northern, middle and southern parts of the Norbotten high mountainous border region.

In the northern and southern parts of the Norrbotten mountain border, the fossiliferous beds of Lower Cambrian zone 4 rest on about a 100 m thick sequence of broadly speaking conformable sediments. In the intermediate zone, the Stora Sjöfallet area, the fossiliferous beds of zone 4 are separated from the Precambrian basement by a thin basal breccia-like conglomerate. Thus, if long periods of erosion did not occur in the areas discussed within the Lower Cambrian, the middle of the Norrbotten mountain area was above sea level for most of Lower Cambrian time, on a peninsula protruding to the west in the Lower Cambrian sea. Thus the trends of the Lower Cambrian transgressions in the Norrbotten area need not have been exactly conformable with the actual trend of the mountain border. The concept of NE–SW trending Lower Cambrian shore-lines along the Caledonian mountain border zone gets no real support from the above-mentioned facts.

The breccia sediment quoted above in the lower part of the autochthonous sediments of the Torneträsk area is not the only one discovered. The mountain border zone north of Torneträsk is the only area in which the breccia sediment of

Vakkejokk type has to date been thoroughly examined and continuously followed over tens of kilometres. Most of the other sedimentary breccias examined are situated at the very bottom of the sedimentary sequence (see Figs. 124–126). The basement rocks beneath the breccias are often weathered to a depth of some metres. The material of the breccias is in general of local origin. In middle and southern parts of the Swedish mountain border zone, coarse breccias in the lowermost part of the autochthonous sediments, looked upon as being of probably Lower Cambrian age, have not so far been discovered.

The Vakkejokk breccia looks in many respects like a tillite and so also do the breccia sediments which occur in other places at the base of the autochthonous deposits. The breccia-in-breccia structure of some parts of the Vakkejokk breccia and the quite different structure and composition of nearby parts of the same breccia are not in line with its interpretation as a real tillite. However, if the Vakkejokk breccia originated in an area with cold climate and seasonal thawing of the soil, the breccia-in-breccia structure of the deposits and the various kinds of structure within the breccia are more easily understood. The writer is of the

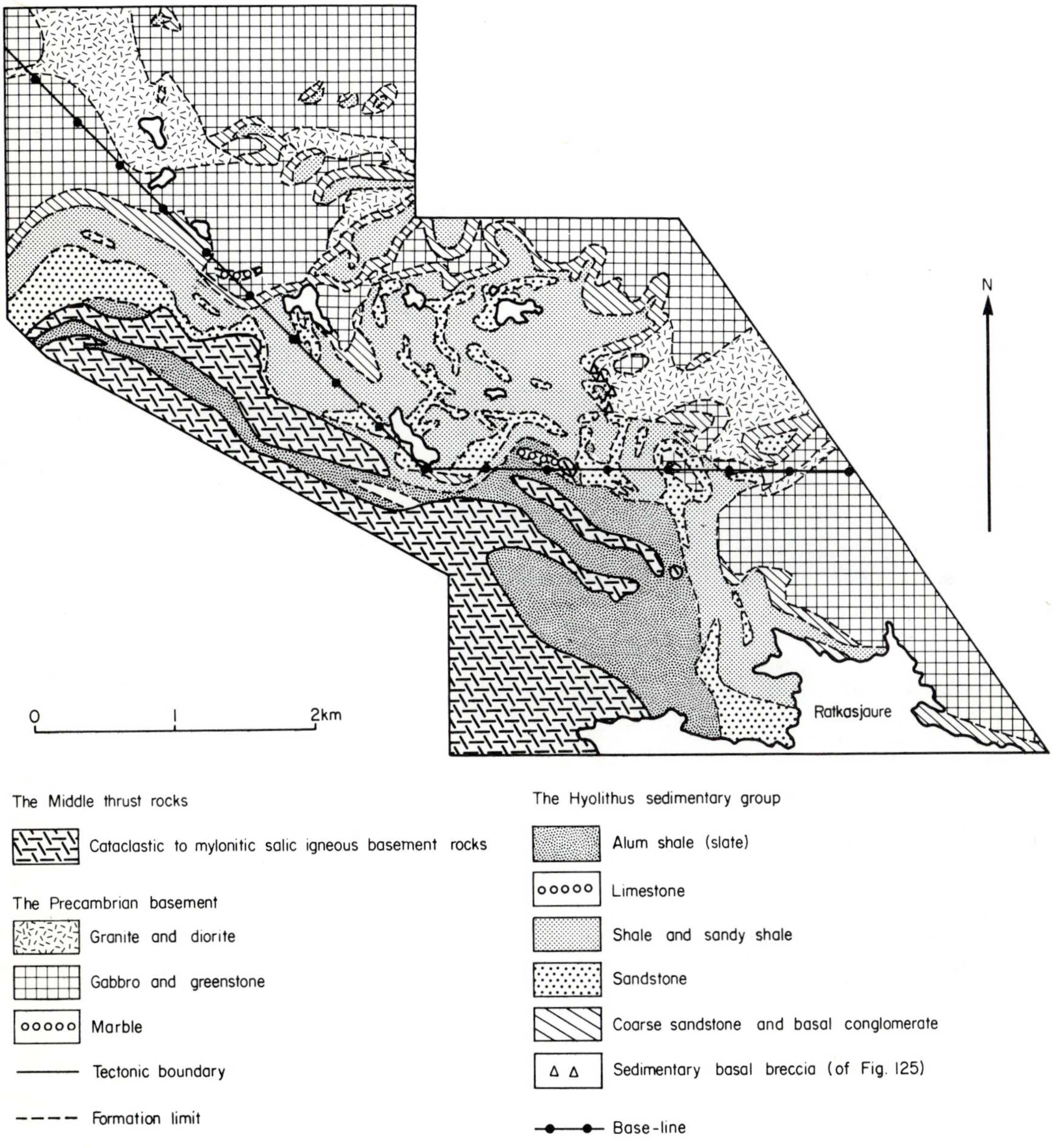

Fig. 124 *Geological map of the area to the northwest of Ratkasjaure at the southern limit of the Sjangeli window.* From O. Kulling (1964)

Fig. 125 *A local block of coarse sedimentary breccia* from the border zone between the Hyolithus group of sediments and the basement on the southern edge of the Sjangeli window, about 2 km northwest of the northwest corner of Lake Ratkasjaure. The breccia passes upwards (to the right of the picture) into a fine-grained, gritty conglomerate without a sharp contact. Photo O. Kulling (1953); see map, Fig. 124

Fig. 126 *A tillite-like basal breccia of the Hyolithus group from Abmojokk*, about 5 km southwest of Lake Sitojaure in the extreme northeasterly part of the southern Norrbotten mountains. The sedimentary breccia lies directly on the old Precambrian crystalline basement; it is 4–5 m thick, very coarse in its lowest part but passes upwards through gritty-grained members to fine-grained dark coloured sandy rocks with scattered light-coloured fragments of rocks and minerals. The shale lying above the clastic rocks is the upper part of the autochthonous sequence of the area. The upper photograph shows the natural rock face of a local erratic boulder. The lower one shows the opposite face of the sample with uncovered, more or less faceted, pebbles. $\times 0.6$. From O. Kulling (1951)

opinion that solifluction processes may have taken part in the creation of the Vakkejokk breccia and also of the breccia rocks at the base of the autochthonous sediments. The climate may have been of a periglacial nature.

Certain stratigraphical facts may be mentioned in favour of the interpretation of the sedimentary breccias as chronostratigraphical counterparts of real tillite rocks of the Varanger ice age. In northernmost Scandinavia there are two areas—the Alta area between the fjords Kvaenangen and Altafjord, and the Tanafjord area—where the stratigraphic positions of the tillite formations have been accurately assessed by Føyn (1938 and 1964). The Alta area—the Alta window with autochthonous sediments on basement rocks—lies some 200–260 km NNE of Lake Torneträsk. The Tanafjord area is situated about 200 km northeast of the Alta area. The table (Fig. 127) consists of the autochthonous sedimentary sequences of the two last-named areas and the Vakkejokk, Luopakte, and other sections of the Torneträsk area.

Føyn (1964) noted a striking resemblance between the sedimentary sequences above the Alta tillite, and those lying on the upper tillite of the Tanafjord area, and as a result of this correlated the Alta tillite with the upper of the two tillite

	TORNETRÄSK AREA Northern Sweden				ALTA AREA Northern Norway	TANAFJORD AREA Northern Norway
	LUOPAKTE SECTION SOUTH OF LAKE TORNETRÄSK	VAKKEJOKK SECTION NORTH OF LAKE TORNETRÄSK	SECTION ABOUT 4 KM EAST OF VAKKEJOKK, NORTH OF LAKE TORNETRÄSK	SECTION ABOUT 7 KM EAST OF VAKKEJOKK, NORTH OF LAKE TORNETRÄSK		
	THRUST ROCKS	THRUST ROCKS	THRUST ROCKS	THRUST ROCKS	THRUST ROCKS	THRUST ROCKS
PROPABLY MIDDLE CAMBRIAN	*Black shale, schistose,* about 72 m	*Black shale, schistose,* about 15 m		*Black shale, schistose,* about 3 m		
LOWER CAMBRIAN	*Shale with Lower Cambrian faunal zone 4 fossils*, 19 m	*Shale*, about 20 m	*Sediments as in the Vakkejokk section*, but thicknesses not measured	*Shale*, 11 m		*Sandstone and shale with Cambrian fossils at some levels* *and* *Lower Cambrian fossils in lowermost part*, 7–800 m
	Sandstone with fossil tracks, 48 m	*Sandstone*, 50–55 m		*Sandstone with fossil tracks*, 27 m		
	Shale with Lower Cambrian faunal zone 2 fossils, 18 m	*Shale*, 17–18 m	*Shale*	*Sandy shale*, 8 m *Shale*, 9 m	*Shale with sandy intercalations*, 20–30 m	*Sandstone, siltstone and shale*, about 1000 m
	Sandstone, with edged to rounded pebbles in its lower part, 2 m		*Sandstone*, 0,5–1 m *Granite conglomerate*, about 0,5 m		*Sandstone with beds of shale and conglomerate*, 10–30 m	
UPPERMOST PROTEROZOIC OR VARANGIAN		*Vakkejokk breccia*, about 6 m	*Vakkejokk breccia*, some m thick	*Vakkejokk breccia, upwards grading into conglomerate*, 2 m	*Alta tillite*, about 10 m	*Upper tillite*, max. 60 m
	Sandy shale with cyclic sedimentation and Spriggia *fossils*, 2 m			*Sandy shale with cyclic sedimentation and* Spriggia *fossils*, 4–5 m		
	Sandstone with fossil tracks, 5 m		*Shale and sandstone with tracks of fossil*	*Sandstone*, about 2 m	*Quartzite and shale*, 100–150 m	*Sandstone and shale*, maximum 400 m
	Sandy shale with cyclic sedimentation and tracks and remains of fossil, about 12 m		about 20 m	*Sandy shale with cyclic sedimentation*, 8–9 m		
	Sandstone with basal arkosic conglomerate, 9 m	*Sandstone with arkosic basal levels*, about 12 m	*Sandstone with arkosic conglomerate to breccia in its lower part*	*Gritty sandstone and arkose with basal arkosic conglomerate*, about 5 m		*Lower tillite*, maximum 60 m
UPPER PROTEROZOIC						*Quartzitic sandstone and arenaceous shale*, about 1200 m
OLD PRECAMBRIAN			*Old Precambrian basement*			*The above bedrock sequence probably lying directly on Old Precambrian basement*

Fig. 127 *An attempted correlation between some Northern Scandinavian Late Precambrian–Cambrian autochthonous sequences*

formations in the Tanafjord area. The quartzite and shale below the Alta tillite are supposed by Føyn to be counterparts of part of the sediments above the lower Tanafjord tillite. The thickness of the deposits of the Tanafjord area suggests that in Late Precambrian–Cambrian time the area was of geosynclinal character, whereas the Alta and Torneträsk areas, with relatively thin sedimentary deposits, would be part of the eastern border of the same depression. If one were to compare the autochthonous sediments of the Torneträsk area with those of the Alta area only, the sedimentary beds below the tillite of the Alta area would be looked upon as counterparts of the sediments below the Vakkejokk breccia in the Torneträsk area. But correlation of the deposits in the two last-named areas with the Tanafjord sequence causes one to consider whether the thick sedimentary formations between the upper and the lower tillite formations of the Tanafjord area are duplicated by deposits in the geosynclinal border zone.

The autochthonous sediments designated as being of Lower Cambrian age south of the Norrbotten mountains, i.e. along the eastern border of the Västerbotten and southern Swedish Mountain Caledonides, differ in some respects from those of the Norrbotten area. Firstly, no fossiliferous sediments of the Lower Cambrian faunal zone 4 have been discovered. Secondly, sediments bearing fossils of Lower Cambrian zone 3 persist along the entire border zone from southernmost Norrbotten to northern Dalarna, where the Caledonian border zone crosses the international border. Exceptionally, sediments with fossils of Lower Cambrian zone 2 are met with. The *Laisberg sandstone* mentioned above in the Laisvall–Laisberg area crosses the Norrbotten–Västerbotten county boundary and runs about 110 km southwards to the Vojmsjö area of middle Västerbotten, where the sandstone beds wedge out and disappear from the eastern border zone. To the south of the Vojmsjö area, half-way between Lake Vojmsjön and Lake Malgomaj, calcareous sandstone with *Volborthella tenuis*, a guide fossil of Lower Cambrian zone 2, occurs. Together with the sandstone, thin formations of phosphorite conglomerate and arkosic sandstone lie on the Precambrian granite basement. However, the sparse exposures examined are not sufficient to work out the stratigraphy. The stratigraphical relation of the Laisberg sandstone to the *Volborthella*-bearing sandstone is unfortunately not fixed. However, both of them disappear from the mountain border zone in the southern part of the Västerbotten mountain border. No autochthonous sediments with fossils of Lower Cambrian zone 2 have so far been discovered along the southern Swedish mountain border—only sedimentary layers with Lower Cambrian zone 3 fossils are met with in this zone. Good exposures of these sediments are, however, quite sparse. The sediments with zone 3 fossils are usually greenish grey to grey shales, marly shales, impure limestones and sandy sediments. No general stratigraphy has been worked out. In the Idre area of northern Dalarna, lead ore deposits were discovered in the autochthonous sediments of the Caledonian border zone. By drilling around Vassbo within the Idre area the stratigraphy of the autochthonous sediments and the extent of lead ore mineralization were studied. According to Tegengren (1962) the Lower Cambrian beds below the Middle Cambrian black shales are about 25 m thick and are composed, from the top, of the following formations (see Fig. 9 in Tegengren's report):

0,1 m conglomerate
9 m quartz sandstone with galena
7 m calcareous sandstone
5 m clayshale
3 m conglomerate and arkose

The basement is composed of Jotnian Dala sandstone with dolerite sills.

In the Clayshale formation a specimen of *Torellella* was found in the shale of a drill core. During the final decades of the last century *Torellella laevigata* and other fossils belonging to zone 3 of Lower Cambrian were found in the sedimentary sequence listed above. However, the calcareous sandstone and bottom conglomerate have to date yielded no fossils.

Middle Cambrian or Paradoxides beds

The Middle Cambrian in Sweden is stratigraphically divided into three stages. These are: 3. the *Forchhammeri* stage, 2. the *Paradoxissimus* stage, and 1. the *Oelandicus* stage. These stages are in turn subdivided into zones. Sediments with fossils from all three stages are found in several parts of the autochthonous sediments of the Caledonian border zone. However, exposures with fossil-bearing Middle Cambrian sediments are extremely few. Most fossil finds are made in local boulders.

In most parts of the eastern border zone of the Norrbotten mountains, schistose black shale rests on Lower Cambrian sedimentary deposits and is followed by thrust rocks. The original thickness of the black shale is never preserved. As the Norrbotten black shale hardly ever contains nodules of stinkstone, in which fossils can be found, consideration of it as Middle Cambrian or partly Middle Cambrian is based on the fact that it lies on sediments with fossils of the uppermost faunal zone of Lower Cambrian. In the southernmost part of the Norrbotten frontier, badly preserved trilobites have been found in stinkstone intercalations in the black shale, which seems to be autochthonous. The fossils belong to the *Oelandicus* stage.

Along the eastern marginal zone of the Västerbotten mountains the autochthonous Middle Cambrian sediments are in black shale facies. The Västerbotten black shale is, however, generally much better preserved than that in Norrbotten. The author examined the collected fossils (KULLING, 1955) and found that most faunal zones within the three Middle Cambrian stages were present. Stratigraphical gaps within the sediments were, however, hard to locate, as most of the fossil-bearing stinkstones were collected as local erratics.

The autochthonous Middle Cambrian sediments in the Jämtland border area are composed of shales and conglomerates. Among the conglomerates, the *Exporrecta* conglomerate of the *Forchhammeri* stage may be mentioned. The lower part of the Middle Cambrian is also made up of conglomerate beds. Thus, the autochthonous Middle Cambrian of the Jämtland border zone is of a littoral facies. Middle Cambrian black shale occurs, however, in the Cambro–Silurian thrust complex to the west. The northern continuation of the littoral Middle Cambrian sediments in Jämtland once covered an area east of the present Västerbotten mountain border. As a result the trend of the littoral zone during Middle Cambrian times was not conformable with the actual NNE–SSW-running erosional limit of the sediments, but was more NE–SW directed. The distribution of the Lower Cambrian Laisberg sandstone in the Västerbotten border zone, described previously, brings about the same conclusion.

In the southern part of the Swedish Caledonides in Härjedalen and northern Dalarne, autochthonous Middle Cambrian rocks occur according to scattered observations. The zone with *Hypagnostus parvifrons* of the *Paradoxissismus* stage, bearing the guide fossil, is located in black shale situated immediately below the base of the lowermost thrust unit of the area.

Upper Cambrian or Olenus beds

The autochthonous Upper Cambrian sediments are restricted to the middle part—the Västerbotten–Jämtland part—of the Swedish Caledonian border zone. In any case, no sediments with Upper Cambrian fossils have been found in the Norrbotten mountain marginal zone or in that of Härjedalen and northern Dalarne. The Upper Cambrian sediments are black shales with lenses of stinkstone. As the Middle Cambrian beds in Jämtland were composed of conglomerates with intercalated clay shale, the deposition of Upper Cambrian black shales in the area suggests a transgression of the sea in an easterly direction. The fauna of the Upper Cambrian black shale is rather monotonous, mainly with representatives of the trilobite families *Olenidae* and *Agnostidae.* The trilobites occur in stinkstone concretions. The stratification of the black shales has been largely disturbed, and their thickness is not exactly known. Most of the examined trilobites are from stinkstone boulders collected in local till.

In Sweden the Upper Cambrian in Olenid shale facies is divided into 8 biostratigraphical zones. The upper 4 zones are called the *Peltoura zone group*, and the 2 lowermost zones the *Olenus zone group.* The table below lists these zones.

Black shale of lower Upper Cambrian age is fairly widely distributed, and occurs between middle Västerbotten in the north and southern Jämtland in the south. The middle and upper parts of the autochthonous Upper Cambrian are reported to be well represented only in selected parts of Jämtland. In middle Västerbotten fossils from the *Peltoura scarabaeoides* zone (see table) have been collected in several localities. However, it must be borne in mind that the black shale areas of the mountain border zone are covered by often thick Quaternary deposits; that most fossils have been collected in till material; and that thrusting has probably carved away very large slices of the crumpled autochthonous sediments. The primary distribution of the different Upper Cambrian faunal zones along the actual mountain

UPPER CAMBRIAN ZONES		VÄSTERBOTTEN	ÅNGERMANLAND	JÄMTLAND
PELTOURA ZONE GROUP	*Acerocare*			?
	Peltoura scarabaeoides	+		+
	Peltoura minor			+
	Protopeltoura praecursor			+
	Leptoplastus and *Eurycare*			+
	Parabolina spinulosa and *Orusia*			+
OLENUS ZONE GROUP	*Olenus* and *Agnostus obesus*	+	+	+
	Agnostus pisiformis	+	+	+

border area is due to these and other circumstances which are little known. According to THORSLUND's investigations (THORSLUND, 1960; see Table on p. 179) a stratigraphical gap of varying magnitude occurs in different parts of the autochthonous Upper Cambrian of Jämtland, between the Upper Cambrian and Lower Ordovician. THORSLUND's field studies are still in progress and only preliminary reports have been published. However, it is clear that the biostratigraphical gap in the later part of the Upper Cambrian indicates that the easterly directed sea transgressions, which took place during the lower and middle Upper Cambrian early in the epoch, were followed by regressions.

ORDOVICIAN

In the eastern autochthonous strata, Silurian sediments are missing. The Upper Ordovician is represented only by the *Tretaspis* shale and limestone, both belonging to uppermost Caradoc. From the northern Swedish mountain border down to the middle part of Västerbotten, Ordovician sediments are lacking. In the area between the large mountain border lakes of Storuman and Vojmsjön in Västerbotten, autochthonous Ordovician sediments are exposed in several places below the Lower and Lowermost thrust rocks, on the long stretch towards and across the international border in northern Dalarne in the SSW. In Jämtland the autochthonous Ordovician is certainly more completely surveyed than in other, northern and southern areas (THORSLUND, 1940 and 1960), and the stratigraphical rock sequence is more completely preserved (Fig. 128).

In the Västerbotten area only the *Limbata* limestone of Upper Arenig age, the Ceratopyge limestone, and the Ceratopyge shale of Tremadoc age can be distinguished. In the Tåsjö area of Ångermanland, the shale containing *Dictyonema* of Tremadoc age is probably autochthonous. Autochthonous Ordovician "Orthoceratite" limestone has long been known in the Härjedalen mountain border region, particularly in the Glöte area. Recent diamond drilling in the Glöte area has penetrated the autochthonous sedimentary sequence. The results of the core examination have not yet been published.

Tremadoc. The *Dictyonema* shale is only found north of the eastern part of Tåsjön. The shale is part of an imbricated sequence of strata, probably of parautochthonous character. It is of interest to note that the Ceratopyge limestone and shale occur in Västerbotten, but not in Jämtland. The Ceratopyge transgression from the west reached the southern part of the Västerbotten marginal zone, but probably not that of Jämtland. The transgression wave seems to be unconformable with the erosional border. The transgression was probably more SE-directed.

Arenig, Llanvirn and Llandeilo. As shown in Fig. 128 the Lower Ordovician in Jämtland starts with the Arenig *Difformis* beds—limy sediments with glauconite-bearing conglomerate at the base. In the Cambro–Silurian thrusts to the west the *Difformis* beds are widely distributed and indicate a western transgression. They make up the lower part of the Hunneberg stage, which increases in thickness in the thrust units to the west with intercalations of shaly beds. The *Planilimbata* limestone following above is succeeded by shales—the lower Didymograptus Shale—of western facies. Limestones of eastern facies form the upper part of the Lower Ordovician and the lowermost part of the Middle Ordovician. Above this limestone sequence, the uppermost part of which is the Llandeilo *schroeteri* limestone, a stratigraphical gap occurs and complements the main part of the Llandeilo. The lower part of the Ordovician strata are in the main well preserved, but not tectonically intact. Good opportunities for studying minor dislocations are offered in the Lower Ordovician limestone quarries near Brunflo at the eastern end of Lake Storsjön. The Lower Ordovician limestones in the Västerbotten marginal zone are all situated immediately below the Lower thrust rocks and as a result are badly preserved.

Caradoc. The Caradoc sedimentary sequence lies with a transgressive contact on an eroded surface

Black Tretaspis Shale Slandrom Limestone	*Caradoc*	UPPER ORDOVICIAN
Limestone Shales Limestone "Loftarstone" Conglomerate Sedimentary Breccia	*Caradoc*	MIDDLE ORDOVICIAN
Stratigraphical gap Schroeteri Limestone	*Llandeilo*	MIDDLE ORDOVICIAN
Platyurus Limestone Vaginatum Limestone	*Llanvirn*	MIDDLE / LOWER ORDOVICIAN
Lepidurus Limestone Limbata Limestone Billingen Lower Didymograptus Shale Hunneberg Planilimbata Limestone Difformis Limestone	*Arenig*	LOWER ORDOVICIAN
Stratigraphical gap	*Tremadoc*	LOWER ORDOVICIAN

Fig. 128 *The autochthonous Ordovician rock sequence in Jämtland*, according to THORSLUND, 1960 (cf. Fig 138)

of underlying Cambro–Ordovician sediments, or on the Precambrian basement, mainly of granite. In the Lockne area, situated south of easternmost Lake Storsjön, it can be shown that the horizontal lithological changes within the same stratigraphical unit are conditioned by sedimentation in the relatively broken topography of the coastal area. The Caradoc sedimentation started with a conglomerate. Sometimes a basal sedimentary breccia separates the conglomerate from the weathered Precambrian basement. Above the conglomerate follows the *Loftarstone*—so-called from the old local name for a dark grey, coarse to fine, stratified calcareous sandstone which represents a transitional deposit between the underlying conglomerate and the limestone above. In places a reef-like locally distributed limestone occurs immediately on top of the weathered basement igneous rocks. The shales and limestone beds following on are closely connected with the underlying sedimentary formations. THORSLUND (1960) reviewed the problems of changes of level and transport of material during the Lower-Middle Caradoc of the Jämtland autochthonous, with special reference to the Lockne area. The beds mentioned above are known as the *Ludibundus* beds. "The gap below the *Ludibundus* beds within the autochthonous is of locally varying magnitude, and shows, from the stratigraphical point of view, a general increase towards the east. The changes of level indicated by it were probably not of exclusively eustatic nature, since certain facts favour the assumption of epeirogenetic movements—a tilting of the eastern Archean with the superimposed Cambrian and Lower Ordovician strata." These facts are gathered from THORSLUND's study (1940) of composition and position of the basal *Ludibundus* conglomerate. The conglomerate contains pebbles not only from the eastern Precambrian and Ordovician beds but also from Cambrian beds in places situated more than 10 km from the regions within the autochthonous which might be considered as their source. Consequently, the Cambrian beds which supplied the fragments were elevated, prior to the transgression, to an exposed position which allowed their disintegration and transport in a westerly direction over stratigraphically higher packets of strata. Thereby also the region of origin of the clastic material in the Middle Ordovician beds of the autochthonous has been indicated.

The top-beds of the autochthonous Ordovician of Jämtland are composed of a so-called "masur" limestone, the *Slandrom limestone*, and black *Tretaspis* shale—the uppermost known autochthonous sediments.

Thus, the autochthonous Cambro–Ordovician sedimentary sequence of Jämtland contains several conglomerates and four verified biostratigraphical gaps, indicating sedimentation in a coastal region or close to a coastline. In several parts of the eastern autochthonous Cambro–Ordovician, proof has been presented concerning trend directions during different Cambro–Ordovician transgressions. While the present eastern erosional limit of the autochthonous sediments runs NNE–SSW, several Cambro–Ordovician coastlines seem to have a more or less NE–SW trend.

CHAPTER XIX

THE LOWERMOST THRUST ROCKS OR THE EAST JÄMTLAND NAPPES

The easternmost units of the thrust rocks in Jämtland occupy a special position because they are composed of sparsely fossiliferous or almost entirely unfossiliferous Cambro–Silurian sediments. The eastern frontal parts of the East Jämtland nappes are not morphologically prominent because the nappe rocks are usually of the same kind as the underlying little-altered sediments. However, for short stretches in which hard rocks, such as quartzites or dragged-along portions of the crystalline basement, compose the front parts of the nappes, erosional cliffs appear. Thus the steep Östberget on Frösön immediately west of Östersund is composed of basement porphyry which forms the lower portion of an East Jämtland nappe.

On the geological map of Sweden (1958) 4 East Jämtland nappes can be distinguished to the south of Lake Storsjön, and 3 to the north of the lake. The lowermost thrusted unit, *the Skute nappe*, appears as a minor nappe unit south of Lake Storsjön. The overlying tectonic unit, *the Bjärme nappe*, is the easternmost front nappe in almost the whole of Jämtland. To the south of Lake Storsjön there follows a quite small nappe of mainly Ordovician limestone. After this comes the *Sunne nappe*; its eastern front runs parallel to the Bjärme nappe across central Jämtland. The thrusted Östberg porphyry west of Östersund belongs to the basal part of this nappe. North of Lake Storsjön the westernmost unit, the *Ås–Huså s nappe* of the East Jämtland nappes, runs in a S–N direction for a distance of about 40 km. The total width of the East Jämtland nappes in their broadest part to the north of Lake Storsjön is about 35 km.

The composition of the bedrock of the East Jämtland nappes differs in several respects from that of the autochthonous sediments. The most conspicuous difference is that not only Cambrian and Ordovician sediments (as in the autochthonous), but also Silurian sediments representing the lower part of Llandovery, contribute to the formation of the East Jämtland nappes.

A detailed stratigraphy of all the East Jämtland nappes has not so far been published. A great many observations on the sedimentary formations of the nappes have been made by THORSLUND and partly recorded in his monograph *On the Chasmops Series of Jämtland and Södermanland* (1940) and in several of his more recent papers (see THORSLUND, 1960). His main scheme for the "Eastern nappes" is shown in Fig. 129.

THORSLUND points out that according to his surveys a continuous lithological change of facies and an increasing thickness of strata can be traced in a SE–NW direction from the autochthonous to the uppermost of the East Jämtland nappes. The stratigraphical gaps in the autochthonous sedimentary sequence were diminished and filled up in the bedrock of the nappes. The lithological change of facies from southeast to northwest is most prominent within the lower to middle part of the Caradoc comprising the Ludibundus beds, but is apparent in most parts of the stratigraphical sequence.

CAMBRIAN

In the Lower Cambrian of the East nappes, quartzite predominates whilst in the autochthonous beds quartzite, calcareous sandstone and conglomerate are present. In the Middle Cambrian the lower part of the stratigraphical sequence is composed of conglomerates, both in the Autochthonous and in the East nappes, while in the upper part, the conglomerate of the Autochthonous has its counterpart to the west in black shale. The sedimentary facies of Upper Cambrian is black shale in both east and west. In the east, however, the autochthonous beds contain large quantities of bituminous limestone and some thin conglomerate layers. Furthermore, erosion has taken away considerable upper portions of the black shale sequence, and the magnitude of the primary stratigraphical gap above the remains of the Upper Cambrian black shale is hard to fix.

ORDOVICIAN

Tremadoc. The Tremadoc *Dictyonema* shale forms a shaly transgression from the west and does not reach the Autochthonous.

Arenig, Llanvirn and Llandeilo. During the Hunneberg stage, i.e. during the lowermost

Lithology	Fossils	Series	System
Limestone Shales with Quartzite Beds Kyrkås: Sandstone, Mudstone, Calcareous Siltstone		*Llandovery*	SILURIAN
Kyrkås: Mudstone, Sandstone Shales with Limestone Inclusion	*Tretaspis latilimbus* *Dicellograptus complanatus*	*Ashgill*	UPPER ORDOVICIAN
Black Tretaspis Shale Limestone			
Triarthrus Shales	*Dicranograptus clingani* *Nemagraptus gracilis*	*Caradoc*	MIDDLE ORDOVICIAN
Robergia	*Climacograptus haddingi*	*Llandeilo*	
		Llanvirn	LOWER ORDOVICIAN
"Orthoceratite Limestone" Lower Didymograptus Shale Difformis Limestone		*Arenig*	
Stratigraphical gap		*Tremadoc*	
Dictyonema Shale			
Stratigraphical gap			
Alum Shale			UPPER CAMBRIAN
Alum Shale Shale Conglomerate Stratigraphical gap			MIDDLE CAMBRIAN
Quartzite Stratigraphical gap			LOWER CAMBRIAN

Fig. 129 *The stratigraphy of the East Jämtland nappes* (according to Thorslund, 1960), cf. Fig. 138

Arenig, a considerable transgression occurred. In the autochthonous beds this event is registered by glauconitiferous and conglomeratic layers, the faunal zone of *Lycophoria difformis*. The basal beds are followed by limestone deposits. In the East Jämtland nappes this limestone becomes thicker and contains a successively increasing number of shaly beds westwards. However, during the Middle Arenig or Billingen stage other conditions prevailed. Thorslund asserts that transport of material from the eastern continental area prevailed during the Billingen stage, because in the Autochthonous and minor nappes in the extreme east the *Phyllograptus* shales are relatively thick, but thin out in more westerly situated thrust units. A comparison of facies conditions and strata thickness of the uppermost Lower Ordovician and the lowermost Middle Ordovician in the East Jämtland stratigraphical sequences is hard to complete as the stratigraphical evidence, based upon fossils, is obtained almost exclusively from the autochthonous beds. Determinable fossils are seldom discovered in the strongly compressed limestone beds of the Allocthonous. As a result the sequence of limestone formations of Upper Arenig and Llanvirn in the East Jämtland nappes is given the collective name of "Orthoceratite limestone".

Above this limestone a shaly sequence with thin limestone beds representing the Llandeilo stage occurs. The marker fossil of the dark shales is *Climacograptus haddingi*, and those of the intercalated limy beds are specimens of the trilobite genus *Robergia*. The sediments fill up the major part of the Llandeilo stratigraphical gap in the Autochthonous.

Caradocian. Within the subsequent Caradoc stage the difference in facies between East Jämtland nappes and autochthonous sedimentary sequences is considerable. The breccia sediment, conglomerates and calcareous sandstone of the Lower Caradoc autochthonous, are replaced by dark shales, particularly by *Nemagraptus gracilis*-bearing shales. Within the Middle Caradoc, the nappe equivalents of the autochthonous limestones with shale intercalations are shales only. However, in the Uppermost Caradoc both the Autochthonous and the East Jämtland nappes are characterized by about the same facies of black shale—the *Tretaspis shale*—and by limestone. The Caradoc of the East Jämtland nappes grades westwards into miogeosynclinal facies.

Ashgillian. As sediments of Ashgillian age are lacking in the Autochthonous there is no possibility of a comparison between the autochthonous and the Eastern nappes beds. The Lower and Middle Ashgillian of the East Jämtland nappes is composed of shales with nodular limestones and corresponds to the biostratigraphical zones with *Dicellograptus complanatus* and *Tretaspis latilimbus.* The topmost beds of Upper Ordovician or Upper Ashgillian are composed of quartzite and mudstone, representing the lower portion of the lithostratigraphic unit known in the literature as the *Kyrkås quartzite.* The Upper Ashgillian is thus the same as the Lower Kyrkås. The Kyrkås beds display littoral facies. A minor stratigraphical gap probably occurs immediately below the quartzite beds.

SILURIAN

The Silurian of the East Jämtland nappes belongs to the Lower and Middle Llandovery, and consists of Upper Kyrkås calcareous siltstone, mudstone and sandstone strata, followed by shale with sandy beds and limestone.

The area of the erosional remains of the East Jämtland nappes described above covers a relatively narrow zone, 35 km at its broadest part, lying mainly in the East Jämtland part of the Caledonides. The length of the area is about 160 km in a N–S direction.

CHAPTER XX

THE LOWER THRUST ROCKS

To the west of the East Jämtland nappes occur the very large thrust units of the Lower thrust rocks. In Jämtland, Ångermanland and southern Lapland the sedimentary rocks from these thrusts are often fossiliferous. In the northern part of Lapland Lower thrust sediments are rarely fossiliferous. In central Jämtland, sediments belonging to the fossiliferous Lower thrust rocks rest immediately above sediments of the Lowermost thrust rocks, i.e. the East Jämtland nappes. ASKLUND, who has surveyed the Jämtland Caledonides, prefers to call all fossiliferous Cambro–Silurian thrusted rocks of Jämtland by the collective name of *Jemtlandian nappes*—not to be confused with the East Jämtland nappes mentioned previously.

The Lower thrust rocks run continuously along the southern half of the Swedish Caledonides and appear intermittently along the northern half. Over the border the thrust masses are widely distributed in southern Norway, but there are few occurrences in the north. Thus, the Lower thrust rocks are an important part of the Scandinavian Caledonides. The bedrock of the thrust masses is of low metamorphic grade, often with schistose or cataclastic rocks. High grade imbrication is frequently a characteristic feature, particularly in eastern parts of the nappe units. The Lower thrust rocks are presumed to have been transported a relatively short distance and the overlying thrust masses which have been transported for a considerable distance seem to have dragged them along. The Lower thrust rocks, using the author's delimitation of them, are supposed by other geologists to be partly autochthonous, partly rock units transported for a considerable distance. In the following description evidence for and against the different schemes of interpretation will be given.

The composition of the Lower thrust rocks is as follows:

Fossiliferous Cambro–Silurian of partly miogeosynclinal facies. Main occurrences are in Jämtland, Ångermanland and southern Västerbotten.

Vemdal quartzite group. Earlier the name "Vemdal quartzite" was used for the thrusted quartzitic rocks to the south of the Storsjö area, and "Ström quartzite" for those to the north of the same area. In this book the widespread quartzite series is called the *Vemdal quartzite group*, in both the Swedish and Norwegian districts. Vemdal quartzite, consisting mainly of quartzitic sandstones with slate intercalations and sometimes with quartzite conglomerate and carbonate beds, occurs in most areas of the eastern marginal zone of the Swedish Caledonides. In the Storsjö area of Jämtland and along the middle of the Norrbotten border area, the Vemdal quartzite is virtually absent. The writer's discovery some years ago of a Late Proterozoic *Spriggia* fossil in sandstone just below the fossiliferous Lower Cambrian sediments in the autochthonous of the Torneträsk area, and the possibility of the tillite-like Vakkejokk breccia of the same area, lying between the fossiliferous Lower Cambrian sediments and the Late Proterozoic ones, being a (periglacial) counterpart of the glacigenous deposits of the Varanger ice age, strengthen his view that the Vemdal quartzite deposits, lying on tillite, should be considered as Lower Cambrian. In an occurrence of Vemdal quartzite in southern Norway fossil tracks of *Scolithus* character have been discovered on the quartzite bedding planes.

The Långmarkberg tillite group. Formations of tillite and varved shale, belonging to the Varanger ice age (often referred to in Scandinavian geological literature as the Eocambrian ice age), constitute the bedrock of the group. The main occurrences of tillite rocks are in northern Jämtland, Ångermanland and southern Västerbotten.

Dolomite, shale and sandstone (scattered occurrences).

The Risbäck Sparagmite group. The Risbäck Sparagmite group is of late Precambrian age. Sparagmite is composed mainly of arkosic rocks of different types. Intercalations of conglomerate, quartzite and shaly sediments occur in the often reddish-coloured coarse clastics.

Precambrian igneous and metamorphic rocks, composed of dragged-along slices of the basement. This thrust basement (overthrusted rocks of early Precambrian age) occurs in several areas as thin slices in the lowest part of the thrust unit. However, the largest areas of thrust basement rocks are in northern Jämtland, and from southern Jämtland as far as northern Härjedalen. Most Lower thrust rock areas along the Swedish Caledonides have been only superficially surveyed. As a result it is possible to give only a general review of the structures and distribution of rocks in selected areas of the thrust masses.

NORRBOTTEN

In the northern Norrbotten mountains, the Lower thrust rocks are called the *Rautas thrust complex*. The low grade metamorphic sediments of the complex are mainly grey to whitish-grey quartzites. Arkose, sedimentary breccia, dolomite, shale and exceptionally limestone are also present. The basement rocks within the Lower thrust rocks are largely composed of granite and syenite. Gneiss, acid to intermediate volcanics and basic eruptives also occur. In the low grade sediments, the only fossils so far discovered have been found in a locality on Mt. Luopakte at the mountain border south of Lake Torneträsk. The fossils present are from faunal zone 4 of the Swedish Lower Cambrian. The Lower thrust rocks are often highly imbricated, but good exposures of imbrication structure are sparse; two examples may be mentioned. One is in the mountain marginal zone north of Lake Rautasjaure, a few tens of kilometres south of Lake Torneträsk, the other west of the well-known tourist station of Saltoluokta near Stora Sjöfallet in the Kerkau mountains.

The first example (Fig. 130) shows the southern slope of Vidjafjället, north of Lake Rautasjaure. The rock slices in the rock wall consist exclusively of sediment. The basement Precambrian slice beneath the sediments seems, however, to be devoid of imbrications. But since the lower part of the thrust rocks is hidden beneath a scree and forest cover, the relationship has not been satisfactorily determined. The sediments within the imbrication nappe are quartzites, phyllites and sandy slates. Lenses of light-coloured dolomite are exposed in the upper parts of the wall. On the imbricate nappe rests cataclastic–mylonitic Precambrian granite to syenite of the Middle thrust rocks.

The second example (Figs. 131–133) occurs west of Saltoluokta on the southwestern side of Lake Langas. In the Kerkau rock wall facing the lake both the upper and the lower contacts of the imbricate nappe are exposed. The imbricate nappe lies on a thin sequence of autochthonous sediments. Below this sequence, the basement is composed of early Precambrian Sjöfall quartzite. Above the imbricate nappe rest erosional remains of the Middle thrust rocks, composed of basement-granite to -syenite, of a somewhat mylonitic

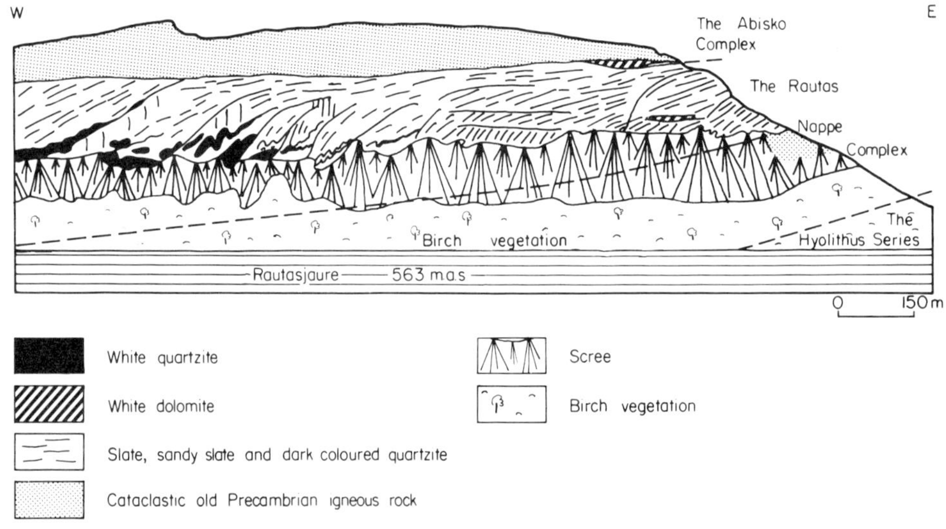

Fig. 130 *The southern aspect of Mt. Vidja to the north of Lake Rautasjaure.* In the mountain scarp the imbricate structure of the sediments of the Rautas Nappe Complex is well demonstrated. The cataclastic old Precambrian igneous rocks outcrop on the eastern slope and compose the crest area. The former one belongs to the Rautas Nappe Complex, and the latter one forms the lower part of the Abisko Complex. Sandstone and shale of the autochthonous Hyolithus group are exposed a short distance to the east of the figured section. From KULLING (1930)

Fig. 131 *Middle portion of the Kerkau rock wall facing Lake Langas.* Photo taken from the eastern shore of Lake Langas. A schematic section through the rocks of the Kerkau rock wall is shown in Fig. 132. The hill further to the right is Kaska Kerkau. The Precambrian Sjöfall quartzite extends up to the lower ledge to the left. The upper cliff with steep structures shows a section through the Lower thrust rocks with well-developed schuppen structure. Between the schuppen nappe and the Precambrian basement is a thin slice of sediments belonging to the Hyolithus sequence. Photo O. Kulling (1946). From O. Kulling (1964)

character. The composition of the imbricated Lower thrust nappe rocks is evident from the figure.

The largest area of exposed lower thrust sediments within the northern Norrbotten mountains lies to the north of Lake Torneträsk (see Fig. 134 in back pocket) in the Kuollejokk window. The northern part of the window is cut off by the international border. The window sediments consist of slate with intercalated dolomite. The white-weathered grey dolomite appears as several SSW–NNE-running formations with intervening slates. The sediments dip rather steeply to WNW; it is quite probable that the window bedrock is imbricated. However, no deep-rooted exposure of window rocks can be found.

One more example of the composition of the Lower thrust rocks in the northern Norrbotten mountains will be given. Mount Jubmotjåkko lies northeast of Saltoluokta tourist station, on the northern side of Lake Langas. The crest of the mountain consists of igneous rocks of the Middle thrust masses. Between the topmost rocks and the autochthonous sediments occur Lower thrust rocks 150–160 m thick, consisting mainly of quartzitic rocks. Intercalated in the quartzites is a sedimentary breccia 12–14 m thick. Most of the fragmented material is quartzite and sandy shale, but granite fragments occur also. The sedimentary breccia is tillite-like and was at first looked upon as a real tillite. The breccia is similar in appearance to a type of Vakkejokk breccia occurring north of Lake Torneträsk.

Along the marginal zone of the southern Norrbotten mountains only a few and small areas of Lower thrust rocks are met with, mainly in the southernmost part.

VÄSTERBOTTEN, ÅNGERMANLAND AND NORTHERN AND MIDDLE JÄMTLAND

In the eastern marginal zone of the Västerbotten mountains the Lower thrust rocks cover large areas in the east and southwest, on both sides of the southerly-protruding Stalon complex of Middle thrust rocks. In Ångermanland, in the northwest Caledonian lobe of the province, the Lower thrust rocks occupy most of the hilly area round Lake Tåsjön. Further south, in northern and central eastern parts of Jämtland, the Lower thrust rocks occur as a NNE–SSW-directed belt. In the southern part (the Storsjö part) of this belt another belt of Lower thrust rocks extends NNW and crosses the international border in the area north of Oldfjällen. Thus, the southerly-protruding Ansätten mountain ridge of Middle thrust rocks in Central Jämtland is circumscribed by Lower thrust rocks in the same way as are the Stalon Middle thrust rocks in southern Västerbotten. In the most westerly part of Jämtland, beyond the western belt of Lower thrust rocks sediments, the substratum of basement rocks outcrops as a N–S-trending anticlinal bedrock area. The bedrock area is usually called the *Mullfjäll anticline.*

During the last few decades several geologists have contributed to our knowledge of the tectonics and stratigraphy of this vast area, in particular

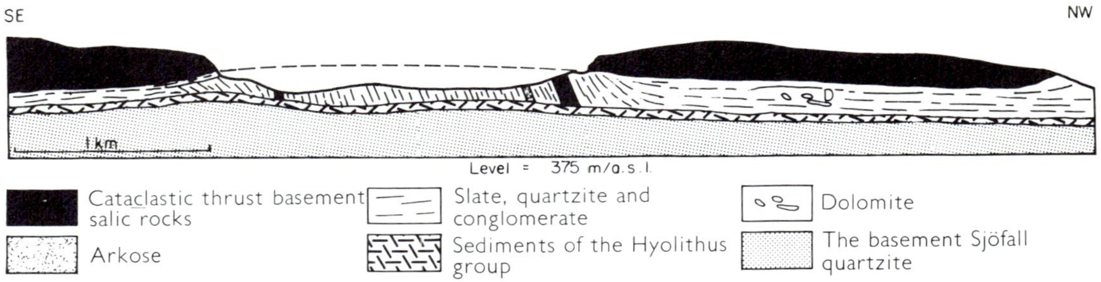

Fig. 132

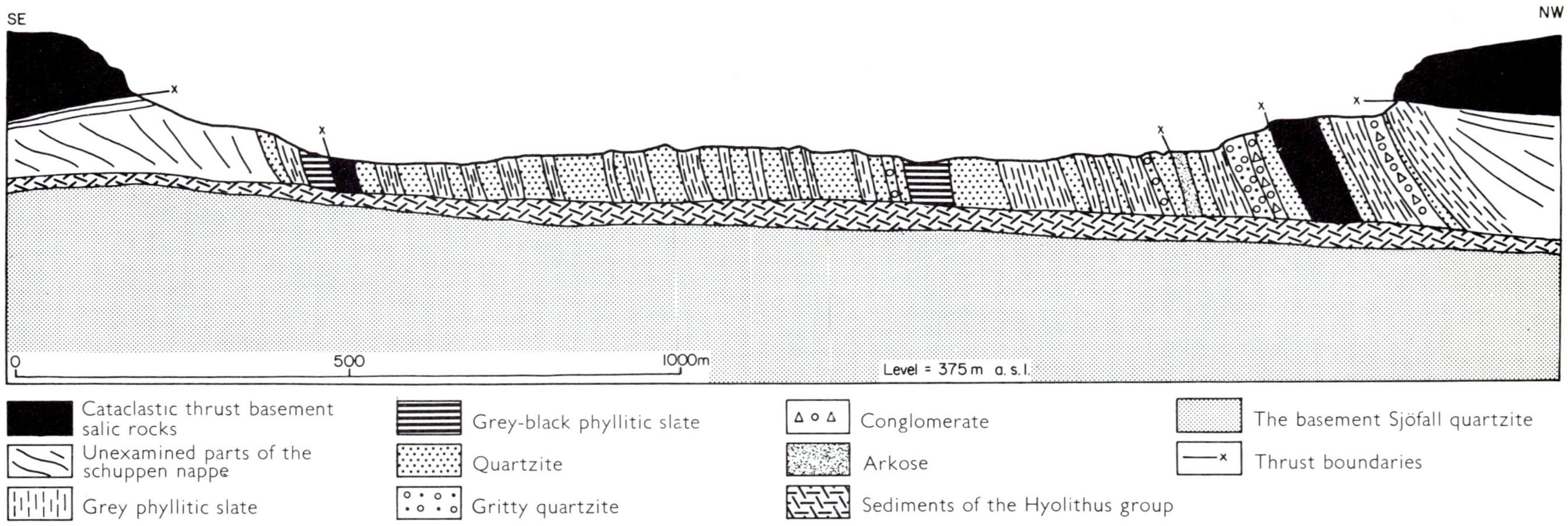

Fig. 133

Figs. 132 and 133 *Sketches of the Kerkau mountain wall facing Lake Langas in the area between the Salto tourist station and Stora Sjöfallet.* Fig. 133 is a more detailed study of the central part of the profile in Fig. 132. On the Sjöfall quartzite of the basement rests some tens of metres of sandstone and shale of the Hyolithus group. A gritty conglomerate-sedimentary breccia has been seen in the lowest part of the sediments. Where, however, the breccia occurs could not be exactly determined as the border between the sedimentary sequence and the basement was not exposed. Lower thrust rocks with schuppen structure rest on the sediments of the Hyolithus group. The rocks have in general been studied along the lip of the rock wall facing Lake Langas. In the rocks of the schuppen nappe the stratigraphy can be discerned. Three presumed schuppen boundaries are shown in the diagram, Fig. 133. In Fig. 132 dolomite lenses are included. These show up well in the rock wall because of their light colour. The Kerkau summit consists of a flat-lying nappe of cataclastic granite and syenite—a part of the Middle thrust rocks

ASKLUND, DU RIETZ, THORSLUND and the writer. Different views on the geology of the Lower thrust rocks have been put forth. The writer's view on the Lower thrust rocks is based mainly on his geological survey of the eastern marginal zone of the Västerbotten mountains. The writer has called all the Lower thrust rocks in this zone by the comprehensive name of the *Blaik nappe complex* (KULLING, 1942 and 1955). In the western part of the rock masses, however, there are several levels of localized thrusting. Many of them are recorded on the geological maps accompanying the two monographs of the writer mentioned above, and seem to be of an overthrust-fold character. The most prominent of the thrust planes runs along the N–S trending border of the thrust-basement granite, occurring in the southwestern part of the Västerbotten marginal zone, west of the southern part of the Stalon Middle thrust-rock complex. As can be seen on the eastern slope of Mt. Gittsfjället, the cataclastic bedrock-granite has been overthrusted onto Vemdal quartzite. In more northerly and southerly parts of the thrust front the granite has disappeared, and the lowest part of the thrust mass is composed of sparagmitic rocks. The writer refers to this western thrust mass as the *Gittsfjäll nappe*. The easterly situated, greater part of the Blaik nappe complex was previously called the *Vemdal nappe* (or in earlier papers the *Ström quartzite nappe*). The bedrock of the Gittsfjäll nappe in Västerbotten, and its southern continuation in northern Jämtland—the very large Svaningen area mainly composed of igneous basement rocks but with several remnants of sparagmite superposed on the igneous rocks—is considered by ASKLUND to be an autochthonous "horst". The Vemdal quartzite belt situated in the eastern mountain border zone in Västerbotten, Ångermanland and northern Jämtland is looked upon by the same author as a belt of klippen with roots far to the west, on the western side of his "horst" zone (ASKLUND, 1960). ASKLUND's view was not, however, in accordance with the data, obtained from the writer's geological mapping. During the last few years the bedrock of the eastern mountain marginal zone of northern Jämtland, Ångermanland and southern Lapland has been electromagnetically surveyed. It has been verified that the Vemdal quartzite of the eastern mountain border zone was not a klippe, but plunged to the west beneath the existing fossiliferous Cambro–Silurian sediments of mostly shaly nature, then reappeared as an eastern member of the "autochthonous horst" zone of ASKLUND. The fossiliferous Cambro–Silurian shales between the eastern and western zones of Vemdal quartzite are the core of a syncline overturned to the east. The folded Lower thrust rocks—the Vemdal nappe—of the actual marginal zone are in addition often highly imbricated, as will be shown later.

The fossiliferous Cambro–Silurian sediments, which according to the writer represent the upper part of the Vemdal nappe, are considered as part of the Föllinge nappe by ASKLUND. This Föllinge complex covers a wide area in northern and central eastern Jämtland and is composed mainly of shales, sandstone and limestones. In the area west and northwest of Lake Storsjön the sediments of the Föllinge complex plunge to the west beneath low grade sediments of the Olden nappe (see ASKLUND, 1960 and THORSLUND, 1960). The writer considers this Olden nappe as the western and upper nappe of the Lower thrust rocks. The nappe is possibly a southwestern counterpart of the Gittsfjäll nappe of southern Västerbotten and the Svaningen complex of northern Jämtland. The distribution of rocks within the parts of the Lower thrust rocks is given in Fig. 135.

WESTERN THRUST PARTS			EASTERN THRUST PART
OLDEN NAPPE	SVANINGEN COMPLEX	GITTSFJÄLL NAPPE	VEMDAL NAPPE
Fossiliferous Ordovician			Fossiliferous Lower Silurian Ordovician Cambrian
	← Vemdal quartzite group →		
	← Långmarkberg tillite group →		
		(Dolomite, shale and sandstone, scattered occurrences)	
	← Risbäck sparagmite (arkose) group →		
Basement granite and porphyry		Basement granite	

Fig. 135 *Major rock groups within different parts of the Lower thrust rocks (the Blaik nappe complex) in southern Lapland, the Tåsjö area of Ångermanland and northern and middle Jämtland*

FOSSILIFEROUS CAMBRO–SILURIAN

According to our present knowledge the Olden nappe consists only of fossiliferous Ordovician sedimentary strata, whereas the Föllinge sequence of the Vemdal nappe contains fossiliferous sediments of Cambrian, Ordovician and Silurian age.

Little is known about the fossiliferous Cambro–Silurian of the Lower thrust rocks in Västerbotten. In the Gittsfjäll nappe, erosion seems to have removed all the fossiliferous sediments. In the

OLDEN NAPPE SEQUENCE	FÖLLINGE SEQUENCE OF THE VEMDAL NAPPE			
MIDDLE WESTERN JÄMTLAND	NORTHERN AND MIDDLE EASTERN JÄMTLAND	ÅNGERMANLAND AND SOUTHERN VÄSTERBOTTEN		
	Ekeberg Greywacke		*Wenlock*	SILURIAN
	Bångåsen Shales { *M. spiralis*, *M. discus*, *M. turriculatus* } Berge Limestone Ede formation Stratigraphical gap		*Llandovery*	SILURIAN
Dark Shale	Shales with Beds of Sandstone, Dark Shale and Limestone	Greywacke and Shale	*Ashgill*	UPPER ORDOVICIAN
Greywacke and Shale	Greywacke Beds — Upper Ogygiocaris Limestone and Shale — Lower Ogygiocaris	Greywacke and Shale	*Caradoc* *Llandeilo*	MIDDLE ORDOVICIAN
Glossograptus Beds	Upper Didymograptus Shale	?	*Llanvirn*	MIDDLE ORDOVICIAN
Vaginatum Limestone Dark Shale Conglomerate	"Orthoceratite Limestone"	Limbata Limestone	*Arenig*	LOWER ORDOVICIAN
	Stratigraphical gap		*Arenig*	LOWER ORDOVICIAN
	Lower Didymograptus Shale Difformis Limestone	Lower Didymograptus Shale Difformis Limestone	*Arenig*	LOWER ORDOVICIAN
	Stratigraphical gap	Stratigraphical gap	*Tremadoc*	LOWER ORDOVICIAN
	Ceratopyge Shales Dictyonema Shale	{ Ceratopyge Limestone and Shale } Dictyonema Shale	*Tremadoc*	LOWER ORDOVICIAN
	Stratigraphical gap	Stratigraphical gap	*Tremadoc*	LOWER ORDOVICIAN
	Alum Shale	Alum Shale		UPPER CAMBRIAN
	Alum Shale Stratigraphical gap	{ Alum Shale, Conglomerate }		MIDDLE CAMBRIAN
	Shale Sandstone Conglomerate	Stratigraphical gap		LOWER CAMBRIAN
	Stratigraphical gap			LOWER CAMBRIAN
	Vemdal Quartzite group	Vemdal Quartzite group		LOWER CAMBRIAN
		Långmarkberg Tillite group		VARANGIAN
		Sparagmite Arkose group		SPARAGMITEAN
Basement Granite and Porphyry	Basement Granite	Basement Granite		OLD PRE-CAMBRIAN

Fig. 136 *Fossiliferous Cambro–Silurian (and its substratum) within the Lower thrust rocks of Southern Västerbotten, Ångermanland and northern and middle Jämtland,* according to Thorslund (cf. Fig. 138)

Vemdal nappe of southern Västerbotten the fossiliferous beds, mainly forming the bedrock in the broad, flat valleys, are almost entirely covered by quaternary deposits. The cursorily examined exposures occur mainly along brooks. In Ångermanland and Jämtland the fossiliferous Cambro–Silurian deposits are fairly well known. THORSLUND (1960) has closely examined the sediments and their fossils over a long period. THORSLUND'S division of the fossiliferous Cambro–Silurian of the Lower thrust rocks is given in the table above (Fig. 136). In the following the term *the Föllinge sequence* will be used when describing the fossiliferous sediments of the Vemdal nappe. The Cambrian strata of the Föllinge sequence consist of basal conglomeratic and sandy beds followed by black shales of Eastern facies. The stratigraphical gaps occurring in the Lower Cambrian, lowermost Middle Cambrian and uppermost Upper Cambrian of the Autochthonous and East Jämtland nappes continue to the west and appear in the deposits of the Föllinge sequence, also. The Olden thrust unit lacks every trace of Cambrian beds. Its oldest sediments—basal breccia and conglomerate deposited on the Precambrian igneous rock-basement—are probably of Upper Arenig age. In the Föllinge sequence the Tremadoc is represented by Dictyonema shale and Ceratopyge shale. The Lower Arenig sediments in the Föllinge sequence have almost the same type of facies as occur in the deposits of the Lowermost thrust units and the Autochthonous. The Upper Arenig beds of the Lower thrust rocks are little-known, since determinable fossils have been extracted only in extremely rare cases from the strongly schistose limestone beds, the "Orthoceratite" beds. In the Llanvirn, Llandeilo and Caradoc, shales and greywackes predominate among the sediments of the Föllinge sequence and the Olden nappe (Fig. 137); they are thus typical miogeosynclinal deposits. The sediments of the western nappes are of another facies in Upper Ordovician time, i.e. in uppermost Caradoc and Ashgill. In the Olden sequence dark shales predominate while in the Föllinge sequence the dark shales occupy only the middle part. The limestone below the shales seems to bear some relation to the Upper Caradoc limestone of the East Jämtland nappes and to the Slandrom limestone of the Autochthonous. The sandstone levels in the shaly beds above the dark shales indicate stratigraphical gaps in sedimentation. No deposits of uppermost Ashgillian occur—this represents a major stratigraphical gap.

After this immersion the first Silurian transgression deposited sandy beds—the Ede quartzite formation, earlier called the *Phacops quartzite*. In the East Jämtland nappes the Upper Kyrkås beds are considered by THORSLUND to be a chronostratigraphic equivalent of the Ede quartzite. The Berge limestone, resting on the Ede quartzite, and earlier called the *Pentamerus limestone*, contains a fauna in which tabulate corals and brachiopods are particularly abundant. Its age is Middle and lowermost Upper Llandovery and is comparable to part of "division 6" of the Norwegian stratigraphical system for the Cambro–Silurian. The thickness of the Berge limestone amounts in some parts of the Föllinge sequence to about 75 m and the limestone is widely distributed. It should be mentioned that an eastern part of the Berge limestone occurs as uppermost beds of the Skute thrust unit of the East Jämtland nappes. Silurian strata younger than the Berge limestone occur in the Föllinge sequence only. Above the Berge limestone there rests a sequence of graptolite-bearing shales black in colour, with calcareous shales and sandy beds in its upper part. The dark shales

Fig. 137 *Local block of graded bedded Föllinge greywacke from the Järpen area of Central Jämtland*, at the mouth of Järpströmmen hydroelectric tunnel. Photo O. KULLING (1943)

are called the *Bångåsen shales* by THORSLUND and their graptolite fauna is of upper Llandovery age. Traces of volcanic activity are very sparse in the miogeosynclinal deposits of Jämtland. However, the lower part of the Bångåsen shales contains thin layers of meta-bentonite. The ash-beds belong to the faunistic zones containing *Monograptus turriculatus* and *M. discus*. The thickest observed ash-layer lies at the base of the graptolite shales and is only 0,5 m thick. The Ekeberg greywacke, which follows above the graptolite shales, is the youngest sedimentary formation so far observed in the Föllinge sequence, i.e. in the entire Cambro–Silurian of Jämtland. The greywacke is considered by THORSLUND to be of lowermost Wenlock age. The only fossils observed lie in an intraformational conglomerate bed and are fragmentary specimens of *Favosites gotlandicus* forma *forbesi*. THORSLUND points out that the Ekeberg greywacke exhibits great lithological similarity to the greywacke of the Middle Ordovician sequence of strata of the western nappes, "which likewise has obviously obtained its material from western continental masses".

In his 1960 review of the Jämtland Cambro–Silurian, THORSLUND concludes his report with the following lines:

"By way of summary we can say that the eastern Cambro–Silurian of Jämtland consists of shallow-water deposits, laid down in a wide trough—a miogeosyncline—which probably became deeper in Middle Ordovician times. It extended along the Caledonian eugeosyncline with its wealth of volcanites, but was separated from the latter by land areas which supplied the material for most of the clastic sediments. This is evident from the predominantly western extension of these sediments and from their thickness, which is great in the west but decreases to the east. By comparison the quantities of material which have been brought from the east, and deposited in the eastern coastal portions of the basin, appear to be small. Here, however, the effects of changes of level are obvious in the succession of strata."

The stratigraphical conditions and facies relationships between the eastern and western parts of the fossiliferous Cambro–Silurian of Jämtland, the "Jämtland Cambro–Silurian field", are summarily illustrated in a schematic section shown in Fig. 138 (see back pocket).

Before finishing this review of the fossiliferous Cambro–Silurian of the Lower thrust rocks of Jämtland a few remarks will be made about the geological position of the so-called *Offerdal conglomerate*. The conglomerate outcrops on the boundary zone between the Middle thrust rocks of the large Ansätten klippe and the fossiliferous Cambro–Silurian north of Lake Storsjön. The conglomerate contains pebbles and boulders of porphyry, granite, quartz and other material from early Precambrian crystalline rocks, but no components from Cambro–Silurian sediments. Thus, the conglomerate is of sparagmite conglomerate type, as described by the writer in the Middle thrust sparagmite complex in the Västerbotten and southern Norrbotten mountains (KULLING, 1934, 1942 and 1955). The Offerdal conglomerate is looked upon by ASKLUND and THORSLUND as the uppermost formation of the Olden nappe sequence, representing the first Silurian transgression of lowermost Llandoverian age. In earlier geological descriptions covering the region (HÖGBOM, 1894 and 1920; TÖRNEBOHM, 1896), however, the Offerdal conglomerate was incorporated in the bedrock of the Ansätten (Offerdal) klippe, as its lowermost portion (see Fig. 154). The rocks immediately above the Offerdal conglomerate were "granite mylonites" and other highly altered thrust-basement rocks, of Middle thrust rock position. The writer has examined specimens of the granite mylonite mentioned above and of the Offerdal conglomerate. Both the granite mylonite and the tectonized conglomerate are of exactly the same metamorphic facies, containing porphyroblasts of epidote and mica minerals. The same state of alteration characterizes the sparagmite, sparagmite conglomerate, and thrust-basement rocks of the Middle thrust rocks (the Stalon complex) of the Västerbotten–southern Norrbotten mountains. The writer is of the opinion that the Offerdal conglomerate, together with sparagmite and other associated rocks, once formed the eastern portion of the Middle thrust rocks of the region. During the overthrust phase the eastern portion of the thrust mass probably became imbricated, and finally overridden, by its western continuation, where early Precambrian acid crystalline rocks formed the lowest part of the thrust mass. (See the description of the Ansätten area, Chapter XXI, pp. 208–209.)

VEMDAL QUARTZITE GROUP

The sediments of the Vemdal quartzite group and those of the autochthonous Lower Cambrian of the eastern border zone are much alike, with light-coloured sandstone beds and multicoloured shales predominating. The low grade thrust sediments are, however, very thick compared with the relatively thin autochthonous beds, and lack a unitary stratigraphic subdivision due to the lack of good marker horizons. The most common quartzites of the Vemdal quartzite group are light grey ones; whitish grey and dark grey quartzites are common also. Grey, red and greenish grey slates comprise important portions of the bedrock of the group. The quartzitic members usually outcrop and the shaly members are covered by quaternary deposits. Arkosic and gravel–conglomeratic beds are present, particularly in the lower part of the group. Exceptionally, thin red and light grey limestone beds occur. The only trace of life in the entire Vemdal quartzite group—probably some kind of protozoan—was discovered in some grey limestone lenses, intercalated in the quartzite rocks. The writer is of the opinion that the Vemdal quartzite group represents a western counterpart of the Lower Cambrian sediments of the Autochthonous.

LÅNGMARKBERG GROUP OF TILLITES AND VARVED SHALES

Traces of the Varanger ice age in Sweden have been discovered in the mountain marginal zone in southern Lapland (Fig. 139), northwestern Ångermanland and northern Jämtland. The name of the group comes from Mt. Långmarkberg in northern Jämtland, where the writer found a well-exposed section of old glacial rocks. On Mt. Långmarkberg, varved glacial shale with scratched pebbles (see Figs. 140 and 141) and two kinds of tillite rocks, each containing block material of different composition, were discovered. The varved shale and its immediate substratum of tillite are greenish grey to purple, and their boulder material is of roughly the same kind. The tillite formation beneath the other glacial rocks is grey to dark grey in colour. In some parts of the area, where the glacial rocks occur, the two tillite rock types have been found. In most localities, however, only one type of tillite represents the glacial deposit. As the glacial rocks rest on sparagmite rocks, sparagmite material is generally present in the tillites. Sediments of igneous and other kinds of rocks are common, also. In the block material of the usually dark grey lower tillite are grey granites, grey sparagmites and sometimes shale and quartzite common rock types. The sediments often predominate in relation to the igneous rocks. The boulder material of the upper tillite formation is to a minor extent composed of sediments. A major portion consists of granite, syenite, porphyry, greenstone, and other igneous rocks. The material of the upper tillite is supposed to be bedrock debris which has been transported for some distance. The sparagmite pebble component of the lower tillite is indicative of material which has been transported over shorter distances. In only one locality, in the most northwesterly part of Ångermanland, the bedrock surface underlying the tillite shows badly preserved ice-scratches, indicating ice movement in a NNW–NW direction. Possibly the land ice

Fig. 139 *Upper side view of varved shale with enclosed block material.* The erratics have dropped from floating icebergs and caused disturbances in the finely varved mud. The edged erratic in the lower right part of the picture is white vein-quartz, and the part of the big rounded boulder shown is grey granite. From about 10 km northeast of Risbäck village in the southern part of the marginal zone of the Västerbotten mountains, × 1. O. Kulling, collected 1936. Photo Inger Signorelli (1966)

was situated in the east. The grey granite pebbles of the lower tillite are of almost the same type, as is common in the old Precambrian Revsund granite east of the Västerbotten–Jämtland mountains. The block material of the upper tillite formation has no known eastern connections. The author is of the opinion that this material has a southerly to southwesterly origin. The finely-varved shale, with scratched and faceted enclosed pebbles, probably obtained its erratics from melting southern ice, producing floating icebergs. In his monographs on the Västerbotten Caledon-

× *1·3*

× *2*

Figs. 140 and 141 *Varved shale with erratic pebbles, and an erratic pebble with striations, faceting and subrounded corners enclosed in the shale.* The enclosed coarse clastic material was deposited in glacial clay from floating icebergs. From the Långmarkberg tillite formation, and the tillite locality on Mt. Långmarkberg in the extreme northeasterly part of Jämtland. From O. KULLING (1955) ides (KULLING, 1942 and 1955) the writer gives tables of boulder calculations and discusses the direction of ice transport during the old glaciation.

SCATTERED OCCURRENCES OF SUB-TILLITE DOLOMITE, SHALE AND FINE-GRAINED SANDSTONE

Between the tillite and underlying sparagmite there sometimes occur relatively thin formations of sediments which cannot be incorporated either in the glacigenous rocks or in the arkose sequence below. These intermediate sediments seem to be remnants of sedimentary formations, which, once deposited on the sparagmite, were partly swept away just before or during the deposition of the tillite rocks. Several types of these sediments occur as boulder or pebble material of the tillite rocks. The *Grytsjö dolomite* will be considered first. It is a light-grey dolomite more than 20 m thick which occurs immediately below the Långmarkberg tillite in an area about 10 km northeast of the eastern end of Lake Kultsjön. The overlying tillite contains scattered pebbles of light-coloured dolomite. In other nearby areas no dolomite has, however, been observed at the stratigraphical level in question. Elsewhere local erratics of dolomite and dolomite conglomerate can be seen in the neighbourhood of outcropping tillite and are considered to lie stratigraphically below the tillite. At Mt. Långmarkberg in northern Jämtland, where the upper tillite is superposed by well-preserved varved shale, the lower tillite is underlain by greyish black to black slates, sandy slates and black fine-grained sandstones. The dark grey lower tillite contains a great many pebbles of precisely the same rock types, as underlie the tillite.

RISBÄCK SPARAGMITE GROUP

The name of the group is taken from the village of Risbäck in the southern part of the Västerbotten mountain marginal zone. Within the sparagmite group arkoses and feldspar-bearing quartzites predominate. Intercalations of conglomerates, sandy shales, shales and sandstones occur. The conglomerate beds in the sparagmite sequence of the Lower thrust rocks are usually fine grained. In the same group of Middle thrust rocks a very coarse conglomerate is a characteristic type of rock. The Sparagmite group probably contains no good marker rocks. Because of the obvious lack of such guide beds, the stratigraphical subdivision of the rocks into well-defined formations cannot generally be carried out. In a selected part of the Gittsfjäll nappe, the author subdivides the Risbäck Sparagmite group as follows: Upper red sparagmite formation, *ca.* 300 m thick; Upper grey quartzite formation, more than 100 m thick; Lower red sparagmite formation, probably about 500 m thick; Lower grey quartzite formation, 100–200 m thick. In the Upper quartzite formation, quartzite and sandy shale predominate. In the Lower quartzite formation, grey quartzite and arkosic quartzite are common. In the sparagmite formations, pebbly conglomerate beds with pebbles up to 1 cm in diameter frequently occur. The material is in most cases badly sorted. Among the sparagmite rocks, a distinction is often made between red and grey sparagmite. However, the colour of the sparagmites changes from deep reddish brown to red, light red, reddish grey, greenish grey to dark grey and in some cases whitish grey. In the red sparagmites, the colour depends on the presence of red feldspar, red porphyry and other red igneous rocks, and on iron oxide impregnation or iron ore films on bedding-surfaces. There are no well-defined sequences of red or grey sparagmite—the different coloured sparagmites grade into each other. In most cases the metamorphic state of the sparagmites influences the colour. The red sparagmites are in the main structurally better preserved than the grey to greenish grey ones. By increasing metamorphism, the diffusely distributed Fe oxide in the sparagmites is concentrated into small crystals; red feldspars and rock fragments become sericitized, etc. Films of new foliated minerals not infrequently hide minute grains of red minerals. However, in some cases sparagmite rocks have initially no percentage of iron oxide or red-coloured minerals and never were of red colour.

PRECAMBRIAN BASEMENT ROCKS

The thrust basement (of overthrusted Precambrian crystalline rocks) of the Olden nappe is the largest area of exposed basement within the Lower thrust rocks of the Swedish Caledonides. The basement is composed of granite—the Olden granite—and porphyry. The basement area was previously called the *Olden anticline*, and the Swedish part of it comprises the southern half of a N–S trending dome-shaped basement area circumscribed on the northern, western, and eastern sides by Middle thrust rocks. Northeast of the Olden anticline, on the other side of the Ansätten ridge of Middle thrust rocks, there is another large outcrop of basement rocks with about the same composition as those in the Olden anticline. However, the outcropping Lower thrust basement in the Mullfjäll anticline in central western

Jämtland, and the small outcrops of such rocks within the Ordovician Olden sediments, are all composed of porphyry. The most common type of Olden granite is an even-grained perthite granite with very few dark minerals, mostly biotite. The cataclastically affected parts of the granite contain chlorite and epidote minerals and are sometimes of augengneissic habit. The granite is believed to be of Early Proterozoic age, with some relationship to the Dala granite. The porphyries of the areas mentioned above are said to be older than the granites and are mostly quartz porphyries of grey to reddish colour.

In southern Västerbotten the Lower thrust-basement rocks are for the main part composed of Revsund granite, a granite of the early Precambrian Svionian cycle. It occurs frequently in the Precambrian east of the Västerbotten mountains. The Revsund granite is usually a coarse porphyritic granite with biotite, and often some hornblende, as dark minerals. The granite is related to migmatite rocks as might be expected. The largest area of thrusted Revsund granite in southern Västerbotten is the Gittsfjäll area, west of Lake Malgomaj; this is the lower part of the Gittsfjäll nappe. Several very small lens-like bodies of Revsund granite are confined to the border zone between the Vemdal quartzite and the underlying autochthonous Cambro-Ordovician sediments. The often highly cataclastic granite shows tectonic limits against both overlying and underlying sediments. The crush rocks are looked upon as remnants of the primary substratum of the Vemdal quartzite.

THE ORMSJÖ AND STALON SECTIONS

The Ormsjö and Stalon sections are cut through parts of the Lower thrust rocks in southern Västerbotten; both have been surveyed in detail.

In the first part of this chapter, the main tectonics of the Lower thrust rocks of southern Västerbotten, southwards to and including middle Jämtland, were described. The account was based on general geological mapping of this vast mountainous area. In recent years two reports of detailed geological surveys within the marginal zone of the southern Västerbotten mountains have become available; one deals with the Ormsjö area (Du Rietz, 1960), and the other with the Stalon area (obtained from Larsson and Moberg). Both surveys have primarily elucidated the imbricate structure of the Lower thrust rocks.

The Ormsjö section. Some decades ago lead ore deposits were discovered west of Lake Ormsjön in the most southeasterly corner of the Västerbotten mountains marginal zone. The ore lay in the Vemdal quartzite of the Vemdal nappe. The Lövstrand area west of Lake Ormsjön was surveyed by detailed mapping and a number of diamond drillings. The Lower thrust rock complex of the area lies, on the whole, almost horizontal; a closer inspection of rocks and structures discloses, however, that the bedrock is highly imbricated (Fig. 142). The section is about 1,5 km in length and at least eight small tectonic units can be recognized. The stratigraphy of the Lower thrust rocks appears to be: basal arkose; quartzite; shale and quartzite; black shale. The autochthonous sedimentary sequence below the imbricate nappe starts with a thin quartz conglomerate, followed by "siltstone-shale with pebbles or layers of calcareous sandstone", partly calcareous siltstone, dark shale (with calcareous layers) and black shale. These autochthonous sediments are, however, somewhat tectonically disturbed in a more detailed sense. It is of interest to note that the stratigraphy of the sediments of the Vemdal nappe is quite different from that of the autochthonous sediments underneath.

The Stalon section. While the Ormsjö section is an example of the structure of the Lower thrust rocks in the extreme eastern border of the Västerbotten mountain marginal zone, the Stalon section clarifies the structure in the western part of the rock complex. It extends from the outlet of Lake Kultsjön in the west towards the east (to S 84° E), is 17 710 m in length, and ends immediately north of the northwestern corner of Lake Malgomaj. The section reproduced here was obtained along the wall of the so-called *Stalon tunnel,* which was blasted for hydroelectric purposes during the last decade. The bedrock walls of the tunnel were examined, several specimens of rocks collected, and almost the entire length of the tunnel photographed. The survey was directed by Moberg of the Swedish State Power Board and the bedrock examination carried out by state geologist Larsson. This unique material has very kindly been put at the disposal of the author.

The Stalon tunnel cuts through the middle of the Arksjö-Krutsjö area; this comprises the southwestern part of the Västerbotten Lower thrust rocks, which lie between the Stalon complex to the east, and the Seve-Köli complex to the west (see the geological map). In its easternmost part the Stalon tunnel has been blasted through rocks of the Stalon complex. The rest of the tunnel cuts through Lower thrust rocks, consisting of folded and imbricated bedrock of the Risbäck sparagmite arkose group, the Långmarkberg tillite group, the Vemdal quartzite group, and fossiliferous Cambrian rocks. No fossils have, however, been found in the sediments designated as fossiliferous Cambrian; but in the eastern part of the tunnel there are black slates of alum slate habit, most probably of Middle and Upper Cambrian age. Nevertheless, the presence of Ordovician sedimentary strata is not wholly precluded. Before blasting the tunnel, a great many diamond drillings were carried out in the area between Lake Kultsjön and Lake Malgomaj. When drawing the diagram of the entire tunnel (Fig. 143) it was considered worthwhile to include those core sections lying less than 100 m from the tunnel. Apart from the diagram of the entire tunnel section, there are four selected parts which have been reproduced on a somewhat bigger scale (Fig. 144). The strike and dip of all fissures and glide planes have been thoroughly measured but not those of the bedding planes. In most parts of the tunnel the bedding planes probably strike in a N–S direction, i.e. at about right angles to the tunnel course. Detailed geological mapping of the bedrock surface above the tunnel has been planned but not yet carried out.

In the selected parts of the tunnel section (shown in Fig. 144) all observed glide planes and fissures have been listed. Within a distance of 2420 m, 61 such planes were measured. Thus, one obviously supposes that the bedrock of the area is highly imbricated. Simple stratigraphical sequences and good formations of primary sedimentary thicknesses are almost absent in areas with such disturbed bedrock.

Almost any group of rocks contains a great variety of rock types. Thus, the Risbäck Sparagmite group consists of red and grey arkoses of varying grain size and percentage of feldspar, with or without intercalated slates of red, green or grey colour. Slates of different colour are also common within the Vemdal quartzite group, within which the quartzites range in colour from light grey to dark grey. In order to get a general view of the main rock distribution within the bedrock around the tunnel numerous rock types have been grouped together and the number reduced to ten, as follows: *Lower thrust rocks*—1. Grey sparagmite 2. Red sparagmite 3. Tillite 4. Limestone (some very thin beds only) 5. Red and green slate 6. Grey slate 7. Quartzite 8. Black slate. *Middle thrust rocks of the Stalon complex*—9. Basal mylonite schist 10. Middle metamorphic Stalon sparagmite.

In Fig. 144, showing the four selected sections of the tunnel, the types of rocks named have been increased by three, namely: grey sparagmite with slate intercalations; red sparagmite with slate intercalations; quartzite with slate intercalations.

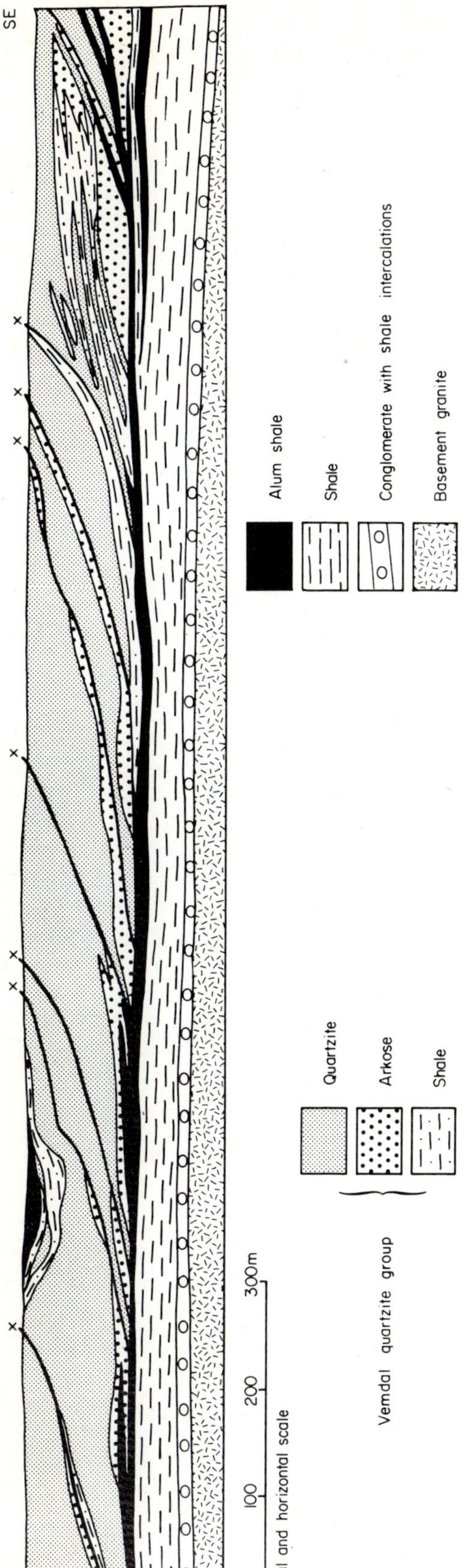

Fig. 142 *Geological section through the Lövstrand area west of Ormsjön. The eastern part of the Västerbotten mountain marginal zone.* Schematically redrawn after T. Du Rietz (1960).

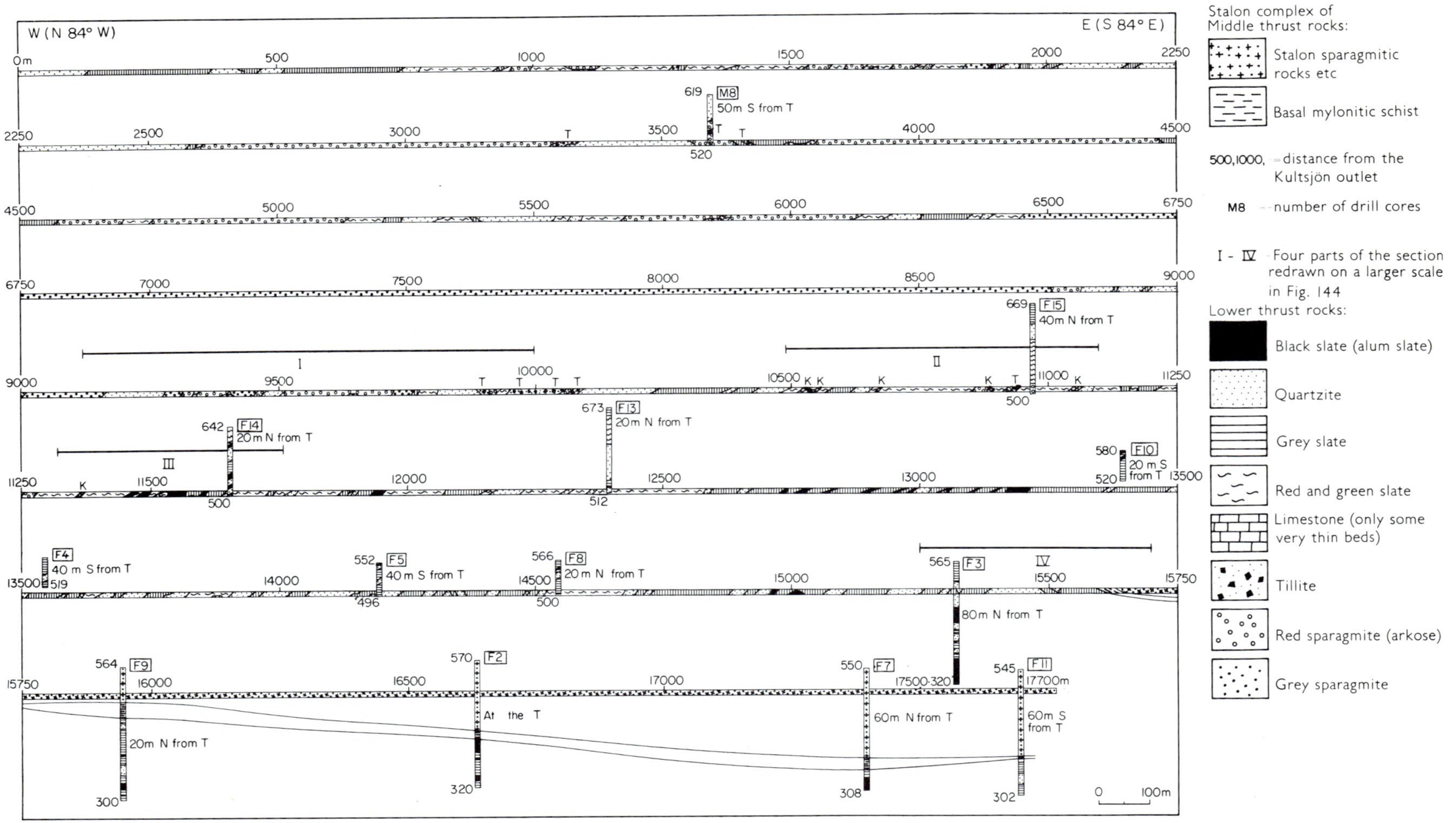

Fig. 143 *The Stalon tunnel, through the middle part of the Arksjö–Krutsjö area in the southwestern part of the Västerbotten Lower thrust rocks.* It goes from the Kultsjön outlet, in the west, eastwards (S 84° E) and ends immediately north of the northwestern corner of Malgomaj. Original by O. Kulling

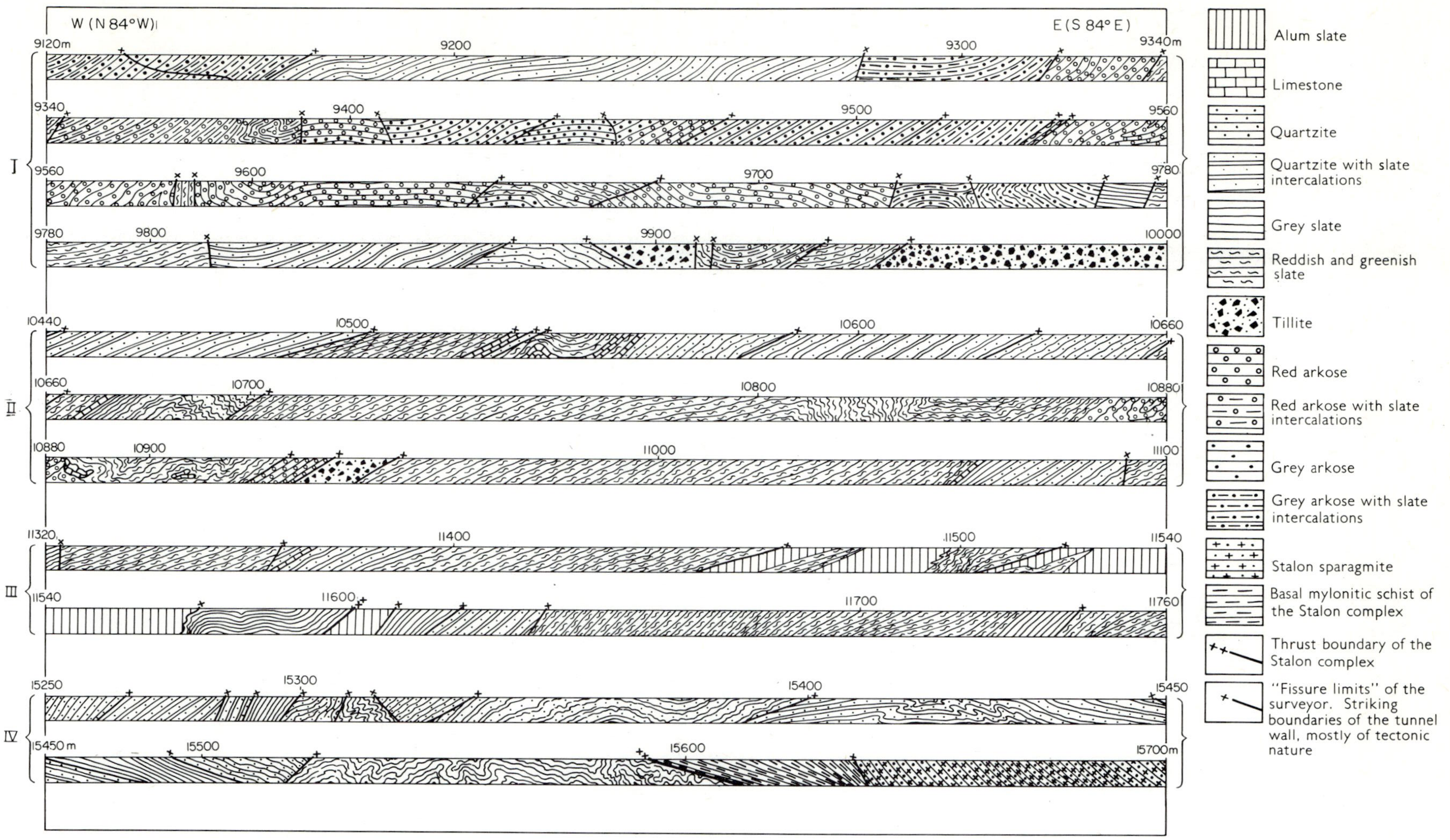

Fig. 144 *Four selected parts, I–IV, of the Stalon tunnel*, elucidating the rock structures more closely than the general tunnel section in Fig. 143. I Between 9120 and 10 000 m; II Between 10 440 and 11 100 m; III Between 11 320 and 11 760 m; IV Between 15 250 and 15 700 m from the Kultsjön outlet

The section through the entire tunnel will be considered first. There are two rather long stretches of grey sparagmite within the tunnel; the Stalon sparagmite from about 15 600 m to and beyond the east end of the tunnel; and grey sparagmite in the middle part of the tunnel from 6450 to 8750 m. The Stalon sparagmite has a bottom mylonitic zone against the underlying Lower thrust rocks. The other occurrence of grey sparagmite seems, according to the figure, to be of anticlinal character. A detailed study of photographs of the tunnel wall demonstrates, however, that the sparagmite often displays rather steep structures and has probably been thrown together in several folds, and in addition extensively imbricated. At a distance of 2,5–3 km to the south of the tunnel section lies the northern end of the overfolded and thrusted granite basement, the granite making up the lowest part of the Gittsfjäll nappe. The extension of the Gittsfjäll nappe to the north of Njakafjället could not be localized during the writer's general survey of the marginal zone of the Västerbotten mountains in 1936 (see the maps in Kulling, 1942 and 1955). It was expected that the thrust level would be clearly registered in the material of the Stalon tunnel but this expectation has not been fulfilled. The forthcoming survey of the bedrock surface above the tunnel may produce further evidence concerning this.

In the writer's monograph of 1942 the relationships of grey and red sparagmites to each other, from a stratigraphic and metamorphic point of view, were treated at some length. The conclusion reached, based on field and laboratory work, was that most of the grey sparagmites were once red ones, which, through processes of regional metamorphic and tectonic character, altered their colour to grey and greenish grey. The imbrication of the bedrock of the tunnel area hampers investigation of the primary stratigraphic relationship, *inter alia*, of the red- and green-coloured sparagmites to each other (see section I, Fig. 144). At about 9400 m and 9650 m on section I there occurs red sparagmite lying on grey sparagmite in two small tectonic packets. In other parts of the same section packets of grey and red sparagmite are bound to adjacent packets of Vemdal quartzite. In most cases, excluding the broad zone of grey sparagmite in the middle part of the tunnel, the sparagmites of the Lower thrust complex of the area are red in colour (Fig. 143). The grey sparagmite mentioned extends for about 2 km southeastwards. Between the grey sparagmite and the upthrusted basement granite 1 km further south is a zone of red to greyish red sparagmite.

Tillite rocks of the Lågmarkberg group have been observed at 5 places in the tunnel wall. Tillite also occurs in a drill core obtained near the tunnel. The tillite at 3300–3340 m will be considered first. The rock is called "tillitic shale" and is bound to red sparagmite in the west and to Vemdal quartzite in the east. The thin tillite bed at 3660–3680 lies, however, in grey shale, and the tillite in drill core "M 8" obtained from close by appears to be intercalated with the uppermost part of the red sparagmite. Of the two last-named occurrences of tillite, probably one or both of them display glide surfaces against their bordering rocks. There are 2 tillite beds at about 9900 and between 10 000 and 10 100 m; one of them lies bound to sparagmite and Vemdal quartzite, the other one is separated from Vemdal quartzite by red and green slate. Most of the contact surfaces of the tillites with the adjacent rocks are tectonic (see the tillite at 9900 m, section I, Fig. 144). The tillite at about 10 930–10 940 m is bound to red sparagmite and red and green slate, and has secondary contact-surfaces with the adjacent rocks. The positions of all tillite rocks of the tunnel area are, but for the last-named one, in accordance with an earlier statement, namely that they lie in the border zone between the Vemdal quartzite group and the Sparagmite group. The tillite (see section II, Fig. 144), together with a thin sheet of red sparagmite, is included in a folded sequence of red and green slates. According to drill core "F 15" (Fig. 143), taken within the neighbourhood of the tillite occurrence, the core slate most probably belongs to the same formation as the slate in the tunnel, and thus lies in the border zone between Vemdal quartzite and sparagmite. Some thin layers of red and white limestone are intercalated in the multicoloured slate (see section II, Fig. 144).

Eastwards of 11 450 m there are repeated occurrences of black slate, most probably of alum slate composition. No finds of fossils and lenses of stinkstone have been reported. The black slates together with adjacent grey slates and quartzite rocks are however looked upon as being of mainly fossiliferous Middle–Upper Cambrian age, as mentioned before.

From 15 600 m to the eastern end of the tunnel the bedrock is part of the Stalon complex. Four drill cores from this area show that the substratum of the Stalon complex is composed of black and grey slates with some intercalated quartzite.

The rocks of the Stalon Middle thrust rocks in the tunnel area seem, according to observations in the tunnel, to be very slightly imbricated. The rock structures generally dip eastwards. In only a few places are westward-dipping zones of crush-rocks found. The often intense imbrication of the Lower thrust rocks, exemplified in the foregoing description, took place during the advance of the Middle–Upper thrust complex—the Main Seve nappe.

During the writer's survey in 1936 of the Lower thrust rocks in the area in which the Stalon tunnel was later blasted, the presence of black slates, situated below dark grey tillite rocks and exposed in a hill south of little Lake Grubbsjön 7 km east of Lake Kultsjön, attracted his attention. These black slates lay in part directly beneath the tillite, partly separated from it by some 10 m of grey slate and quartzite. The lowest part of the dark grey tillite was cataclastic. Among its boulder material was black siltstone and slate, besides other sediments and igneous rocks—mainly granite and quartz-porphyry. The black slate lying between the Lågmarkberg tillite and the underlying sparagmite can be seen at only a few places, but, because of its presence in several occurrences of tillite rocks, it is presumed to represent erosional remains of a formerly (in late-pretillitic times) widespread formation of black fines.

During investigations for construction of the Stalon tunnel, several diamond boreholes were drilled in the area north of Lake Grubbsjön which now lies between the lake and the tunnel in the north. Fig. 145 gives the positions of seven boreholes and the composition of the cores obtained. Within the cores the following rocks were discernible: 1. Quartzite 2. Interbedded quartzite and slate 3. Grey slate 4. Green slate 5. Red slate 6. Black slate 7. Sparagmite 8. Sparagmite with shaly intercalations 9. Sparagmite with quartzitic intercalations. (B indicates brecciated rocks and C a Conglomeratic layer; the colour of the sparagmite is not mentioned.)

A few words will be said about the composition of the cores. In the two most southerly lying cores, M 23 and M 14, sparagmite occurs in the lower part of the cores, superposed by black slate. Possibly the sequence of black slates and quartzite rocks represents late pre-tillitic sediments; no tillite rocks have, however, been reported in the cores. In core M 23, sparagmite lies above the upper of the two black slates. The lowest part of this sparagmite is brecciated, and possibly the whole of the upper sparagmite is overthrusted. The upper portions of the cores M 19, M 22 and M 20 consist of quartzite and shale which are looked upon as representatives of the Vemdal group. Only in core M 20 does black slate occur between the Vemdal sediments and the lower-lying sparagmite rocks. In core M 22, black slates and quartzite rocks comprise the lowermost part of the core, lying beneath the sparagmite. The quartzitic portion of these basal sediments is reported to be brecciated. Possibly the bedrock above the black slates in question is overthrusted. In core M 21 there are some beds of quartzite and shale in the middle part of the thick sparagmite sequence. Whether the core sequence is tectonically generated or not is uncertain. In the upper part of core M 10 there is a 15 m thick formation of sparagmite, intercalated in a thick sequence of shale and quartzite. The "sparagmite" may be a feldspar–quartzitic part of the Vemdal group. The writer in a monograph of 1942 described these coarse feldspar-bearing quartzite intercalations in the Vemdal group.

The above comments are not claimed to clear up all the geological questions of this area. A detailed geological survey has not been carried out in the area north of Grubbsjön, but is highly desirable.

SOUTHERN JÄMTLAND, HÄRJEDALEN AND NORTHERN DALARNE

The Lower thrust rocks of the southern part of the Swedish mountains, extending from south of the Storsjön area to the point at which the Caledonide rock complex crosses the Swedish border in northern Dalarne, are almost devoid of fossiliferous Cambro–Silurian sediments as they have been eroded away. Sediments of the Vemdal quartzite group, the Hede sparagmite group, and the thrust basement granite remain. However, very large parts of the Lower thrust rocks in the southern mountains have hardly been surveyed at all.

Along the eastern border of the southern mountains Vemdal quartzite forms the Caledonian front. Sections with imbricated quartzite occur very occasionally. However, when Asklund (1960) gave a review of the southern part of the Caledonides, he described steep-dipping structures in part of the Lower thrust rocks and called them "collar" structures. As an example he quoted Mt. Hoverberget at the southern corner of Lake Storsjön. The majority of the mountain consists of thrusted porphyry. Steep-dipping Vemdal quartzite has been observed on some sides of the mountain. Asklund is of the opinion that the Vemdal quartzite was once deposited around a Precambrian mountain. The writer interprets such Hoverberg structures as products of rock imbrication of the same type as occurs in more northern parts of the Lower thrust rocks. The Vemdal quartzite with attached slices of Precambrian basement rocks, constituting the Vemdal quartzite nappe, disappears to the west beneath rocks belonging to the Hede–Rogen sparagmite complex, to thrust-basement granite, or to some eastern protrusions of Middle thrust rocks. The two first-named rock complexes constituting the

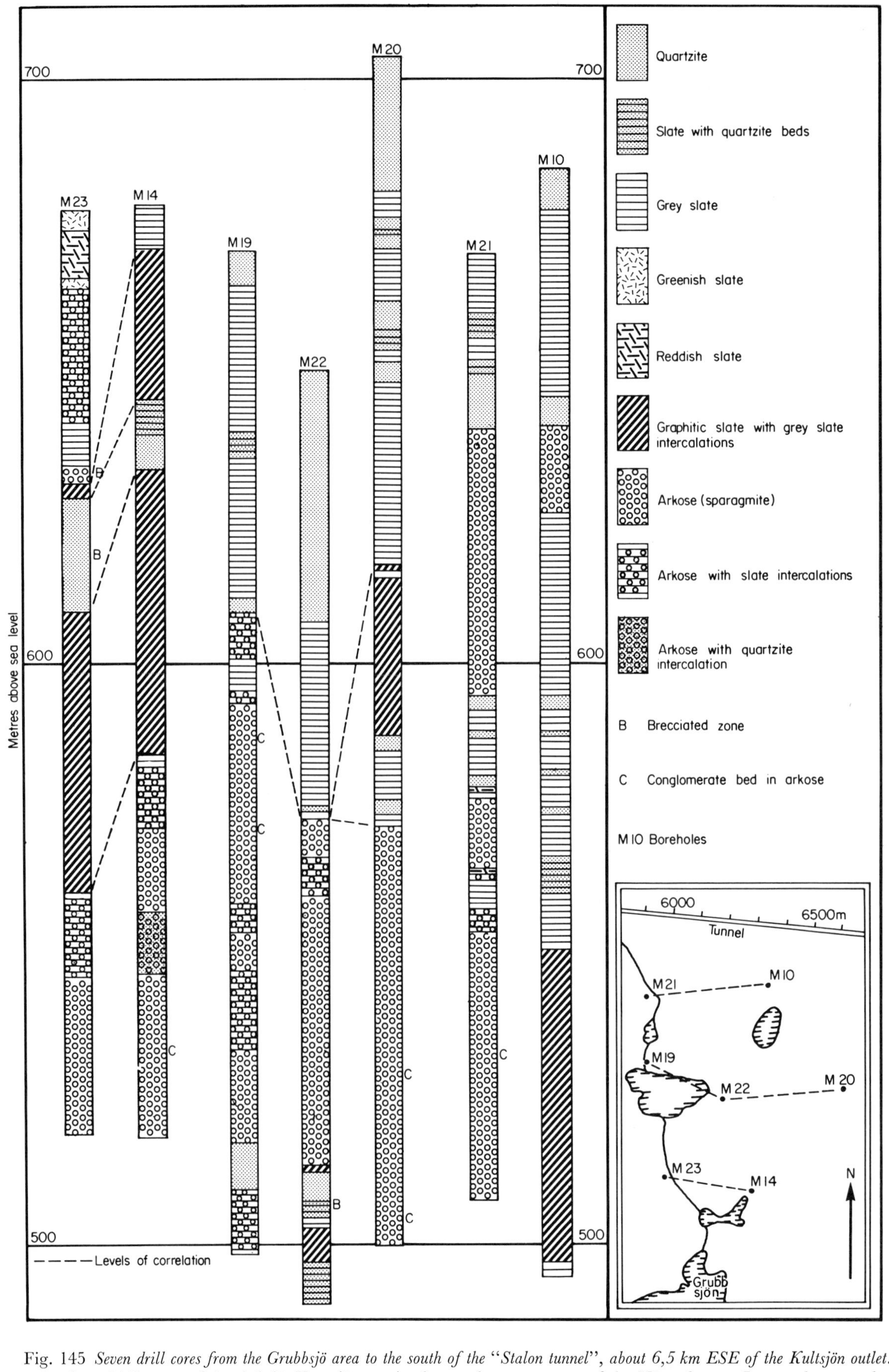

Fig. 145 *Seven drill cores from the Grubbsjö area to the south of the "Stalon tunnel", about 6,5 km ESE of the Kultsjön outlet.* In the middle part of the Arksjö–Krutsjö area in the southwestern part of the Västerbotten Lower thrust rocks. See map of the area

western and upper parts of the Lower thrust rocks in the southern mountains are described below.

In the northeastern corner of the Hede–Rogen sparagmite region—in the *Hede area*—the tectonic and stratigraphical properties of the Hede sparagmite group and the Vemdal quartzite are well known from the survey carried out by STÅLHÖS (1956). His stratigraphy of the "Sparagmite group" of the Hede area is (from the top) as follows:

Red sparagmite formation. Purple to red and green arkoses and light feldspathic quartzites with intercalations of siltstone and coarse polymictic conglomerate. Thickness up to about 200 m.

Grey Sparagmite formation. Grey to light grey arkoses and feldspathic quartzites with siltstone and conglomeratic intercalations. Thickness from 0 to about 100 m.

Hede limestone formation. Light to dark grey calcareous sandstones, arkoses, stratified calcareous siltstones, shales, and occasionally intraformational conglomerates and sedimentary breccias, the latter grading into tillite-like "boulder clay". Thickness from 0 to about 30 m.

The Hockla basal formation. Greywacke sandstone, dark grey shale and basal conglomerate. Thickness from 0 to 15 m.

Within the Hede area the red sparagmite has no superposed bedrock unit, but in other parts of the extensive Hede–Rogen sparagmite region superposed rocks are always part of the Middle thrust units.

Outside this small portion of the Hede–Rogen sparagmite region, information on its bedrock is very sparse. As is seen from the above table and from STÅLHÖS's geological map, the various formations of the Sparagmite group are rather unevenly distributed within the Hede area. In some parts of the area all four formations rest directly upon the basement granite. Thus, the Hockla basal conglomerate can be seen resting on the granite in the type locality at Hockla, in the western part of the Hede area. The "boulder clay" of the Hede limestone formation is deposited on the granite basement in the eastern parts of the area. The grey sparagmite and above all the red sparagmite are often deposited directly on the basement. The basement granite is strongly weathered in places and, being a lower part of a thrust unit, it is often exposed to weaker or stronger tectonic influence. The cataclastic granite has zones of mylonitization. The original red colour of the granite has changed in tectonized zones to grey and greenish grey.

On TÖRNEBOHM's geological map of the Middle Scandinavian Caledonides (1896) the red and grey sparagmite rocks, as well as the Hede limestone and basement granite of the Hede area, were distinguished. However, he called the limestone the *Biri limestone*, suggesting that the limestone at Hede is lithostratigraphically and probably chronostratigraphically the same as the Biri limestone in the southern part of the large sparagmite region in southeastern Norway. The Hede–Rogen sparagmite region and the Norwegian sparagmite region together represent one continuous sparagmite district.

With regard to the correlation across the international border of the Hede sparagmite group in the Hede area and its Norwegian counterpart in the Mjösa area, STÅLHÖS adheres to TÖRNEBOHM's use of the Biri–Hede limestone as a reference horizon. He points out that "both exhibit remarkable lithological analogies, such as rhythmical banding and local boulder clay facies" and gives the following correlation between the "Sparagmite" group of sediments:

	HEDE AREA	MJÖSA AREA
	Not observed. (Probably eroded away)	Ekre shale Moelv tillite
HEDE SPARAGMITE GROUP	Hede sparagmite Hede limestone Hockla basal formation	Moelv sparagmite Biri limestone Biri conglomerate and Bröttum sparagmite

A remarkable fact is the abundance of carbonaceous material in the matrix of the Hockla basal beds and coal enrichment in some of the Bröttum sparagmite strata. This together with their similar stratigraphical positions strengthens the correlation between the Hockla and Bröttum beds.

The Vemdal quartzite sediments are also dealt with by STÅLHÖS. He distinguished a lower division of quartzite rich in feldspar, and an upper one of quartzite in which feldspars were sparse. This is in accordance with the well-known stratigraphy of the Vemdal quartzite on the Norwegian side of the border, where the lower portion is called the *Vardal sparagmite* and the upper portion the *Ringsåken quartzite*.

To the north of Hede, large areas of thrust-basement granite form the substratum of the Sparagmite group. In the Hede area a small portion of the same granite is thrusted locally to the east above the sparagmite rocks. The northern and western limits of the small granite areas are considered as fault lines by STÅLHÖS. The eastern

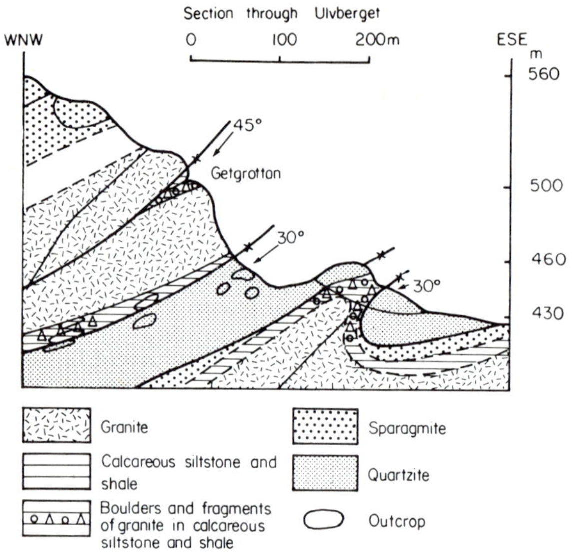

Fig. 146 *Section through the imbricated bedrock of Ulvberget about 5 km east of Hede village in Härjedalen.* From G. STÅLHÖS (1956)

frontal part of the Sparagmite–granite nappes has an imbricated structure, as shown in Fig. 146.

Before leaving the problem of the Lower thrust rocks of the southern Swedish mountains, a few remarks will be made about the *Ovik quartzite* in the area 40–80 km north of Hede and immediately south of the Storsjön area. In the northeastern part of Mt. Oviksfjällen the basement porphyry substratum of the quartzite outcrops. To the north, the Ovik quartzite disappears beneath shales and greywackes which, according to the geological map of Sweden (1958), are of Ordovician age. The Ovik quartzite is considered on the map as a southern part of the Olden nappe. The quartzite differs from the usually whitish to grey Vemdal quartzite in more easterly areas by its often darker, blue-black to dark grey colour. Many years ago (see WIMAN, 1893) Upper Cambrian trilobites were discovered in black shale intercalations in the Ovik quartzite beds. Probably the entire Ovik quartzite sequence represents Cambrian sediments deposited near a shore. On the geological map of 1958 and in the accompanying description (ASKLUND, 1960) the Ovik quartzite is considered as being older than fossiliferous Cambrian.

CHAPTER XXI

THE MIDDLE THRUST ROCKS

In his standard monograph on the Caledonides of Central Scandinavia, TÖRNEBOHM (1896) gave an account of the formation of the old mountain chain. He demonstrated the extent of large over-thrusting and distinguished between thrusting of first and second order. TÖRNEBOHM's term *the main nappe of large thrusting* (Swedish: "Den stora överskjutningens huvudskolla") for the distant thrusted rocks in the Central Scandinavian Caledonides corresponds principally to the concept of *the main Seve nappe* in this paper on the Middle and Upper thrust rocks taken together. After surveying the bedrock of the high Sarek mountains in northern Lapland, HAMBERG (1910) arrived at the conclusion that the overthrust masses were composed of an upper and a lower nappe. On Sarek the upper one is to a large extent composed of metamorphic basic igneous rocks and HAMBERG called it *the Amphibolite nappe*; the lower one with a bedrock mainly of syenite was called *the Syenite nappe*. Taken together, the upper and lower thrust units of HAMBERG correspond to TÖRNEBOHM's Main nappe. The Sarek syenite nappe is thus the same as the writer's Middle thrust rocks of that area.

The concept of large thrusting, still generally accepted in 1910 (the year of the XI International Geological Congress in Sweden), was abandoned for some decades by Scandinavian geologists shortly afterwards, but regained its influence on geological thought in the thirties (KULLING, 1930; ZENZÉN, 1932; ASKLUND, 1933). Both TÖRNEBOHM's and HAMBERG's concepts have had their advocates during the last thirty years.

The author is of the opinion that parts of the Middle thrust rocks, together with the Upper thrust rocks, have their roots in the eugeosyncline. The former are a slice of the Precambrian basement below the eugeosyncline, often together with the very lowest sedimentary units resting upon it. The latter are geosynclinal rocks of eugeosynclinal type. During the phase of overthrusting a movement zone arose within the middle portion of the thrust complex, due to different abilities to transmit pressure in the upper and lower parts of the thrust rocks. Later, a more or less tectonic boundary arose between the upper and lower parts of the large thrust mass, and the upper limit of the Middle thrust-rock complex was formed.

The regional metamorphic alteration of the Middle thrust rocks is usually of middle grade, biotite facies, but their tectonization is often very strong. Mylonites, cataclasites and tectonically banded rocks are common constituents of the rock complex.

Middle thrust rocks occur along the whole of the Swedish Caledonides. In some segments of the mountain chain, the rock complex mainly composes the eastern portion; in other segments, it is common in both eastern and western parts of the Swedish mountains.

The most common constituents of the Middle thrust rocks are acid igneous early Precambrian (Archean) rocks such as granite, syenite and gneissic rocks. In some segments, large slices of sparagmite (arkosic) rocks, quartzites and sediments of other composition occur. The Middle thrust rocks are frequently composed of an upper sedimentary portion and a lower granite–syenite portion. The Middle thrust rocks have been (with varying degrees of justification) subdivided into different tectonic units in some areas.

The Middle thrust rocks in different parts of the Swedish Caledonides often have local names, alluding to the geography or to the composition of the bedrock.

The description of the Middle thrust rocks below will proceed from north to south. Every district has contributed something to the general picture.

PÄLTSA AREA

In the small *Pältsa area*, situated in northernmost Norrbotten on the Norwegian–Finnish

Figs. 147–148 *Two views of the eastern front of the Caledonides in the Torneträsk area*

Fig. 147 *shows the southern aspect of the lower part of the southern slope of Mt. Vaivantjåkko, to the north of Lake Torneträsk.* The mountainous, steep slope in the background is composed of mylonitic rocks of the Abisko complex. In the middle of the picture the overthrust limit between a western and an eastern schuppen may be seen. The basal part of the eastern schuppen is registered by a low wall. A very thin sheet of quartzite of Lower thrust rock type divides the Abisko complex from the autochthonous sediments, which form the lower part of the slope. The Vakkejokk breccia outcrops in a low ledge behind, i.e. to the north of, the light-looking zone short of the trees in the lower part of the slope. The levelled bedrock in the foreground represents the old Precambrian crystalline basement. Photo O. Kulling (1952)

border, a lobe of the Caledonides extends to the west of Lake Kilpisjärvi. The Middle thrust rocks are composed mainly of granite, syenite and fine-grained gneiss. The rocks are cataclastic to mylonitic. A thin sheet of fine-grained gneiss with fresh feldspars separates the tectonized igneous rocks from the superposed garnet-bearing mica schists of the Upper thrust rocks.

TORNETRÄSK AREA

In the next area to the south, the *Torneträsk area*, the Middle thrust rocks are composed mostly of a lower part of cataclastic to mylonitic acidic igneous rocks, and an upper part of banded quartzitic and phyllitic schists with intercalations of light dolomite. This was called the *Abisko*

Fig. 148 *The Luopakte–Kaisepakte portion of the Caledonian front south of Lake Torneträsk, viewed from N 40° W, from the shore of a lagoon on the northern side of the lake.* The mountain wall of Kaisepakte to the right is composed of Abisko mylonites, as are the slopes of Luopakte in the background, except for the most easterly one, which consists of quartzite of a Lower thrust rock position, cf. Fig. 110. The autochthonous sequence lies immediately below the Lower thrust rocks and its lower limit coincides roughly with the lower boundary of the newly fallen snow. Photo O. Kulling (1965)

complex (the Abisko nappe) by the writer after a type area of the Middle thrust complex to the south of Lake Torneträsk, around Abisko. The border between the Abisko complex and the underlying bedrock is usually easy to trace (Figs. 147–149). The strongly tectonized Precambrian basement rocks lie directly upon the often contorted sediments of the Lower thrust rocks (Fig. 149). At the upper border of the Abisko complex as well as at the border between the upper and lower portions of the complex, however, different conditions prevail. The reason for this is that the banded tectonites by their very nature have tectonic-looking boundaries with both the overlying and the underlying bedrock. In some areas Abisko complex type are exemplified on the map of the Påtjojaure area in the mountains southwest of Lake Torneträsk (Fig. 150). North of Lake Torneträsk occurs the beautifully folded dolomite formation in the Middle thrust rocks, visible on the steep southeast slope of Mt. Tidnotjåkko (Fig. 151), situated in the easternmost corner of the mountain marginal zone. Thanks to the white-weathering dolomite, the transversal folding which is so characteristic of the Middle thrust rocks is well displayed. In the southwestern part of the section the layers have been overfolded to the northeast. The often recumbently folded beds of hardschists are, however, just as often folded over to the southwest as to the northeast. To the west

Fig. 149 *Richly quartz-veined and contorted portion of the uppermost part of the quartzite-schist sequence of the Lower thrust rocks in the area north of Lake Torneträsk, on the mountain slope immediately west of Tuoptevaratjak.* The cataclastic–mylonitic granite–syenite of the Middle thrust rocks follows immediately above the varatjak, quartzite–schist rocks. From O. Kulling (1964)

mapping has indicated that a structural unconformity occurred between parts of the bedrock of the Abisko complex and the Upper thrust rocks (Fig. 150). Within other parts of the same rock sequence, crystalline schists of the Upper thrust rocks follow, apparently conformably, on the banded quartzites of the Abisko complex. The crystalline schists of the Upper thrust rocks show somewhat higher regional metamorphism of garnet grade and higher, in contrast to the biotite grade characteristic of the banded schists—the so-called hardschists.

Any real stratigraphy can hardly be applied to the sedimentary upper part of the Abisko rock complex because of the tectonic banding of the sediments, their intricate folding, and the numerous zones of gliding. However, a light-coloured dolomite in the hardschist beds is available as a type of marker rock. The complicated tectonics of the Middle thrust rocks of the of Torneträsk, the Middle thrust rocks consist mainly of hardschists, which have been separated from the autochthonous basement granite of the Sjangeli window by a thin sequence of sandy sediments. According to a borehole through the banded to streaky rocks of the thrust complex, layers of dolomite of various thicknesses and graphitic schists are interbedded with the hardschists. Because of strong tectonization, however, no conclusion can be drawn regarding the primary bedding.

SITASJAURE AREA

Examples of the composition and mode of occurrence of the Middle thrust rocks in the *Sitasjaure area* at the southern edge of the Sjangeli window will now be considered. Among the metasediments, the typical hardschists of the northern

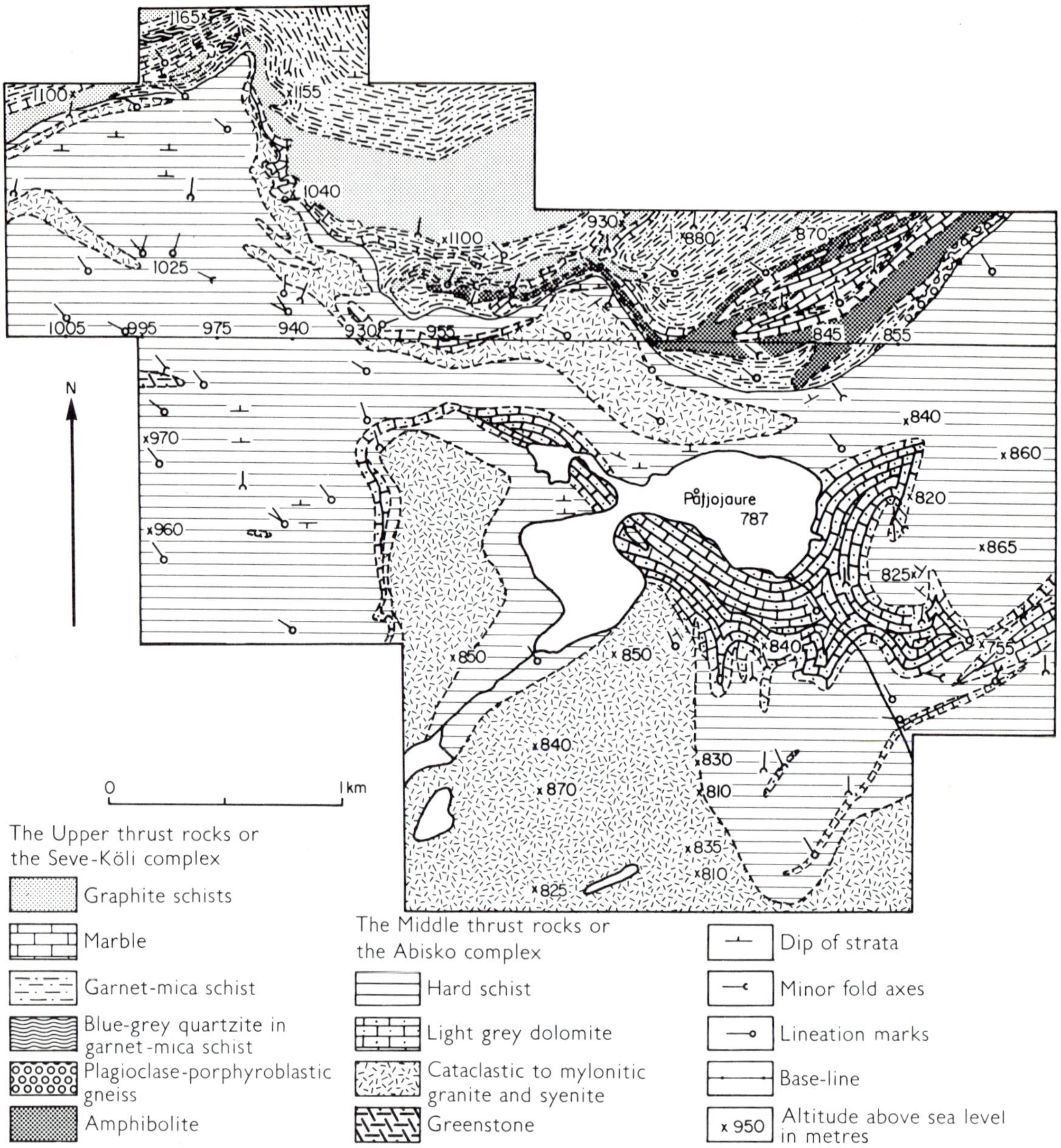

Fig. 150 *Geological map of the Påtjojaure area on the southern slope of the Western Abisko high mountains.* From O. Kulling (1964)

areas have been replaced partly by granular quartzites and biotitic mica schists around Lake Sitasjaure. For this reason the boundary between the Middle and the Upper thrust rocks has been difficult to fix in several places. The overthrust basement is largely made up of fine-grained gneissose granites and gneisses. To the east these rocks pass into mylonitic–cataclastic acid eruptive rocks which predominate in the easterly parts of the Middle thrust rocks of the mountain chain. On Mt. Muokkeris to the north of western Lake Sitasjaure, the Middle thrust rocks display imbricate structures.

AKKAJAURE AREA

The Middle thrust rocks of the *Akkajaure area* in the central Norrbotten mountains represent the widest area of cataclastic, overthrust-basement rocks in Lapland. Hardschists of Abisko type seem to be sparse in the upper parts of the Middle thrust rocks around Lake Akkajaure. Slices of low grade sediments, partly with attached little-altered basement rocks, occur in the cataclastic rock complex, and may be interpreted as units of Lower thrust rocks, tectonically included in the complex. The writer's knowledge of the

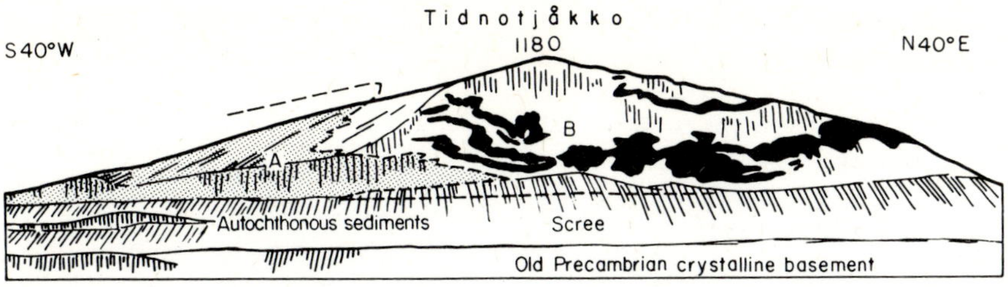

Fig. 151 *Sketch of Tidnotjåkko seen from the southeast.* The mountain forms the southeast corner of the Caledonide rocks north of Lake Torneträsk. On a base-levelled Precambrian rock surface rests the Hyolithus sedimentary sequence. The upper part of the autochthonous rocks is covered by scree. In the rock face above the scree Middle thrust rocks are exposed. The readily visible dolomite in the rock face is pale, grey-white weathered and demonstrates clearly the transverse folding and crumpling in the sediments of the upper part of the Middle thrust rocks. From O. KULLING (1960a). A Cataclastic–mylonitic granite and syenite of Precambrian age; B Hardschist and dolomite (the latter black in the sketch)

large Akkajaure thrust-basement complex is, however, very incomplete and additional mapping must be carried out.

The Akkajaure thrust-basement complex is the northern part of HAMBERG's Syenite nappe in the Sarek area, mentioned above. According to HAMBERG the major part of the igneous thrust rocks consists of syenite. In the western part of the Sarek high mountains and in still more westerly situated mountains of Upper thrust rocks, small windows of Middle thrust igneous rocks are exposed on the bottom of the valleys. The westernmost of these windows lies near the southeastern corner of Lake Virihaure and its bedrock is composed of anorthosite.

RUOUTEVARE AREA

The *Ruoutevare area* of the Middle thrust rocks lies about 50 km south of the eastern end of Lake Akkajaure, and is the only portion of the Middle thrust rocks in Lapland which holds known valuable ore deposits, namely titaniferous iron ore. This ore is a thin cake-shaped segregation occurring in anorthosite to anorthosite-gabbro. In places these igneous rocks pass into potassium-bearing rocks of monzonitic, syenitic and granitic composition and into ultrabasic rocks such as peridotite, pyroxenite and hornblendite. The ore minerals present are titaniferous magnetite, ilmenite and iron spinel. Instead of spinel the ore often contains corundum—sometimes more than 10%. Boreholes show that the cake-shaped ore is some tenths of metres thick. The high grade ore is estimated to cover an area of about 135 000 m^2, and the entire ore-containing Ruoutevare area outcrops over 150 km^2.

Occurrences of anorthosite and allied rocks have been met with in some of the "windows" mentioned above, and also in the vicinity of Kisouris in the southern Akkajaure area. Here, low grade titaniferous iron ore occurs together with the anorthosite rocks. The anorthosite rock sequence with or without iron ore segregations probably occurs widely in the Middle thrust-basement complex below the Upper thrust rocks to the northwest and NNW of the Ruoutevare ore field.

TJEGGELVAS–SAGGAT AREA

A narrow zone of Middle thrust rocks is exposed on the eastern mountain border from the mountain marginal zone east of Ruoutevare, to the north of Lake Hornavan 70–80 km southwards. The thrust rocks are called the *Tjeggelvas–Saggat complex.* The northern part of the complex is composed of Late Precambrian sparagmite rocks with conglomerate intercalations, and its middle and southern parts of as yet insufficiently surveyed schists and gneissose rocks of assumed Precambrian age.

STALON COMPLEX AREAS

A very large portion of the Middle thrust rocks, called the *Stalon complex*, extends for some 200 km along the Caledonian marginal zone, from south of Lake Hornavan in southern Norrbotten to south of Lake Malgomaj in southern Västerbotten. Within the Upper Caledonian thrust rocks west of the Stalon rock complex, "windows" of the complex are exposed. The main constituent of the complex is sparagmite with intercalated or basally situated coarse conglomerate. The material of the conglomerates is mainly early Precambrian igneous rocks. In the sparagmite area tillite-like breccias and light-coloured quartzites have been noticed and probably constitute upper portions of the sedimentary sequence, equal in age to the tillite of the Långmarkberg glacial deposits and the quartzite of the Vemdal clastics. Over wide areas, particularly along the eastern front, the sparagmite rocks descend to the base of the Middle thrust rocks. In the northern, southern and western parts of the Stalon rock complex the bottom portion of the thrust rocks consists of early Precambrian igneous rocks. In the south, mainly highly schistose greenstones, possibly to a great extent of volcanic origin, occur and in the north syenite rocks similar to those of the Syenite thrust complex in the middle Norrbotten mountain border zone. There are five "windows" of Middle thrust rocks within Upper thrust rocks in northern Västerbotten and southern Norrbotten. Sparagmite rocks occur in four of the five exposures. In the fifth—the Bångfjäll rock complex northeast of Tärnaby—the bedrock is composed of syenite and granite, with a thin quartzite sheet at the top.

The above synopsis on the Stalon rock complex reflects the writer's conception of the complex proper (see KULLING, 1942 and 1955). In a monograph (1942) on the bedrock of the eastern marginal zone of the Västerbotten mountains the western limit of the Västerbotten marginal zone was discussed, and it was pointed out that in the Lake Storuman area a boundary could not be drawn between the Stalon sparagmite and the micaceous schists to the west—the sparagmite was gradually transformed into mica schists. The writer's subsequent survey of the geological zone is very incomplete but has strengthened his interpretation of the so-called eastern mica schists as western metamorphic equivalents of the Stalon sparagmite rocks.

The sparagmite rocks of the Stalon complex are usually of middle grade, whereas those of the Lower thrust rocks are, as mentioned earlier, of low grade. The latter are arkoses containing chlorite–sericite; the former are characterized by biotite and epidote, and sometimes by garnet. The colour of the Stalon sparagmite is usually greenish grey to grey. Sometimes coarse sparagmite conglomerates, and exceptionally some very easterly-lying Stalon sparagmites of slight schistosity, are red in colour. There is a very gradual increase in metamorphic alteration of the Stalon sparagmites from east to west—from little-altered arkoses in some most easterly areas to garnet-bearing mica schists in the west. In the middle part of the Västerbotten marginal zone the crystalline schist-bearing western continuation of the Stalon sparagmite rocks has been examined only along some E–W-running sections. The primary composition of the frequently garnet-bearing schists is in many cases hard to determine.

BORGA AREA

The southern Västerbotten–northern Jämtland part of the crystalline schist-bearing Middle thrust rocks is relatively well known. The first general survey of the bedrock of the Västerbotten marginal rock zone was given by the author (1942). STAARGAARD (1955) and VAN DER HARST (1956) then carried out detailed geological mapping of the mountainous Borga area, where the Middle thrust rocks mentioned are an important part of the bedrock. In the table (Fig. 152) the formations in the Borga region, according to

GENERAL DIVISION	STRATIGRAPHICAL–TECTONIC DIVISION
UPPER THRUST ROCKS	*Amphibolite–mica schist formation* with ultrabasic rocks. Zones of strong migmatization
	Amphibolite–metamorphic arkose formation with partially metamorphic diabase
	Major thrust plane
MIDDLE THRUST ROCKS	*Quartzite* — / *Mica schist formation* with partially metamorphosed diabase } Imbricate zone
	Hardschist-bearing meta-gneiss formation with partially metamorphosed diabase
	Thrust plane
	Metamorphic arkose formation (with partially metamorphosed diabase outside the map area)
	Thrust plane
	Greenschists
LOWER THRUST ROCKS	*Phyllite formation*
	Thrust plane
	Feldspar quartzite

Fig. 152 *Stratigraphical–tectonic scheme for the bedrock of the southern, i.e. Jämtland part of the Borga area,* according to VAN DER HARST (1956)

VAN DER HARST, are listed. VAN DER HARST's subdivision of the bedrock of the Borga area is the most detailed one. The basis of classification is in accordance with an earlier one of STAARGAARD. The main tectonic division is according to the author's interpretation.

During a reconnaissance survey in the twenties and thirties, well-preserved to partially metamorphosed diabases were observed in the bedrock zone of southern Västerbotten, here considered as the western metamorphic part of the Middle thrust rocks. The diabases were conformably to discordantly intruded into a sequence of relatively highly metamorphosed sediments of mainly arkose–sandstone–sandy shale origin. The primary sedimentary composition of the often garnetiferous schists was hard to settle. In places metamorphic sparagmite-arkoses of Stalon complex type were detected. White and grey micaceous quartzite rocks reminiscent of the Vemdal quartzite were also present.

The various formations listed in VAN DER HARST's scheme will be mentioned, with reference to experience gained by the author during a bedrock survey of the southern Västerbotten areas.

The lowermost units in Fig. 152—low grade phyllite and feldspar quartzite—represent a western part of the writer's Gittsfjäll nappe of Lower thrust rocks. The feldspar quartzite belongs to the sparagmite rocks and the phyllite is most probably a lower shaly part of the Vemdal quartzite.

The formations in the table up to the "major thrust plane" are Middle thrust rocks. The greenschists are a sheet of overthrusted basic rocks and the metamorphic arkose formation is a micaceous western part of the Stalon sparagmite. The hardschist-bearing meta-gneiss formation, the mica schist formation and the quartzite formation are possibly middle metamorphic western equivalents of the Vemdal quartzite rocks. Within the Middle thrust rocks, the hardschists of the Torneträsk area seem to be of about the same composition as the hardschist-bearing meta-gneiss formation in the Borga area. The partially metamorphosed diabases intruded in the Middle and Upper thrust rocks are of particular interest. In the southern Swedish mountains diabases of the same kind intruded in the Middle thrust rocks have been called the *Ottfjäll diabase* for many years; the name is taken from Mt. Ottfjället, 60–70 km west of Lake Storsjön in Jämtland. For example, on TÖRNEBOHM's 1896 map of the Central Scandinavian Caledonides the distribution of Ottfjäll diabase is shown. In the Borga area the diabase is discordantly to conformably injected and represents probably one and the same phase of intrusion. All grades of metamorphic alteration are present among the diabases, from slightly transformed diabase with preserved ophitic structure, through uralitized types to schistose garnet-bearing amphibolite. The igneous structure is often preserved in central parts of the diabase sills and lenses, whereas the border zones are of schistose amphibolite, with a gradual transition zone between the two. The banded hardschists are sometimes intersected by diabases. The intrusion of basic magma must thus be later than the movements responsible for the tectonic banding. The intrusion of the diabase seems to have taken place during a primorogenic phase of the Caledonian orogeny before the main thrusting. There is, however, a noticeable difference in the distribution of the diabase rocks in the southern Swedish mountains. In this district the diabase is restricted to the upper part of the Middle thrust rocks. Diabase sills are common immediately below the base of the Upper thrust rocks but do not penetrate them. In the Borga area VAN DER HARST has observed several diabase dykes in the amphibolite–metamorphic arkose formation of the Upper thrust rocks. In the southern mountains, the absence of diabase dykes in the Upper thrust rocks could be taken as evidence for the presence of a major thrust plane against the diabase sill-bearing Middle thrust rocks below. In the Borga area, the opposite conclusion to that of a big thrust limit between the Upper and Middle thrust masses is reached. The two upper formations, namely the quartzite and the mica schist formations of the Middle thrust rocks in the Borga area, display imbricate structure and indicate the presence of a thrust zone. In accordance with his general views, the writer believes in a thrust zone of moderate magnitude between the Middle and the Upper thrust rocks, but not in thrusting of such a major character that the Upper and the Middle thrust rocks could have advanced to their present positions from different root-areas.

VAN DER HARST in his valuable publication mentioned thrust planes and strong differential movements in several zones of the rock sequence, and featured them on his geological map of the Borga area. The thrust planes are generally found at the junction of rock units of different lithology. The thrust zones have developed particularly within the more incompetent rock masses formed by sandy and silty rocks. The major thrust plane beneath the Middle thrust rocks, which according to the writer lies beneath the greenschists (see Fig. 152), is, however, not distinguished by VAN DER HARST. A schematic section across the Borgafjäll area is given in Fig. 153.

BÅGEDE AREA

A NNE–SSW-trending belt of mylonitic thrust-basement rocks extends from south of the Borga area to Lake Hotagen, farther south. The highly altered basement rocks are mainly of granitic composition. The southern continuation of the Middle thrust rocks of the Borga area with metamorphic arkoses and other metamorphic sediments probably occurs above and to the west of the thrust-basement rocks but seems to be included in the large western crystalline thrust mass of the region.

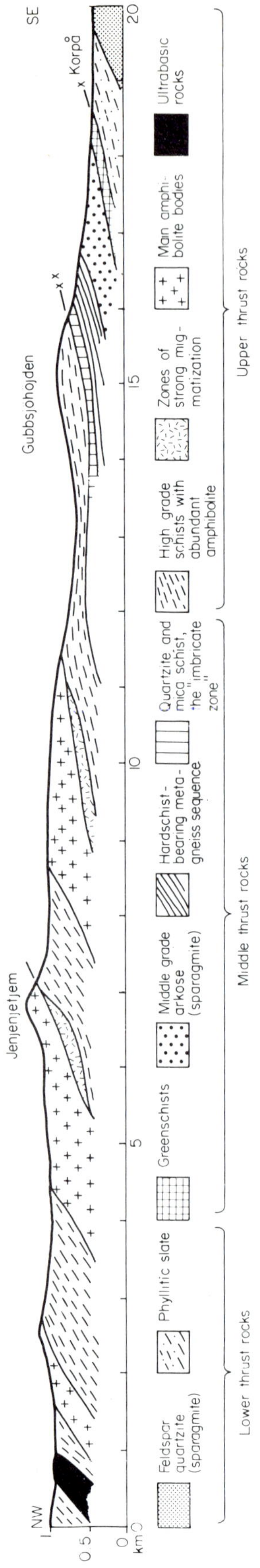

Du Rietz (1960) has called the thrust-basement rocks the "Lower mylonite nappe" and the rocks west of them the "Seve nappe". In the southern half of the zone a thin "Upper mylonite nappe" is distinguishable between the "Lower mylonite nappe" and "the Seve Nappe". In his description of the geology of the southern Swedish Caledonides Asklund (1960) refers the Upper mylonite to his Serv nappe and the Lower mylonite nappe to his Granite–mylonite nappe. The Serv nappe of the belt is, according to Asklund, composed of quartzites, arkoses, mica schists and limestones, the last-named rocks being the lowermost formation of the sedimentary sequence of the nappe.

ANSÄTTEN AREA

The Middle thrust rocks mentioned above extend from Hotagen in the north to Offerdal, about 20 km north of Lake Storsjön in the south. Törnebohm called this isolated thrust complex the *Ansätten nappe*, and looked upon it as a klippe of the Main thrust nappe; he gave local names to all large outlying parts of the Main thrust nappe. The Ansätten thrust complex is a high mountainous area, surrounded to the east, south and west by low-lying areas of fossiliferous Cambro–Silurian sediments belonging to the Vemdal and Olden nappes. The southern, lower part of the Ansätten complex is composed of so-called "hällskiffer"—a hard, micaceous schist easily cleaved into plane slabs. The rock is of economic value and is used for building purposes as flagstones, etc. The hard micaceous schist has no relict structure. In northern parts of the Ansätten complex the lower part of the bedrock sequence is composed of amphibolite rocks and granite–mylonite with relict structure. On the thrust-basement rocks, the upper central part of the Ansätten complex consists of fine-grained mica schists, greenschists and white micaceous quartzites. In the quartzite rocks diabase sills of Ottfjäll type occur (Stålhös, personal communication).

In the chapter on the Cambro–Silurian of the Lower thrust rocks of Jämtland, the Offerdal conglomerate, lying between the folded fossiliferous sediments and the "hällskiffer" of the Ansätten Middle thrust rocks, was mentioned. According to Högbom (1894 and 1920) and Törnebohm (1896) the polymictic coarse conglomerate belongs to the lowest part of the Ansätten klippe. The conglomerate sometimes occurs together with altered sparagmite rocks, and

Fig. 153 *Section across the Borgafjäll area in the most southerly part of the Västerbotten mountains.* Drawn by O. Kulling, partly after T. H. van der Harst

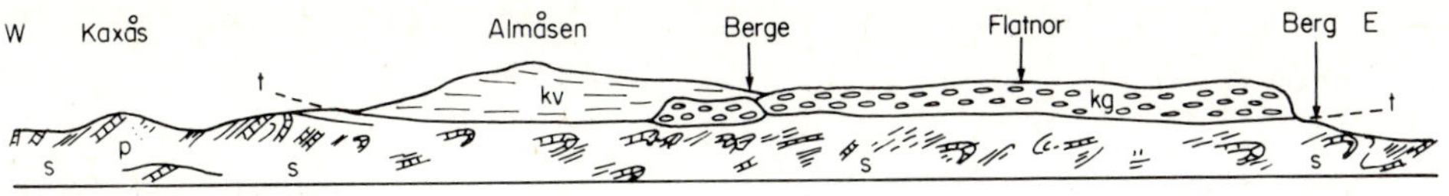

Fig. 154 *The southern aspect of the Ansätten outlier*. Sketch by A. G. Högbom (see Högbom, 1894), reproduced by A. E. Törnebohm (1896). According to Högbom the section is about 20 km long. *kv* Micaceous quartzitic schist; *kg* Offerdal conglomerate; *t* the thrust plane of the Middle thrust rocks; *s* folded-together Cambrosilurian, mostly Silurian limestone and shale; *p* Precambrian quartz porphyry

is considered as a Stalon sparagmite conglomerate, with the same geological position and state of alteration as the typical Stalon conglomerates occurring in the Västerbotten and southern Norrbotten Stalon sparagmite complex (see Fig. 154).

To the south of the Ansätten high mountainous area lies the small *Alsen* or *Landverk hill*, a klippe of Middle thrust rocks, composed mainly of "hällskiffer", with sills of Ottfjäll diabase.

SOUTHERN SWEDISH MOUNTAIN AREA

In the southern Swedish part of the Caledonian range, in southern Jämtland and most of Härjedalen, the Middle thrust rocks occupy a very large continuous area. Their western and northern portions are partly covered by Upper thrust rocks. East of the large Middle thrust complex lies the *Fuda klippe* and to the south are two rather small klippen. The biggest of them is called the *Frönberg klippe* and it is cut by the international border.

Before entering upon a description of the Middle thrust rocks of the southern Swedish mountains, a few comments will be made on the connection between this southern thrust complex and the Middle thrust rocks north of the Storsjö area. The Lower thrust rocks of the Olden anticlinal zone run from the Olden area about 40 km northwards on the Norwegian side of the border up to Bugvassfjella, and from there westwards to some 10 km southeast of Grong, where the Lower thrust-basement rocks (at this point forming the western part of the Olden nappe) disappear to the north, west and south beneath Middle thrust-basement rocks. However, the Middle thrust-basement rocks are an integrated part of the west coast basement of southern Norway. Thus, the Norwegian west coast basement rocks, together with the eugeosynclinal Trondhjem–Bergen–Stavanger rock belt, may be a western part of the Main Seve nappe.

The bedrock of the southern Swedish mountains was very carefully surveyed during the last decade of the nineteenth century by Högbom (1894) and Törnebohm (1896). Törnebohm's geological map of 1896 contains only a few errors compared with up-to-date maps; he considered the Upper and the Middle thrust rocks of the author as the main distant thrusted nappe, and divided the lower part of the thrust masses, here called the Middle thrust rocks, into an upper part of Sparagmite rocks grading upwards into crystalline schists with sills of Ottfjäll diabase, and a lower part consisting of the thrust basement of augen gneiss and granitic rocks.

During the last decade the southern Jämtland–northern Härjedalen region of the southern Swedish mountains was geologically remapped by Strömberg (1955, 1961). In his first paper on the bedrock of "Middle Härjedalen" he subdivided Törnebohm's Main thrust rocks into three nappes: the *Åreschist nappe* with crystalline schists; the *Serv nappe* with metamorphic sparagmites, intruded by Ottfjäll diabase, with a "basal series"; and the *Augen gneiss nappe*.

In his paper of 1961 on the bedrock of southern Jämtland and the northern Härjedalen mountains, Strömberg altered the name of the crystalline schist complex from Åreschist nappe to *Helags nappe* and divided the bedrock of the Särv nappe into several rock units (see Fig. 155).

Strömberg divided the augen gneiss and allied rocks in the lower part of the Middle thrust rocks in the following way: augen gneiss and blasto-mylonite; granite–mylonite and blasto-phyllonite.

Strömberg adheres to Törnebohm's belief in the thrust-basement nature of the augen gneiss, and expounds the view that the augen gneiss was formed metamorphically by a late porphyryblastesis on a mylonitized mixture of granites and basic rocks. The main part of the Serv nappe is composed of feldspathic quartzites and greywackes, with transitions to real quartzites. Törnebohm called these rocks sparagmites. Strömberg in his first paper adheres to Törnebohm's terminology. After a closer examination, he subdivided the sedimentary rocks of the Serv complex into two large groups—the feldspathic quartzite beds and the greywacke beds. The latter is to a great extent dark, fine-grained siltstone. In the southern part of the map area, quartzitic rocks are almost the only ones present; in the northern part both types of rocks occur side by side.

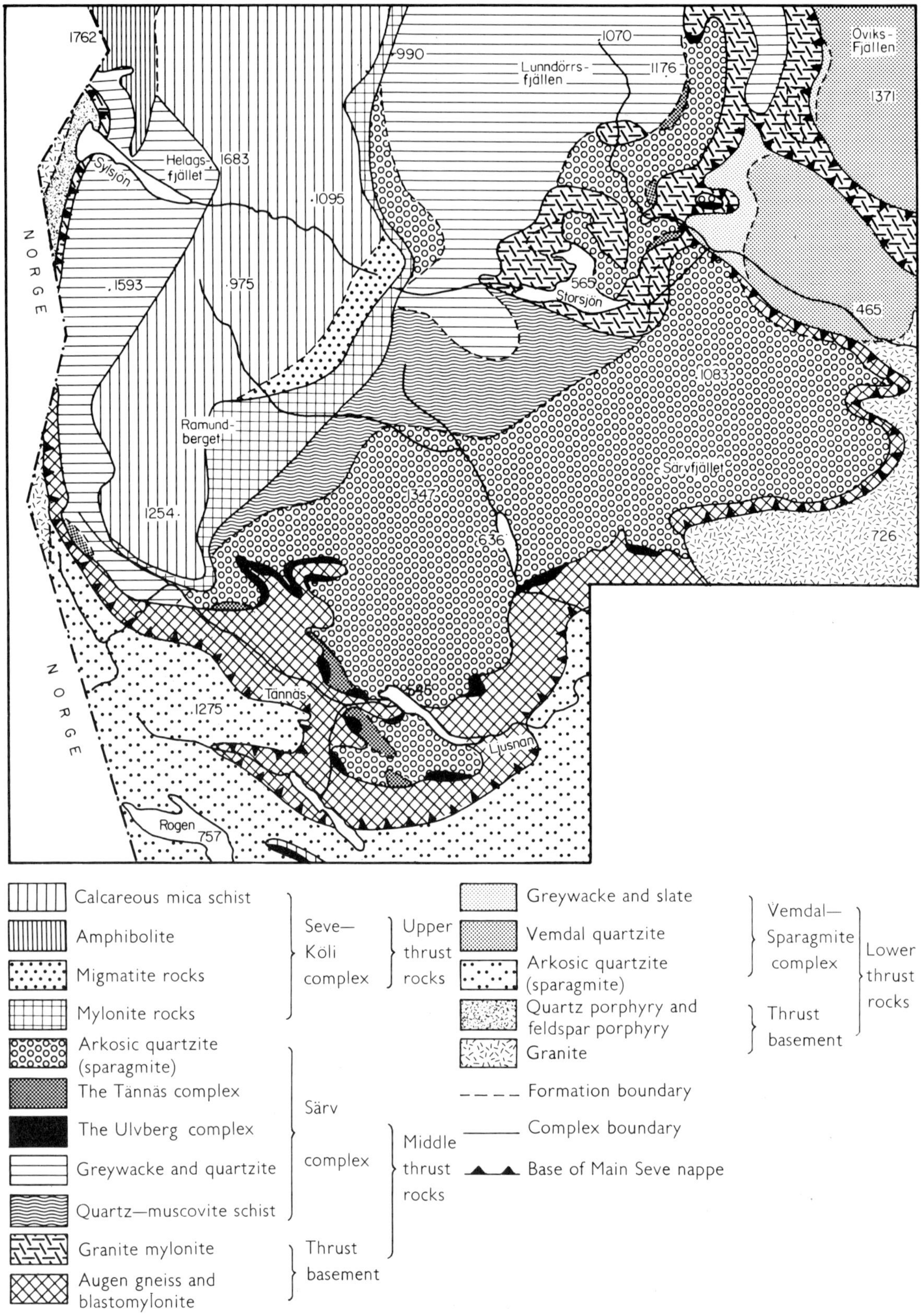

Fig. 155 *Geological map of the southwestern part of the county of Jämtland.* By A. STRÖMBERG (1961). Some minor alterations of geological and tectonic boundaries performed by O. KULLING (1965)

STRÖMBERG has pointed out that according to apparently very early tectonic disturbances and strong schistosity, the age relationships of the two series can hardly be worked out in an unambiguous way. The basal group in STRÖMBERG's first scheme —the rocks immediately above the "augen gneiss"—were divided into three formations: the *Ulvberg complex*, the *Tännäs complex* and the *Grönstack complex*. Furthermore, he considered the Sylarna amphibolite, a granite mylonite sheet immediately below the Helags nappe, and a quartz–muscovite schist formation as parts of his Serv nappe. Some comments on the rock complexes listed above will be made.

The *Ulvberg complex* contains dolomite-marble and white quartzite to feldspathic quartzite and occurs as thin lens-shaped bodies between the Serv nappe proper and the Mylonite nappe. In the eastern boundary zone of the Serv nappe the Ulvberg sediments are missing. In many localities the mode of occurrence of the rock complex bears witness to very intense tectonic disturbances and in many places the sedimentary rocks seem to have been more or less wedged out. The author is of the opinion that the Ulvberg sedimentary rocks, as they are always found lying directly on the mylonite–augen gneiss rocks, were deposited originally on the basement rock surface. Subsequent thrusting has disturbed the original sedimentary contact and caused actual tectonic contact and regional tectonization.

The rocks of the *Tännäs complex* lie in places near Tännäs in the southern part of the mountain district immediately above the sediments of the Ulvberg complex. In other parts of the district, the Tännäs rocks have been traced to the boundary zone between the augen gneiss and the lightly-coloured feldspathic quartzites, and in one locality to the north the Tännäs rocks lie as a lens-shaped body on the border between the feldspathic quartzite beds and the greywacke beds. The Tännäs complex is composed of felsitic and amphibolitic to anorthositic rocks. According to STRÖMBERG the Tännäs rocks were deformed in Precaledonian time and are similar in composition to some members of the Norwegian Jotun group of igneous rocks. The Tännäs rock complex should not be confused with the Tännäs gneiss, a name which is often used for the augen gneiss. Ottfjäll diabase has not been found in the Tännäs complex rocks but it occurs, however, in close connection with rocks of the complex in the neighbourhood of Tännäs. STRÖMBERG states that it is difficult to distinguish in the field between deformed Ottfjäll diabase and deformed Tännäs gabbroid rocks; the contact relationships are not known.

The *Grönstack complex*, also called the basal fragmental rocks, occurs only in the Tännäs area as a thin deposit on the Tännäs complex. The breccia is composed of fragmental material from the underlying rocks. STRÖMBERG considers that the breccia is most likely a sedimentary deposit.

The quartz–muscovite schists bear some resemblance to the quartzitic hardschists in the Middle thrust rocks of the Torneträsk area. The schists occur between the feldspathic quartzite formation and the granite–mylonite sheet, which according to STRÖMBERG represents the uppermost part of the Serv nappe proper.

The *Sylarna amphibolite*, lying in the south-westernmost part of Jämtland, is crossed by the international border; it was thought by TÖRNEBOHM to be a lower but integrated part of the crystalline schists. STRÖMBERG noted that the Sylarna amphibolite lay between the Helags crystalline schists and the greywackes of the Serv nappe proper, and because of some similarities in composition between an amphibolite dyke in a western portion of the Serv greywacke, and the Sylarna amphibolite, he regarded the latter as an upper part of the Serv nappe. Later he thought that the Sylarna amphibolite most probably belonged to the Helags complex—the Upper thrust rocks.

The stratigraphy of the sedimentary formations in the Serv nappe proper has not, according to the above facts, been wholly solved. The three formations—the Ulvberg, Tännäs and Grönstack complexes—lie in the southern part of the map area, in the boundary zone between the feldspathic quartzite and the underlying augen gneiss. In the northern part of the map area the bedding planes in the eastern and western parts of the greywacke formation dip beneath the feldspathic quartzites, which thus seem to represent the younger formation. In the eastern border zone of the greywacke formation a lens-shaped body of rocks of the Tännäs complex occurs bound to the east by feldspathic quartzite. Some kilometres to the north of the Tännäs rock lens, in the same zone, lies a small body of Ulvberg dolomite. In the middle of the greywacke area there is an occurrence of carbonate rocks. If the rocks of the Tännäs and Ulvberg types mentioned above are chronostratigraphically equal to the ones in more southerly parts of the area of the Swedish Caledonides under consideration, the greywacke beds are the lowest-lying sedimentary formation of the Serv nappe proper, and must thus represent a northern sedimentary facies which thins out towards the south.

The distribution of the Ottfjäll diabase in the distant thrusted rocks of the southern Swedish mountains is very peculiar. Diabase sills occur in

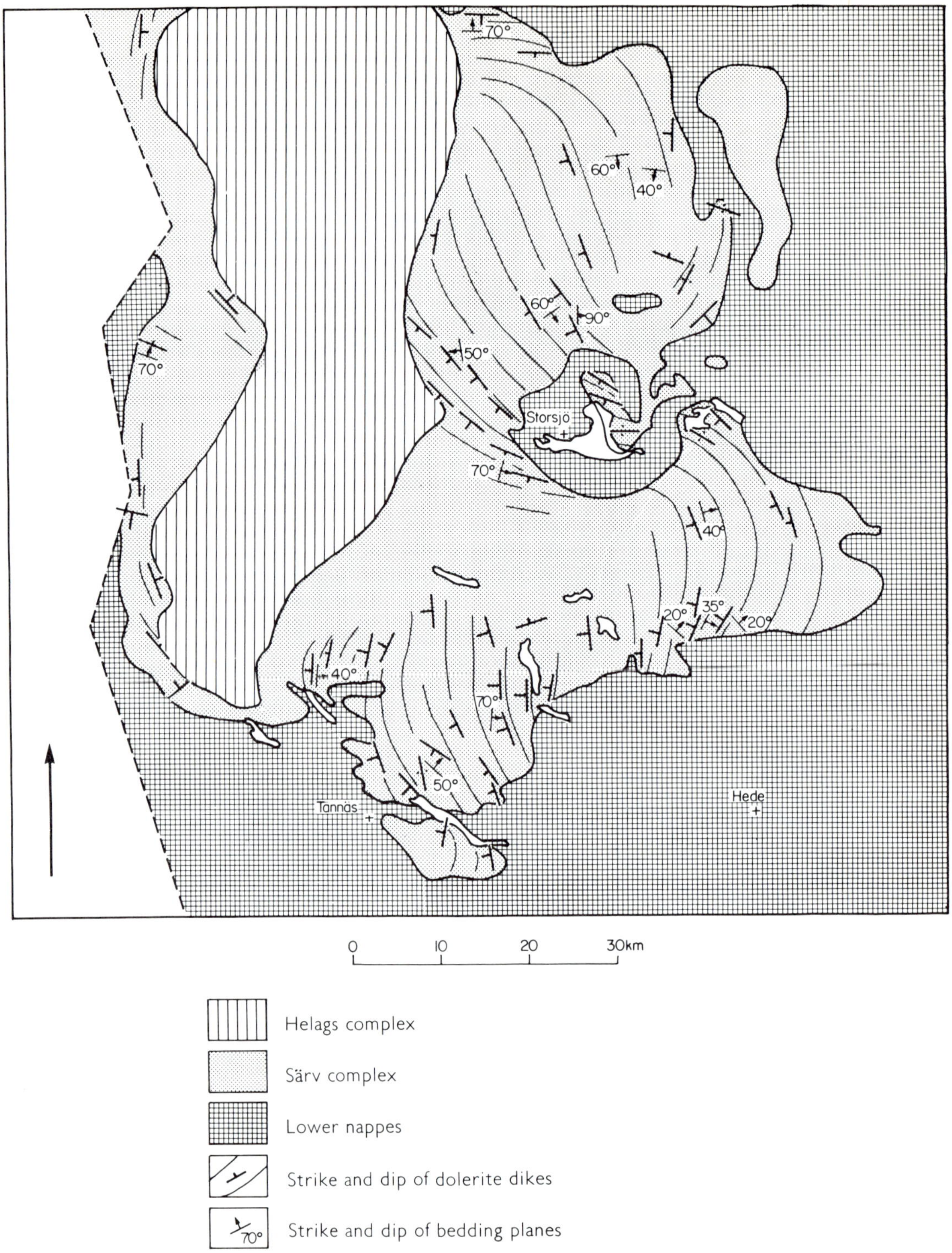

Fig. 156 *Distribution, strike and dip of dolerite dykes of the Särv complex in the southwestern part of the county of Jämtland.* By A. Strömberg (1966)

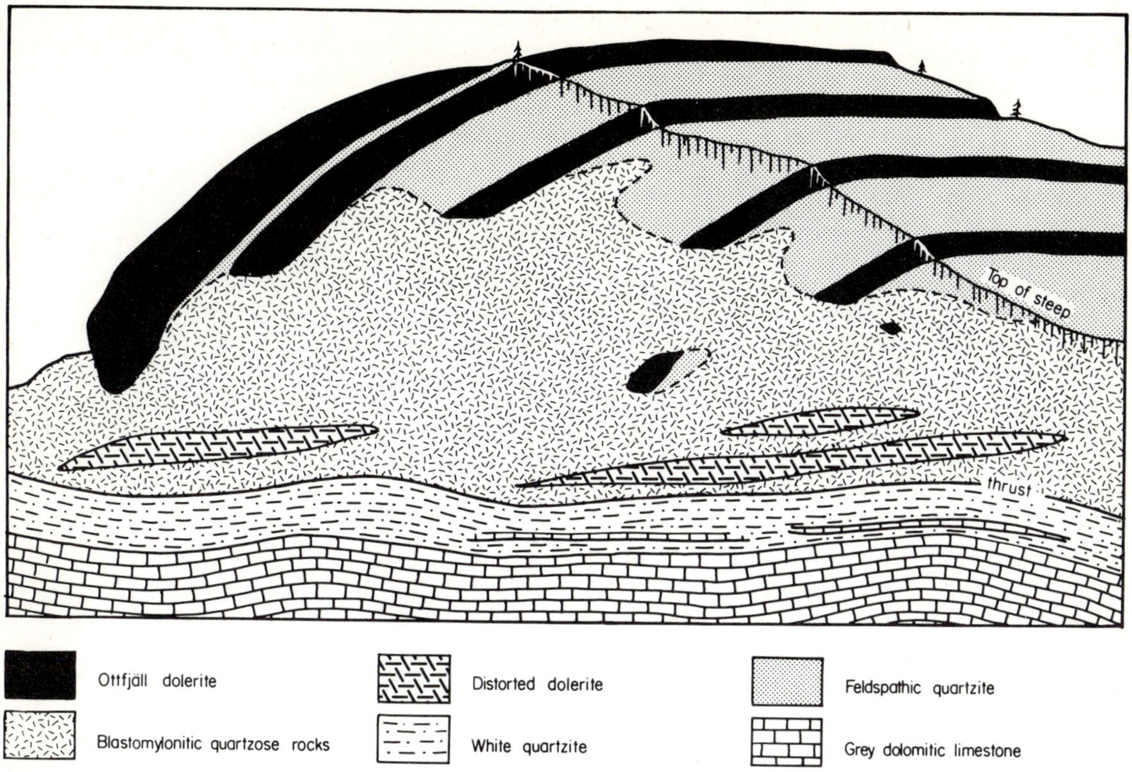

Fig. 157 *A schematic sketch of the southern steep slope of Ulvberget.* Ulvberget hill lies in the eastern boundary zone of the Serv sparagmite complex, about 13 km east of Tännäs village in the middle part of Härjedalen. The sketch shows a zone of thrusting in the lower part of the dolerite-penetrated sparagmite (the "feldspathic quartzite") and the underlying Ulvberget complex, which is composed of quartzite and limestone. Redrawn from a sketch by A. Strömberg; see A. Strömberg (1955)

the feldspathic quartzites and greywacke beds of the Serv complex Middle thrust rocks. No traces of Ottfjäll diabase have, however, been located, either in the thrust-basement rocks of the mylonite, Ulvberg, Tännäs, Grönstack and the quartz–muscovite schist complexes of the Serv nappe proper, or in the Upper thrust rocks of the Helags nappe. The partially metamorphosed Ottfjäll diabase thus occurs in a middle zone of the large Middle thrust rock masses. No sources of the diabase magma have been observed in the rock complexes lying above and below. The orientation of the Ottfjäll diabase in every minor area is very uniform and the diabase occurs in swarms. On Törnebohm's geological map of 1896 the trend of the Ottfjäll diabase sills is mainly N–S, with some deviations. E–W-running diabases occur in a limited area, but this orientation seems to be exceptional. Strömberg has studied the orientation of the diabase sills in relation to the bedding planes of the sedimentary rocks intruded by them (see Fig. 156). His conclusion is that the sedimentary rocks often lie at an angle of almost 90° to the diabase sills, and suggests that the system of joints into which the diabase is intruded was at right angles to the sedimentary bedding-planes irrespective of whether the latter were undisturbed or had been folded prior to the intrusion. However, observations of sedimentary bedding planes in the strongly schistose quartzite and siltstone beds in question are relatively few. On Strömberg's geological map the trends of the diabase sills are to a large extent about the same as on Törnebohm's map. More detailed examination of the orientation of the diabase sills by Strömberg in the remapped area shows, however, regular deviation from the trend (see Fig. 156). Strömberg has surveyed the boundary zone between the diabase-penetrated sediments of the Serv complex and the underlying Ulvberg complex. A sketch of Mt. Ulvberget is given in Fig. 157. The Ulvberget complex of the area is composed of white quartzite on grey dolomitic limestone. Above the quartzite lie greenish-coloured blastomylonitic rocks with long horizontal lenses of distorted diabase; also present are some small fragments of feldspathic quartzite and better preserved diabase of the same kind as composes the upper part of the section. Strömberg's survey has shown that the mylonitic rock must have been formed from the diabase-penetrated feldspathic quartzite and that the greenish-coloured highly altered rocks thus represent a zone of thrusting. This thrust zone is probably of minor importance and registers a zone of crushing and tectonic mobility between the upper and lower rock units of the Middle thrust complex of the region. The lowest portion of the Middle thrust rocks, below the Ulvberg complex, consists of "augen gneiss and blastomylonite".

Some additional remarks will be made on the klippen mentioned above. On the eastern border of the southern mountains, south of Lake Storsjön, lies the *Fuda klippe*. TÖRNEBOHM, who named the klippe, has described its bedrock; this consists mostly of greenish, strongly schistose sparagmite. He also mentions chloritic schists, striped gneisses, striped porphyries, other tectonized rocks of supposed thrust-basement nature and mylonites lacking relict structure. It is of particular interest to note that coarse polymict conglomerates and sedimentary breccias were discovered in the easternmost part of the thrust rocks; the coarse sediments were obviously beds in sparagmitic rocks. On TÖRNEBOHM's geological map these coarse clastics have a somewhat analogous position to the Offerdal conglomerate of the Ansätten "klippe" to the north of Lake Storsjön.

In the southernmost part of the Swedish Caledonides there are *four very small klippen* composed of augen gneiss and a somewhat bigger one which continues across the international border, called the *Frönberg klippe* by TÖRNEBOHM. The bedrock of the thrust unit consists of augen gneiss, striped sparagmite rocks, and also representatives of the mylonites and the Serv complex. On the Norwegian side of the border, the klippe is called the *Kvitvola klippe* by TÖRNEBOHM.

CHAPTER XXII

THE UPPER THRUST ROCKS

As described in Chapter XVII the writer has divided the Upper thrust rocks as follows: the *Seve–Köli rock complex*, which occupies the lower and major part of the thrust masses and is the tectonic backbone of the entire Scandinavian Caledonides; the *Virihaure nappe* in the western part of the middle Norrbotten mountains; and the *Storfjäll nappe complex* in the northwestern part of the Västerbotten mountains.

THE SEVE–KÖLI COMPLEX
INTRODUCTION

The composition of the Seve–Köli complex is often quite different in eastern and western portions, which in parts denotes the lower and upper portions. In the Swedish mountains, four continuous districts of the Seve–Köli complex can be distinguished. 1. The *Pältsa area*, in northernmost Norrbotten at the Finnish–Norwegian border, representing an isolated protrusion from the Norwegian side. 2. The *northern Norrbotten area* down to north of Lake Akkajaure, bound to the west by the Sjangeli basement anticline; south of the Torneträsk area there are several erosional "klippen" of the Upper thrust rocks. 3. The *largest continuous area of the Seve–Köli complex, extending from south of Lake Akkajaure* in the north, to the *Oldfjäll area* in middle-west Jämtland, where the Upper thrust rocks cross the international border. 4. The western part of the southern Swedish mountains—the *western Jämtland–Härjedalen mountains.*

Of the four areas, the bedrock of the first, second and fourth is composed mainly of highly or relatively highly metamorphosed rocks, representing lower parts of the Seve–Köli complex. Their altered sediments are almost wholly devoid of fossils. To date, some very badly preserved stem segments of Pelmatozoans have been found at one locality only—at Handöl in the northwestern part of the area 4. In the third, large area, fossils of the type mentioned above were discovered in some scattered localities as early as the last decades of the nineteenth century. After that, no fossils were found until 1925, when the author discovered relatively well-preserved ones in two sedimentary formations in the middle of the Västerbotten mountains. These two fossiliferous formations, the *Slätdal limestone* of Uppermost Ashgillian, and the *Broken graptolite shale* of Middle Llandovery, were located in the middle of a sequence of formations with characteristic sediments and volcanic rocks. During the continuation of a survey of the Lapland part of the Caledonides, the writer discovered fairly well-preserved fossils in some other areas. These fossils have not yet been described in a paper. They were all found in carbonaceous rocks in the middle of the Norrbotten mountains and a preliminary examination has dated them as Upper Ordovician, partly chronostratigraphically equal to the Slätdal limestone.

Sediments and lavas of typical eugeosynclinal facies predominate in the Seve–Köli rocks. Calcareous shales, greywackes, conglomerates, sandstones to calcareous sandstone, limestones, basic lavas and tuffs are the most common rocks. Lavas of acid to intermediate composition have also been met with. Among sediments of peculiar composition, the serpentine conglomerate and serpentine sandstone rocks should be mentioned; they comprise a formation in the lower part of the eugeosynclinal stratigraphic sequence.

Igneous rocks of basic and ultrabasic composition, considered as intrusive bodies, occur also. Very large areas in the eastern part of the Seve–Köli complex are composed of amphibolite rocks, collectively designated as the *Eastern amphibolite* (Fig. 158). This amphibolite is on the whole conformable with the enclosing rocks. Conformable alternation of rocks of different lithology with amphibolite of eastern type is frequently seen. However, in places within the Eastern amphibolite area, the writer has observed sills of Eastern amphibolite type discordantly cutting across layered rocks with the primary bedding

Fig. 158 *The eastern part of the southern front of the high Kebnekaise mountains*, looking WNW from the southern slope of the Ladtjovagge U-shaped valley. The Kebnekaise tourist station to the right lies on the southern slope of Mt. Kebnetjåkko; to the left is the peak Tuolpagorni. The Eastern amphibolite forms the dark-looking upper part of the mountain mass. Feldspathized and other high grade schists compose the lower (partly scree-covered) portion of Tuolpagorni and its eastern continuation. Below the schists follows a sheet of mylonitic thrust-basement crystalline rock. All the rock masses thrusted from afar lie directly on autochthonous sediments of the Hyolithus group or are divided from these by a thin slab of parautochthonous slate–quartzite. The old Precambrian crystalline basement of the area outcrops in the area around the tourist station. Photo V. Schytt (1965)

preserved. No traces of lava structure have been reported in the high grade altered rocks of the Eastern amphibolite zone.

Forty or fifty years ago, during the period when Törnebohm's overthrust theory was not accepted, the main view held on the mode of origin of the Eastern amphibolite was as follows: basic magma had intruded into the eugeosynclinal deposits in some zone to the west of the international border, and from there had conformably forced its way into the sediments to the east, as far as the zone of its present occurrence in which it had solidified. This view does not seem to be realistic. The author feels that the basic magma forced its way up into that part of the eugeosyncline from which the Seve–Köli rocks were overthrusted, directly beneath that part of the eugeosyncline in which the basic rocks were of greatest thickness. If, or to what extent, the basic magma reached the floor of the geosynclinal sea, is still an open question.

EASTERN GNEISS–MIGMATITE AND AMPHIBOLITE ZONE

The crystalline schists intercalated in or situated immediately below the Eastern amphibolite are for the most part highly altered and transformed to gneissic and migmatitic rocks. Because of this the eastern part of the Seve–Köli complex is sometimes called the Eastern amphibolite–migmatite zone, or by a similar term. However, there are some parts of the zone in which amphibolite, but no highly feldspathic schists, occur. Other parts contain highly altered schists without or with only a few thin amphibolite bodies. Thus, there is no immediate connection between the basic rocks and the state of metamorphic alteration of the schists. In these parts of the Seve–Köli complex, in which the lower part of the complex can be followed in a westerly direction, both the Eastern amphibolite and the highly transformed schists thin out and disappear. It must be borne in mind that the migmatite rocks in the east have no western continuation.

Examples will be given of the nature and mode of the occurrence of the Eastern gneiss–migmatite rocks of the Seve–Köli complex—the Upper thrust rocks—starting in the *Birtavarre area*, a Norwegian area north of the Norrbotten mountains, because of the primary importance of the results obtained in this area (Padget, 1955).

Padget has divided the long-range thrusted rocks as follows:

Birtavarre series {a. non-granitized schists
b. granitized schists

Sparagmitic schists

The sparagmitic schists are the same as the thrust-basement unit of the author's Middle thrust rocks. The greater part of the Birtavarre series seems to be a northern continuation of the Seve–Köli complex and the granitized lower part of it is equivalent to the writer's Eastern gneiss and migmatite rocks. The lowest part of the non-granitized portion of the Birtavarre series consists of carbonate rocks—the "big limestone series". Thus, there are no amphibolite rocks immediately above the granitized schists, which look partly like gneiss and partly like augen gneiss. Lens-shaped bodies of amphibolite, are, however, present in the schists. It has not been possible to subdivide the granitized rocks as their composition varies too much from one area to another. In all parts of the granitized rocks examined, there is abundant evidence of the later growth of quartz–feldspathic material. The predominance of the alkali feldspars orthoclase, microcline and albite is always noticeable.

The next example concerning the composition of the lowermost part of the Seve–Köli complex refers to the *Pältsa area* in the extreme northerly corner of Sweden, where a small portion of the Upper thrust rocks extends into Swedish territory. Directly above the cataclastic–mylonitic thrust-basement rocks of the Middle thrust zone, a thin bed of fine-grained gneiss with fresh feldspars divides the Middle thrust rocks from the overlying garnet–mica schists of the Upper thrust rocks. The fine-grained gneiss is thus the only equivalent, in the Pältsa area, of the granitized schists farther north.

In the *Torneträsk area* about 100 km south of the Pältsa area the feldspathized lower part of the Seve–Köli complex lies above the hardschist of the Middle thrust rocks and below the Eastern amphibolite. The granitization of the schists is not very pronounced. Common rock types are feldspar–porphyroblastic gneiss, gneissose mica schists of various types and mica schists with large porphyroblastic muscovite grains etc. Quartz–feldspathic lenses present are pseudo-conformable and are mostly thin. Towards the west, both the Eastern amphibolite and the porphyroblastic gneissic rocks thin out. In the western part of the Torneträsk area none of the rocks mentioned are present in the basal zone of the Seve–Köli complex. The geological map of the Påtjojaure area in the high Western Abisko mountains southwest of Lake Torneträsk (Fig. 150) shows the zone in which the porphyroblastic gneiss and amphibolite thin out and disappear.

In the *southern part of the northern Norrbotten mountains,* from south of the Torneträsk area in the north to Lake Akkajaure in the south, the composition of the lowest part of the Seve–Köli complex is roughly the same as that of the Torneträsk area. The gneissose schists and hardschists below them, however, become thinner towards the south. In the Akkajaure area the feldspathic schists generally lie directly on mylonitic thrust-basement rocks. The same is true of the large *Sarek–Pårtefjäll–Tarrekaise high mountain complex* to the south of Lake Akkajaure, where the Eastern amphibolite predominates. In more southerly parts of the Norrbotten mountains, the role of the Eastern amphibolite diminishes, and the scattered amphibolite bodies are enclosed in garnetiferous and feldspathic schists. However, most of this large high grade metamorphic schist area has been very superficially surveyed. The same is true of its southern continuation in *northern Västerbotten.* Geological mapping of this extensive area was carried out about forty years ago and a reassessment is required.

Valuable contributions to the geology of the eastern high grade metamorphic portion of the Seve–Köli complex in the *middle and southern Västerbotten areas and in northern Jämtland,* have been made by the following: Du Rietz (1938) on the Borka area in Västerbotten and the Muruhatten area in northern Jämtland; Faddegon (1940) on the Rivovardo area in Västerbotten: Michel (1950) on the Borkafjäll area in Västerbotten; Staargaard (1955) on southeast Borgafjäll in northern Jämtland; and van der Harst (1956) on the Borga area in southern Västerbotten. (Faddegon's monograph is in Dutch, with a German summary; all the other accounts are in English.) It may be added that in the writer's monograph on the Caledonides of the Västerbotten mountains (1955, Swedish with English summary) the problems of the eastern, highly metamorphic rock zone were of course dealt with at some length. In central southern Västerbotten and northern Jämtland the Eastern amphibolite occupies an important part of the highly metamorphic rock belt of the Seve–Köli complex. Migmatite rocks form a central zone of the rock belt which is thus bound to the west and east by amphibolite. Peridotite is a common constituent of the rock belt, usually more closely connected with amphibolite rocks than with schists. The migmatite (see Fig. 159) is typical K migmatite, with microcline as the principal feldspar. The highest metamorphic facies observed within the migmatite zone is the almandine–diopside–hornblende-subfacies of the amphibolite facies.

Fig. 159 *Migmatite type from the Lower part of the Seve–Köli complex.* North of the western part of Lake Gardsjön in the middle eastern part of the Västerbotten mountains. From O. KULLING (1955)

The basic igneous rocks of the eastern, highly metamorphic belt within and bounding the migmatized zone are on the whole migmatized to only a small extent themselves but are penetrated by conformable as well as discordant pegmatitic veins, mostly of trondhjemitic character. Diffuse feldspathization to migmatization has, however, sometimes been encountered in the amphibolite rocks. Banded amphibolite types have also been observed.

As mentioned earlier, the Eastern amphibolite is mostly superposed on the migmatitic schists. Crystalline schists above this amphibolite do not usually display any remarkable feldspathization. When no amphibolite rocks are present, the migmatite rocks grade upwards into garnet–mica schists. In their papers DU RIETZ, FADDEGON and MICHEL describe the transition zone towards the west and upwards, from the migmatitic rocks via garnet–mica schists to phyllitic schists of middle grade.

The Eastern amphibolite and the zone of migmatization, according to what has been said before, thin out to the west. What evidence for this is to be found in the Västerbotten–northern Jämtland area? To the west of the eastern high grade zone of the Seve–Köli complex there are three areas in which the lower part of the rock complex is uncovered. These are: along the eastern front of the antiform-shaped Middle thrust rocks of the Bångfjället area in the centre of the northern Västerbotten mountains; in the Fjällfjället antiform in the southwestern Västerbotten mountains; and along the eastern front of the Börgefjell window at the international border west of Mt. Fjällfjällen.

The *Bångfjället thrust-basement rocks* are superposed to the east by a thin quartzite formation, and this in its turn by phyllite rocks belonging to the middle grade rocks of the Seve–Köli complex. The phyllite rocks are supposed to have been in thrust contact with the quartzite, which is considered as an upper member of the Middle thrust rocks. No high grade schists or amphibolite of Eastern type are present in the lowermost part of the Seve–Köli complex.

The *Fjällfjällen antiform* is composed of *feldspar-rich quartzite* of arkosic composition, probably representing the western portion of the arkosic Sparagmite group, which has been described earlier for both the Lower and the Middle thrust rock complexes. On the geological map of Västerbotten county (1955), the writer designated the quartzite rocks in the Bångfjället area and the Fjällfjällen arkose mentioned above as representatives of the oldest quartzite formation in the entire Västerbotten Seve–Köli bedrock-area. Both these quartzite rocks often show tectonic banding. A paper on the area between Mt. Fjällfjällen and the international border with Norway was recently published by ZACHRISSON (1964). He states that the eugeosynclinal sequence of strata has a primary depositional contact with the Fjällfjäll arkose.

Between the Fjällfjäll antiform and the Börgefjell basement window on the international border to the west is the synclinally formed Seve–Köli rock area, which ZACHRISSON (1964) called the *Remdalen syncline.* The basement is composed of granite, and is considered as autochthonous by ZACHRISSON. On the granite lies a sequence of mainly sandstone beds of primary superposition. Above the window rocks follow the thrust complex of Seve–Köli rocks. The lowest formation of thrust rocks—the Fjällfjäll arkose—is superposed by the same formations of geosynclinal facies as are registered in the eastern limb of the syncline. In the Remdalen syncline area no highly metamorphic rocks have been found in the lowest part of the Seve–Köli rocks. The lowest sedimentary members of the Seve–Köli rocks are of middle metamorphic grade and are called phyllites.

In the three areas of the Västerbotten mountains mentioned above no high grade rocks or migmatite components occur in the lowest part of the Seve–Köli complex, nor is amphibolite of eastern type met with. The reason for the absence of high grade rocks in the western, lower part of the Seve–Köli complex will be dealt with in a later section.

The terms "Eastern gneiss and migmatite" and "Eastern amphibolite" have been used for high grade rocks in the eastern part of the Seve–Köli complex, and at the same time in the eastern part of the Lapland mountains. In the *southern Swedish mountains*, erosional remains of the Seve–Köli complex are met with only in the western to extreme western part of the mountain chain, and the terms above are not used. In Törnebohm's monograph of 1896 on the Central Scandinavian Caledonides, with the southern Swedish mountains as an eastern part, the high grade metamorphic rocks of the Main thrust nappe were called the *Åre gneiss* and the amphibolite gneiss and schist. On Törnebohm's geological map, the high grade metamorphic rocks mentioned occur in the high Åreskutan mountain belt of western Jämtland and in the mountainous international border zone in westernmost Jämtland. The "Eastern amphibolite" is often called the *Sylarna amphibolite* in the southern mountains. Mt. Sylarna, consisting largely of amphibolite, lies in the southwestern corner of Jämtland, right on the international border. According to Törnebohm, there are amphibolite intercalations in the Åre gneiss. Generally the gneiss seems to lie above the amphibolite rocks. Subsequent examinations of the high grade rock zones of western Jämtland have neglected the high grade rocks of the area. In his account of the southern part of the Swedish Caledonides, Asklund (1960) states that the garnetiferous gneissose rocks and amphibolite in the high grade rock portion of the Seve–Köli complex—called the Seve nappe proper by Asklund—seem to be Precambrian metamorphic rocks of mainly Archean age.

THE CENTRAL WESTERN SEVE–KÖLI AREA

In the introduction to the Seve–Köli bedrock, the Swedish part of the Seve–Köli complex was divided into four areas. In the largest of these, covering most of southern Norrbotten, Västerbotten and northern Jämtland, relatively well-preserved fossils have been found in several localities, and these, being concentrated in one area, provide a basis for establishing the Seve–Köli stratigraphy.

Björkvattnet–Virisen Area (Central Västerbotten mountains)

This area contains two stratigraphically well-defined fossiliferous formations, to date the only ones in the eugeosynclinal complex of the Swedish Caledonides. The fossiliferous formations—a coral bearing limestone and a graptolite slate—occur in a stratigraphic sequence containing several good marker horizons. This sequence was later found in other parts of Swedish Lapland.

In 1924, the writer surveyed the central part of the Björkvattnet–Virisen area. During a visit to the area in 1925, fossils were discovered in two of the sedimentary formations, and this stimulated further investigation. After presenting the geological results in a monograph (Kulling, 1933), the writer did not revisit the area until 1963 and 1964, when more graptolite fossils were discovered in a slate overlying the earlier graptolite occurrence. During this period some minor areas were surveyed in greater detail, and an examination was made of drill cores from the area west of Björkvattnet, the cores being made available by Kungl. Vattenfallsstyrelsen.

The new data supports the previously established stratigraphy. The mapped area can conveniently be divided into an eastern part—the area between Björkvattnet and Virisen—and a western part, comprising Vardofjället and Arefjället. In the former, sedimentary structures including graded bedding, current bedding and fossils are relatively well preserved. In the latter, the lithologies are generally more deformed and are characterized by tectonic banding. The tectonic boundary between the two bedrock areas is verified from west of Lake Björkvattnet southwards to Mt. Stora Grokas. The northern and southern continuation of the boundary is little known. The writer used to name the western bedrock area the Grokas complex. The stratigraphy of the eastern area will be dealt with first.

The *stratigraphy of the Björkvattnet–Virisen area* is as given at the top of the following page.

Peridotite, the lowest member of the bedrock, outcrops in the easternmost part of the area. The rocks are to some extent of primary habit—dunite and pyroxenite—and to some extent transformed

Viris group Lövfjäll group Broken group	SILURIAN
Slätdal group Vojtja group Gilliks group Seima group (Quartzite conglomerate at Seimajaure) Serpentine conglomerate or Rotik conglomerate Peridotite rocks	ORDOVICIAN

into serpentine and soapstone. In the north some peridotite outcrops have a breccia structure, of somewhat doubtful origin.

Serpentine conglomerate. During the twenties the writer observed serpentine conglomerates in some isolated outcrops in the eastern belt of ultrabasic rocks, east of Virisen.

Quartzite conglomerate at Seimajaure. The conglomerate lies between the ultrabasic rock and greenstone of the Seima group and has been found on either side of the narrow middle part of Lake Seimajaure only.

The Seima group. This is composed of normal basic volcanites with some intercalations of a slaty nature. In the Seimajaure area the volcanites are structurally speaking relatively well preserved and may be with or without carbonate amygdaloids. Greenstone beds with agglomeratic to conglomeratic structure have been found. So far as is known, no keratophyre lavas are present in the Seima greenstone belt. In the western part of the Björkvattnet–Virisen area, the basic volcanic rocks present are mainly greenschists lacking preserved primary structures.

The Gilliks group. Greywacke conglomerate (see Figs. 160–162), greywacke-feldspar-bearing quartzite and black slate are the main constituents of this group. Greenschist intercalations occur. Stratigraphically, the Gilliks sediments seem to be the conformable successors to the Seima volcanic rocks. Coarse clastic rocks usually comprise the lower part of the group sequence and black slates the upper part. In the western part of the area, the greywacke conglomerate beds dominate the fine-grained clastic rocks of the group. In the east the reverse conditions prevail. The composition of the block material in the greywacke conglomerate is quite remarkable. Na granite and Na keratophyre occur both as pebbles and as large scattered blocks. Dark oolitic dolomite and several types of sandstone also occur in the conglomerate material. Exceptionally small grains of amphibolite (epidote-amphibolite) can be seen. The source of the material forming the Gilliks clastic rocks is most probably entirely within the eugeosynclinal area. None of the conglomerate block material appears to be Precambrian.

The Vojtja group. Quartzite conglomerate, often very coarse, and well-washed quartzite are the main rock types of this group. Calciferous quartzite occasionally constitutes a transition bed with the Slätdal limestone. In the Björkvattnet–Virisen area, quartzitic facies of the Vojtja group predominate in the west, and conglomeratic facies in

Figs. 160–162 *Three types of Gilliks greywacke conglomerate from the central part of the Västerbotten mountains.* From O. KULLING (1955)

Fig. 160 *From the northern slope of Mt. Gillikstjåkko.* × 2

Fig. 161 *From the steep northern slope of Mt. Brokfjället.* The enclosed boulder to the right is of dolomite, 0,5 m in diameter

the east. In an isolated outcrop of coarse quartzite conglomerate looked upon as a member of the Vojtja group, in the southeastern part of the area, pebbles of greenstone were observed. The greenstones lacked primary structures but in composition seemed to be of the Seima greenstone kindred.

The sediments of the Vojtja group lie with a large stratigraphical unconformity upon the sediments of the Gilliks group and are followed without any sedimentary break by limy strata of the Slätdal group.

In the *Slätdal group*, blue-grey limestone and grey to dark grey marly limestone are common. Calcareous quartzite, marl and limestone-breccias are also met with as well as black to dark grey slates. The Slätdal limestone is often richly fossiliferous, and the marly limestone usually contains well-preserved fossils. The fauna, described by the

Fig. 162 *From the lower part of Slätdalsbäcken brook.* The light-coloured layer to the left of the hammer has a greenschist matrix

writer (1933), is of uppermost Ashgillian and rich in tabulate corals. In a calciferous quartzite of the group the large thick-shelled brachiopode *Holorhynchus giganteus* Kiaer was discovered—this is a good marker fossil of uppermost Ashgillian, equivalent to stage 5 b in the Oslo area.

The Broken group. This has two major formations —the Broken quartzite and the Broken slate, the latter superposed on the former. Below the Broken quartzite formation lies a sequence of dark slates with thin beds of phyllitic quartzites, comprising the lowermost beds of the Broken group. The Broken quartzite is light grey and partly calcareous. It successively grades into the Broken slate through numerous thin slaty beds in its uppermost part. The graptolite-bearing part of the slate measured in 1925 was only 1,7 m thick and lay 4,5 m above the Broken quartzite. As mentioned previously the graptolites belong to the

Fig. 163 *An illustration of rhythmical sedimentation, sometimes of graded bedded nature, in the Lövfjäll group calcareous sandstone–sandy slate sequence.* The 5 m sample is from drill core F 4 from the Geumå area to the west of Björkvattnet. The part of the drill core shown is from the upper limb of a recumbent syncline; the sediments are inverted and are younger downwards (to the left) (cf. the section of 8 drill cores in Fig. 166)

Petalograptus folium and *Cephalograptus cometa* zones of Middle Llandovery. The graptolite-bearing slate observed in 1963 in the upper part of the Broken slate lies in the neighbourhood of that measured in 1925. The graptolite remains collected in 1963 are however badly preserved. SKOGLUND has very kindly examined the fossils, and according to his preliminary determination most of the graptolites are large species of the genus *Petalograptus.* He also found a *?Glyptograptus* and generically indeterminable diplograptidae. The writer shares SKOGLUND's view that the slate containing large species of genus *Petalograptus* probably represents the *Monograptus sedgwicki* zone, the lowest zone of Upper Llandovery.

The Lövfjäll group. Common rock types are lime-bearing phyllites, lime-bearing quartzite, greywacke and slate. Light grey to dark grey slates are intercalated. The sediments of the group often display well-developed graded bedding. There are usually thin bedded interlayerings of sequences of phyllitic slate and limy quartzite,

Fig. 164 *Current bedding in partly graded bedded Lövfjäll greywacke.* Southeast of Lake Vesken in the central part of the Björkvattnet–Virisen area. From O. KULLING (1933)

quartzite and limy quartzite, and slate and limy phyllite in the sediments with partly graded bedding. Fig. 163 illustrates the rhythmical structure of the sediments, and Fig. 164 shows current bedding in a partly graded bedded greywacke of the Lövfjäll sequence. In the lower part of the Lövfjäll group, thin beds of greenstone to greenschist can be seen (see the geological map of the northwestern part of Virfjället, Fig. 165). The igneous rocks do not have a distinct volcanic structure.

The Viris group. The sediments of this group are composed of mainly grey fine-grained quartzite—the Viris quartzite. The subordinately occurring feldspar is mostly an acid plagioclase, and microcline-bearing Viris quartzite is sparse. In the lower part of the Viris group rock types characteristic of the Lövfjäll group occur as intercalations. Quartzite conglomerates are met with in the transition zone between the Lövfjäll and Viris groups. One of them, the *Gimja conglomerate,* occurs in sediments of Lövfjäll facies. The Viris quartzite group is the topmost sedimentary unit known in the Swedish part of the Caledonian eugeosynclinal sequence.

In the Björkvattnet–Virisen area, the Viris quartzite forms the core of a N–S trending syncline, or more correctly a synclinorium overfolded to the east. The author's view of the tectonics of the central Västerbotten mountain area is given on a block diagram in his monograph of 1933. A further contribution to the interpretation of the tectonics of the area is given in Fig. 166; this shows the composition of diamond drill cores taken from the area west of Lake Björkvattnet. Seven of the eight cores are taken along an E–W-running line. Core F 4 is taken from a point 400 m south of the line. Diamond drilling was carried out in connection with a projected hydro-electric scheme and the cores, after geotechnical examination, were placed at the writer's disposal. Cores B 2 and B 10 come from the western inverted

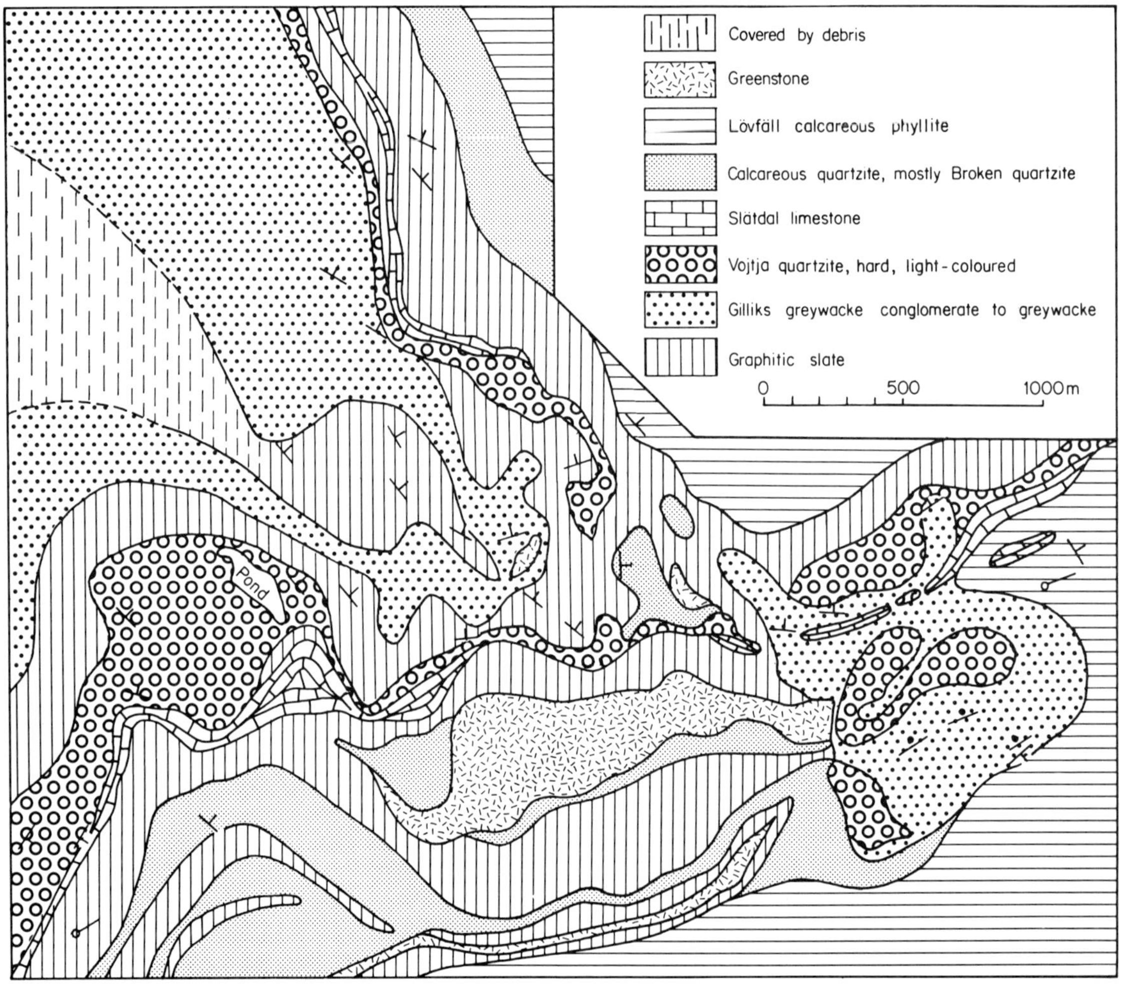

Fig. 165 *Geological map of the northwestern part of Virfjället, lying north of the western part of Lake Virisen.* Original by O. Kulling, assisted by G. Fredrikson

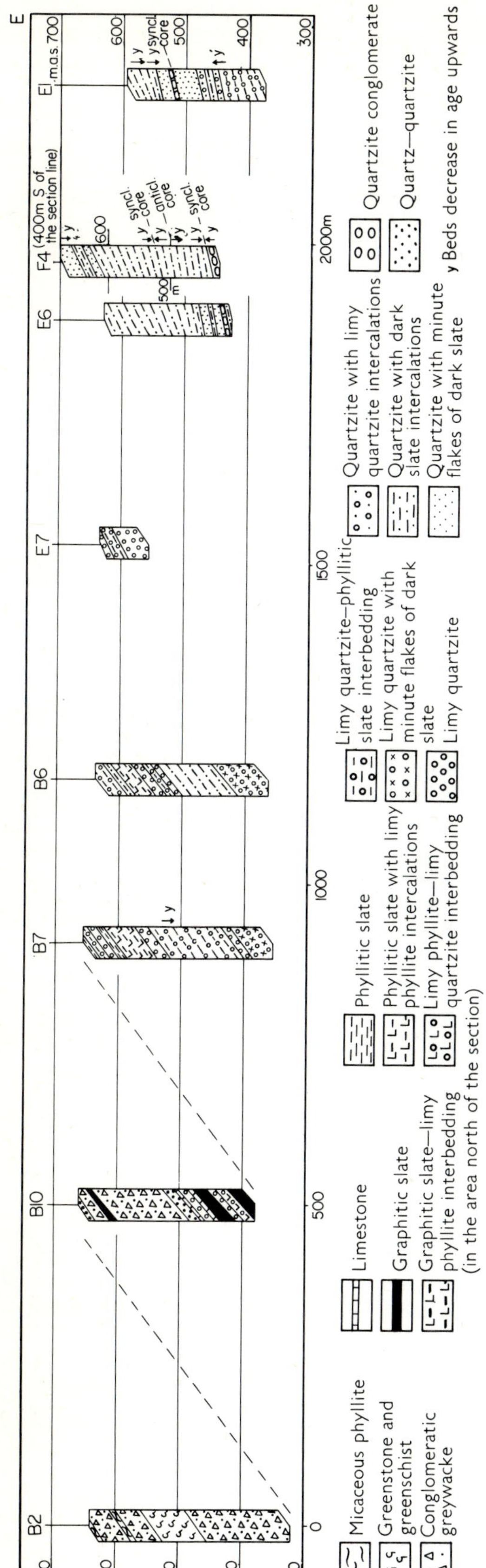

limb of the syncline. Gilliks conglomeratic greywacke forms the greater part of the two western cores. In the lower part of core B 10, sediments of the Gilliks group are followed by Vojtja quartzite and Slätdal limestone with intercalated limy quartzite; also by the Broken group sediments of black slate, limy quartzite–limy phyllite and more black slate. The Lövfjäll group sediments have been investigated mainly by means of cores B 7, B 6 and E 7, although core B 6 has not been examined by the writer, and no evidence as to the original order of succession of the sedimentary rocks has been obtained from it. In the three eastern cores the Viris group and Lövfjäll group sediments are represented. The quartzite conglomerate occurring in all the cores is probably the northern equivalent of the Gimja conglomerate. In cores F 4 and E 1, studies of graded bedding show that the sedimentary sequence of rocks is isoclinally folded. Examination of cores obtained from east of the area of the numbered cores revealed that the same type of folding occurring in the eastern part of the above section prevailed further east as a sequence of isoclinal folds overfolded to the east.

The anticlinal rock zone which follows to the west of the N–S-trending synclinorium beyond Lake Björkvattnet has an imbricated structure. This western bedrock zone has, however, been very superficially surveyed. The anticlinal zone, with outcropping Gilliks greywacke conglomerates and greenschists, trends southwards. In the area northwest of Lake Virisen the zone bends to the east and the rocks mentioned above plunge beneath the over-lying formations of the sequence. The geological map of the northwestern part of Mt. Virfjället given in Fig. 165 shows the area in which the structure of the plunging bedrock anticline has been studied. The purpose of the survey was, *inter alia*, to locate the stratigraphical position of some greenstones, observed by the author during reconnaissance mapping of the area in the twenties. In the northern part of the map area, the greywacke conglomerate and black slates of the Gilliks group outcrop. Above, to the east and south follows the Vojtja quartzite, and to the south the Slätdal limestone. The black slates and limy quartzite of the Broken group which follow enclose the lower of the two greenstone bodies. The upper greenstone lies on the boundary between the Broken sediments and the Lövfjäll limy phyllite–quartzite. No distinct lava structures have been found in the greenstone rocks.

The easternmost part of the map area is an upthrusted rock unit containing Gilliks greywacke

Fig. 166 *Eight drill cores from the Geumå area to the west of Lake Björkvattnet.* Original by O. Kulling conglomerate, Vojtja quartzite and Slätdal limestone. The Vojtja quartzite in part forms low hills, surrounded by lower-lying Gilliks clastics. The Slätdal limestone is partially surrounded by the Gilliks rocks. The boundaries of the different rocks are probably not exposed.

Before leaving the Björkvattnet–Virisen area the thickness of the rocks present will be considered. Tectonic deformations have changed the primary thicknesses of these rocks everywhere. Rocks of highly differing grades of competence form the bedrock. Black slate for example behaves in a way other than quartzite, etc. The rock sequences of the Viris group, the Lövfjäll group and the Gilliks group are all several hundred metres thick. The Broken group sediments are often 70–100 m thick; the Slätdal group sediments vary from 10–50 m, and the Vojtja clastics from 10–160 m in thickness; these figures are taken from the author's estimates in a monograph of 1933.

The bedrock of the northern and southern continuation of the Viris synclinorium has been summarily surveyed in comparison with that of the Björkvattnet–Virisen area; Seima volcanic rocks are, however, probably absent. The Gilliks group appears to be represented by quartzites and slates. Vojtja and Slätdal sediments occur in several places in the area between Lake Virisen and Lake Kultsjön to the south; Broken group sediments are, however, not known beyond the Björkvattnet–Virisen area. The Lövfjäll group sediments cover wide areas to the south of Lake Virisen—of their distribution north of the Björkvattnet–Virisen area little is known. The Viris quartzite occurs only within the area immediately to the south of Lake Virisen.

Remdalen Syncline Area

The Virisen synclinorium which has just been described is followed to the west by the Fjällfjällen anticline with its northern continuation, the Rotikken anticline. The N–S-directed Fjällfjällen anticlinal zone is composed of arkoses to feldspar-bearing quartzites, and is thought by the writer to be a protruding western stratigraphic continuation in the Seve–Köli complex of the Sparagmite group arkoses belonging to the Middle thrust rocks of the mountain marginal zone. To the south, in the area west of Lake Kultsjön, the Fjällfjällen anticline consisting of arkosic rocks plunges down beneath phyllitic rocks. To the north the Fjällfjällen arkose also disappears beneath phyllites in the eastern part of Mt. Vardofjällen. However, according to available geological maps covering the Fjällfjällen area (Högbom, 1925 and Du Rietz, 1941), the bedrock of the area has been disturbed by thrusting or by complicated folding and there is not a clear picture of the sequence of tectonic events; further to the north in the Rotikken anticlinal zone of Mt. Arefjället the core of the anticlinal zone under consideration is composed of peridotite.

The Remdalen syncline area is of particular geological interest because of its important ore deposits. Sulphide ores have been known for many years in the Scandinavian Caledonides, but most of them have been found on the Norwegian side of the international border. In 1918, the geological survey of Sweden commenced investigations in the northern Jämtland–Southwestern Västerbotten mountains. In 1918 and 1919 the Stekenjokk and the Remdalen sulphide ore deposits were discovered. Subsequent work has confirmed the great economic value of the deposits, particularly of Stekenjokk ore. Detailed prospecting in the area should eventually result in mining operations.

Three important papers on the ore deposits mentioned above and their surroundings have been published—two by Högbom and Du Rietz (already mentioned) and the third by Zachrisson (1964). The last-named paper—"The Remdalen syncline, stratigraphy and tectonics"—is based on very detailed geological mapping with aerial photographs on a 1:10 000 scale and on numerous boreholes within the ore deposits. The published geological map of the Remdalen syncline on a 1:25 000 scale resulting from this represents considerable progress in the geological mapping of the Swedish Caledonides. Zachrisson's stratigraphical scheme for the Remdalen syncline is given in Fig. 167. The somewhat different composition of rocks in the western and eastern limbs of the syncline is shown on the diagram. Data on thicknesses have not been included since each formation shows large variations which are probably in part of primary origin and in part due to deformative movements during folding and thrusting. According to Zachrisson the complete stratigraphical sequence of the syncline probably amounts to more than 4000 metres.

Quartzite conglomerates occur at two different levels, generally in association with limestones. The content of acid and basic volcanic rocks (quartz-keratophyres and greenstones to greenschists) is very high. Three different quartz–keratophyre-bearing formations can be distinguished. There seems to be a large stratigraphical gap within the sequence of strata at the base of the Remdalen group, where a minor unconformity can be found.

The intrusive rocks within the map area are of two types. Most of the intrusive rocks are rather light-coloured saussuritic hornblende gabbros. The contact zones are generally strongly schistose

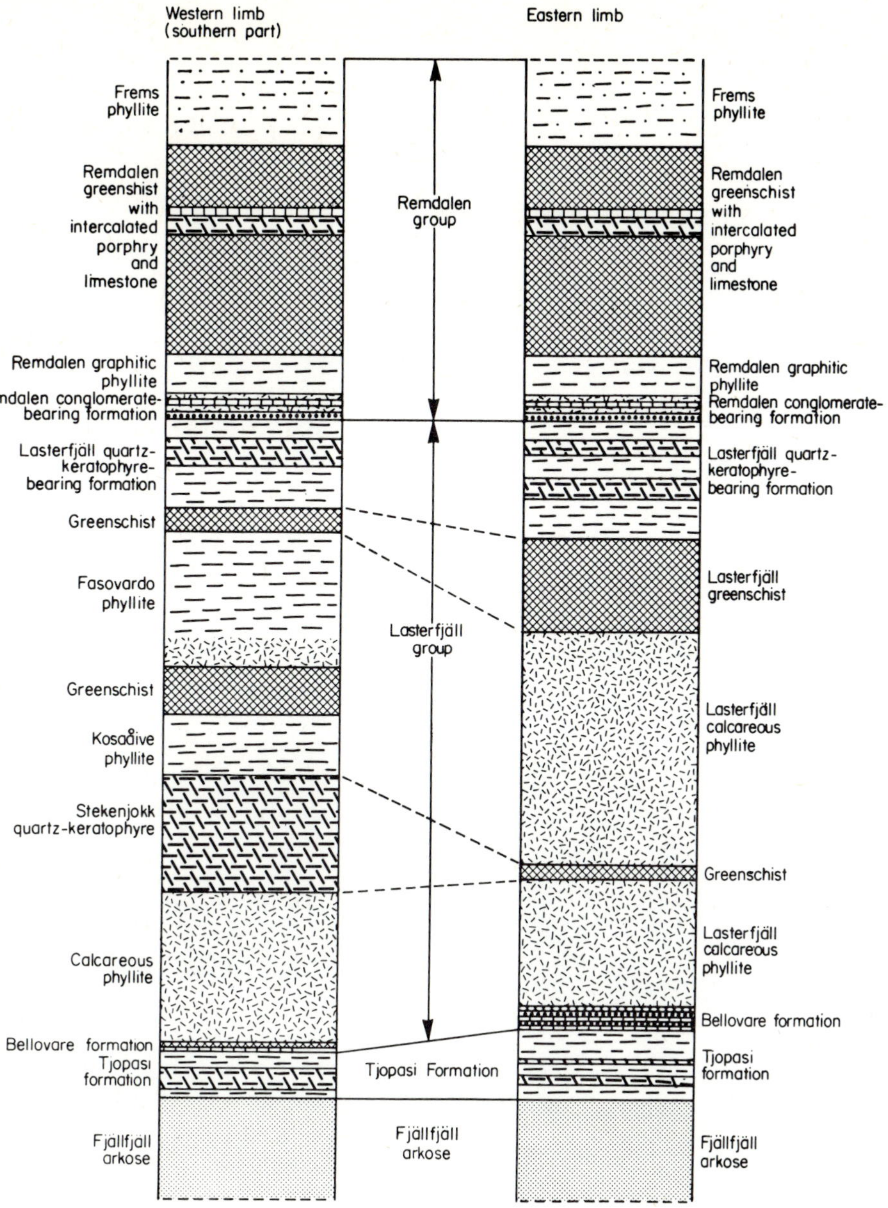

Fig. 167 *Stratigraphy of the Remdalen syncline. Comparison between western and eastern limbs.* Taken from ZACHRISSON (1964)

and often difficult to distinguish from the fine-grained basic volcanic rocks. ZACHRISSON points out that gabbro lenses are a characteristic feature, particularly of the calcareous phyllite rocks, although the entire sequence up to its topmost formation—the Frems phyllite—has been intruded. He further maintains that the intrusions are pre- or synkinematic, since they have been folded and sheared in the same way as the surrounding rocks. The minor group of intrusive bodies consists of serpentines, hornblendites and generally garnet-bearing amphibolites. As is well known in other parts of the eugeosynclinal sequence, the ultrabasic rocks are concentrated at a rather low stratigraphic level—in the Remdalen area at the top of the Fjällfjällen arkose and in the Tjopasi phyllite. North of the Remdalen area they are associated with a well-developed serpentine conglomerate, the Rotik conglomerate (KULLING, 1933). It is of great interest to note that ZACHRISSON discovered minor bodies of ultrabasic rocks situated much higher in the stratigraphical column. Thus small serpentine lenses have been found in the lower part of the Stekenjokk quartz keratophyre and in the middle of the Remdalen group rocks.

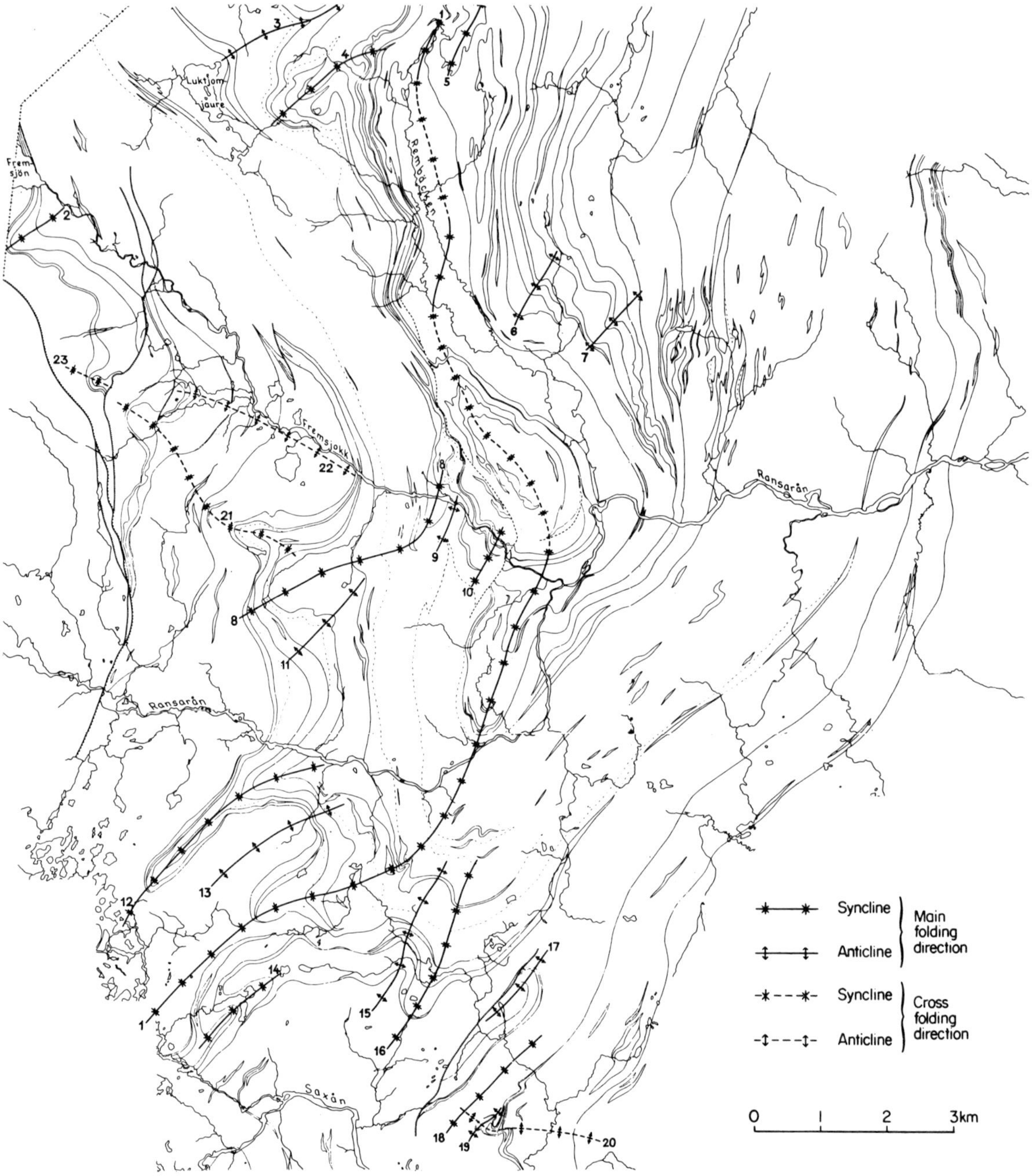

Fig. 169 *Folding in the Remdalen Syncline.* From E. Zachrisson (1964)

Another result of importance, obtained as a consequence of detailed geological mapping, was that thinning out or thickening of rocks within the various formations could be satisfactorily demonstrated. The geological map and published cross sections (Fig. 168 in back pocket) illustrate for example that the Stekenjokk quartz keratophyre occurs in the southwest and probably has a volcanic centre in that direction. The Lasterfjäll calcareous phyllite is much thicker in the eastern limb of the syncline than in the western one. The equivalent to the upper part of the calcareous phyllite in the western limb is a group of sediments of a different facies with a high percentage of volcanic rocks. The cross sections in Fig. 168 show that the Remdalen synclinal sequence of rocks has a thrust-contact with the underlying St. Börgefjell basement complex in the west.

The main Caledonian axes of folding, trending NNE, are generally responsible for the structures of the Remdalen syncline. This is particularly true in the southern, eastern and northernmost parts of the Remdalen area mapped by ZACHRISSON (see Fig. 169, illustrating folding). In some parts of the mapped area there is folding around northwesterly fold axes. In the most southeasterly part of the mapped area cross-folding of this type occurs also. ZACHRISSON has pointed out that the main folds and cross-folds seem to be connected with each other by successive transitions, and that some observations possibly indicate that the cross-folding is later than the main folding; this, however, need not imply that they belong to different phases.

Blåsjö Area

Within the Seve–Köli complex there is another small area which has been geologically mapped in detail and the results recently published—namely the Blåsjö area of northern Jämtland (NILSSON, 1964). Lake Blåsjön is situated about 35 km SSW of the southern end of the Remdalen area. The region described extends from the international border in the west to the northwestern part of Lake Stora Blåsjön in the east. In the area between the two areas geologically mapped in detail a thorough geological survey is in progress. However, the results already obtained permit of a correlation between some formations of the two areas, as shown below.

The bedrock forms a syncline the fold axis of which generally trends N 50° E, and is flat to gently undulating; this is shown in the two sections in Fig. 170 (see back pocket). The geological map, Fig. 171 (see back pocket) shows the northern part of the mapped area. NILSSON has established the following stratigraphical succession for the Blåsjö area:

NORTHERN PART	SOUTHERN PART
BRAKKFJÄLL PHYLLITE	
Graphitic phyllites, quartzites, greenstones and greenschists Greywackes with polymict, partly jasper-bearing conglomerates Bjurälv limestone Graphitic phyllites, quartzites, greenstones and greenschists	Graphitic phyllites, quartzites, greenstones and greenschists
Portfjäll conglomerate	
Basalt–quartz keratophyre formation	
Blåsjö phyllite with gabbroic intrusions	

The lowest formation in the succession, the Blåsjö phyllite, is very thick and calcareous. This calcareous phyllite can be continuously followed to the north as far as the Remdalen area, where its counterpart appears to be the lower part of the Lasterfjäll calcareous phyllite, situated between the Stekenjokk quartz keratophyre and the lower-lying Bellovare formation (compare this with ZACHRISSON's stratigraphical scheme). The basalt–quartz keratophyre formation of the Blåsjö area is the southern continuation of the Stekenjokk quartz keratophyre in the Remdalen area. The Portfjäll conglomerate, superposed on the volcanic beds, is a monomict quartzite conglomerate. NILSSON has pointed out that interbedded graphitic layers in the conglomerate indicate that it is of marine origin. To the north the conglomerate can be followed as far as south of the Remdalen area. However, a counterpart of a conglomeratic nature has not been found in that area. In the Remdalen syncline succession there are interstratifications of quartzite in the lower part of the Kosaåive phyllite formation, which ZACHRISSON feels are counterparts of the Portfjäll clastics. The Portfjäll conglomerate extends from Lake Stora Blåsjön, 30 km to the SSW, to Mt. Portfjället on the international border and then southwards some 10 km on the Norwegian side. Thus, the quartzite conglomerate is a good stratigraphical marker over a large area.

A thick phyllitic complex with intercalated quartzites and basic volcanic rocks rests on the Portfjäll conglomerate of the Blåsjö area. In the northern part of the area a limestone formation—the *Bjurälv limestone*, followed by greywackes and polymict conglomerates—forms the middle part of the phyllitic complex. The Bjurälv limestone represents a northwestern facies and occurs widely on the Norwegian side of the border. The greywackes are coarse, sometimes with interbedded layers of graphitic phyllite and polymict conglomerate. The pebble material of the conglomerate consists mainly of keratophyre, but pebbles of quartz keratophyre, greenschist and quartzite also occur; one conglomerate bed contains jasper pebbles. This jasper material in the conglomerate beds has been found only in the lowest part of the eugeosynclinal sequence of strata in the Norwegian Trondheim region and is used by Norwegian geologists as a criterion for assigning the rock to the Lower Ordovician.

The phyllitic complex with its intercalations is followed by the Brakkfjäll calciferous phyllite; according to NILSSON this is the youngest sedimentary formation within the Blåsjö area.

The Blåsjö area, of small extent, is the most southerly area of the entire Swedish Seve–Köli complex within which the bedrock has been satisfactorily mapped. Our knowledge of the low grade Seve–Köli rocks of the southern Swedish mountains—i.e. the Tännfors region of western

Jämtland—is to date very superficial; geological mapping has, however, recently started.

Sulphide ore deposits of economic value occur in the Blåsjö area, as shown on the geological map. Both the Blåsjö area and the Remdalen synclinal area are situated in the SSW–NNE-striking chalcopyrite–sphalerite ore belt in the western part of the Swedish Caledonides (see Chapter XXVI on economically valuable rocks and minerals). Ore prospecting is continually in progress.

After the discovery of the ore outcrops the area was subjected to a geoelectrical survey. The so-called Slingram method (*Loop frame* method) is the most common ground method in use; this indicates quickly and at moderate cost any superficial electrically conducting bodies, particularly sulphide ores and graphite-bearing slates. No quantitative calculations could, however, be made from the survey. The side-dip of the conducting ore body or graphitic layer can often be determined (see Parasnis, 1962). Fig. 172 (see back pocket) is a map showing the geoelectrical anomalies in the northern part of the Blåsjö area, as found by the Slingram method. It can be compared with the geological map in Fig. 171. The graphitic phyllite of the area is shown perfectly on the map of electromagnetic anomalies. There is a distance of 80 m between the measured sections and 20 m between every measured point. Above a conducting layer the electromagnetical anomaly is always negative (red curves on the map).

Rotikken Anticlinal Area

The Rotikken anticlinal area lies north of the Blåsjö area, beyond the Remdalen synclinal area described above, and west of the Björkvattnet–Virisen area. It includes the eastern part of Mt. Arefjället—Mt. Vardofjällen, which are low mountains. The northeastern part of the area was surveyed by the writer during the twenties and its bedrock described (Kulling, 1933). The northern continuation of the Remdalen–Fjällfjällen area as far as west of Lake Virisen was superficially mapped by Du Rietz (1941). The bedrock of the area differs structurally from that of the Björkvattnet–Virisen area, as most rocks in the westerly area are tectonically banded. The structure of the schists is exemplified in Fig. 173. It is often difficult to say which types of well-preseved rocks in the easterly area have real stratigraphic counterparts among the banded rocks of the westerly area. Several marker rocks such as conglomerates, greywackes, limestones, graphitic slates and greenstones are, however, discernible. North of the Fjällfjällen anticline of arkosic sediments is the continuation of the anticlinal zone, known as the Rotikken anticline; its core is composed of peridotite. According to the author's account the stratigraphies of the eastern and western limbs of the anticline are different, see table below.

WESTERN LIMB	EASTERN LIMB
Greenstone with gabbroid lenses, dark slate and some thin beds of conglomerate and quartz keratophyre	Greenstone, quartz keratophyre and dark slate
Grey banded phyllite	Grey banded phyllite
Svälva quartzite conglomerate and hard quartzite	Hard quartzite
Dark slate and greywacke Greenstone with quartz keratophyre beds Limestone–greywacke–dark slate beds, *the Grey group* Greenstone with quartz keratophyre beds Dark slate	Greenstone with quartz keratophyre beds
Rotik serpentine conglomerate	*Rotik serpentine conglomerate*
Anticline core of peridotite	

Above the volcanics of the eastern limb comes a somewhat tectonized facies of the Gilliks, Votja, Slätdal, Broken and Lövfjäll group rocks. It is not certain if the correlation in the table above is right, or whether, as supposed by the writer in 1933, the Svälva quartzite conglomerate and the greywacke beds are stratigraphical counterparts to the Votja quartzite conglomerate and Gilliks clastics. But there are marked differences between the pebble material of the western greywackes and the typical greywackes of the Gilliks group; furthermore, the formations above the Svälva conglomerate with their banded phyllites, very thick dark slates and volcanic rocks differ in composition from the rock sequence above the Vojtja clastic beds of the Björkvattnet–Virisen area. The Seima greenstone of the Björkvattnet–Virisen area is the eastern counterpart to the upper part of, or possibly to the whole of, the eastern volcanics, which differ from the Seima volcanics in the presence of quartz keratophyre rocks. The Rotik serpentine conglomerate (see Figs. 174 and 175) is the lowest sedimentary formation in both limbs of the Rotikken anticline; the conglomerate rests on the peridotite core of the anticline. Greywacke beds and limestones intercalated in the sequence of volcanic rocks occur in the surveyed part of the western limb of the anticline. In addition the Svälva conglomerate, a quartzite

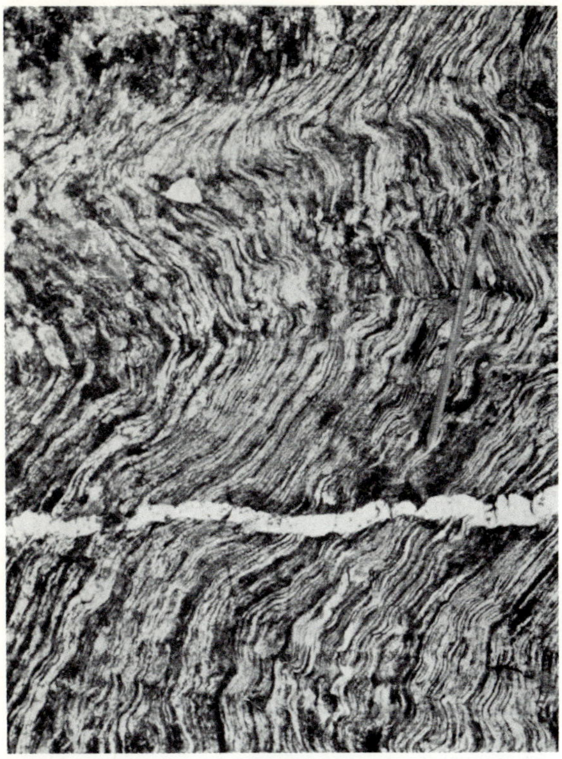

Fig. 173 *Tectonically banded phyllitic schist from the southwestern slope of Mt. Gillikstjåkko*

conglomerate, occurs in the upper part of the sequence in the western limb.

The distribution of the volcanic formations within the district is difficult to resolve. Most volcanic rocks have been transformed into monotonous greenschist–greenstone rocks without preservation of primary structures. In the review on the Remdalen area it was mentioned that the Stekenjokk quartz keratophyre within the Lasterfjäll group occurred in the southwest. In the western limb of the Remdalen syncline, the quartz keratophyre is very thick in the south but thins out towards the north and disappears on the northern side of the Fremsjokk valley. This disappearance is considered by ZACHRISSON to be a combination of primary decrease and tectonic disturbance. In the eastern limb of the syncline the stratigraphical counterpart of the quartz keratophyre seems to be a thin greenschist bed.

The geological maps and papers of DU RIETZ (1941) and ZACHRISSON (1964) show a tentative correlation between the stratigraphical sequence of the Remdalen area and that of the Rotikken anticline area. The Rotik serpentine conglomerate has not been found in the Remdalen area, but in an adjacent area (to the north) it has been found in the northern continuation of the Tjopasi phyllite formation. The greywacke sediments of the Rotikken area do not occur in the Remdalen area.

In the Björkvattnet–Virisen area the Gilliks group sediments display different facies in the east and west, with the coarse western greywacke conglomerate changing eastwards into a more quartzitic facies. In the Blåsjö area, greywacke beds and other sediments occur in northern and northwestern parts. No greywacke beds have been found in the southern part of the Blåsjö area. Generally speaking, greywacke sediments within the part of the Seve–Köli complex under consideration seem to have a westerly source. However, only a few selected areas have been more thoroughly surveyed and it is too early to discuss the distribution of the various sedimentary formations in the district.

The Svälva quartzite conglomerate of Arefjället is possibly a northern continuation of the quartzite conglomerate of the Bellovare formation. Greywacke beds of Gilliks facies have not been found within the Remdalen area. The Remdalen quartzite conglomerate with overlying calcareous phyllite and limestone layers is possibly a southern counterpart of the Vojtja conglomerate—Slätdal limestone of the Björkvattnet–Virisen area. The upper part of the Remdalen group rocks seems to be partly comparable with the rocks of the Broken and Lövfjäll groups in the northern area.

The area west of the Rotikken anticline is at present being mapped. This work will provide a basis for more detailed reassessment of the western stratigraphy and its relationship to that in the eastern limb as it is found in the type area of Björkvattnet–Virisen.

Aivo–Sarvas Area

The area lies on both sides of the Västerbotten–Norrbotten county boundary and on the international border with Norway. The name Aivo–Sarvas given by the author alludes to two mountains—Mt. Aivotjåkko in Västerbotten and Mt. Sarvastjåkko in Norrbotten. Most groups of rocks examined and delimited in the Björkvattnet–Virisen area have also been found in the Aivo–Sarvas area. This northern continuation of the stratigraphical sequence of central Västerbotten forms the upper and middle part of the Seve–Köli rocks in the Aivo–Sarvas area. The rock-groups of the lower part of the Seve–Köli thrust complex of the area are thus a valuable complement to the main stratigraphy.

The Seve–Köli thrust complex of the Aivo–Sarvas area borders to the north on the Nasafjäll window rocks which are composed of early Precambrian granite and gneiss; a thin group of sediments occurs between the basement rocks and the Seve–Köli thrust mass. East and west of that part of the Aivo–Sarvas area examined the

Fig. 174 *Local boulder from south of Rotikken, immediately south of the River Vapstälven.* From O. KULLING (1955)

Fig. 175 *Conglomerate type with sandy-grained lenses.* From the southwestern slope of Mt. Rotikken. Photo by O. KULLING (1942)

Figs. 174–175 *Serpentine conglomerate from the Rotikken area in the middlewestern region of the Västerbotten mountains*

Seve–Köli bedrock is relatively unknown and will not be discussed in this description. To the southwest the Seve–Köli rocks plunge beneath mica schist rocks of the Storfjäll nappe.

During reconnaissance-mapping of the Västerbotten mountains in the twenties QUENSEL observed a quartzite conglomerate and some other rocks of marker character in the central part of the Aivo–Sarvas area, and suggested that the quartzite conglomerate was a northern counterpart of the quartzite conglomerate at Vojtjajaure in the central Västerbotten mountains. In 1934 the writer examined some parts of the Aivo–Sarvas area and came to the same conclusion, namely that the rock sequence which had been surveyed could be stratigraphically connected with that of the Björkvattnet–Virisen area. From 1948 to 1951 the Boliden Mining company carried out ore prospecting in the Aivo–Sarvas area and its environs. The geological results of this survey were collected in a report (MARKLUND, 1952). Recently the writer revisited the Aivo–Sarvas area, and in a limestone immediately above the continuous quartzite conglomerate of the area, poorly preserved fossils were found, but some recrystallized tabulate corals only could be identified. It was clear however that the limestone represented the Slätdal limestone and the conglomerate the Vojtja conglomerate. The division of rocks in the part of the Aivo–Sarvas area surveyed is given in Fig. 176 (see also map, Fig. 177).

As appears from Fig. 176, the upper and major part of the Seve–Köli rocks of the Aivo–Sarvas area is closely connected with the stratigraphical sequence of the Björkvattnet–Virisen area. The grey quartzite on the Aivotjåkko crest is probably a northern counterpart of the Viris quartzite of central Västerbotten. The lowest of the three

THE STORFJÄLL NAPPE COMPLEX		
15	Mica schist.	
THE SEVE–KÖLI COMPLEX		
14	Quartzite	The Viris group
13	Calcareous sandy sediments with slate intercalations (rhythmical sedimentation, see Figs. 178–179)	The Lövfjäll group and the Broken group
12	with three formations of quartzite conglomerate (uppermost is the *Aivo conglomerate*)	
11	Limestone with scattered fossil remains	The Slätdal group
10	Quartzite conglomerate and quartzite	The Vojtja group
9	Limestone, partly with conglomerate structure	The Tjanga limestone
8	Quartzite with intercalations of polymict conglomerate, greywacke and slate. Scattered occurrences of greenstone	The Gilliks group
7	Quartzite	The Vuortna quartzite
6	Greenstone	The Sarvas greenstone
5	Limestone with greenstone intercalations	The Pieske group
4	Metamorphic sediments and volcanic rocks	The lowest part of the Seve–Köli complex
3	Banded tectonite rocks	
THE AUTOCHTHONOUS ROCKS		
2	Sedimentary rocks considered as western equivalents of the Hyolithus group	
1	Archean basement. The Nasafjäll window	

Fig. 176 *Division of rocks in part of the Aivo–Sarvas area*

intercalated quartzite conglomerates in the thick subjacent sequence of calcareous sandy Lövfjäll phyllites is considered to be a stratigraphic counterpart of the Broken quartzite of the Björkvattnet–Virisen area. The Slätdal limestone and Vojtja quartzite conglomerate are the two best marker beds in the Aivo–Sarvas area. The Tjanga limestone appears to be connected with the overlying clastic beds. In the Caledonian eugeosyncline not only quartzite and conglomerate but limestone also occur indicating shallow water deposition. The arenaceous sediments, considered as Gilliks sediments, have intercalations of typical Gilliks greywacke conglomerate, and their quartzitic rocks contain a blue clastic quartz which is a typical marker mineral for the Gilliks clastics of central Västerbotten.

Beneath the Gilliks group sediments follow the light grey, hard Vuortna quartzite, the Sarvas greenstone and the Pieske limestone. Greenstone also occurs both in the limestone and in the metamorphic sedimentary rocks beneath it. These lower rock units of the Seve–Köli stratigraphical sequence in the Aivo–Sarvas area do not have typical counterparts in or around the Björkvattnet–Virisen area of central Västerbotten. The Sarvas greenstone probably records volcanic activity which took place at an earlier date than that which formed the Seima basic volcanic rocks of the Björkvattnet–Virisen area. About 8 km south of the Aivo–Sarvas map area, peridotite occurs as lens-shaped bodies in high grade metamorphic rocks forming the lower part of the Seve–Köli complex. The entire district south of the Aivo–Sarvas area is, geologically speaking, almost unknown.

The lowest part of the Seve–Köli complex consists of banded tectonite rocks. Below the Seve–Köli complex the autochthonous sediments are of middle metamorphic grade. Alum slate, grey slate and quartzite are the main constituents of this probable western counterpart of the so-called Hyolithus sediments at the eastern mountain border.

Mavas Area

The Mavas area lies 70–80 km NNE of the Aivo–Sarvas area and stretches from Lake Ikesjaure in the south to Lake Peskehaure in the north and along the Norwegian border. The Mavas bedrock area occupies the same position north of the Nasafjäll basement window as the Aivo–Sarvas bedrock area does south of the window. The high grade lower part of the Seve–Köli complex east of the Mavas area covers a wide area but has not been surveyed very much. West of Lake Ikesjaure the middle metamorphic grade Seve–Köli rocks plunge westwards beneath higher grade metamorphic rocks, composed of a lower and an upper thrust sheet. These rock sheets, lying in the international border zone, are considered to be northern tectonic counterparts—the lower one mainly of greenstone to the Storfjäll nappe, and the upper one of high grade schists to the Rödingsfjäll nappe. The distribution of these thrust rocks on the Norwegian side of the international border is not very well known.

The southernmost part of the Mavas area, in the mountains between Lake Ikesjaure and Lake Mavasjaure, was surveyed by the writer in 1934. On that occasion tabulate corals in limestone lenses, intercalated in typical coarse Gilliks greywacke conglomerate, were discovered. Several colonies of *Halysites* and a specimen of cf. *Columnaria vacua* Foerste were collected. As the oldest Tabulata are known to be early Middle Ordovician, the Gilliks clastics could not be older than Middle Ordovician. The pebble material of the Gilliks greywacke conglomerate is mainly of the same type of suite as characterizes the Gilliks conglomerate rocks of the central Västerbotten mountains, i.e. pebble types of Na granite, Na keratophyre, sandstone and phyllitic rocks.

On the same occasion the Upper Ordovician marker sequence of fossiliferous Slätdal limestone

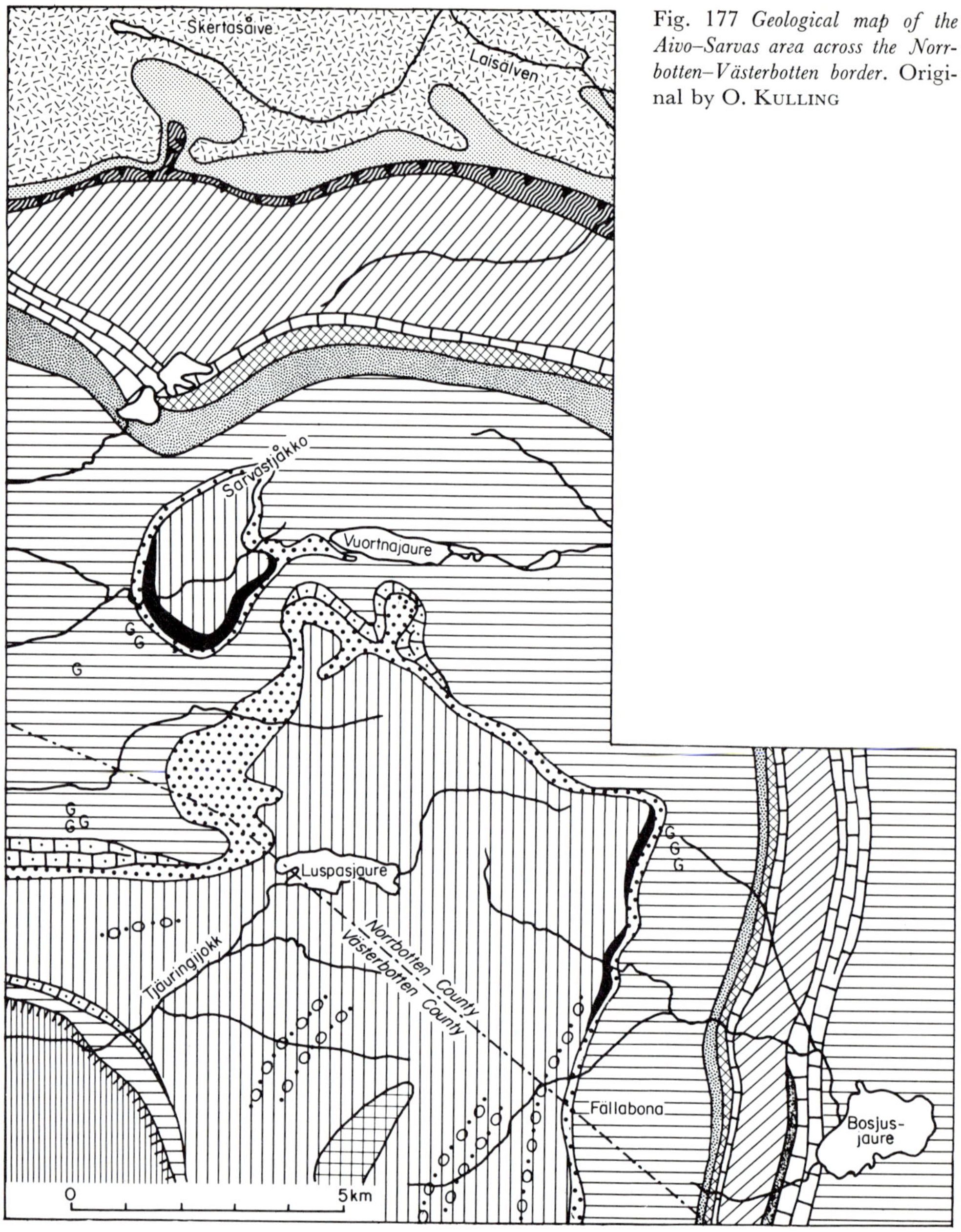

Fig. 177 *Geological map of the Aivo–Sarvas area across the Norrbotten–Västerbotten border.* Original by O. Kulling

The Storfjäll nappe complex
- Mica schist

The Seve-Köli complex
- Quartzite of the Viris group sediments
- Calcareous, sandy Lövfjäll sediments with
- three observed levels of quartzite conglomerate
- Limestone of the Slätdal group
- Quartzite conglomerate and quartzite of the Vojtja group
- Tjanga limestone, partly with conglomerate structure
- Quartzite with intercalated polymictic conglomerate, greywacke and slate of the Gilliks group
  G Observed greenstone.
- Quartzite of the Vuortna group
- Greenstone of the Sarvas group
- Limestone with greenstone intercalations. The Pieske group
- Quartzite west of Bosjusjaure
- Little-surveyed metamorphic sediments and volcanic rocks
- Banded tectonite rocks constituting the lowest part of the Seve-Köli complex

Autochthonous rocks
- Slates and quartzites. A western counterpart of the Hyolithus group
- Old Precambrian crystalline basement. The Nasafjäll window
- Lower limit of the Storfjäll nappe
- Lower limit of the Seve-Köli complex

Fig. 178

Fig. 179

Figs. 178–179 *Noteworthy example of rhythmical sedimentation, partly of graded bedded type.* The calcareous sandy phyllites of the Lövfjäll group in the Aivo–Sarvas area. About 3 km ESE of Luspasjaure on the Västerbotten–Norrbotten border. Photos by O. Kulling (1963)

on Vojtja quartzite conglomerate was located in the rock sequence above the Gilliks clastics. Among the fossils collected from the Slätdal limestone was the first cephalopod ever found in sediments of the Swedish part of the Caledonian eugeosyncline. The fragmental fossil found was an orthoconic cephalopod with a cyrtochoanitic siphonal tube.

In the lower part of the surveyed bedrock series a thick sequence of limestone and calcareous slates was discovered. The limy sediments could be followed from the middle part of Lake Mavasjaure, southwest over the eastern end of Lake Ikesjaure and further southwest across the Norwegian border. The limy group was referred to as the *Pieske limestone.* Later the limestone was followed northwards to and beyond Lake Peskehaure (Pieskejaure) from which it got its name.

The northern continuation of the area up to Lake Peskehaure was reconnaissance-mapped by the author in 1957 and 1958. According to this survey the stratigraphy of the *Seve–Köli rocks of the Mavas area* is as shown in Fig. 180.

Above the uppermost group of sedimentary rocks to the north of Lake Ikesjaure lies the Schistose greenstone. At first this greenstone was considered to be the uppermost geological formation of the Seve–Köli sequence. However, to the south of the lake, greenstone of about the same type as in the north apparently lay just above the

10	Grey and black slate with grey calcareous slate intercalations.	The Lövfjäll group and the Broken group
9	Fossiliferous limestone and marly slate (only in the southern part of the area).	The Slätdal group
8	Quartzite conglomerate and quartzite	The Vojtja group
7	Greenstone with agglomerate beds.	The Gilliks group
6	Grey calcareous slates, grey and black slates and greywacke conglomerate to quartzite. Intercalations of greenstone.	
5	Grey to black slates with intercalations of calcareous slate, limestone, quartzite, quartz keratophyre and greenstone.	The Mavas group
4	Limestone with intercalations of schist and calcareous schist.	The Pieske group
3	Schists of various kinds	
2	Dark dolomitic limestone with schist intercalations	The Keddåive group
1	Partially surveyed sequence of schist with intercalations of limestone and greenstone. Its lowermost part is somewhat feldspathic and underlain by banded so-called hardschist with the local name of Juron quartzite.	

Fig. 180 *Stratigraphy of the Seve–Köli rocks of the Mavas area*

Pieske limestone, and above the greenstone were high grade metamorphic schists quite clearly thrusted onto it. The regional survey of the area showed that it was quite probable that the greenstone was of thrust nature also.

The Seve–Köli sequence of the Mavas area contains several very good marker formations, such as the fossiliferous Slätdal limestone, the Vojtja quartzite conglomerate, the Gilliks conglomeratic greywacke, the Pieske limestone, and the Keddåive dark dolomitic limestone. The dolomitic limestone was first noted on Mt. Keddåive in the mountainous area northeast of Lake Peskehaure, where an almost black carbonate formation lies just below the Pieske limestone of that area. The Mavas group of rocks, with various kinds of sedimentary rocks and acid and basic igneous rocks probably of lava type, seems to occupy about the same position as the Vuortna quartzite in the Aivo–Sarvas area. The sedimentary rocks of the Gilliks group in this area contain a much higher percentage of clastic material than do the rocks of the same group in the northern area. It seems probable, in view of the material present, that during pre-Vojtja time at least the Mavas area occupied a much more central position within the eugeosynclinal trough than did the Aivo–Sarvas area.

Area North of Lake Peskehaure

The rock sequence described in the Mavas area extends across the northwestern part of Lake Peskehaure and from there strikes northeast. As the upper part of the Pieske limestone outcrops along the northwestern shore of Lake Peskehaure, which is about 1,5 km from the international border, the greater part of the rock sequence northwest of Lake Peskehaure lies on the Norwegian side of the border. North of the lake, however, the entire middle grade metamorphic rock belt reenters Swedish territory. In Norway the entire Seve–Köli rock sequence above the Pieske limestone, which used to be called the Muorki schist (Vogt, 1927), is some 10 km broad.

The Seve–Köli rock sequence north of Lake Peskehaure has been examined from the Pieske limestone in the southeast up to the base of the overlying thrust complex; this is called the Virihaure nappe and comprises the upper part of the Upper thrust rocks in the Sulitelma–Salohaure district. The writer's investigations have, however, been restricted to some narrow sections across the Seve–Köli complex on both sides of the international border.

The lowest group of rocks examined, *the Pieske limestone*, contains intercalations of limy schists, quartzite rocks, and, in the east, greenstone beds.

The *Mavas group* which follows is composed of grey to black slates, with intercalations of sandy slates, calcareous slates, quartzites, quartz keratophyres and greenstones. In some eastern sections quartzite to calcareous quartzite lies immediately above the Pieske limestone; this is probably a thin northern counterpart of the Vuortna quartzite in the Aivo–Sarvas area.

The sequence of rocks north of Lake Peskehaure, referred to as the *Gilliks group*, contains several beds of coarse greywacke rocks of Gilliks type, but the major part of the group is composed of grey to black slates and fine-grained sandy slates. Greenstone intercalations also occur.

Above the Gilliks group follow black slates with some thin fossiliferous limestone beds in the upper part. The area north of Lake Peskehaure, particularly the northern part on the southern slope of Mt. Sulitelma, is geologically well known for its fossil content. The first discoveries—some pelmatozoan stem segments—were made by the renowned Swedish fossil hunter von Schmalensee in 1898. He was sent there by Sjögren, who was the leader of the geological survey of the Sulitelma ore deposits from 1893 to 1898. (See Sjögren, 1900.) The fossils were found on both sides of the international border about 6 km north of the northern shore of Lake Peskehaure. When the Sulitelma ore deposits were remapped by Vogt

(1919–1921) he visited the fossil locality and discovered some Bryozoan colonies in thin limestone lenses (Vogt, 1927). According to Bassler of the U.S. National Museum, Washington, the fossil bryozoans are of one species, *Dianulithes petropolitanus* Dybowski. The vertical distribution of this bryzoan species has been discussed a great deal. The fossil seems to indicate upper Middle Ordovician. In Vogt's monograph of 1927 there is a special geological map, scale 1:30 000, of the fossil-bearing area on the southern slope of Mt. Sulitelma. The E–W-lying fossil-bearing limestone crosses the international boundary at cairn 239, the "Isberg cairn". On Vogt's map there are two rock units which may be dealt with here; both of them lie in the stratigraphical sequence above the fossil-bearing limestone. The lower-lying of them is a Na keratophyre agglomerate, probably more than 100 m thick, apparently consisting of lens-shaped inclusions in slates; the slates were called chlorite granulite by Vogt. In the thick overlying formation of black schists there occurs the other of Vogt's lens-shaped bodies, consisting of chlorite granulite. The writer's survey of the area mentions among other things that the stratigraphical rock-sequence of the area is strongly disturbed, and a reliable stratigraphy is not likely to be worked out. To the author it seems probable that both the black slates with *Dianulithes petropolitanus*-bearing limestone layers and the overlying slates with Na keratophyre volcanic rocks belong to the pre-Vojtja part of the bedrock, and thus probably form the uppermost part of the Gilliks group of rocks. Vojtja type quartzite conglomerate and Slätdal limestone have not, however, been found in the area. Above the slates with volcanic intercalations there occur graded bedded quartzite to slate rocks and grey slate to grey calcareous slate rocks of about the same facies as prevails in the Lower Silurian sediments (Lövfjäll facies) above the Upper Ordovician Vojtja–Slätdal group of sediments in the central part of the Västerbotten mountains. The sequence of rocks described above disappears to the northwest, beneath the rather high grade metamorphic rocks of the Virihaure nappe. On an aerial photograph of the southeastern slope of Mt. Sulitelma the thrust limit of the Virihaure nappe is quite easily discernible for a stretch of some 15 km.

Eastern Part of the Virihaure Area

The next area to the north in the middle grade portion of the Seve–Köli complex, within which determinable fossils have been obtained and within which the stratigraphical sequence of rocks includes both Ordovician and Silurian formations, is that going from the eastern surroundings of Lake Virihaure northwards to the area round the eastern part of Lake Vastenjaure. Particular attention will be paid to the southern part of the area, i.e. to the eastern surroundings of Lake Virihaure (see geological map, Fig. 181).

Numerous fossils have been found in limestone beds in the area east of Lake Vastenjaure. The stratigraphy of this area has been worked out and described (Kulling, 1947). This new stratigraphical column was compared with, among others, that of the Seve–Köli rocks of the central Västerbotten mountains (according to Kulling, 1933) and a rather good correspondence between the main stratigraphical development of the bedrock of the two areas which lie 220 to 240 km apart was obtained. The correlation shown covered the Silurian and uppermost Ordovician part of the stratigraphical column, from the Lövfjäll group down to and including the Vojtja group of clastic rocks. In the sequence beneath this sedimentary formation no conglomerate greywacke of Gilliks facies could be found. On a later occasion the writer followed the sequence of rocks southwards with the fossiliferous limestone as a marker formation. In the area to the north of the eastern part of Lake Virihaure, the fossiliferous limestone was relatively well exposed and its fossils fairly well preserved. The fossil localities on the mountain slopes north of Arasluokte bay on Lake Virihaure have yielded the richest harvest of coral fossils obtained to date in the eugeosynclinal part of the Swedish Caledonides.

The geology of the area lying between the Peskehaure and Virihaure areas will be briefly mentioned before entering upon a description of the area around the eastern part of Lake Virihaure. In the eastern part of the intervening area the Pieske limestone is widely distributed, and also the Keddåive dolomitic limestone beneath it. Other rocks of value as markers within the area are black slates, quartzites and greenstone formations. The bedrock of the entire area is of a more mica schistose nature than the bedrock of the area north of Lake Peskehaure, and represents the lower part of the Seve–Köli stratigraphical sequence of rocks. The upper part of the stratigraphical column—the Upper Ordovician–Silurian deposits—does not appear until east of Lake Virihaure.

The bedrock area east of Lake Virihaure has a broad NNW-plunging synformal structure. At the northeastern corner of Lake Virihaure the otherwise broad synform has a locally isoclinally folded core, plunging NNW. The compositions of the stratigraphical sequence of rocks in the eastern and western limbs of the synform differ from each other in some respects (see Fig. 181). In the

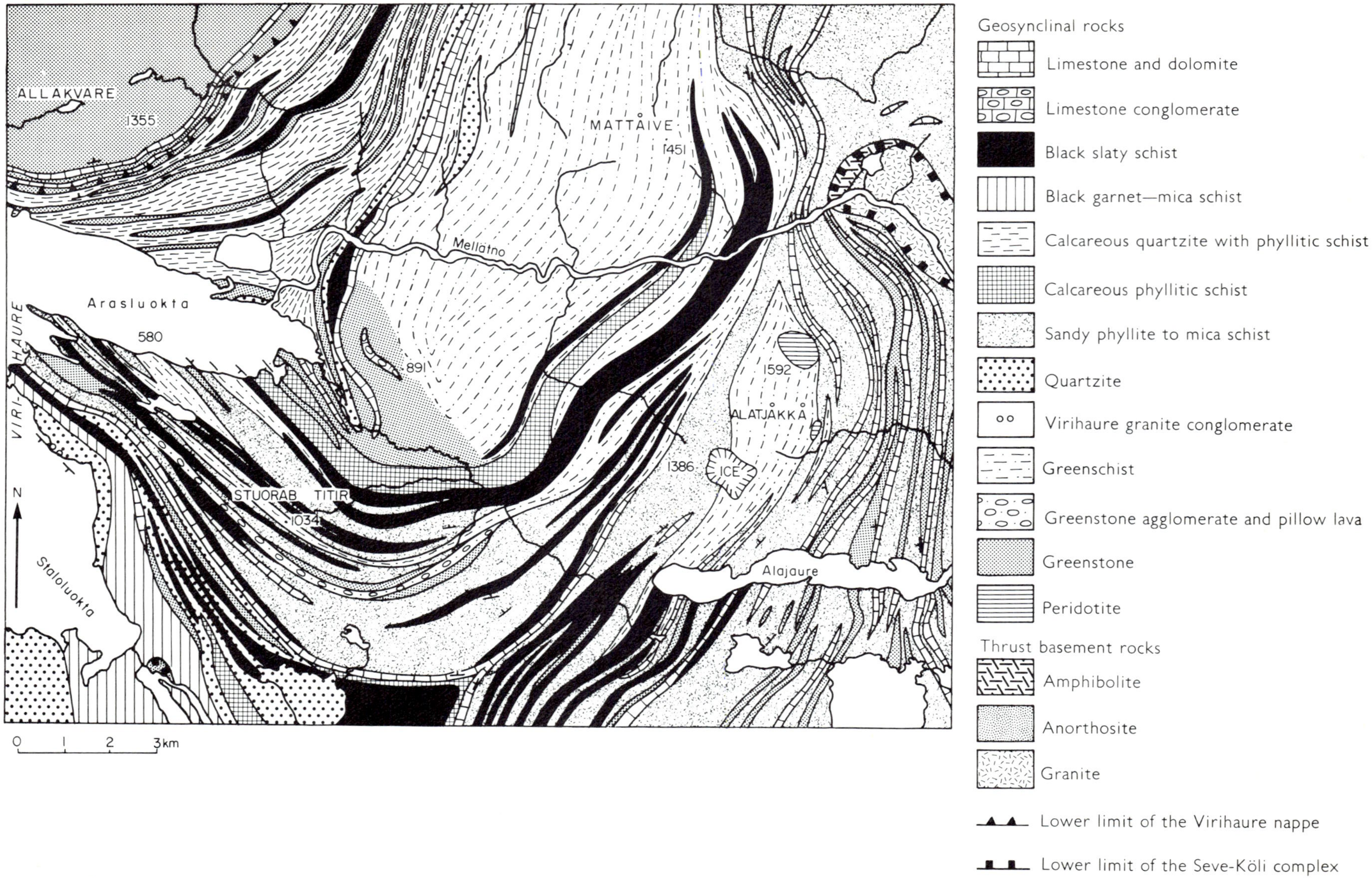

Fig. 181 *Geological map of the area east of Lake Virihaure.* Original by O. KULLING, assisted by A. FREDRIKSON

brief description of the stratigraphy which follows the main facts will be outlined.

The *Eastern amphibolite* of the Swedish Caledonides lies immediately to the east, below the lowest sedimentary formation of the eastern limb. Within the Eastern amphibolite there occur some conformable beds or flat lenses of micaceous schist and marble, i.e. magnesitic to Mg-rich dolomite lenses.

Mica schistose quartzite rocks follow above the Eastern amphibolite. Within the sedimentary rocks there are amphibolite beds of the same types as characterize the Eastern amphibolite.

The *Virihaure quartzite* with granite-conglomerate beds is probably the western counterpart of the quartzite mentioned above, within the eastern limb. The Virihaure quartzite occurs around the southernmost part—at Staloluokta bay on Lake Virihaure. Granite-conglomerate beds within the quartzite group have been found both south and east of Staloluokta bay. The substratum of argillaceous rocks is not exposed.

In the eastern limb the *Keddåive dolomite* lies above the quartzite rocks. The dark-coloured dolomite group contains repeated beds of amphibolite, blackschist and other schists. The Keddåive dolomite is followed by *mica schists* of various kinds and then by the *Pieske limestone.* This carbonate group also occurs together with amphibolite rocks in a conformable position. The micaceous rocks above the Pieske limestone are mostly quartzitic. They probably represent a northern continuation of the Vuortna quartzite in the Aivo–Sarvas area previously described.

The rocks of the western limb will now be dealt with. An example of the composition of the rock sequence between the Virihaure quartzite and the higher-lying Pieske limestone of the western limb in an area about 4 km north of the southeast corner of Lake Virihaure could be located on Fig. 181. The lowest rock member in the mapped area, black garnet–mica schist, has quartzite intercalations of Virihaure quartzite type going downwards. Carbonate rocks occur in the sequence of strata between the black garnet–mica schist and the higher lying Pieske limestone. The carbonate rocks are probably a western counterpart of the Keddåive dolomite further east, but are of limestone composition. Beds of quartzite, calcareous quartzite, black schist and greenschist form the bedrock between the lower and upper groups of carbonate rocks.

The occurrence of peridotite, anorthosite and granite rocks within the area mapped east of Lake Virihaure is of particular interest. The granite and anorthosite rocks represent upheavals of parts of the Middle thrust rocks and are considered by the writer as thrust-basement rocks. The anorthosite occurrence near the southeast corner of Lake Virihaure is surrounded by black garnet–mica schists with intercalated quartzitic rocks. East of the area under discussion the writer found the same anorthositic type of rock within the Middle thrust rocks. A titaniferous iron ore deposit, which lies in the eastern mountain border area of the Middle thrust rocks some 50 km southeast of Lake Virihaure, contains, among other igneous rocks, anorthosite of the type mentioned above. The small peridotite occurrences lie within the lower part of the Seve–Köli complex, stratigraphically in the vicinity of the Pieske limestone. The two peridotite occurrences on the special map (Fig. 181) lie to the north of Lake Alajaure within a N–S-trending bedrock zone, where the intensely folded Pieske limestone appears as repeated streaks together with amphibolite and quartzitic mica schist.

Titirnjarka peninsula, lying northwest on the eastern part of Lake Virihaure, has a ridgy character and, in the main, a well-exposed bedrock. The Pieske limestone serves as a marker bed in the middle of the rock formations which strike northwest. Immediately above the Pieske limestone of the peninsular area follow quartzitic schists, as in the eastern limb of the synform mentioned above. Intercalations of black schists and limy beds occur. Above the sandy group of rocks lies a greenstone formation with agglomerate beds and pillow-lava horizons. Further upwards comes a sequence of alternating sandy schists, black schists and greenschists, crowned by a formation of calcareous phyllitic schists. This rather thick sequence of sandy sedimentary beds could possibly be looked upon as a northern, smaller characteristic counterpart of the Gilliks clastics in several more southerly-lying parts of the Seve–Köli complex.

In the isoclinal core zone of the synform at the eastern end of Arasluokta bay (the easterly-directed bay on Lake Virihaure) local disturbances have made stratigraphical examination difficult. Light-coloured dolomite and light- and dark-coloured limestone appear to form a stratigraphic lower zone with overlying quartzite; fine conglomeratic beds constitute the core zone itself. Greenstone to greenschist and black to dark grey slates are other components of the core zone. A glance at the geological map (Fig. 181) discloses remarkable behaviour by the greenschist–greenstone rocks in the core zone and to the east of it. In the east the schistose basic volcanic rocks of the eastern limb occupy a very broad zone 5 km in width in places. Going over to the western limb, the basic rocks dwindle to a thin greenstone bed. This behaviour of the basic rock formation is somewhat difficult to understand; it is connected with the intensity of tectonic deformation within the bedrock area.

The stratigraphy of the upper part of the geological sequence in the eastern Virihaure area has largely been examined in the northerly-directed valley north of the eastern end of Lake Virihaure. On the slopes of the valley, and along brooks within the area, the writer has obtained a very large collection of fossils from the outcropping limestone formations; these formations are certainly of Upper Ordovician age and are a northern stratigraphical counterpart of the Slätdal limestone. North of Virihaure the Seve–Köli sequence of rocks is tectonically overlain by amphibolite rocks, comprising the lower part of the Virihaure nappe. The tectonic boundary with the Virihaure nappe rocks is almost invisible, as the rocks on both sides of the boundary are of roughly the same metamorphic grade.

Stratigraphy of the area north of Arasluokta bay

10 The upper, major part of the sequence is composed of calcareous sedimentary rocks, ranging from calcareous phyllite to calcareous quartzite. Dark phyllitic slate and greenstone occur as repeated intercalations. The whole group of sediments are, in the author's view, a northern counterpart of the Lövfjäll group, well-known in the central and northern parts of the Västerbotten mountains. The Lövfjäll group sediments north of Lake Virihaure are several hundred metres thick. The underlying sedimentary formations are all relatively thin and of uneven thickness due to tectonic disturbances.

9 Dark grey phyllitic slate and black slate. A few metres thick.

8 Marly limestone with scattered fossils. Some tens of metres thick.

7 Grey quartzite. Probably 10 m thick.

6 Dark grey slate with some marly beds. Less than 10 m thick.

5 Fossil-rich limestone formation. About 40–50 m thick. The upper part is grey, marly and very rich in well-preserved fossils (Fig. 182). The lower part is dark grey in colour. In the lowest limestone beds, scattered edgy pebbles of light grey dolomite and other rocks of the immediately subjacent sedimentary formations occur, together with fossil remains (Fig. 183).

4 Quartzite, only a few metres thick.

3 Dolomite breccia, sometimes with edgy material, but also with a more conglomeratic structure. Probably 10 m thick.

2 White and grey dolomite, often well-bedded. Probably more than 100 m thick.

1 Quartzite, some metres thick.

There follows a very thick formation of greenschist and greenstone. In these tectonized basic rocks, intercalations of green quartzite, dolomite and limestone have been found by the writer.

The question remains as to how the sedimentary formations listed above can be correlated with the Slätdal–Vojtja sequence—the stratigraphic backbone of the Seve–Köli series of formations in southern and middle Lapland. The large gap in the stratigraphical column occurs above the dolomite formation. The dolomite breccia, overlying quartzite and scattered pebbles within the lowest part of the thick fossiliferous limestone all indicate a large stratigraphical gap followed by marine transgression with limy precipitation within the coastal zone. Thus, the thin quartzite formation represents the Vojtja clastics and the thick fossil-rich limestone the Slätdal limestone. Formations 7 to 9 in the above stratigraphical column are possibly northern counterparts of the Broken group.

Vertsajokkots Section east of Lake Vastenjaure

The most northerly occurrence of the fossiliferous Slätdal limestone in Swedish Lapland outcrops along Vertsajokkots brook, which has its source on the southern slope of Mt. Låtats and from there flows westwards to the southeast corner of Lake Vastenjaure. The writer discovered the fossiliferous limestone in 1947 and in the same year published (Kulling, 1947) a report on, *inter alia*, the Upper Ordovician–Lower Silurian stratigraphy of the area east of Lake Vastenjaure. The northern part of the area east of Lake Vastenjaure consists of Virihaure nappe bedrock. The lower part of the nappe is composed of the following: pillow lava (the so-called Låtats lava); quartz keratophyre; and micaceous phyllite with a layer of unfossiliferous impure limestone. As in the Virihaure area the tectonic boundary between the Virihaure nappe rocks and the underlying Seve–Köli rocks is hard to locate. The upper part of the Seve–Köli rocks belongs to the Lövfjäll group of sediments and has been examined in scattered outcrops on the mountain slopes north of Vertsajokkots and along the westernmost part of the brook. The calcareous phyllite to calcareous quartzite of the group is several hundred metres thick, and greenstone beds occur as conformable intercalations. In the middle part of Vertsajokkots the stratigraphical succession (Fig. 184) has been measured, ranging downwards from the lowest part of the Lövfjäll group to basic

Fig. 182 *Fossiliferous limestone with parallel-lying elongated conical tubes of a fossil species of undetermined systematic position.* No septa or other cephalopode characteristics have been observed. The parallel arrangement of the tubes possibly indicates near-shore deposition of the carbonate material. Slätdal limestone in the area north of the eastern corner of Lake Virihaure. Photo O. Kulling (1964)

volcanic rocks in greenschist facies below the Vojtja group sediments.

Below this sedimentary sequence amounting to 263 m there follows to the east a very thick sequence of greenschists with scattered beds of light dolomite.

The sequence of sedimentary formations east of Lake Vastenjaure listed in Fig. 184 is the northern continuation of the sequence to the north of the easternmost part of Lake Virihaure. In the sequence just described the 32 m thick quartzite (no. 2 in the table) is considered to constitute the Vojtja group clastics, and nos. 3 to 5 constitute the Slätdal group of calcareous beds. The whole sequence, starting with no. 6 (quartzite) followed upwards by slates and limestones with quartzite beds up to and including no. 11 (slate), is probably a counterpart of the Broken group sediments in southerly-lying areas. Similarly no. 12 (quartzite) and the following fines represent the lowest part

Fig. 183 *Richly fossiliferous limestone with an edgy inclusion of light grey dolomite.* The limestone is considered as a northern counterpart of the Slätdal limestone in central Västerbotten and is thus of Upper Ashgillian age. The area north of the eastern corner of Lake Virihaure. Collection O. Kulling. Photo Inger Signorelli. Natural size

14	15 m	Dark grey phyllite with beds of calcareous phyllite. Lowermost a bed of black slate.
13	7 m	Marly slate and limestone with fossil remains.
12	12 m	Light grey quartzite and calcareous quartzite with marly beds. Lowermost a bed of fine conglomerate.
11	45 m	Upper and major part of dark grey to blackish grey phyllitic slate with scattered quartzitic beds. The lowermost 10 m is of black slate.
10	8 m	Uppermost and lowermost parts are of marly limestone with fossils. The middle part is of black slate.
9	10 m	Light grey quartzite and calcareous quartzite.
8	25 m	Grey phyllitic slate with beds of blackish grey slate.
7	12 m	Grey marly slate and limestone with fossils.
6	5 m	Light grey quartzite and calcareous quartzite.
5	30 m	Dark grey to blackish grey limestone, richly fossiliferous. Intercalations of black slate with limestone lenses.
4	15 m	Blackish grey to black slate.
3	12 m	Upper part of bluish grey limestone. The lower part of grey marly slate contains fossiliferous limestone beds. At the base is a thin bed of white dolomite.
2	32 m	Grey to bluish grey quartzite with some thin intercalated fine conglomeratic beds.
1	35 m	The upper part is composed of grey marly slate with limestone beds and dark grey phyllitic slate. The limy beds are fossiliferous. The lower part is dark grey slate with marly slate.

Fig. 184 *Stratigraphical succession of middle Vertsajokkots*

of the Lövfjäll group. The distance between the two areas with fossiliferous bedrock sequences is about 8 km. The difference in composition between the two stratigraphical columns presented is partly caused by primary facies differences. However, the bedrock of the two areas shows many signs of tectonization; sufficient knowledge of it cannot be gained without detailed mapping of the entire geological zone.

In the area east of Lake Vastenjaure, the upper part of the Seve–Köli sequence, including the fossiliferous formations, disappears towards the north and northeast beneath the Virihaure nappe rocks. To the east, the middle and lower formations of the Seve–Köli sequence are about the same as described in the area around the eastern part of Lake Virihaure and in the adjacent area to the south. Thus, the lava formation, with agglomerates and pillow lava beds, and the Pieske limestone are well developed. Because of the style of folding, the limestone is repeated and occurs as narrow zones with a N–S trend. A peridotite body occurs and lies conformably with one of the limestone bodies.

In the preceding pages all the relatively well-mapped bedrock areas within the Swedish part of the Seve–Köli complex in which, with exception of the Blåsjö area in northern Jämtland, the bedrock contains not only Ordovician but also Silurian deposits have been dealt with. The areas considered lie in southern Norrbotten, Västerbotten and northernmost Jämtland. The remaining part of the Swedish mountains includes the northern Norrbotten mountains, the middle and southern Jämtland mountains and the Härjedalen and northernmost Dalarna mountains. The Seve–Köli stratigraphy of the remaining mountain areas will now be considered.

Northern Norrbotten Mountain Area

The Seve–Köli rocks of the northern Norrbotten mountains have been dealt with in the writer's description accompanying a map of the mountain area (1964, in Swedish with English summary). In the northern Norrbotten mountains, and in all other parts of the Swedish mountains in which the upper part of the Seve–Köli column has been eroded away and only the high grade metamorphic lower part of the stratigraphical column is left, primary rock structures are relatively sparse. The rock transformation within some zones of Seve–Köli rocks in the northern Norrbotten mountains is remarkable. In westernmost parts of the mountain area the rocks show slightly increasing metamorphism upwards. In the eastern part of the mountains however the metamorphic grade in the bedrock beneath the Eastern amphibolite is considerably higher than it is above the sheet of basic igneous rocks.

The uppermost group of rocks within the northern Norrbotten Seve–Köli sequence is the *Gicce gneiss*, a migmatitic schist complex occurring only in the international border zone southwest of Lake Sitasjaure. In the adjacent part of Norway the Gicce gneiss occurs widely and is pseudo-stratigraphically located. The rock name was proposed by Foslie (1941).

In the international border zone southwest of Lake Sitasjaure, the Gicce gneiss lies above amphibolite rocks, in the upper western parts of which sparse pegmatite veins occur.

To the east of these amphibolite rocks there follow three isolated erosional remains of the same amphibolite formation. The most easterly of the amphibolite bodies contains relict lava structures. The amphibolite formation has been named the *Patta greenstone lava* by the writer (after Mt. Pattatjåkko). The rocks immediately below the Patta greenstone are of different composition in the east and west. In the east, the greenish grey *Patta quartzite* lies beneath the greenstone, while in more westerly areas, in particular southwest of Lake Sitasjaure, *calcareous mica schist* with garnet and garben hornblende occupy the same position. Judging from the scant observational material available to the writer, the Patta quartzite is an eastern shoreline deposit, whilst the limy schists to the west denote deposition at a greater distance from land.

Fig. 185 *Calciferous Nuolja marble with numerous fragments of harder rocks which were originally layers in the marble.* The more brittle layers were broken during tectonic deformation. The fragments then drifted apart and rotated in the plastic surroundings. The southwestern part of the westerly Abisko high mountains about 3 km east of Lake Håjkanjaure. From O. Kulling (1950)

Where the Patta quartzite is of greatest width, in the mountains around Mt. Pattatjåkko, a thin marble formation occurs directly beneath it; its occurrence is, however, intermittent. Below the Patta quartzite or the marble formation lie *quartzite and quartzitic schist* with interbeddings of greenschist and mica schist. Below this rock complex comes a *graphitic schist formation* followed by a *marble formation*—the *Nuolja marble*. This marble formation is looked upon by the writer as the northern Norrbotten counterpart of the Pieske limestone. In the southern parts of the northern Norrbotten mountains the Nuolja marble is rather intermittently distributed, probably due to strong tectonic disturbances (see Fig. 185). In the northwestern part of the northern Norrbotten area, around the western part of Lake Torneträsk, there are at least two formations of marble with mica schist in between. In the high western Abisko mountains at the western end of Lake Torneträsk the upper marble formation, occurring on Mt. Nuolja among others, is strictly speaking the Nuolja marble. The lower marble formation occurring in the north is often of more or less dolomitic composition and dark in colour. In the southern part of the northern Norrbotten mountains a thin quartzite formation often lies immediately below the Nuolja marble and has a relict gritty structure in some places. In the same area the lowest, schistose part of the rock sequence down to the Eastern amphibolite has not yet been satisfactorily worked out.

The above description of Seve–Köli stratigraphy in the northern Norrbotten mountains is based mainly on field observations. A closer petrological and chemical examination of collected rock specimens has not yet been carried out, neither has a planned detailed geological survey of the high mountainous area around Mt. Pattatjåkko. The stratigraphical column of the northern Norrbotten mountains appears to be a northern counterpart of the Pieske limestone and some of its superposed rock formations in the southern Norrbotten mountains. The Upper Ordovician and Silurian deposits of the southern mountains are completely eroded away in the northern Norrbotten mountains. The topmost geological formation discernible in the northern mountains—the *Patta greenstone lava*—is possibly a northern counterpart of the Seima greenstone lava in the central Västerbotten mountains.

Southern Swedish Mountain Area

The Seve–Köli stratigraphy of the southern part of the Swedish Caledonides is not very well known. On the bedrock map of Sweden (1958) the subdivision of rocks within the middle metamorphic grade Seve–Köli area of the southern mountains, i.e. within the Tännfors region of westernmost Jämtland, is the same as on TÖRNEBOHM's old geological map (1896). Three N–S-directed rock units are discernible; these are the eastern or phyllite zone, the middle or Stuedal schist zone and the western or garben schist zone. In the phyllite zone TÖRNEBOHM mentioned phyllite rocks with intercalations of limestone, quartzite, arkose and conglomerate. "Stuedal schist" is an old Norwegian term for greenish grey biotitic porphyroblastic, mostly sandy mica schist rocks. The garben schist is dark grey and often contains very large hornblende porphyroblasts.

The Tännfors rock complex is underlain and surrounded by high grade metamorphic schists—called *Åre gneiss* by TÖRNEBOHM—and by mica schist with amphibolitic intercalations. At the border between the Stuedal schist and the high grade metamorphic schists in the south some elongated bodies of serpentine occur. The serpentine at Handöl is of economic value and is used commercially.

After the geological survey of the Tännfors region by TÖRNEBOHM the area was remapped some decades later by FRÖDIN (1922). FRÖDIN, who like most Scandinavian geologists in the twenties did not believe in large thrusting, came across a N–S-directed thrust plane between the phyllite mentioned above and the Stuedal schist west of it. He considered the overthrust plane and the tectonization of rocks in the thrust-zone as signs of quite local dislocation in a small easterly-directed overturned antiform. In the western half of the Tännfors region FRÖDIN also noticed a N–S-trending overthrust plane with a slab of high grade schists forming the lowest part of the bedrock immediately above the plane. FRÖDIN's supposition was that the overthrust zone represented a zone of magmatic intrusion, combined with some kind of dislocation. To the writer it seems probable that the central part of the Tännfors region represents an isolated part of a nappe—a klippe with a western continuation on the Norwegian side of the border. In the eastern and western border zones of the Tännfors region FRÖDIN observed arenaceous rocks—"sparagmitic schists"—forming the lowest part of the Tännfors rock complex, and intercalated with greenstone which he considered to be intrusions. Although FRÖDIN as a representative of the geological thinking of his time did not believe in large thrusting, his geological mapping disclosed a great many tectonic results of importance. These results have been wholly overlooked in most geological reports on the southern Swedish Caledonides during the last three decades.

In 1964, geological remapping of the Tännfors region and its surroundings was started by STRÖMBERG who in the same year wrote a paper on the geology of the area. He stated that the rocks of the most easterly part of the region contained, *inter alia*, greywackes followed below by a coarse conglomerate with pebbles of serpentine and occasionally of porphyry and limy quartzite. This sedimentary sequence of rocks is overlain to the west by the thrust unit previously mentioned, with tectonically banded schists in the lowest part.

Badly preserved stem segments of Pelmatozoa were collected in the Tännfors region by HÖGBOM, in a marble layer in the neighbourhood of the Handöl serpentine quarry. These fossils are the only ones that have been discovered in the eugeosynclinal rocks of the southern Swedish mountains. The Tännfors region is a stratigraphically low-lying part of the Seve–Köli rock sequence.

Conglomerates with serpentine pebbles of the type mentioned above had already been described in TÖRNEBOHM's monograph of 1896 on the Central Scandinavian Caledonides. The conglomerates occurred together with serpentine bodies, in a thin S–N-extending middle metamorphic zone of Tännfors bedrock type, the bedrock zone being situated along the international border northwest of the Tännfors region. As in the Tännfors region the ultrabasic rocks and serpentine conglomerates were situated in the very zone in which the middle grade metamorphic bedrock graded into high grade metamorphic rocks (to the east). TÖRNEBOHM gave several examples of the stratigraphy of the conglomerate- and serpentine-bearing bedrock. A stratigraphical sequence could not however be worked out. The serpentine and serpentine conglomerate occur together with amphibolite, garben schist, garnet–mica schist, greenstone conglomerate and polymict conglomerate, with pebbles of greenstone, quartzite, schist, etc.

COMMENTS ON THE DISTRIBUTION OF THE SEVE–KÖLI GROUP ROCKS

As has been described, erosional remains of the upper part of the Seve–Köli sequence, down to and including the Gilliks group sedimentary rocks, have been found from the southern part of the Västerbotten mountains to the middle of the Norrbotten mountains, a stretch of about 290 km.

The uppermost sedimentary unit, the *Viris quartzite*, occurs mainly in the longitudinally-running synform of the central Västerbotten mountains, and a small erosional remnant of the quartzite forms the summit of Mt. Aivotjåkko in the northernmost part of the mountains. These two quartzite deposits are probably the youngest sediments within the Scandinavian Caledonide eugeosyncline.

The *Lövfjäll group* of calcareous sandy facies seems to be of about the same composition in southern and northern occurrences, with the exception of the international border area across lakes Ikesjaure, Mavasjaure and Peskehaure, where strata considered as part of the Lövfjäll group are mainly developed as slates and calcareous slates, i.e. without pronounced sandy members.

Typical *Broken group rocks*—the light grey Broken quartzite and the dark graptolite-bearing slate—have been found only in the central part of the Västerbotten mountains. In other areas quartzites, conglomeratic beds and slates may or may not be real counterparts of the Broken group. In the Aivo–Sarvas area on the Västerbotten–Norrbotten border, the lower of the two conglomerate formations lying higher up than the Vojtja quartzite conglomerate is considered by the writer to be of Broken group age. In the area around the eastern parts of Lake Virihaure and Lake Vastenjaure a rather thin quartzite formation above the fossiliferous Slätdal limestone is thought by the writer to be a northern counterpart of the Broken quartzite.

The sequence *fossiliferous Slätdal limestone–Vojtja quartzite conglomerate–quartzite* has been met with in all the above-mentioned areas except that north of Lake Peskehaure, where as a result dating of the rocks is somewhat uncertain. The Vojtja conglomerate and quartzite represent a major transgression within the eugeosyncline. It may be pointed out that in the Aivo–Sarvas area and in the vicinity of Lake Virihaure–Lake Vastenjaure, limestone occurs not only above the clastic rocks mentioned but also immediately below them. In the area north of Lake Virihaure the lowest part of the Slätdal limestone contains scattered pebbles and also angular rock material. The Slätdal limestone and also the limestone beneath the Vojtja clastic rocks are both deposits which originated near a shore. The limestone lying immediately beneath the Vojtja clastic rocks is in the writer's view some kind of transgression sediment in a very flat environment. The *Vojtja transgression* was preceded by a pronounced upheaval of land within the eugeosyncline. During the *pre-Vojtja regression*, erosion carried away unknown portions of the uppermost eugeosynclinal deposits and the Vojtja transgression rocks were laid down on an eroded surface. However, the sediments below and above the transgression surface show the same metamorphic state of preservation; no traces of Upper Ordovician orogenesis have been found.

The *Gilliks group rocks*. The deposits immediately below the Vojtja clastic rocks consist, as pointed out above, of rock types such as coarse greywacke conglomerate, greywacke to quartzite, black slate, greenstone lava and greenschist. In the central part of the Västerbotten mountains, where the first dated fossils in the entire Seve–Köli complex were found (this chapter, page 219), the Gilliks group of sediments below the Vojtja clastics contain the characteristic Gilliks "conglomerate slate" i.e. unsorted greywacke conglomerate with a slate-bearing matrix. The boulder and pebble material of the coarse deposits—Na quartz keratophyre, Na granite, dark- to light-coloured dolomite, quartzite and exceptionally amphibolite—suggests an intra-eugeosynclinal source. The distribution of this characteristic coarse greywacke type of rock and of its conglomeratic, quartzitic and arkosic equivalents in the Seve–Köli areas is of particular interest.

The distribution of Gilliks clastic rocks within the Västerbotten mountains will be considered first. The coarse Gilliks rock types of central Västerbotten appear to be absent in southern Västerbotten. In the southeastern part of the region the Gilliks clastic rocks are represented by greywacke quartzite and quartz phyllite (Du Rietz, 1935). In the southwestern part, no sediments of Gilliks type occur beneath the Vojtja clastics (locally called the Remdalen conglomerate). Were such sediments once present the pre-Vojtja epoch of erosion has since swept them away. In the northernmost part of Västerbotten—the Aivo–Sarvas area—the Gilliks group is composed mainly of feldspar-bearing quartzites with some beds of polymictic conglomerate in its western part. The facts presented may indicate a northwestern to western source for the Gilliks clastic rocks. The distribution of the Gilliks clastic rocks in the southern to middle Norrbotten area may suggest this also as there are coarse Gilliks greywacke conglomerates in the Ikisjaure–Mavasjaure district; in the district north of Lake Peskehaure lying NNE of the Ikisjaure–Mavasjaure area, however, the Gilliks clastic rocks seem to be mainly of quartzitic composition.

The magnitude of the stratigraphical break below the Vojtja clastic rocks and the age of the Gilliks group rocks are hard to establish because no real marker fossils have so far been found in the Gilliks or older Swedish Seve–Köli sedimentary rocks. The discoveries mentioned previously of *Halysites* and other tabulate corals in limy lenses within the Gilliks greywacke conglomerate, and the discovery of the bryozoan colony *Dianulithes petropolitanus* in fine-grained rocks which are probably an upper portion of the Gilliks group, lead to the single conclusion that the Gilliks group rocks are probably not older than Caradocian. It may be mentioned that the Slätdal–Vojtja rocks above the Gilliks group rocks are of Upper Ashgillian.

In this connection it is of interest to mention two renowned coarse clastic groups of rocks on the Norwegian side of the international border. In the Hölonda–Horg area in the western part of the Trondheim region, fossil discoveries have been reported in relatively low grade eugeosynclinal rocks for many years. According to VOGT (1945) a polymict conglomerate, the *Volla conglomerate*, containing pebble material with a composition reminiscent of that of the coarse Gilliks clastics, is immediately superposed on a graptolite-bearing dark slate—the *Dicranograptus* slate of Upper Caradoc age. The polymictic conglomerate is separated from a quartzite conglomerate—the Lyngestein conglomerate—situated higher up in the sequence, by sandstone and lavas. The Lyngestein conglomerate is considered by the writer to be a stratigraphical counterpart of the Vojtja conglomerate.

The other polymictic conglomerate of interest is the so-called *Moberg conglomerate* in the western Norwegian Bergen region where eugeosynclinal deposits occur (KIAER, 1929). The Moberg polymictic conglomerate is of greywacke type; above the clastic group lies a fossiliferous limestone with a coral fauna of Lower Ashgillian age. A little higher up in the sequence there occurs quartzite and a quartzite conglomerate—the *Skarpefjell conglomerate*—probably a stratigraphical equivalent of the Lyngestein and Vojtja conglomerates in the extreme northerly parts of the eugeosyncline. It may be mentioned that slate and limestone of Lower Llandoverian age lie above the Skarpefjell conglomerate. If no objections can be raised to the correlations proposed above, it would appear that there was regional greywacke sedimentation within the eugeosyncline during the Uppermost Caradocian and that the stratigraphical gap beneath the Vojtja clastics probably accounts for only part of the Lower Ashgillian.

The *Seima group of mainly greenstone lava.* In the Seimajaure area in the eastern part of the central Västerbotten mountains, the basic Seima lava formation lies directly beneath the Gilliks clastic rocks. In the upper part of the lava sequence the greenstone is amygdaloidal, and agglomerate beds occur. In the lower, eastern part of the formation the increase in metamorphism eastwards blots out primary structures and renders further stratigraphical examination difficult. In the Remdalen area of southwestern Västerbotten the *Lasterfjäll greenschist* has been considered as the western counterpart of the Seima volcanic rocks. In the Remdalen area the Lasterfjäll basic volcanic rocks occur mainly in the eastern limb of the N–S-trending synform; this behaviour indicates that the centre of the extrusive rocks is in the northeast, as the lava rocks in the eastern limb thicken to the north. The lava zone across Lake Seimajaure is probably a proximal portion of the lava eruptions.

Going northwards from the central Västerbotten mountains to the Aivo–Sarvas area, the Seima volcanic rocks seem to dwindle out. The thick formation of greenstone lava in that area—the *Sarvas greenstone*—does not occur at the base of the Gilliks clastic rocks, but between the underlying Vuortna quartzite formation and the Pieske limestone. Greenstone beds within the limestone sequence also seem to be connected with the greenstone immediately above the limestone.

The next area to the north which will be considered is the Titirnjarka peninsular area on the eastern side of Lake Virihaure. In the sequence of strata between the Slätdal–Vojtja sequence of rocks and the Pieske limestone there are several, possibly six, greenstone to greenschist "formations". Some of them occur as large lens-shaped bodies, others are continuous formations. The uppermost green rocks in the area are probably a result of volcanic activity during Gilliks time. The sandy rocks below these upper green rocks are not however of typical Gilliks facies. In the lower part of the sequence there occurs a formation of greenstone—the *Titir lava*—with agglomeratic beds and well-preserved layers of pillow lava. This lava formation is probably a northern equivalent of the Seima lava in central Västerbotten. The lava formation is separated from the lower-lying Pieske limestone by a sequence of sandy rocks, which is probably a northern equivalent of the Vuortna quartzite mentioned in the foregoing description. The sandy formation has thin intercalations of black slate and limestone. In the area east of Lake Virihaure, greenstone beds are sparse immediately above the Pieske limestone. To the southeast however greenstone occurs partly in strata above the Pieske limestone and partly as layers within the limestone sequence; it thus has the same stratigraphical mode of distribution as the Sarvas greenstone in the Aivo–Sarvas area. Before concluding this discussion on the distribution of the basic volcanic rocks, particularly of the Seima lava, mention should be made of the *Patta lava* in the southwestern part of the northern Norrbotten mountains (KULLING, 1964). In this greenstone formation on Mt. Pattatjåkko, there are levels with primary lava structures; the greenstone is the uppermost stratigraphical unit of the area and occurs as isolated erosional remnants in both the southwestern part of the Norrbotten mountains and in adjacent areas on the Norwegian side of the border. On Mt. Pattatjåkko the *Patta quartzite* lies immediately below the volcanic formation; it is thus probably a northern counterpart of the Vuortna quartzite in the Aivo–Sarvas area.

In this review, stratigraphic correlations between the Seve–Köli rock sequences of different areas have been made for the upper part of the stratigraphic column, comprising the Silurian and Upper to Middle Ordovician groups of rocks. The lower part of the Seve–Köli group rocks is almost devoid of fossils, and good marker rocks are sparse. The southern Norrbotten Pieske limestone and its northern Norrbotten counterpart, the Nuolja limestone, serve, however, as a lower Seve–Köli marker formation of prime value.

The Swedish Seve–Köli sequence of formations is an eastern portion of the Caledonian eugeosyncline. As exemplified in the preceding description the bulk of the sediments are shales, sandy to limy shales, sandstones to limy sandstones, greywackes, and conglomerates. Limestone and dolomite constitute a minor part of the deposits and occur mostly as rather thin formations. In more westerly parts of the eugeosyncline carbonate rocks occupy considerable portions of the stratigraphical column. This is particularly true of the Norwegian Seve–Köli zone west of the Norrbotten mountains. The best example of this western limy facies is probably in the Håfjell synform in the Ofoten area—the "Håfjellmulden" west of Narvik. The Håfjell succession is about 5000 m thick, and about a third of this sequence of rocks is limestone and dolomitic limestone. The age of the sequence is, according to the writer, probably Middle–Upper Ordovician and Silurian. The sequence has three arenaceous formations; the lowest is a polymictic conglomerate, considered to be a northern counterpart of the Gilliks clastics; the two upper formations are composed of quartzitic sandstone and can possibly be correlated with, respectively, the Vojtja and the Broken clastic rocks of central Västerbotten. In districts lacking any traces of fossil remains, the geologist often has to be rather dogmatic in his assertions.

Before concluding this section on Seve–Köli stratigraphy, the distribution of ultrabasic, peridotite rocks and serpentine conglomerate and breccias will be discussed. The peridotite rocks serve as almost a guiding level in the lower part of the eugeosynclinal sequence and are occasionally followed by serpentine breccias to serpentine conglomerates. In places, the serpentine breccias are of a somewhat agglomeratic appearance. It is probable that the rising ultrabasic magma made its way into the unconsolidated mud at the bottom of the eugeosyncline and remained there. The only known fossiliferous serpentine conglomerate within the Scandinavian Caledonides is the Otta conglomerate in the southern corner of the Norwegian Trondheim region. The fauna of the conglomerate is surprisingly well preserved; an examination of the trilobites (by Hedström, 1930) indicated that the conglomerate was of uppermost Arenigian age, but according to a later examination of the gastropods (by Yochelson, 1963) lowermost Middle Ordovician age seemed the most plausible. In 1933 the writer suggested that the Otta conglomerate might be of roughly the same age as the serpentine conglomerates discovered in the central part of the Västerbotten mountains. It is of interest to note that Bailey and McCallien (1952) expressed the opinion that the peridotite body enclosed in Lower Ordovician basic lava in the extreme southwesterly part of the Scottish Caledonides probably consisted of effusive rocks.

STORFJÄLL NAPPE COMPLEX

The delimitation of the Storfjäll nappe complex against the underlying Seve–Köli rocks is admittedly based on relatively little field evidence. The writer's experience of the vast nappe area is insufficient, and the papers published on selected parts of its bedrock do not give a general synopsis of the geology of the entire area. The thrust complex may be divided into the southern or Södra Storfjället–Tängvattnet area, and the northern or Norra Storfjället–Älsvattnen area.

Södra Storfjället–Tängvattnet Area

The *Södra Storfjället area* was geologically surveyed during the twenties by Beskow (1929), and his monograph on the mountainous area contains a great deal of information. Beskow's stratigraphical division of the bedrock of Mt. Södra Storfjället is based on his interpretation of the bedrock as isoclinally folded; a basaltic lava formation, called the *Mesket formation*, is the uppermost stratigraphical unit and he suggested that it forms the core of three NE–SW-lying synforms. Below the Mesket formation Beskow's stratigraphy includes a sequence of mainly quartzitic rocks with some thin formations of limestone and conglomerate, with a thin, intermittent basic lava bed above the lowest limestone formation. After some revisional field work within the area and a reexamination of the geological facts available, the writer in a monograph on the Caledonides of Västerbotten county (1955) interpreted the Södra Storfjället area as a NE–SW-directed synform. The formations in the upper part of the synformal area seem to bear close relationships to, and can be correlated with, the formations within the central Västerbotten Björkvattnet–Virisen area. The stratigraphy of the Södra Storfjället area is in the writer's view as shown below; the stratigraphical column refers mainly to the southeastern limb of the synform.

13 *Sandy calcareous phyllites and quartzites* representing the core of the synform.

12 *Phyllitic slates.* At mountain summit point 736, the writer has noticed calcareous beds and thin conglomeratic levels within slaty rocks, probably referable to the lower part of the "phyllitic slates".

11 *Limestone* more than 16 m thick observed in only one locality in the area SSE of summit point 736.

10 *Quartzite conglomerate to quartzite.* In the limestone locality mentioned above it is about 70 m thick. According to Beskow's geological map the conglomerate can be followed for more than 10 km. In the northwestern limb no conglomerate has been observed at the same stratigraphical level—only a thin quartzite bed.

9 *Mesket greenstone lava formation.* These basic volcanic rocks with agglomeratic and tuffaceous beds contain intercalations of slaty sediments and quartz keratophyre. The high-lying area of Södra Storfjället with the crests of Mt. Mortotjåkko and Mt. Njåsats is formed, *inter alia,* of typical Mesket lava.

8 *Morto quartzite conglomerate,* which occurs intermittently. In places the coarse clastic rocks are 25–30 m thick.

7 *Phyllitic slates,* "biotite phyllite", with quartzitic intercalations. In the western part of the formation the slates grade into calcareous sandy phyllites.

6 *Polymictic conglomerate* with intercalations of phyllitic slates.

5 *Ropen quartzite conglomerate* occurring intermittently.

4 *Greenstone lava.* A thin formation, occurring intermittently.

3 *Ropen limestone,* this is continuous and can be followed right round Mt. Södra Storfjället and is therefore a good marker formation in the northwestern limb also.

2 *Phyllitic to sandy schists* with intercalations of conglomeratic greywacke, quartzite and greenstone.

1 *Ropen granite,* the lowest rock unit within the Södra Storfjället thrust rocks.

Beskow is of the opinion that the igneous rock unit was intruded during the overthrust period. However, the igneous rocks and the overlying sedimentary rocks apparently show the same grade of metamorphism and tectonization and the granite body seems to lie quite conformably beneath the sediments. The granite is a salic quartz–albite type and can probably be interpreted as a recrystallized and later cataclastically altered quartz keratophyric lava. Several small elongated peridotite bodies outcrop on the northern slope of Mt. Södra Storfjället. They all lie in the lowest part of the stratigraphical sequence given, in the schists below the Ropen limestone. Serpentine conglomerate of Rotik type is not met with.

As has been mentioned, the stratigraphical column given above is that of the south-eastern limb of the Södra Storfjället synform. In the northwestern limb the coarse clastic and sandy beds of the southeastern limb are replaced by shaly deposits to a large extent. Thus, shaly facies seems to predominate in the northwest and the sandy, near shore facies in the southeast.

Sedimentary formations nos. 13–10 form the actual core of the Södra Storfjället synform and can probably be considered as northwestern counterparts of the Lövfjäll, Broken, Slätdal and Vojtja sediments of the Björkvattnet–Virisen area. The isolated remnant of the limestone formation within the area of the synform core indicates relatively strong tectonization within the bedrock area. If, as supposed by the writer, the Mesket lava of Mt. Södra Storfjället is the northwestern continuation of the Seima lava of the Björkvattnet–Virisen area, then the Gilliks sediments with their characteristic greywacke members intercalated between the Vojtja conglomerate and the Seima volcanic rocks do not occur in the Södra Storfjället sequence. Formations nos. 8–4 of the Södra Storfjället sequence, with upper and lower quartzite conglomerates and intermediate conglomerates of polymictic character, are hard to correlate with formations in adjacent parts of the Caledonian eugeosyncline, as the polymictic clastic rocks seem to be of rather local facies, i.e. they often pass rapidly into fines. The persistent limy formation in the lower part of the Södra Storfjället sequence is known as the *Ropen limestone*; in northern Västerbotten and in Norrbotten the *Pieske limestone* is a persistent limestone occurring in the lower part of the Seve–Köli stratigraphical column; these two limestone deposits are probably chronostratigraphically comparable.

The *Tängvattnet area*—the area lying between Mt. Södra Storfjället in the south and Mt. Artfjället and Mt. Norra Storfjället in the north—was superficially mapped during the twenties;

the northern part was later remapped by the writer (Kulling, 1955), and the southern part by Murris (1957). Most types of rocks occurring in the Södra Storfjället area are found in the Tängvattnet area. However, rock formations suitable as markers are sparse. The bedrock is partly imbricated and zones of strong tectonic movement have been found. The material available does not permit of working out a reliable stratigraphical column for the entire Tängvattnet area. Within the area, grey calcareous slates and black graphitic slates with quartzite intercalations predominate. Limestone, dolomite and peridotite occur as small, scattered, mostly lens-shaped bodies. The lowest group of rocks within the area is, according to Murris, the greenschist-bearing quartzite–mica schist group situated in the southernmost part of the area, along the E–W-directed Jovatten valley, and the topmost group is the greenstone formation in the international border zone to the west of Lake Tängvattnet. Murris is of the opinion that the greenstone body has been thrusted above the underlying graphitic slates. Northwest of Tängvattnet the writer has examined the northern continuation of this greenstone body, which borders on limestone, peridotite and slates. The whole sequence of rocks is folded in a complicated fashion and disappears to the northwest beneath the migmatitized rocks of the Rödingsfjäll nappe. Murris has formulated the following tentative stratigraphical column for the Gieravardo–Jofjället area: calcschist formation; graphite schist formation; greenschist formation; and the lowest unit, the quartzite–mica schist formation. The writer has cursorily examined the western continuation of the bedrock sequence of the Gieravardo–Jofjället area at Lake Favnvandet in Norway, some 10 km to the west of the international border. The sequence obtained from this examination is given below. From the top:

6 Phyllitic slate with levels of calcareous slate and graphitic slate.
5 Quartzite.
4 Greenstone.
3 Phyllitic slate, partly graphitic.
2 Calcareous phyllitic slate.
1 Graphitic slate.

The relatively thick formation of calcareous phyllitic slates is a western continuation of Murris's "calcschist formation". The continuation of the quartzite formation mentioned above on the Swedish side of the international border—a stretch of about 8 km—was drawn on the geological map of the Västerbotten mountains of 1929 but is missing on Murris's map. Where the western outcrop of the quartzite was drawn on the 1929 map at the summit of Mt. Girtotjåkko, Murris's map has a thin formation of "metakeratophyre". Immediately to the west of this rock unit, at about the same stratigraphic level, there occur thin beds of dolomite, dolomite conglomerate and calcareous slate.

Before leaving this stratigraphically little-known area some information will be given about Tjålt hill some 20 km north of Tängvattnet. The following formations have been listed (Kulling, 1955):

5 Quartzite (the Tjålt formation)
4 Greenstone
3 Limestone (possibly a northern counterpart of the Ropen limestone)
2 Calcareous slate with dolomite intercalations (the Rödingsnäs formation)
1 graphitic slate with levels of quartzite (the Murberg formation).

If the quartzite formations observed in different parts of the Tängvattnet area are of one and the same stratigraphic unit, and the sequence quartzite–greenstone–limestone on Tjålt is a northern counterpart of the Ropen quartzite–conglomerate–greenstone formation and Ropen limestone in the Södra Storfjället area, future geological mapping has a stratigraphical marker sequence from which to start.

The rocks of the Södra Storfjället–Tängvattnet area are, generally speaking, tectonically affected to a greater extent and are of a somewhat higher grade of regional metamorphism than the underlying fossiliferous Seve–Köli rocks in the southeast. This rock sequence is mostly in the sericite–chlorite and biotite–chlorite subfacies of the greenschist facies, whereas the rocks of the overthrusted Södra Storfjället rock complex are in epidote–amphibolite facies. The lower part of the rock complex is garnet-bearing and levels of staurolite–mica schist occur. Retrogressive metamorphism increases to the northwest and all grades of conversion of granite, hornblende and biotite into chlorite are common features; complete pseudomorphs often occur. In the Tängvattnet area, the rock sequence above the stratigraphically low-lying garnet-bearing quartzite–mica schist formation is, according to Murris, in greenschist facies. Metamorphism increases towards Mt. Norra Storfjället in the northeast and the slaty schists are followed by garnet–mica schists with garnets increasing in number and size.

Norra Storfjället–Älsvattnen Area

The northern, major portion of the Swedish part of the Storfjäll nappe complex extends from northeast of the Tängvattnet area in the south to the northern corner of the Västerbotten mountains, where the thrust rocks cross over into Norway. The eastern erosional boundary of the thrust complex generally follows the N–S-trending Tärnasjön–Tärnaån valley. In the northwest the Storfjäll thrust complex disappears beneath the migmatite rocks of the Rödingsfjäll nappe. Mt. Norra Storfjället occupies the southeastern part of the thrust complex with peaks up to 1593 and 1764 m a.s.l. Lake Laisan at the southern foot of the mountain lies at a height of 452 m. Other mountains within the complex are Eastern Mt. Artfjället to the west of Mt. Norra Storfjället, the northern continuation of Mt. Norra Storfjället including the peaks Tjutetjåkko, Gabbi and Mesket, and in the northern part of the complex the low mountainous area around the Älsvattnen lakes. The bedrock of the Norra Storfjället–Älsvattnen area was superficially mapped during the twenties (Quensel and Backlund, 1929). Their map shows the main distribution of the rocks, with the lower grade metamorphic rocks as a northwestern belt and the higher grade metamorphic rocks forming the main southeasterly part of the area. During the thirties and forties the writer made excursions and carried out revisional work within the area (Kulling, 1955). Most parts of Mt. Norra Storfjället and surrounding areas were geologically remapped by Dutch students during the same period. Field work and scientific assessment of the collected material under the experienced leadership of Brouwer resulted in a number of published theses at the university of Amsterdam.

The relatively low grade northwestern belt of the thrust complex will be considered first; this extends from the eastern part of Lake Överuman up to and including the Älsvattnen area in the northeast, which is the northern continuation of the Tängvattnet chlorite grade bedrock area. In the western part of the belt, the bedrock immediately below the Rödingsfjäll migmatite is composed of limestone, dolomitic limestone, greenstone and phyllitic, partly calcareous schists. As the carbonate rocks lie immediately below a major thrust plane, they are highly tectonized and represent a zone of differential movement. In several places the carbonate rocks are rich in variously sized fragments of resistant rocks such as amphibolite and quartz–feldspar-rich rocks. The amphibolite may occur as lava beds or intrusions within the carbonate rocks. The acid rock fragments can be proved to emanate partially from the *Murivardo granite*, of which is composed Mt. Murivardo to the west of Lake Övre Älsvattnet on the international border. They occur in this vicinity as variously sized inclusions or involutions in the locally broad zone of carbonate rocks. The age of the Murivardo granite is somewhat dubious—it may be a tectonic rock wedge of thrust-basement rocks. In the Övre Älsvattnet area there are some peridotite bodies within the limestone regions. The Ruopsokvare peridotite ridge east of Mt. Saretjåkko outcrops as the core of a limestone antiform. The limestone in the Övre Älsvattnet area and in other parts of the northwestern belt is felt by the writer to be the northern counterpart of the Ropen limestone. The writer's stay in the so-called northwestern belt was not sufficiently long to permit him to work out a proper stratigraphical table for the bedrock.

The major part of the Storfjäll nappe complex in the most northerly part of Västerbotten consists of high grade rocks and lies immediately east of the limestone-bearing bedrock belt just described. The boundary zone between higher and lower grade metamorphic rocks appears to be the site of strong differential movements. However, it is uncertain whether there is a real overthrust level in the zone. In his monograph, "The Mesket–Gabbi region", Aleva (1950) draws attention to this bedrock zone and calls it "the zone of tectonic mixing". His summary of the nature of the bedrock zone is as follows: "This belt is in general made up of more or less biotite-, carbonate- and sometimes garnet-bearing foliated amphibolites and hornblende schists, and dark grey to black, mostly garnet-rich, platy and fissile schists. Local and subordinate to these are platy to laminated chlorite–muscovite schists and hardschiefer-like rocks, chlorite-rich schists, albite–hornblende garben schists, garnet–mica schists and staurolite–mica schists." He points out that rocks of the "tectonic mixing zone" to the northwest gradually pass into rocks of the lower grade metamorphic zone. The mountain area to the south of Aleva's "Mesket–Gabbi region" i.e. the northern part of Mt. Norra Storfjället, is dealt with by Mulder (1951). He is of the opinion that the zone of differential movements becomes indistinct to the south and the transition from higher grade to lower grade metamorphic rocks is rather gradual.

At the beginning of the twenties Quensel and his co-workers came across a great many rocks which aroused their interest, when mapping Mt. Norra Storfjället and adjacent mountain massifs. Quensel discovered a basic lava formation—the *Mesket lava*—and a quartzite conglomerate, which he considered to be a counterpart (in a high metamorphic environment) of the Vojtja conglomerate in relatively low grade rocks of the central Västerbotten mountains. Bodies of norite, gabbro and granite and areas of migmatite rocks were also found in the middle of the high grade mountain schists. The geologists mentioned above called the conglomerate occurrences the *Vojtja conglomerate* (as did the writer in his report of 1938, which dealt with, among other things, the stratigraphy of the Mesket–Gabbi area). Later in a monograph on the Caledonides of Västerbotten county (1955) the author stressed, *inter alia*, that quartzite conglomerate occurred at several levels within the eugeosynclinal sequence, and that a stratigraphic correlation was acceptable and obviously strengthened if the bedrock areas compared had a complete sequence of marker formations, were of about the same metamorphic grade, and were situated within one and the same nappe unit. The two marker formations within the actual high grade metamorphic area are the quartzite conglomerate mentioned above and the Mesket lava formation. The geological map of QUENSEL and BACKLUND (1929), those in the papers of ALEVA and MULDER, and the writer's Västerbotten bedrock map of 1955, all show that the two formations are not conformable with one another. Thus, the quartzite conglomerate extends from Mt. Mesket 7 km directly southwest, whereas the Mesket formation curves directly from the eastern slope of Mt. Mesket some 6 km southeast and then curves back westwards to immediately east of the southern part of the conglomerate. Another type of occurrence of quartzite conglomerate within the high grade bedrock area is as large lenses folded together in mica schists of variable composition, for example on Mt. Valle in eastern Artfjället.

According to ALEVA the high grade metamorphic rocks within the Mesket–Gabbi area can be divided into the following units:

Upper schists
Vojtja conglomerate
Middle schists
Mesket amphibolites
Lower schists
Albite–tonalite and transitional rocks of the schists, exposed in the area north of ALEVA's mapped area and there in contact with the lower schists.
Intrusive basic igneous rocks, mainly norite (in the amphibolite massif of Mt. Laivatjåkko north of the mapped area, beneath the Mesket amphibolite); and amphibolite (the amphibolite massif of Mt. Laivatjåkko north of the mapped area).
Zones of fragment-bearing gneissose rocks occurring in close connection with norite in the southern part of the mapped area.
Metasomatic and migmatitic rocks in the southern part of the area.

Aleva feels that the upper, middle and lower schists are of about the same primary composition, and that the occurrence of thin quartzite beds in all three formations points to very similar sedimentation conditions. The garnet–mica schists of the formations contain staurolite, kyanite and sillimanite. On the southern slope of Mt. Mesket, the writer (KULLING, 1938) observed thick calcareous schists within the schist formation, called the *Middle schist* in ALEVA's table above. In a section about 6 km west of Mt. Gabbi the Middle schists contain two beds of polymictic conglomerate in a sandy mica schist environment.

In some scattered occurrences of polymictic conglomerate in other parts of the Middle schist formation the writer noticed pebbles of trondhjemitic to quartz keratophyric composition, i.e. of the same type as is common in the Gilliks greywacke conglomerate of the Björkvattnet–Virisen area in the central part of the Västerbotten mountains. However, in most respects, and of course to some extent depending on the high grade metamorphic transformation of the Middle schists, they are very unlike the typical coarse Gilliks clastics. The "Vojtja conglomerate" in the Mesket–Gabbi area is polymictic, and according to ALEVA the following types of pebbles occur in recrystallized matrix transformed to medium-grained garnet–mica schist or hornblende quartzite: quartzite, plagioclase quartzite, hornblende–plagioclase quartzite and amphibolite. If the conglomerate and the Mesket lava are the stratigraphical counterparts (within the Storfjäll nappe) of the Vojtja quartzite conglomerate and the Seima lava of the Seve–Köli complex in central Västerbotten, then the Gilliks sediments between the two formations just mentioned and the Middle schists of the Mesket–Gabbi area may be considered as at least partly contemporary deposits. The stratigraphical gap below the conglomerate, together with the often intricate folding and the levels of differential movement within the rock sequence, probably caused the unconformable nature of the conglomerate with the underlying bedrock. In the high grade stratigraphical sequence no marble occurs either above the conglomerate or in the lower part of the sequence. In the high grade bedrock area the Ropen limestone ought to be found in the region in which albite tonalite and norite occur.

The *Mesket amphibolite* is a thick sequence of basic lava rocks in amphibolite facies. The rocks are metamorphic derivatives of lava beds with agglomerates, tuffs and local admixtures of argillaceous and calcareous sediments. Primary structures are mostly obscure. Thus, agglomerate rocks may have been transformed into anything from amphibolite rich in ellipsoid bodies of varying size to amphibolite with indistinct lenticular structure. The thickness of the Mesket volcanic formation in its central part is probably more than 1000 m. However, the complicated tectonics and strong deformation of the rocks within the area do not allow a reliable estimate to be made.

The *acid igneous rocks* which are called albite tonalite by ALEVA are known in other parts of the Storfjäll nappe area as *albite granite* and *meta-quartz keratophyre*; they seem to appear as conformable bodies. In the area north of Mesket–Gabbi transitional rocks going to schists occur with the albite tonalite and because of this the igneous rock sequence may be an acid lava formation with pyroclastic rocks and reworked material. The Na-rich igneous rocks seem to be restricted to the lower parts of the eugeosynclinal deposits.

In the easternmost part of the Storfjäll nappe, and west of the middle part of Lake Tärnasjön, there occurs a relatively extensive granite mass, the *Juomotjåkko granite*. The granite body about 6 km long is surrounded by Mesket amphibolite and is thought to be an intercalation in the lava formation. The light-coloured microcline granite contrasts strongly with the surrounding dark amphibolite. Elongated zones or lenses of Mesket amphibolite are found in several places within the granite mass. The microcline granite is biotite-bearing, but the biotite content varies from place to place. The lenses or slides of amphibolite in the granite are often intersected by leuco-trondhjemitic veins. MULDER described the Juomotjåkko granite in the paper mentioned above; he states, *inter alia*: "The granite mass as a whole may be assumed to be more or less lenticular or basin-shaped, with a slight elongation in a WNW direction. Although the contacts may be locally of a secondary nature, and the western contacts are irregular, the bedding observed in the Mesket amphibolite series in the neighbourhood of the contacts suggests a mainly concordant position for the granite mass. Near the contacts the dip is towards the granite and exposures of the upper contact have not been found anywhere. In several localities in the Mesket amphibolite series similar small massive biotite granitic to trondhjemitic masses occur. They commonly do not exceed several tenths of a metre in size, while some of these masses are of a somewhat different character, on account of their slightly foliated appearance, their high content of epidote and the fact that microcline is nearly always absent. Thin leuco-trondhjemitic veins intersect these masses locally. In some places intensive later shearing has mylonitized the rocks."

In connection with the above description of the Juomotjåkko granite within the high grade part of the Storfjäll nappe complex it may be mentioned that several small granite bodies occur within the so-called "lower metamorphic" western part of the Storfjäll nappe; the granite bodies are situated north of the eastern part of Lake Överuman. The writer has examined granite bodies, which are distinctly discordant with the surrounding schists. In most cases, however, the granites seem to be concordant intrusions. According to MULDER the acid intrusions vary in composition from biotite–microcline granite and tonalite to quartz-bearing hornblende diorite. The salic members of the igneous rocks cut through the more basic ones.

Basic igneous rocks of noritic to gabbroic character outcrop in four places within the high grade metamorphic bedrock area of the northern Västerbotten mountains. In the following description "norites" refers to these basic rocks. The largest area of norite occurs west of the northern part of Lake Tärnasjön as an E–W-lying body about 10 km in length, immediately beneath the Mesket amphibolite formation. The next largest body lies south of the southeastern part of Lake Överuman and appears as the 5 km broad circular core of a dome-shaped antiform. The norite is surrounded by high grade schists. The two other outcropping norite bodies are very small; it may be pointed out, however, that all four norite bodies are bound to bedrock areas of migmatitic composition. Parts of the basic rocks are of quite original habit and composition. There are all gradations from augite norite to garnet-bearing biotite norite, and from gabbro to uralite gabbro and amphibolite rocks. In the generally occurring medium- to fine-grained rather massic rocks there are coarse-grained types of pegmatitic habit. The writer has noticed that coarse-grained olivine gabbro penetrates fine-grained norite in places, and that the basic rock in its turn is older than the penetrating rocks of trondhjemitic composition. The trondhjemitic sills have caused amphibolitization of the surrounding gabbro–norite rocks. The border zones of the basic rocks with the surrounding rocks are mainly amphibolitized. Remarkable *fragment-bearing gneissose rocks* are always found in the vicinity of the norite. ALEVA, who distinguished the rock group, summarized the origin of the gneissose rocks as follows: "After the intrusion of the norite and probably helped by the heat of it, the schists of the higher metamorphic series suffered an ultra-metamorphic reconstitution or nearly complete anatectic remelting, probably facilitated by addition of alkalis, especially sodium." The fragment-bearing gneissose rocks were characterized as a garnet–mica gneiss with porphyroblastic plagioclase enclosing a variable number of fragments, mostly of amphibolite and granulite-like rocks.

Migmatite rocks within the area of the Storfjäll nappe complex. In the Gabbi–Mesket area described by ALEVA, zones of metasomatic and migmatitic rocks occur, mainly in the rock zone around the norite. Some of the schists, Mesket amphibolites and fragment-bearing rocks have been migmatitically altered in this way. The migmatized schists are veined and soaked by a white quartz–Na-feldspar rock. Locally the leucocratic material thickens to large lenses. In the centre of migmatization, strike and dip of the rocks are hard to measure and undirected, very large muscovite porphyroblasts have been developed. At some distance from the centre of migmatization the intensity of veining and soaking decreases, and the rocks pass into normal mica schists, amphibolites etc.

On Mt. Norra Storfjället and northeast Mt. Artfjäll, large areas of migmatite rocks occur as the topmost rock units of the mountains. The migmatite rocks mentioned are described in a paper on the geology of northeast Artfjäll by BUDDING (1951) and in another on some opaque minerals in the amphibolite rocks of Norra Storfjället by UYTENBOGAARDT (1953). UYTENBOGAARDT includes the migmatite rocks of the top area of Norra Storfjället in the upper part of his "Syter Series" of high grade rocks, which also contains high grade mica schists with conglomeratic layers; at the boundary between the migmatite and schist there is a small norite body. BUDDING after describing migmatite rocks in northeast Artfjäll stated that no sharp boundary could be observed between the migmatites and the surrounding high grade mica schists. Some years ago LINDSTRÖM visited a restricted part of the northern border zone in the area of migmatization on Mt. Norra Storfjället—the area north of hamlet Umasjö and east of Lake Överuman. His view is that the migmatite rocks are probably overthrust units and klippen of the Rödingsfjäll nappe. As the writer has had almost no field experience of the migmatite areas and of their relationships to adjacent high grade schists it has not been possible for him to evaluate these conflicting statements. BUDDING and UYTENBOGAARDT carried out detailed mapping of their areas, and LINDSTRÖM's observations were made during a short visit to the area east of Lake Överuman.

VIRIHAURE NAPPE

The name of this thrust complex is taken from Lake Virihaure which lies near the international border in the middle western part of the Norrbotten mountains. The Swedish portion of the nappe extends from the area around the northwestern part of Lake Virihaure as a relatively narrow belt, and goes northeast along Lake Vastenjaure and the river Vuojatädno to Lake Kutjaure, where it thins out; in the opposite direction it extends southward to Mt. Sulitelma, where the thrust-mass crosses the international border. South of Mt. Sulitelma, the eastern part of the thrust complex reenters Swedish territory north of Lake Ikesjaure, runs about 15 km southwards and then crosses the international border again.

Before entering upon a description of the Virihaure nappe bedrock, KAUTSKY's tectonic scheme for the Sulitelma–Sallohaure region will be outlined. He divided the mountain schists of the region into four large nappes: the Gasak, Salo, Vasten and Pieske nappes. The writer's division of these mountain schists, however, forms three thrust units which, as stated before, are: the outlying middle western Norrbotten portions of the Rödingsfjäll nappe (including the Stipok, Gasakjaure, Sulitelma and Vuoksa klippen); the Virihaure nappe; and the lowest thrust unit, the large Seve–Köli complex (which corresponds to the major part of KAUTSKY's Pieske nappe). The Norrbotten portions of the Rödingsfjäll nappe are given the collective name of the *Stipok nappe*, after the Stipok klippe—its most northerly klippe (KULLING, 1964). The Stipok nappe consists of parts of KAUTSKY's Gasak, Vasten and Pieske nappes and the whole of his Salo nappe. The Virihaure nappe includes the eastern part of the Vasten nappe mentioned above. Some parts of KAUTSKY's Pieske and Gasak nappes are considered by the writer to be parts of the Virihaure nappe.

The boundary between the upper part of the Seve–Köli rocks and the rocks of the Virihaure nappe is hard to locate in most places at first sight. The first rocks in the zone are mainly of biotite grade, whereas the overthrusted rocks contain scattered garnet porphyroblasts or, if the composition was appropriate to its formation, garben schist. The geological field map of the area southwest of Lake Virihaure shows for example that the bedrock formations of the Virihaure nappe lie unconformably upon the Seve–Köli rock complex.

At present, a general stratigraphical table for the Virihaure nappe cannot be established. Intricate folding, squeezing out or thickening of strata, and zones of gliding make conclusions as to the stratigraphy difficult. However, the principal stratigraphical features can be mentioned. In the area southwest of Lake Virihaure, the Virihaure nappe contains the following group of rocks: calcareous mica schist and crystalline limestone, greenstone, peridotite, trondhjemite

Fig. 186

Fig. 187

Figs. 186 and 187 *Låtats lava with pillow structure* (Fig. 186), and *Columnar jointing* (Fig. 187). Mt. Låtats, east of Lake Vastenjaure. Photos by O. KULLING (1946 and 1955)

and granite conglomerate. The peridotite occurs as isolated or agglomerated lenses in the greenstone, or at the boundary between the greenstone and limestone. The trondhjemite is generally attached to the limestone rocks as pseudo-conformable lenses. The entire Virihaure nappe sequence is folded in a complicated fashion. The crystalline limestone–calcareous mica schist group constitutes the upper formation of the stratigraphical sequence, the greenstone volcanic rocks the middle formation and probably the granite conglomerate—the Sårjås conglomerate—the lower formation. The central part—the limestone part of the N–S-trending Virihaure sequence—is thought to be a disrupted synform; the Sårjås conglomerate in the western border zone of the complex is considered to be part of the core of an imbricated antiform. In lower parts of the coarse clastic rocks the conglomerate structure passes into breccia structure. According to KAUTSKY (1953), who first described the conglomerate formation, the westernmost part can possibly be interpreted as a brecciated granite—the primary substratum of the breccia–conglomerate rocks. In the graphitic schists considered as eastern members of the Rödingsfjäll nappe, which lies to the west above the Virihaure nappe, there are several tectonic lenses of Sårjås conglomerate of various sizes.

To the north of this area, northwest of Lake Virihaure and around the western part of Lake Vastenjaure, the upper portion of the Virihaure nappe is composed of the greenstone formation which occurs in the area southwest of Lake Virihaure. The greenstone is often well preserved and of plagioclase–porphyritic nature. Beneath the lava formation lies a sequence of calcareous sediments, usually developed as garben schists and biotite-rich mica schists.

During the summer of 1966 the writer surveyed the area west of lakes Virihaure and Vastenjaure in the southern part from Lake Virihaure in the east to the Norwegian border in the west, and divided its bedrock into the following geological units: greenstone formation and underlying calcareous mica schists, referred to the Virihaure nappe; and dark grey mica schists with amphibolite bodies and light-coloured quartz–mica schists of Virihaure quartzite type, lying on garnet-rich tectonite rocks. This sequence of rocks could be a western part of the Seve–Köli complex. The tectonite unit has tectonic boundaries with both the overlying and underlying rocks. Below the tectonite is a light-coloured quartzite formation with some calcareous quartzite beds; the sediments lie on a gneiss formation. The quartzite and the underlying gneiss are included in the Middle thrust rocks. To the west, below the gneiss unit comes the so-called *Tysfjord granite*; this covers very large areas on the Norwegian side of the border. A protrusion of the granite crosses the border northwest of Lake Virihaure, forming a small area south of border cairn 242 A. This granite may form the autochthonous basement of early Precambrian rocks below the Caledonian rocks. The area on the Norwegian side of the border has been geologically surveyed by Foslie (1942). According to Foslie the Tysfjord granite is "a rather homogeneous granite with abundant microcline-perthite, characterized by hastingsitic amphibole, lepidomelane and generally some fluorite and epidote–orthite."

The plagioclase–porphyritic greenstone formation mentioned above in the area around the western part of Lake Vastenjaure extends eastwards across and south of the lake to Mt. Låtats. Here there are very fine pillow and columnar structures in some beds of the volcanic formation (see Figs. 186 and 187). Beds of quartz keratophyre also occur. There are several intercalations of dark grey slates, micaceous slates and quartzite beds within the lava succession, and similar sediments lie above the Låtats lava rocks. Beneath the lava sequence, calcareous slates with intercalations of coarse greywacke and beds of impure recrystallized limestone have been found. At present, however, a general stratigraphy for the Låtats area cannot be presented.

This description of the rocks of the Virihaure nappe is necessarily brief as our knowledge of the rock complex is rather superficial.

CHAPTER XXIII

THE UPPERMOST THRUST ROCKS OR THE RÖDINGSFJÄLL NAPPE

Rödingsfjället mountain lies on the international border zone in the northwestern part of the Västerbotten mountains. It consists of high grade rocks of migmatitic type. During field work on a geological map of the Västerbotten mountains in 1945 the writer noted that the high grade rocks in the international border zone were underlain to the east by middle grade rocks, and that the boundary between the two rock complexes was of a tectonic nature. The writer suggested that the high grade rock complex along the international border, lying on both sides of Lake Överuman, be called the *Rödingsfjäll nappe*; in a lecture some years later he described it. However, the main part of the Rödingsfjäll nappe lies on the Norwegian side of the border. In Norway, the eastern front of the nappe runs for at least 150 km in a SSW direction from Mt. Rödingsfjället. The Swedish part of the nappe front is about 40 km long; from its northern corner it can be followed another 40 km to the north on the Norwegian side of the border, to east of the Dunderland valley. The further course of the thrust boundary has not yet been investigated.

In the international border zone of the Norrbotten mountains, there are four areas of rocks which the writer feels are eastern protrusions or parts of klippen of the Rödingsfjäll nappe. These are: the *Juoksa thrust region* southwest of Lake Ikesjaure continuing to Mt. Svajefjäll in Norway; the *Sulitelma thrust region* on both sides of the international border, with the Baldoaive mountain area west of Sulitelma as a western part of the thrust rocks; the *Gasakjaure thrust region* immediately north of Mt. Sulitelma; and the *Stipok klippe*, extending from north of Lake Vastenjaure to W of the NW part of Lake Akkajaure.

NORTHWESTERN VÄSTERBOTTEN PART

Returning to the Västerbotten part of the Rödingsfjäll nappe, the stratigraphical subdivision of the bedrock of the eastern part of the thrust complex is from the top, as follows: garnet–mica schist formation; marble formation; migmatized mica schist formation with upper and lower beds of amphibolite (see Fig. 188, northwestern part). In places a thin zone of quartzose mylonite rock forms the boundary zone next to the underlying middle grade rocks of the tectonic substratum beneath the Rödingsfjäll nappe (see Fig. 189). The migmatite rocks are often migmatized in zones. Figures 190 and 191 exemplify the usual migmatite types. The mobilized and precipitated material is mostly of trondhjemitic composition but material of granitic type occurs, too. The feldspathized mica schist rocks are pervaded by discordant pegmatitic veins, which are occasionally ptygmatically folded. The conformably occurring amphibolite beds are probably of lava type; amphibolite dykes occur however. They are in fact older than the trondhjemitization (migmatization) of the rocks (Fig. 192). Geographically the Rödingsfjäll nappe bedrock can be divided into three NE–SW-trending zones: the eastern or migmatite zone, the middle or marble-bearing synclinorium zone and the western garnet–mica schist to gneissose rock zone with discordant leucocratic dykes. The dykes run in a direction parallel to N 65° W joints and are generally of trondhjemitic composition, but granitic dykes occasionally occur. The dykes vary in thickness from a few centimetres up to more than 100 m. Some dykes can be followed for long stretches. The rocks of the dykes are generally of porphyritic structure, and feldspar crystals in them may attain a length of up to 20 cm. Quartzose dykes often intersect the feldspar-rich dykes. In the leucocratic dykes inclusions of micaceous schists have been found, and the schistosity of these enclosed schists is nearly always oriented parallel to that of the surrounding rocks. Valuable observations on the rocks and tectonics of the Västerbotten part of the Rödingsfjäll nappe are given in two geological theses submitted to the university of Amsterdam—one

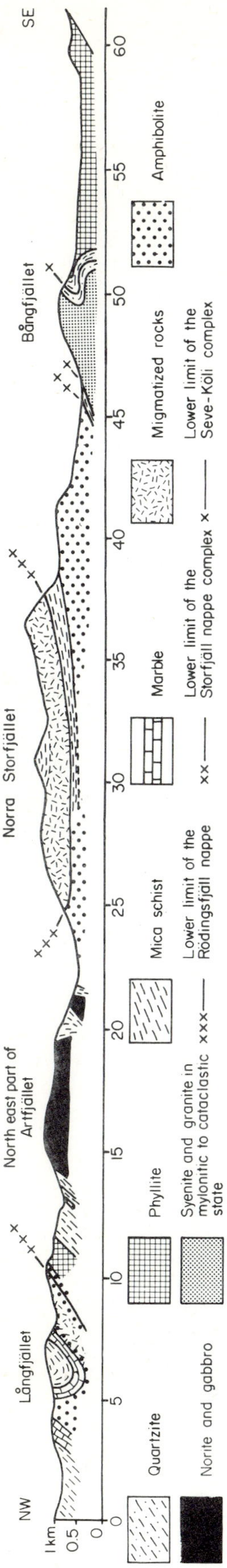

by De Munck (1951) on the area south of Lake Överuman, and the other by Lakeman (1952) on the area northeast of this lake. The writer has described the thrust rocks in a monograph of 1955.

WESTERN NORRBOTTEN PARTS

The four thrust regions of the Rödingsfjäll nappe in the Norrbotten mountains—the *Juoksa, Sulitelma, Gasakjaure and Stipok*—lie in the southwestern part of the mountains and extend beyond the Norwegian side of the border.

The *Juoksa thrust complex* lies southwest of Lake Ikesjaure; on the Swedish side of the border it consists mainly of staurolite–mica schists.

The *Sulitelma thrust complex* forms the upper part of Mt. Sulitelma and continues northwards to Lake Nedre Sårjåsjaure. The upper part of the thrust unit is composed of olivine gabbro—the *Sulitelma gabbro.* Below the gabbro follow metamorphic basic igneous rocks with marble formations and included peridotite bodies. In the lower part of the thrust complex, thin sheets of gneissose rocks occur. The olivine gabbro shows no signs of metamorphic alteration and is looked upon by the writer as being of post-tectonic origin. The Sulitelma gabbro has been thoroughly described in a monograph on the geology of the Norwegian part of the Sulitelma area (Vogt, 1927). Vogt however did not support the theory of overthrusting, and he interpreted the metamorphic basic rocks below the gabbro as metamorphic parts of the Sulitelma gabbro.

The *Gasakjaure part of the Rödingsfjäll nappe* is the international border area from Lake Nedre Sårjåsjaure in the south to west of Sirkasluokte bay on Lake Virihaure in the north. This rock complex consists for the most part of mica schists and calcareous quartzites. In the rather complicated, folded sequence of strata there are intercalations of graphitic schist, greywacke conglomerate, graded bedded sediments, quartz keratophyre and greenstone. In the northwestern part of the thrust complex, some minor bodies of Sulitelma gabbro have been found. The metamorphic grades of the rocks increase westwards and downwards. A characteristic staurolite–mica schist occurs in the international border zone; on the Norwegian side the schists are followed by gneissose rocks.

The *Stipok klippe* is the northernmost of the four tectonically high lying, high grade metamorphic units in the most westerly part of the middle

Fig. 188 *Section across the northwestern part of the Västerbotten mountains*

Fig. 189 *The east boundary of the Rödingsfjäll nappe on the northeast slope of Mt. Rödingsfjället on the interstate boundary of northwestern Västerbotten about 20 km south of Överuman.* The hard rock of the nappe consists of migmatite. Its bottom zone is rather quartzitic and considerably tectonized. The footwall consists of soft, black, lenticularly schistose phyllitic rocks. Farther south calcareous phyllitic rocks and limestone reach the boundary of the Rödingsfjäll nappe. Photo taken looking northwest. From O. KULLING (1955)

Norrbotten mountains. The major part of the Stipok klippe lies on the Swedish side of the international border. Figure 193 is a general geological map of most of the thrust complex, and Fig. 194 is a special map of Mt. Stipok. The Stipok thrust complex is bound in the north and northeast to thrust-basement rocks of the Middle thrust complex, and in the south and southwest to rocks considered as part of the Virihaure nappe. On the Norwegian side of the international border, the geological outlines are taken from FOSLIE's geological map of Hellemobotn and Linnajavrre (FOSLIE, 1942).

The igneous rocks of the Stipok complex—the quartz keratophyre, trondhjemite, greenstone and peridotite—are all of a conformable to pseudo-conformable nature. The quartz keratophyre and greenstone constitute formations in the stratigraphical sequence. The trondhjemite appears as lenses of various sizes within or along the boundaries of crystalline limestone. The peridotite occurs in two ways: partly as long and relatively thin conformable bodies and partly as small lenses, often

Figs. 190–191 *Examples of migmatite rocks in the lower part of the Rödingsfjäll nappe in the northwestern part of the Västerbotten mountains*

Fig. 190 *From the southern slope of Brantsfjället north of Överuman.* Photo by O. KULLING (1944)

Fig. 191 *From Olfjället on the international border with Norway, about 25 km south of Överuman.* From O. Kulling (1955)

occurring in swarms. The high grade sediments of the Stipok complex can be subdivided in the following way: black garnet–mica schist to graphitic schist, grey garnet–mica schist, garbenschist, calcareous garbenschist, calcareous mica schist, limestone and dolomite, and biotite gneiss, a feldspar-rich schist. The high grade sediments are often very coarse-grained, with large hornblende and garnet porphyroblasts. Scattered lenses of kyanite occur in Al-rich schists which are particularly noticeable in the Mt. Stipok area.

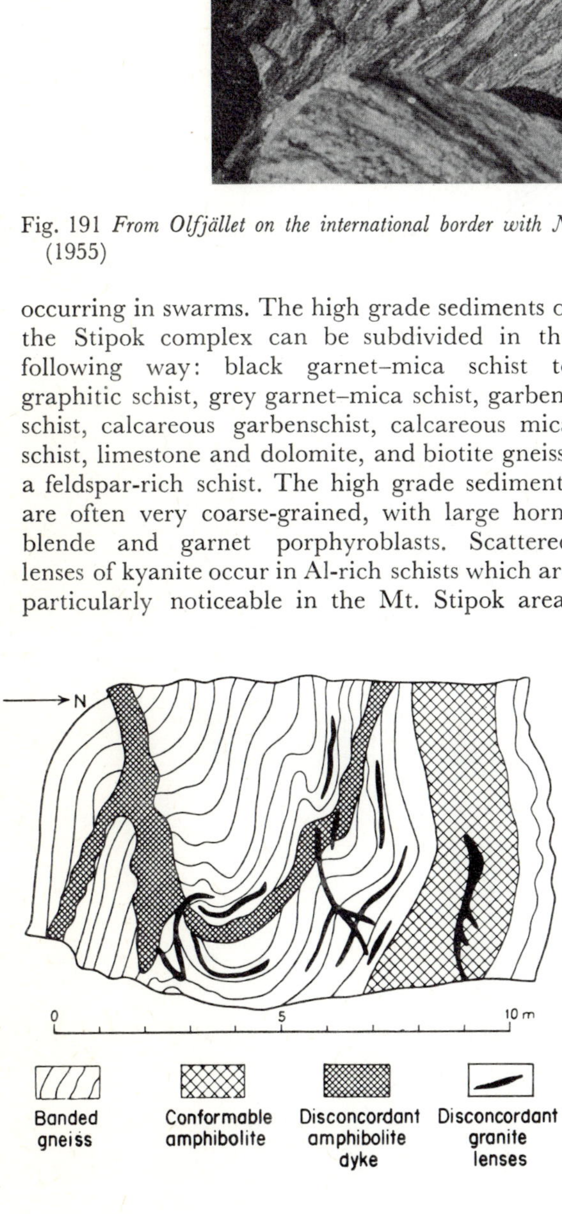

Fig. 192 *Sketch from an outcrop on the shore of Lake Överuman about 3 km northwest of the farm Strimasund.* The sketch exemplifies the geology of the migmatite zone in the lower part of the Rödingsfjäll nappe. From O. Kulling (1955).

Not only typical garbenschists but also hornblende–mica schists with scattered hornblende, lacking real garben structure, have been included with the garbenschists. A subdivision of the various hornblende-bearing schists was attempted but this was only partly completed.

It is difficult to obtain an unequivocal interpretation of the prethrust stratigraphy of the Stipok nappe bedrock. The rock formations of the nappe have been folded and refolded and the primary thicknesses of these plastically deformed formations have changed. Thus, a continuous rock formation could be transformed into a chain of lenses or it could vanish. In order to get detailed information on the geology of part of the Stipok nappe, the writer has worked out a special map of the Mt. Stipok area (see Fig. 194). Mt. Stipok is an isolated mountain in the middle part of the nappe area, with its summit 1240 m above sea level. Lake Slappejaure, 4 km southwest of the summit, lies at an altitude of 725 m. The rocks on the upper part of Mt. Stipok are relatively flat-lying, and dip towards the centre of the mountain. The summit of the mountain consists of black garnet–mica schist, the topmost formation in a rock sequence of synformal nature. East and west of the southern part of Lake Slappejaure, black schists of the same kind as on Mt. Stipok occur, enclosing a garbenschist formation to the east which is interpreted as the core of a synform. The antiformal zone between the two synformal parts of the bedrock is indicated by southerly-directed "protrusions" of calcareous mica schists on both sides of Mt. Stipok (see map, Fig. 194). The bedrock stratigraphy of the Mt. Stipok area is as follows (from the top):

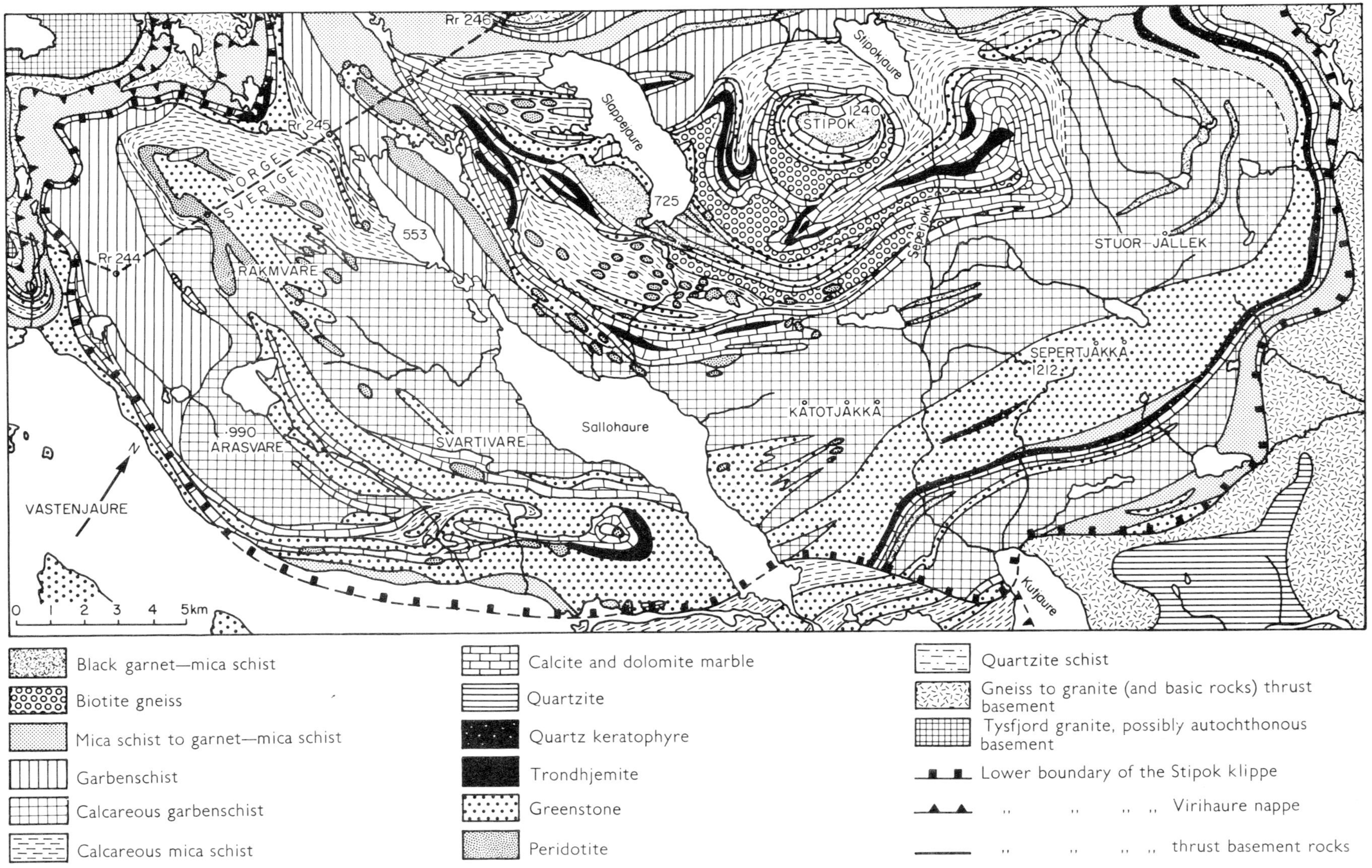

Fig. 193 *Geological map of the Sallohaure area in the middlewestern part of the Norrbotten mountains.* Original by O. KULLING. The geology on the Norwegian side of the international border is from S. FOSLIE's geological map, Linnajavrre (1942) (Map completed 1936, publ. 1937 and description publ. 1942)

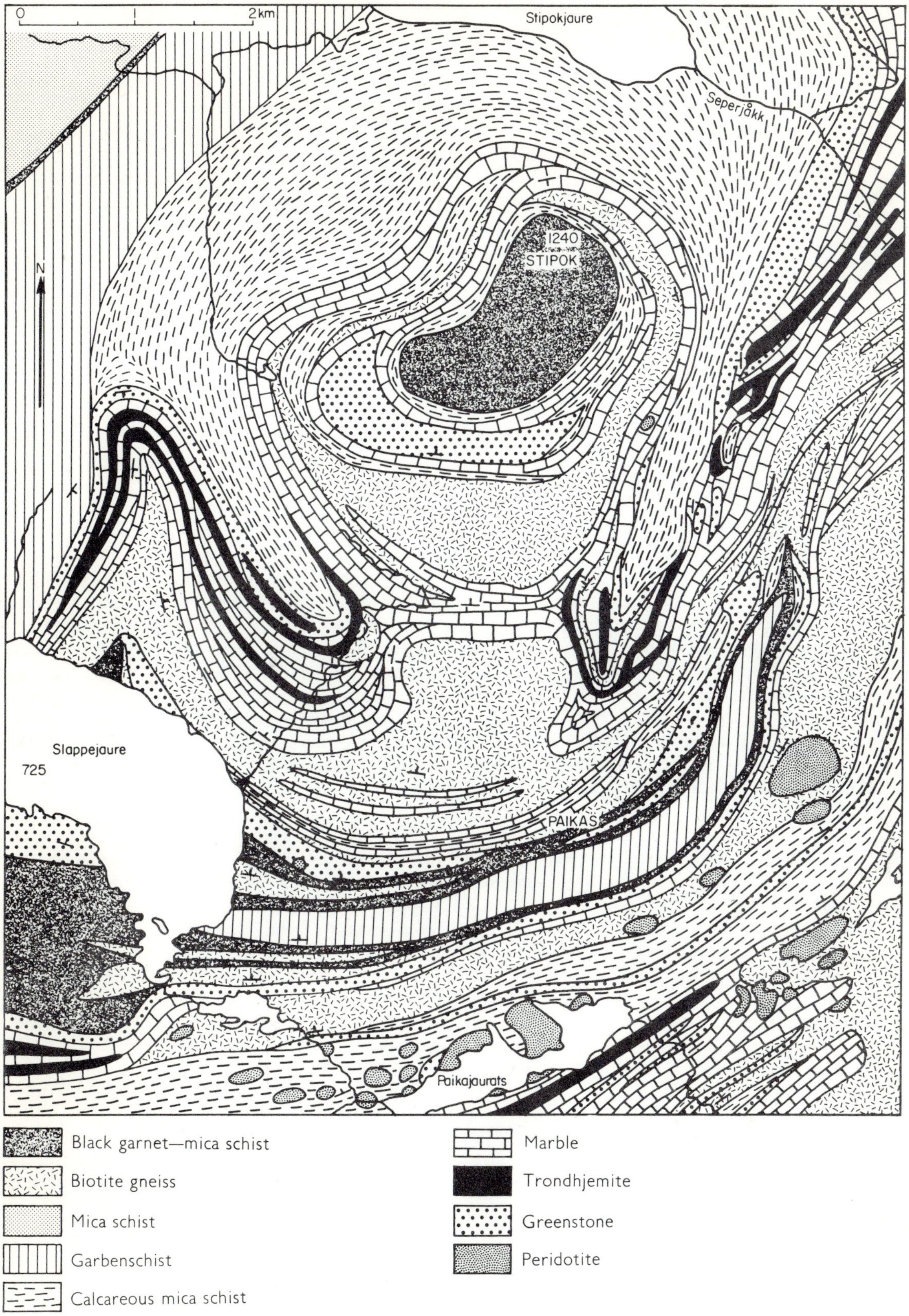

Fig. 194 *Geological map of the Stipok area in the northern part of the Sallohaure area, middlewestern Norrbotten mountains.* Original by O. KULLING, assisted by G. FREDRIKSON

Garbenschist (east of Lake Slappejaure)
Black garnet–mica schist—the so-called *Stipok schist*
Calcareous mica schist and limestone (on Mt. Stipok)
Amphibolite
Limestone, biotite gneiss and calcareous mica schist
Biotite gneiss to biotite–mica schist
Main limestone with pseudoconformable trondhjemite sills and lenses
Amphibolite
Calcareous mica schist
Garbenschist
Black schist
Mica schist—the lowest formation within the Mt. Stipok map area.

Peridotite bodies occur chiefly within the formations below the main limestone, but exceptionally within the main limestone and amphibolite beds. The amphibolites are probably lava rocks. The biotite gneiss occurs in all gradations to biotite–mica schist and as a result the feldspar porphyroblastic rock is considered to be a paragneiss of greywacke origin. Examination of the various rocks of the area is still in progress.

The main map of the Stipok nappe area will now be considered (Fig. 193). The geology of the Mt. Stipok area is representative of the northwestern part of the Stipok nappe. The limestone of the Mt. Stipok area extends 6 km northeast of the mountain, right to the international border southwest of Lake Slappejaure. The two large areas of calcareous mica schist containing several peridotite lenses, south and southeast of Lake Slappejaure, are considered by the writer to be antiformal parts of the bedrock, but the calcareous sediments in them do not contain hornblende porphyroblasts, which, on the other hand, are common in the schist area south of and below the limestone formation. The large area south and southwest of Lake Sallohaure—the southern part of the Stipok nappe—differs in composition from the Mt. Stipok–Slappejaure area. The bedrock of the southern area consists mostly of garbenschists, often with a pronounced percentage of calcareous material.

Large amphibolite bodies are present south of Lake Sallohaure and in the international border zone west of the lake. Their boundaries with the adjacent rocks are partly of a lobate nature. Carbonate rocks occur along the boundaries of the amphibolite south of Lake Sallohaure, or lie, together with trondhjemite and peridotite, in the amphibolite region. The western amphibolite is mostly bounded by peridotite rocks. The limestone connected with the trondhjemite south of Lake Sallohaure appears to be a southern counterpart of the main limestone in the Mt. Stipok area. The limestone formation runs westwards and forms the boundary with the substratum of the Stipok nappe throughout the southwestern part of the map area; it is partly dolomitic. The carbonate formation running along the northern border of the amphibolite regions south of Lake Sallohaure is mostly light-coloured dolomite or dolomitic schist. The Mt. Svartivare peridotite occurs in the middle of the carbonate formation.

In the northeastern part of the Stipok nappe the lowest formations of the nappe sequence occur, from a stratigraphical point of view. A sequence of amphibolite–quartz keratophyre–limestone forms the stratigraphical backbone, suitable as a marker bed, within the bedrock area. The lowest formations within the area are calcareous garbenschist and garnet–mica schist. A number of thin amphibolite levels occur but have not been recorded on the map. The lowest levels of the nappe are often plagioclase porphyroblastic gneissic schists.

The author's geological survey of the Stipok klippe and its environs is still under way. The rocks of the klippe and those of the Virihaure nappe (the portion southwest of Lake Virihaure) have several similar characteristics, such as richness in peridotite and trondhjemite. This is particularly true of zones containing marbles. However, other geological factors indicate that these rock complexes occupy different tectonic positions.

CHAPTER XXIV

COMMENTS ON FACIES AND ROCK DISTRIBUTION

THE STRATIGRAPHICAL RELATIONSHIPS BETWEEN THE DIFFERENT GEOSYNCLINAL BORDER AND GEOSYNCLINAL DEPOSITS

The autochthonous and Lowermost thrust sediments are of shelf facies; the Lower thrust sediments are mainly of miogeosynclinal, and the Upper and Uppermost thrust sediments of eugeosynclinal facies. The sedimentary rocks from the continental border zone have mostly been paleontologically well dated. In the miogeosynclinal sediments, formations with well-preserved faunas are sparse. However, the age of the formations within the stratigraphical sequences has been established in most cases. The Upper thrust sediments—the Seve–Köli sedimentary deposits—have yielded determinable fossils of stratigraphical value in only a few places and in a rather restricted part of the stratigraphical column. As a result it is hard to give a satisfactory review of the depositional assembly within the geosynclinal area and on its eastern border. Fortunately, however, the determinable fossils found in the eugeosynclinal sediments were discovered in formations in the middle part of the stratigraphical column, and in areas with a discernible stratigraphy. Some badly preserved fossils have been found in the Uppermost thrust rocks in scattered localities. The only conclusion that can be drawn from these finds is, however, that the fossil-bearing formations are younger than Cambrian. In the Upper thrust rocks the only dated formation of Silurian age is the Broken slate—part of the Broken group in the Central Västerbotten mountains—with graptolites of upper Middle Llandovery age. In the miogeosynclinal or Lower thrust rocks no graptolite shales of that age have been found to date. However, the *Berge* or *Pentamerus limestone* of the Föllinge sequence in western Jämtland is chronostratigraphically of upper Middle Llandovery age. In Central Västerbotten, in the Broken area and in the Föllinge sequence of Jämtland, the fossiliferous formations mentioned are underlain by sandstone formations—by the Broken quartzite in central Västerbotten and by the Ede or Phacops quartzite in Jämtland. The Ede or Phacops quartzite formation is of Lower Llandovery age. The magnitude of the stratigraphical gap beneath the transgression clastic rocks is not exactly fixed however as the lowest part of the Ede quartzite seems to be unfossiliferous. As it is, the Ede clastic rocks as well as the Broken clastic rocks record the first Silurian transgression in the miogeosynclinal and eugeosynclinal regions. In the East Jämtland nappes also, in the eastern border zone of the miogeosyncline, there are chronostratigraphic equivalents of the Berge limestone and the Ede quartzite (the Ede quartzite is called the *Upper Kyrkås quartzite* by THORSLUND). The Broken–Ede–Upper Kyrkås quartzite rocks represent the *first Silurian transgression*, in the eugeosynclinal, miogeosynclinal, and continental border area.

The preceding immersion period is called the *Västerbotten regression* by the writer.

The Upper Ordovician sediments are well documented in the eugeosynclinal deposits, as well as in the miogeosynclinal and continental border beds. The Slätdal limestone in Central Västerbotten has been dated as Upper Ashgillian; it passes downwards into sandstone and coarse quartzite conglomerate—the Vojtja clastic rocks. These are the best known and most widely distributed transgression deposits in the entire eugeosyncline, documented in both the Västerbotten and southern Norrbotten mountains. In all probability they occur on the Norwegian side of the border in the Kjölihaugen quartzite conglomerate of the northern Trondheim area and in the Lyngestein conglomerate of the western Trondheim area. There are also conglomerate and quartzite deposits which probably register the *Vojtja transgression*, in the southwesternmost part of the eugeosyncline to the south of Bergen.

In the miogeosynclinal Föllinge thrust rocks of Jämtland, the Upper to Middle Ashgillian is represented by shales with beds of quartzite, and in the East Jämtland miogeosynclinal border nappes, the Lower Kyrkås quartzite formation is dated as Upper Ashgillian. In the autochthonous beds of Jämtland the stratigraphical column reaches the upper Caradocian. Ashgillian deposits, if they were ever present, have, however, been eroded away. In southern Norway in the autochthonous sequence of the Oslo region, the uppermost Ashgillian is represented by a transgressive calcareous sandstone; this clastic formation lies on Middle Ashgillian gastropede limestone.

The Vojtja and Lower Kyrkås clastic rocks are uppermost Ordovician transgression deposits within the eugeosynclinal and synclinal border areas; the pre-Vojtja regression is called the *Lapland regression* by the writer.

In the eugeosynclinal deposits of the Björkvattnet–Virisen area in Central Västerbotten and the Aivo–Sarvas area in the Västerbotten–Norrbotten border zone, coarse clastic rocks occur in the uppermost part of the stratigraphical column. These are quartzite conglomerates younger than the Vojtja conglomerate described above—namely the Gimja conglomerate in Central Västerbotten and the Aivo conglomerate in the Aivo–Sarvas area. Both lie in calcareous quartzite and (Viris) quartzite which form the uppermost part of the stratigraphic column in the areas. The youngest transgression beds have not yet been dated. In this connection, however, the composition of the uppermost miogeosynclinal deposits of the Föllinge sequence may be mentioned. Above the Berge limestone, there occur shales with graptolites of Upper Llandovery age. On top of these Bångåsen graptolite shales lies the Ekeberg greywacke, composed of greywacke-like sandstone with layers of shale and conglomerate beds of intraformational character. THORSLUND states that the presence of *Favosites gotlandicus forma forbesi* suggestes lowermost Wenlock age.

In Jämtland there are thick greywacke deposits in the pre-Silurian part of the Olden and Föllinge miogeosynclinal sequences. According to THORSLUND's survey the greywackes are mostly of Middle Caradocian age—the zones of *Nemagraptus gracilis* and *Dicranograptus clingani*. In the eugeosynclinal sequence of Västerbotten and southern Norrbotten, the Gilliks greywackes and greywacke conglomerates, lying immediately or almost immediately below the Vojtja transgression clastic rocks, are probably in part of the same age as the greywackes of Jämtland mentioned above. In both Västerbotten and southern Norrbotten the Gilliks rocks in the west contain typical greywacke rocks but towards the east quartzites occur. These quartzitic rocks may be somewhat younger than the western greywackes. In the western part of the Trondheim region, the polymictic Hopla or Volla conglomerate, the pebble material of which is about the same as that of the Gilliks greywacke conglomerate, lies immediately above graptolite shale—the *Dicranograptus shale*—of Middle to Upper Caradoc age. Thus, the Hopla–Volla conglomerate seems to have some stratigraphical connections with the Gilliks clastics. At this point, mention can be made of the Gilliks greywacke conglomerate in the south Norrbotten Ikesjaure area, in which the pebble-bearing greywacke sediment contains limestone lenses with *Halysites* and other tabulate corals, which demonstrate the undoubted marine character of the western facies of the Gilliks group rocks.

The transgression within the eugeosyncline seems to be of Upper Caradoc age, and the preceding period of unrest may be called the *Gilliks disturbance*; this probably occurred in lower to middle Caradoc times.

In this description of the Upper and Uppermost thrust rocks a great many conglomerates and quartzites within eugeosynclinal sequences older than those mentioned above, i.e. in strata of lower Middle Ordovician and earlier in the stratigraphical column, have been mentioned. Some of the clastic formations are shown to occur widely; others are found only in restricted areas. No fossils enabling dating are known. As a result, comparisons with the lower to middle parts of the eugeosynclinal deposits and the fossiliferous miogeosynclinal and continental border sedimentary sequences are hard to carry out.

The autochthonous of Jämtland in which the upper part of the Caradoc, constituting the youngest erosional remains, is composed of shales and limestones will be considered first. The lower part of the Caradoc is an inundation sequence, starting with a sedimentary basal breccia, and followed upwards by conglomerate, then by calcareous sandstone—the Loftarstone. The whole sequence, including the superposed *Ludibundus* limestone, is called the *Ludibundus* beds, and the transgression is known as the *Ludibundus* transgression. In the eastern marginal zone of the Jämtland Cambro–Silurian the *Ludibundus* clastic rocks lie upon different portions of Lower Ordovician sediments, upon Cambrian strata, or upon a weathered old Precambrian crystalline basement. The stratigraphical gap beneath the *Ludibundus* beds is of Middle–Upper Llandeilo age. The pre-*Ludibundus* regression is called the *Jämtland regression.* In the Föllinge and Olden sequences which lie in the position of the Lower thrust rocks, the lower part of the greywacke beds, discussed in the foregoing, may be a western counterpart of the *Ludibundus* beds. The *Ludibundus* transgression of the autochthonous seems to have coincided with the greywacke facies development of the Föllinge–Olden area and with its development into a miogeosyncline.

In the eugeosynclinal Seve–Köli sequence in northern Västerbotten and southern Norrbotten a light-coloured hard quartzite formation—the *Vuortna quartzite*—occurs beneath the dark-coloured Gilliks rocks. It has not been verified that a real stratigraphical gap between them exists however. The Vuortna quartzite is separated from the lower-lying Pieske limestone by a mostly thin greenstone formation. If the rock sequence described above were characteristic of large areas within the eugeosynclinal sequence, the Vuortna clastic rocks could possibly be interpreted as connected with the Gilliks clastic rocks and as being a eugeosynclinal counterpart of the geosynclinal-border *Ludibundus* transgression rocks of Lower Caradoc age. Elsewhere in the eugeosynclinal deposits, however, for example in more southerly parts, thick deposits of sedimentary rocks and volcanic rocks occur, lying between the rock sequence and a limestone formation with a closely following or directly superposed quartzitic to conglomeratic formation; this sequence is supposed to be a lithostratigraphical equivalent of the Vuortna–Pieske sequence. The matter is somewhat complicated; there is possibly a stratigraphical gap between the Gilliks and Vuortna rocks in this area.

In the above, five groups of clastic rocks within the eugeosynclinal Seve–Köli sequence were compared with probable stratigraphic counterparts in the central Jämtland Cambro–Silurian region of miogeosynclinal and continental border sequences, but several conglomerate and quartzite formations in the eugeosynclinal Upper and Uppermost thrust rocks were not mentioned in this comparison. Considering the stratigraphical columns of different areas described, attention may be drawn to the following quartzite conglomerates: the *Portfjäll conglomerate* of the Jämtland Blåsjö area; the *Bellovare conglomerate* lying on the Bellovare limestone in the southwestern Västerbotten Remdalen area; the *Svälva conglomerate* in the western Västerbotten Arefjället area; and the *Ropen* and *Morto conglomerates* in the Västerbotten Södra Storfjället area. Two granite conglomerates may be mentioned—the *Virihaure conglomerate* and the *Sårjås conglomerate*, in the western Norrbotten Virihaure area. Lastly, the *Rotik serpentine conglomerate* can be mentioned. The Ropen and the Morto conglomerates lie in the Storfjäll nappe, the Sårjås conglomerate in the Virihaure nappe, and all the other conglomerates mentioned in the Seve–Köli complex.

The writer feels that the Ropen and Bellovare limestones, both superposed by quartzite or quartzite conglomerate, are southern lithostratigraphic counterparts of the Pieske and Nuolja limestones, which themselves are often overlain by quartzite deposits.

The Bellovare conglomerate seems to be a southern continuation of the Svälva conglomerate. According to ZACHRISSON (1964) the Portfjäll conglomerate is at a somewhat higher level than the Bellovare conglomerate, but it does not occur in the Remdalen area, where its probable stratigraphic equivalent, a slaty rock with sandy intercalations, lies in the lower part of the Kosaâive phyllite formation. In the Södra Storfjället part of the Storfjäll nappe, the Morto conglomerate lies stratigraphically higher up than the Ropen conglomerate; if the Ropen conglomerate is a lithostratigraphic counterpart of the Svälva conglomerate, the Morto conglomerate may be a stratigraphical equivalent of the Portfjäll conglomerate. The writer is of the opinion that there are at least two main levels of well-washed shore deposits within the lower part of the eugeosynclinal sequences, and that they lie below the Gilliks clastics.

The Rotik serpentine conglomerate of the Rotikken area in western Västerbotten lies directly on partly serpentinized peridotite. The peridotite rocks are almost a guiding level in the lower part of the eugeosynclinal sequence. It seems probable that ultrabasic magma made its way into the unconsolidated mud of the sea bottom and remained there. On emerging, the soft sediments above the peridotite were swept away, and by abrasion a conglomerate with monomictic material was produced from the peridotite. The only fossiliferous serpentine conglomerate known in the Scandinavian Caledonides is the Ordovician *Otta conglomerate* which occurs in the southernmost or Otta part of the Trondheim region. According to HEDSTRÖM (1930) the trilobites of the fauna suggest late Arenig, but the gastropods seem to suggest a somewhat younger age. The Rotik and the Otta serpentine conglomerates could be of about the same age. If this is so, all the coarse clastic rocks mentioned above as being younger than the serpentine conglomerate must be younger than Arenig, and the Nuolja, Pieske, Ropen and other limestone rocks within the lower part of the geosynclinal sequence are probably of Middle Ordovician age.

The Virihaure granite conglomerate lies in the lowest part of the Seve–Köli sequence in the Virihaure area. A thick sequence of strata separates the conglomerate from the higher-lying Pieske limestone of the area; the conglomerate is probably the lowest-lying coarse clastic deposit known in the eugeosyncline. The Sårjås granite conglomerate outcrops at the thrust boundary between the Virihaure nappe and the Rödingsfjäll nappe, and appears to be the lowest member of a disrupted anticline in the western part of the Virihaure nappe. A limestone, considered as a counterpart in the Virihaure nappe of the Pieske limestone in the Seve–Köli complex, is separated from the underlying granite conglomerate by a greenstone formation. According to the writer's interpretation of the stratigraphic data, the Sårjås conglomerate is younger than the Virihaure conglomerate. The Sårjås conglomerate may thus record a later stage of transgression within the eugeosyncline than the Virihaure conglomerate does, but may possibly be a western lithostratigraphic continuation of it. In their areas of deposition, the Precambrian basement appears to have been exposed.

From this summing-up of the stratigraphic distribution of coarse clastic rocks within the lower-middle parts of the eugeosyncline it is clear that an attempt to compare conglomerates in sequences lacking dated fossils, with sandy deposits within the stratigraphical column of the fossiliferous miogeosynclinal and continental border deposits, runs into difficulties. Before reviewing the stratigraphical gaps and transgression beds in the deposits east of the eugeosyncline the paleogeographical conditions within the Lapland part of the old eugeosynclinal trough will be briefly described.

A straight line going in a NNE direction from the Björkvattnet–Virisen area in Central Västerbotten crosses the Aivo–Sarvas area at the Norrbotten border and also the Mavas and Virihaure areas of southern Norrbotten. If the trend of the old geosyncline was conformable with the general trend of the Caledonian chain, the facies conditions registered in the deposits of these four areas would be very similar in character. But, as the data obtained from the areas indicates, the sequence of strata differs in composition in each area. The Aivo–Sarvas area will be considered first. The pre-Vojtja sequence down to the Pieske limestone is, disregarding the volcanic rocks, composed of quartzitic rocks—the Gilliks quartzite and the Vuortna quartzite. In the Mavas area to the NNE, however, considerable portions of the deposits at corresponding stratigraphical positions consist of grey calcareous slates and black slates, and here the clastic part of the Gilliks sediments is for the most part of conglomeratic greywacke type. In the Björkvattnet–Virisen area to the SSW the Gilliks clastic rocks are mostly of the same kind as are found in the Mavas area. In the area around the eastern part of Lake Virihaure there is a very thick sequence of strata between the Pieske limestone and the higher-lying Vojtja clastic rocks. No quartzites and pebble-bearing greywackes of Gilliks facies have been found in this strata sequence, which contains alternating calcareous schists, black schists, sandy schists, volcanic rocks, and at the top a dolomite formation with superposed dolomite breccia immediately below the Vojtja clastic rocks.

It is clear from the foregoing account that the Aivo–Sarvas area was fairly near a shore during the post-Pieske–pre-Vojtja period, while all the other areas during the same time were farther from the shore. Before the thrusting took place, the eastern border of the eugeosyncline may have been situated not too far east of the Aivo–Sarvas area. Within the Vojtja and later periods, i.e. within the Uppermost Ordovician and the lower part of the Silurian, the Aivo–Sarvas and other Lapland areas in the Seve–Köli development seem to be of about the same facies, but the configuration of the eugeosynclinal troughs is not the same as before the pre-Vojtja disturbance.

From the almost wholly unfossiliferous deposits of the middle-lower parts of the eugeosynclinal sequence, we turn to their probable counterparts in the fossiliferous miosynclinal and continental border deposits.

In the autochthonous and the lowest nappes of Jämtland there is a large stratigraphical gap between the Cambrian and the Lowest Ordovician. In the west, the gap is partly filled by shale—*Dictyonema* shale and *Ceratopyge* shale of Middle Tremadoc age. The first extensive Ordovician inundation occurred in Lowermost Arenig times. The transgression deposits—the richly glauconitiferous basal portion of the Hunneberg stage (the beds with *Lycoporia difformis*)—are conglomeratic in the autochthonous but westwards, in the nappes, pass into glauconitiferous limestone with shaly beds. According to THORSLUND there is a stratigraphic break of Middle Arenig age in the Föllinge sequence. The gap widens to the west and eventually occupies possibly the whole of the lower Arenig. No conglomeratic beds are known at the base of the Upper Arenig "Orthoceratite limestone"; in the Olden sequence, however, conglomerate and quartzite beds form the transgression deposits probably of Upper Arenig age, lying on an Old Precambrian basement. Directly above the basal beds follow dark shales and a limestone —the *Vaginatum* limestone—which belongs to the lowermost Llanvirn.

In the allochthonous of the East Jämtland nappes and in the Föllinge sequence there are Upper and Middle Cambrian black (alum) shales, and lower-Middle and Lower Cambrian conglomerates and sandstones. In the author's view the black shale facies belong to the continental border facies. The Föllinge sequence, which is of miogeosynclinal facies from Middle Ordovician and upwards, seems to be of shallow shelf facies within the Cambrian and Lower Ordovician. In the Olden Lower thrust rocks, no Cambrian or lowermost Lower Ordovician occur; they have probably been eroded away. In the Föllinge sequence, in the East Jämtland nappes and in the autochthonous deposits, there is a pronounced stratigraphical gap between the Lowermost Ordovician and the Upper Cambrian beds. In the Autochthonous, as has been verified, erosion has swept away considerable parts of the once present upper portion of the Upper Cambrian sediments. Comparing the clastic beds of the lower-middle parts of the eusynclinal, miosynclinal and continental border deposits mentioned above, it must be admitted that no valuable connecting link has to date been found. If the serpentine conglomerate situated in the lower part of the eugeosynclinal Seve–Köli sequence is of Upper Arenig age, the strata beneath the conglomerate may possibly belong to the Lower Ordovician, and the Virihaure granite conglomerate in the Virihaure area is possibly an eugeosynclinal counterpart of the Arenig basal clastic rocks in the miogeosynclinal Olden sequence. The lower eugeosynclinal rocks in the eastern part of the Seve–Köli complex are often very highly metamorphosed; they are migmatized and lack any primary structures. As a result the age relationships of the lowest deposits within the eastern part of the eugeosyncline are not very well known. Before concluding the section, the stratigraphical relationships within the lowest part of the Seve–Köli complex in the Remdalen synclinal area of southwestern Västerbotten (see the writer's earlier description, and ZACHRISSON's paper of 1964) will be described. The Fjällfjäll arkose, forming the core of the Fjällfjäll anticline to the east of the Remdalen syncline, can be interpreted as a metamorphic western stratigraphical counterpart of the Sparagmite formation in more easterly and lower-lying tectonic complexes. The Fjällfjäll arkoses may thus be looked upon as eugeosynclinal basement rocks of Late Proterozoic age. In the lowest formation of the eugeosynclinal sequence—the *Tjopasi graphitic phyllite*—there are both quartz keratophyric volcanic rocks and lenses of peridotite. Outside the mapped Remdalen area the Tjopasi formation also contains a serpentine conglomerate level. The peridotite rocks are probably of Lower Ordovician age. If this assumption is correct, Cambrian sediments are probably absent in this part of the eugeosyncline, and the coarse clastic rocks in the lowermost part of the Remdalen sequence are thus within the lower part of Ordovician.

In his paper of 1964 ZACHRISSON discussed *inter alia* the relationship between volcanism and sedimentary facies. He gave evidence of a certain regularity of facies variations, so that the facies of the sediments seemed to be influenced by volcanic activity. Thus acid volcanism was usually accompanied by strong graphitic facies, and immediately after a volcanic event conglomerate-bearing strata followed. ZACHRISSON concluded by saying: "A possible explanation is that the strongly explosive volcanism which yielded the quartz keratophyres also caused changes leading to regressions of the sea and to changed conditions of sedimentation."

THE DISTRIBUTION OF VARANGIAN ROCKS IN THE SCANDINAVIAN CALEDONIDES

In a monograph on the marginal zone of the Caledonides of southern Swedish Lapland (1942) the old glacial rocks, earlier called Eocambrian glacial rocks, were renamed the *Varanger glacial* rocks by the writer. This name is taken from the Varangerfjord area of northernmost Norway, where the glacial nature of the old rocks was proved for the first time (REUSCH, 1891). REUSCH discovered an ice-scratched substratum below the tillitiferous rock in a locality in the Bigganjargga area north of the most westerly part of Varangerfjord. Varanger tillite with faceted and ice-scratched pebbles and accompanying varved shale with included erratics was discovered by the writer in northern Sweden, at Mt. Långmarkberg in northern Jämtland (KULLING, 1938, 1942). In the monograph of 1942, the writer reviewed the distribution of Varangian remains in Scandinavia, East Greenland and Northeast Land. The Varangian remains in East Greenland were discovered by the writer and POULSEN in 1929 (see KULLING, 1930 and POULSEN, 1930) and the old glacial rocks in Northeast Land by the writer in 1931 (KULLING, 1932; 1934). Within recent years a great many observations on Varangian glacial rocks and sedimentary breccias of tillite-like character in the eastern border zone of the Scandinavian Caledonides have been described. However, most of the tillite-like rocks, said to be true glacial rocks, lack tillite qualities such as faceted and ice-scratched block material, striated surfaces of the tillite substratum, and accompanying varved shale. Sometimes the tillite-like rocks are composed of blocks of material of rather local origin and foreign material transported over long distances is absent. Thus a great many sedimentary breccias, thought to be of glacial origin, need to be critically reexamined. The writer feels that several tillite-like sedimentary breccias in the Varangian deposits are *most probably deposits of a periglacial nature, formed by frost weathering and solifluction.*

The Varangian in Sweden. In the Swedish part of the Caledonides, rocks known as Varangian glacial rocks have been found in autochthonous sequences, in Lower thrust rocks and, exceptionally, in Middle thrust rocks. In a report of 1951 the writer divided the breccia type rocks, looked upon as Varangian glacial rocks, into two main groups —the *Långmarkberg tillite formation*, deposited on an eroded surface of the Late-Proterozoic Sparagmite group of mainly arkosic rocks, and the *Sito tillite formation*, lying directly on or just above old Precambrian basement rocks. In autochthonous areas the Sito breccia is followed upwards by sedimentary beds, which contain fossil remains of Lower Cambrian age in places. The Långmarkberg glacial rocks have been divided into an upper and a lower tillite formation, both of which contain ice-scratched pebbles and other glacial features, but with obvious differences in pebble composition and colour. After the writer's re-examination of the Sito group of rocks, they were reclassed as sedimentary breccias probably of a periglacial nature.

As described previously, a somewhat tillite-like sedimentary breccia—the Vakkejokk breccia—occurs intercalated in the lower part of the autochthonous sedimentary deposits north of Torneträsk in the northern Norrbotten mountains. It has a "breccia-in-breccia" structure (breccia inclusions of the same composition as parts of the main breccia) and other nontillitic qualities. The breccia lies below shaly sediments which (in other parts of the Torneträsk area) contain a fauna of Lower Cambrian zone 2. The writer's view on the stratigraphical position of the Vakkejokk breccia is elucidated in Fig. 127, a correlation between some well-known Northern Scandinavian autochthonous sedimentary sequences. It is of interest to note that the formation of sandy shales with cyclic sedimentation below the Vakkejokk breccia bears fossils of Late Proterozoic age. Fossil tracks of this type have also been noted in fragments of sandy shale or sandstone in the breccia. In the Luopakte section south of Lake Torneträsk, where no real Vakkejokk breccia is present in the autochthonous sedimentary deposits, the chronostratigraphic site of the breccia is considered as being the stratigraphical gap between the Late Proterozoic fossiliferous *Spriggia* shale and the overlying transgressive sandstone formation (see Fig. 115). The author feels inclined to consider the Vakkejokk breccia and the *Spriggia* bed with its underlying sediments down to the crystalline Precambrian basement as members of the Varangian stage. The Vakkejokk breccia in the Torneträsk area and the basally situated Sito breccia in other parts of the autochthonous sediments of the Norrbotten mountains are possibly periglacial counterparts of the Långmarkberg tillites; the upper Långmarkberg tillite can be correlated with the Vakkejokk breccia and the lower tillite with the Sito sedimentary breccia.

If the correlation table, Fig. 127, and what has been said above are valid, the *Spriggia*-bearing shale in the Luopakte section of the Torneträsk area thus forms a horizon in a Varangian interglacial sedimentary sequence. Most of these sediments are probably of marine nature as the marine *Spriggia*-bearing fauna for example could migrate into the upper part of the deposits.

The Varangian in Norway. In the southern-Norwegian Sparagmite region the Moelv tillite appears to be the southern counterpart of the lower tillite formation in the Långmarkberg glacial rocks. Pebbles with glacial striae have not been found however in the Moelv tillitiferous rocks. The Moelv tillite lies on the top of the Sparagmite group rocks and another breccia type rock, called the basal tillite, is found lying directly on or just above the old Precambrian basement rocks, but below the Cambrian sediments (in certain parts of the Sparagmite region). According to published descriptions the basal tillites seem to be of about the same general composition as the Sito breccias in northern Sweden. In the Tanafjord–Varangerfjord region in the northernmost parts of Norway an upper and a lower tillite formation occur, separated from each other by a rather thick sandstone–shale sequence, and underlain by quartzitic sandstone and arenaceous shales of great thickness. There are some layers with Cambrian fossils in this sequence of sandstones and shales above the upper tillite.

Our knowledge of the deposits classed as Varangian is very incomplete, and that of the positions of the glaciated areas during the lower and the upper glaciation stage almost non-existent. Probably no very large land ices of Quaternary type existed, but a great many glaciated centres of moderate magnitude did, and around them wide regions with a periglacial climate. It is doubtful whether a glaciated area of high mountain to Alpine type existed. The shaly deposits above the tillite rocks are, as far as is known, very seldom of varved shale type. Probably most of the Varangian ice masses went into the sea. The varved Långmarkberg shale mentioned previously, and shown in Figs. 140 and 141, is finely varved and packed with material from melting icebergs. This type of deposit of so-called distal fluvioglacial nature is not common as a type of deposit in submerged areas in front of a rapidly melting land ice, but probably intimates deposits in a lake or a fresh water bay before a stagnant glacier, perhaps indicating a low degree of precipitation and a low temperature. The widespread appearance of fossils in the Lower Cambrian immediately after the end of the Varangian ice-age seems to have a causal connection. Attention has been drawn to this in earlier reports of the writer. Probably the air-charged glacier water from the melting land ices and the quantity of fresh fines have a creative influence. It is for example a well-known fact that marine life in front of the Antarctic continental ice is much richer than in other parts of the oceans.

THE DISTRIBUTION OF IGNEOUS ROCKS WITHIN THE EUGEOSYNCLINAL DEPOSITS

Most igneous rocks within the eugeosynclinal deposits were recrystallized during the period of regional metamorphism and were sheared and folded in the same way as the surrounding rocks. Thus, it is often difficult to determine whether the igneous rocks are of effusive or intrusive character. This is particularly true of basic rocks, and as a result neutral words such as greenstone or amphibolite are usually used for schistose igneous rocks of this type.

A prominent feature of the eugeosynclinal Seve–Köli sequence is the so-called *Eastern amphibolite*, situated in the lowermost eastern part of the sequence, and occupying very large areas. Parts of this amphibolite—usually the central parts—have an ophitic texture, but most of it is wholly recrystallized into schistose amphibolite. Usually the amphibolite is conformable with the surrounding schists, but in places the basic rock cuts across adjacent sediments. Lenses of gabbro norite and peridotite of various sizes are included in the Eastern amphibolite. No volcanic structures have been found in the upper parts of the amphibolite. The Eastern amphibolite may be a conformable intrusion in the lowest part of the eugeosynclinal sequence. It is probably more plausible to think of this huge rock complex as constituting areal eruptions over the geosynclinal floor.

The *gabbro* and *norite* rocks in both the Upper and Uppermost thrust complexes lie in the lower part of the sequence and are all relatively small. The gabbro of Sulitelma type was formed after the period of regional metamorphism and thus belongs to a late- or post-orogenic phase.

The *peridotite* rocks, sometimes accompanied by hornblendites, are concentrated at a very low level, which has been known as the "peridotite level" for many years; they probably penetrated the eugeosynclinal floor of the Seve–Köli sequence during Lower Ordovician times. In the Remdalen synclinal area in the southwestern part of the Västerbotten mountains, some very small peridotite bodies have been found in the upper part of the stratigraphical sequence also. Correlation of the surrounding rocks shows that they are probably of Lower Silurian age. In the Virihaure nappe sequence in the southern Norrbotten mountains, peridotite rocks also occur. The Virihaure nappe is probably a sequence of rocks from the oldest part of the eugeosynclinal stratigraphical column.

Caledonian intrusive rocks of *granitic composition* are very rare within the Swedish parts of the eugeosynclinal deposits. The only notable occurrences lie in the northernmost part of the Västerbotten mountains, between lakes Överuman and Tärnasjön, within the Storfjäll nappe complex of the Upper thrust rocks. The largest granite body is known as the *Juomotjåkko granite* and is intercalated in a volcanic greenstone formation, the Mesket greenstone lava. Veins of microcline granite and leuco-trondhjemite occur in the greenstone around the granite. In an adjacent area the acid intrusive rocks are called mica–quartz diorites. The acid intrusive rocks mentioned were formed after regional metamorphism occurred and are probably a little younger than the migmatitic state of metamorphism.

There are two kinds of Na feldspar-bearing acid igneous rocks—*the albite granites* and *the trondhjemites*—occurring in the highly metamorphosed parts of the eugeosynclinal deposits. The trondhjemites are always small, elongated bodies of sill or lens shape, occurring partly as lenses of various sizes in or within the neighbourhood of marble, partly as fillings of tension fissures, and partly as discordant veins. The albite granites are granite-grained, schistose igneous rocks occurring in the lower part of the stratigraphical sequences in both Upper and Uppermost thrust rocks, but have a rather restricted distribution. In some earlier papers on the Caledonides the acid igneous rocks were looked upon as intrusive rocks, but the writer intended to convey that the rocks were probably recrystallized quartz keratophyres. No primary structures have been found in the mostly gneissic rocks.

Volcanic rocks are not uniformly distributed in the eugeosynclinal deposits. Most of the lava rocks lie in the pre-Silurian part of the stratigraphical column, they are probably of Ordovician age. Conformably-lying igneous rocks occur in some Silurian sequences and are considered to be partly lavas. Indisputable lava formations such as pillow structure, agglomeratic structure and so on have been verified in the Ordovician lavas only. Greenstone lavas occur mostly; acid lavas of quartz keratophyric type are sometimes found in the middle and lower parts of the pre-Silurian sequences. In the Remdalen synclinal area, however, four acid volcanic formations occur. Three of them are in the pre-Silurian part of the sequence and are of a quartz keratophyric nature; the fourth lies in the middle of the thick Silurian Remdalen greenstone, and is a thin formation of porphyry composition. In the same syncline there are seven greenschist–greenstone beds occurring in the area, which has been surveyed in detail. As an example of the non-uniform distribution in the eugeosynclinal area of a particular lava formation reference may be made to the review on the extension of the Seima group of mainly greenstone lavas in Chapter XXII, p. 244, dealing with the distribution of the Seve–Köli rocks.

CHAPTER XXV

COMMENTS ON TECTONICS, METAMORPHISM AND LINEATION

TECTONICS AND METAMORPHISM

The most striking feature of the geology of the eastern part of the Swedish Caledonides is the inversion of zones of regional metamorphism. This reverse metamorphism is a result of overthrusting. In the central and western parts of the Caledonian chain in Swedish territory both normal and reverse metamorphism can be seen. Normal metamorphism prevails in most parts of the distant thrusted rock complexes, but inverted metamorphism seems to be intimately bound up with zones of migmatization, and characterizes the bedrock of the lower parts of such altered zones.

The Lower and Lowermost thrust complexes have their roots in the miogeosynclinal and geosynclinal border parts of the geotectonically affected region and display low grade metamorphism. The higher lying distant thrusted rock complexes consist of typical eugeosynclinal deposits in their middle and upper parts. The lower portion of these thrust rocks—the Middle thrust complex—consists of altered basement crystalline rocks, often attached to a sheet of the arenaceous deposits lying immediately above. The Middle thrust complexes are possibly dragged-along parts of the eugeosynclinal substratum, with the immediately overlying sediments of mainly arenaceous type. The Middle thrust rock complexes form very competent rock units, whereas the eugeosynclinal rock masses are, taken as a whole, of a rather incompetent nature. During the overthrusting phase a zone of movement arose in the boundary zone between the rock masses. The arenaceous zone above the thrust-basement rocks in particular became tectonically affected and transformed into banded quartzose tectonites. The thrust-basement crystalline rocks became mylonitic in eastern parts. In westerly regions however the transformation ended in a more gneissic state. The regional metamorphism of the Middle thrust rocks can usually be classed as middle grade, whereas the metamorphism of the Upper thrust rocks immediately above is always high grade and their lower parts show feldspathization and often real migmatization. This state of affairs exists in the eastern parts of the thrust-masses but towards the west migmatization disappears. Thus, there is no visible connection between the zone of migmatization and a source of supply beneath it or in the axial zone of the orogene. The creation of migmatite rocks seems to be favoured by intense material displacement in deeper sections of the thrust rocks and by the presence of rocks which arrest material at the migmatite stage.

What has been said regarding the migmatization of the lower eastern parts of the Upper thrust rocks also applies to regional metamorphism in the entire Swedish Caledonides; namely, that it has not been verified that the source of heat or magmatic material was below or in the axial zone to the west. Regional metamorphism seems to be an inner Caledonian event caused by material displacement.

Regional metamorphism of the different *étages* of the Swedish Caledonides is discussed briefly below:

The *autochthonous sediments* in the eastern border zone of the Caledonides have been little-altered and well-preserved sedimentary structures and fossils occur, particularly in their lower parts, and in competent layers. There are minor disturbances and secondary structures against the overlying thrust rocks. In the Norrbotten mountains the topmost formation of the autochthonous sediments usually consists of black shale. This incompetent type of rock is always of pronounced schistosity and slaty habit; it often constitutes the soapy material on which thrust rocks of various kinds have slipped away. In places, seams and lenses of black slate have been incorporated in the advancing thrust masses.

In the window areas in the western parts of the Swedish Caledonides the autochthonous sediments are not so well preserved as in the eastern border zone of the chain. Usually only the arkosic to conglomeratic bottom layer of the sedimentary sequence contains well-preserved primary structures. Westwards, the autochthonous sediments have increasing schistocity of a sericitic type.

Thus, the western or window parts of the autochthonous sediments are sometimes structurally more changed than the sediments of the Lower thrust masses. Parts of the autochthonous sediments are metasomatically altered (see Chapter XXVI on ore deposits).

The *Lowermost thrust rocks*—the East Jämtland nappes, of mainly fossiliferous sediments, are relatively little altered. Their fossils are to a great extent paleontologically easily determinable. The imbrication structures of the rocks seem to be less steep than those of the superposed Lower thrust rocks.

The *Lower thrust rocks* of sedimentary composition, with or without attached parts of their crystalline basement rocks, are often characterized by steeply-dipping imbrications. The sedimentary rocks have been changed into quartzites, slates, contorted limestones etc., and most of the fossils in the sediments are indeterminable. Thus, a detailed stratigraphical survey in these rock sequences is difficult. Stages characterized by sericite and sericite and chlorite, and stages with greyish green mica, have been discerned. In more tectonically affected zones the feldspar (mostly K feldspar) in arkoses and other clastic rocks has been partly sericitized. Sericite films often disguise the red pigment of the feldspars in the rocks and create the so-called *grey metamorphism*. Diaphthoritic and metasomatic processes play a subordinate role. Quartz–chlorite material frequently fills veins and fissures. The thrust-basement crystalline rocks are partly cataclastic, and mylonitic zones occur.

The *Middle thrust rocks*—the lower portion of the distant thrusted rocks—have different primary composition in northern and southern, as well as often in upper and lower, parts. As a result, different parts react in different ways under the affect of metamorphic processes.

The crystalline basement of the complex, composed of granites, syenites etc., has largely been transformed into mylonite rocks. In the western Swedish mountains these mylonites pass into a gneissic state; streaks and tectonic lenses of rocks with relict primary structures occur in the mylonite rocks.

Large parts of the Middle thrust rocks in the southern Swedish mountains and in southern Lapland consist of Late Precambrian arkosic rocks known as the *Sparagmite rocks*. In the Lapland area the Sparagmite rocks with their underlying thrust-basement crystalline rocks were called the *Stalon complex* (or nappe). It is of interest to note that the metamorphism of the Stalon arkosic sediments increases from east to west. In some areas in the northeast metamorphism appears in chlorite–biotite facies, but west of these districts metamorphic alteration is characterized by chlorite–biotite–epidote with or without garnet. Further to the west the epidote-rich metamorphic arkoses change to garnet–feldspar-bearing mica schists. Thus, the westwards increase in metamorphism of the Stalon sparagmite rocks starts in the east with a rock alteration comparable to that of the Lower thrust rocks, and ends in the west with rather high grade rocks (almost of Upper thrust rock facies). The middle metamorphic arkoses have usually lost their primary red colour and gained a greenish grey colour. Only some intercalated coarse conglomerates with red granite and syenite pebbles and some extreme easterly-lying arkoses are still reddish in colour.

The *Upper thrust rocks*—the major part of the distant transported rocks of the chain—show normal metamorphism in most parts of the southern Swedish mountains and also in southern Lapland.

The formations with determinable fossils within the Seve–Köli sequence are situated in the upper half of the normal metamorphic rock sequence, some distance above the garnet isograde. It could be said that the fossil-bearing bedrock areas mentioned are situated in the eastern syncline of the Seve–Köli complex. In the southwestern part of the Västerbotten mountains there is also a western syncline present, divided from the eastern syncline by the Fjällfjällen anticline. The western syncline—the Remdalen syncline—is characterized by often banded metamorphism of the so-called *Grokas tectonic facies*, and the few fossil remains observed are mostly in a rather poor state of preservation.

In the eastern part of the eastern syncline mentioned, regional metamorphism is of normal character (Kulling, 1933; Du Rietz, 1938; Faddegon, 1940). However in the most easterly parts of the Seve–Köli complex, in the so-called eastern gneiss–migmatite and amphibolite zone, the metamorphic conditions are rather complicated, and have been thoroughly investigated in only a few areas. The migmatite and feldspar porphyroblastic rocks usually occur beneath the large eastern amphibolite, and the writer's view of this is that the damming effect of the basic rocks caused a certain concentration of the mobilized quartz–alkali–rich material to build up. This false contact metamorphism has been dubbed *metamorphism by damming* (Kulling, 1955). Van der Harst (1956) in a monograph on the geology of the Borga area in the southernmost Västerbotten mountains described, *inter alia*, the plagioclase compositions in Lower, Middle and Upper thrust rocks. The bedrock of the Borga area is an extreme westerly part of the Middle thrust rock complex, with arkoses in albite–epidote–amphibolite facies, and with an overlying "hard schist-bearing gneiss formation" and mica schists in amphibolite facies. This rock complex is separated from the overlying high grade metamorphic rocks of amphibolite and kyanite–staurolite facies by a zone of intense imbrication. This overlying rock complex is considered to be part of the Seve–Köli complex by the writer.

According to VAN DER HARST the metamorphic arkoses (sparagmites) are characterized by albite. The overlying rocks, up to and including the lower part of the amphibolite–mica schist zone, contain oligoclase. All the overlying amphibolite–mica schist–mica gneiss rocks have andesine as the plagioclase. There are small areas of strong migmatization in both the oligoclase and the andesine zones, the migmatite rocks usually being situated immediately below the amphibolite bodies. VAN DER HARST surveyed the plagioclase with normal and inverse zoning in the area. He concluded that strictly speaking the formation of the inversely zoned plagioclases appeared to have taken place in the period of regional metamorphism whilst normal zoning was related to migmatization phenomena of a later date. The peculiar inversely zoned plagioclases seem to reflect a rise in temperature and may have been "produced during progressive regional metamorphism". The mountain region west of the Borga area is geologically speaking almost unknown. In this area regional metamorphism gradually decreases and the plagioclases change to oligoclase and later albite; the albite metamorphic zone is in the central part of the eastern syncline mentioned previously, where some levels with preserved fossils have been found, and regional metamorphism is normal.

In northern Norrbotten, where from a stratigraphical viewpoint only the lower part of the Seve–Köli rock complex exists, regional metamorphism slowly increases upwards and migmatite rocks occupy the topmost part of the thrust complex, situated in the international border area northwest of Lake Akkajaure. On the Norwegian side of the border the migmatite—the so-called Gicce gneiss—is widespread, and is roughly conformable with the stratigraphy. In the Narvik–Ofoten area the gneiss corresponds to the migmatized upper part of the Narvik schists (VOGT, 1950). Above the Gicce migmatite comes a sequence of high grade sediments more than 5000 m thick—the so-called Håfjell sequence; and then another bedrock zone of migmatization, the *Niingen migmatite zone*. A very brief survey of the Narvik–Ofoten area did not disclose any major thrust levels beneath the two zones of migmatization; consequently all the crystalline schists of the area have been included in the Seve–Köli complex. Thus, according to the above, there are three zones of migmatization in the complex: the upper or *Niingen migmatite zone*, the middle or *Gicce zone*, and the lower or *Marsfjäll migmatite zone*. Mount Marsfjället lies in the southern part of the Västerbotten mountains, the part in which the eastern migmatite zone is rather broad and well developed.

The relationship between metamorphism and tectonics in the Storfjäll nappe complex is rather complicated. In the northern part of the complex there is high grade metamorphism in the east and middle grade metamorphism in the west, but the point of metamorphic alteration cannot be exactly fixed; it lies near the boundary between the greenschist facies and the albite–epidote–amphibolite facies. Between the eastern and western bedrock areas there is a zone of strong differential movement (ALEVA, 1950; MULDER, 1951). In the writer's monograph on the Västerbotten part of the Caledonides (KULLING, 1955) this zone was said to represent a real thrust zone. A planned, critical reexamination of the rocks and tectonics of the Storfjäll nappe complex has not yet been carried out, however, and many problems concerning the complex are still unsolved.

The Uppermost thrust rocks. The erosional remains of the Uppermost thrust rocks (the Rödingsfjäll nappe) in the Swedish Caledonides all lie in the most westerly parts of Lapland and occur as protuberances from the Norwegian side of the international border. The geographical connections between the different offshoots can be fixed on the Norwegian side of the border. Norwegian and British geologists have mapped the distribution of the Uppermost thrust rocks west of Swedish Lapland, from Namdal in the south to southeast of Westfjorden in the north. The thrust rock complex has a great many klippen. One of the biggest of these klippen lies west of the Nasafjäll granite window and bears the local name of the *Beiarn nappe* (NICHOLSON and RUTLAND, 1965); it stretches from Svartisen in the south to north of Skjerstadfjorden in the north.

The southern Lapland part of the Rödingsfjäll nappe and most of its Norwegian counterparts consist of a lower migmatized portion and an upper portion of high grade schists and marbles. However, the Stipok klippe in the middle western Norrbotten mountains, considered by the writer as belonging to the Rödingsfjäll thrust rocks, shows no really migmatized lower part, but is

Fig. 195 *Quartzitic hardschist with linear structure.* Eastern shore of Abiskojokk, about 0,5 km southwest of Abisko tourist station. Picture taken looking southeast. The linear structure trends NW–SE. A weaker linear structure, not visible in the picture, lies in the direction N 70° W. From O. Kulling (1964)

rich in pseudo-conformable trondhjemite sheets and lenses in and around the marble formations, where levels of plagioclase porphyroblastic gneissic rocks are also present. In the middle western Norrbotten mountains, trondhjemite sheets and lenses also occur in the marble-bearing portion of the Virihaure nappe.

LINEAR STRUCTURES IN THE SWEDISH CALEDONIDES

All the tectonic complexes in the Swedish Caledonides are relatively flat lying. The rock formations of every thrust sheet have been folded and disrupted in multifarious ways. Imbrications, schistosity and other tectonic structures often make it difficult to obtain an acceptable picture of the primary relationships and thicknesses of the rocks. The Swedish part of the Scandinavian Caledonides trends SSW–NNE, as do most of its tectonic complexes. Axes of the large-scale folds within the various complexes are also mainly conformable to the trend of the chain. However, small scale folds and other linear structures run in a great many directions. In large parts of the distant overthrusted rock complexes, the most prominent linear structures are transversally directed, representing lineation in a *Quartzitic hardschist* (see Fig. 195). Besides this main structure one can often observe weaker impressed linear structures, which have been formed at other times and in other directions of movement to the main one. In fine-grained, banded tectonites in the Middle thrust rocks, for example, such superimposed linear structures are rather easily found. The nature of the transversally

Fig. 196 *Phyllitic mica schists with lineation, striking N 75° E, older than folds striking N 5° E.* Northern shore of Balvatn on the Norwegian side of the international border, northwest of Lake Mavasjaure in Sweden. Looking S 75° W. From M. Lindström (1961)

directed lineations and cross-folding has been much debated, but has still not been resolved.

Information on Caledonian linear structures is given in the following sources: Geological survey descriptions to geological maps of the Västerbotten and northern Norrbotten mountains (KULLING, 1955 and 1964); papers on the geology of the western Västerbotten mountains (HÖGBOM, 1925; DU RIETZ, 1941; ALEVA, 1950; DE MUNCK, 1951; MULDER, 1951; KIEFT, 1952; LAKEMAN, 1952; ZACHRISSON, 1964); a paper on the Caledonian tectonics of the southern Jämtland–Härjedalen mountains (STRÖMBERG, 1961); and several special papers by LINDSTRÖM on lineation in selected areas of the northwestern Norrbotten mountains (LINDSTRÖM, 1958 and 1961) (Fig. 196).

In his study on lineation in the northern Norrbotten mountains mentioned above (1958), LINDSTRÖM paid particular attention to transverse folds and emphasized that they were closely associated with and parallel to typical lineation and that they were just as often overfolded to the northwest as to the southeast (see Fig. 151, first published by the writer in a report of 1930). As regards the mutual age relationships of the different transverse structural directions, LINDSTRÖM claimed (see LINDSTRÖM in KULLING, 1962) that the oldest lineation direction was about S 50° E, followed by lineations at about S 60° E and later by lineations at roughly S 75° E.

In conclusion, something will be said about the directions of penetrative lineations in the western part of the Västerbotten mountains. In the uppermost thrust rocks—the Rödingsfjäll nappe—the upper and middle parts of the complex are characterized by small fold axes, in almost a N–S direction, while the lower, migmatized part of the nappe, according to measurements taken in the northeastern part of the nappe, shows a penetrative lineation striking about N 60° W. In the thrust masses beneath the Rödingsfjäll nappe in the Storfjäll nappe complex, the chief lineation structure of the lower and middle, high grade metamorphic parts is mainly conformable with the transverse lineation of the Rödingsfjäll nappe, while in the upper, relatively low grade part of the Storfjäll complex the trends of lineation are rather complicated, and include both longitudinal and transverse linear directions. The Seve–Köli complex extends southeast of the Storfjäll thrust complex. In the low grade metamorphic upper part with longitudinally directed eastern and western synclines, and in the Fjällfjällen anticline lying in between, the direction of the small-scale fold axes is usually conformable to the general trend of the large-scale axes, but minor areas with cross fold axes do occur. ZACHRISSON, who has made a very detailed geological map of the southern part of the western or Remdalen syncline (1964), states that "the two fold systems cannot be shown to belong to different folding phases, and their age relationship consequently cannot be decided upon". He also concludes that "the piling up of minor nappes and the formation of imbricate structures preceded or occurred simultaneously with the major thrusting eastwards". In the northeastern continuation of the Fjällfjällen anticline, part of the Middle thrust rocks overfolded to the east—the Bångfjäll anticline of thrust-basement crystalline rocks—outcrop. In the gneissic to mylonitic rocks, transverse small-scale fold axes are more common than longitudinal ones. In the eastern high grade part of the Seve–Köli complex in the Västerbotten mountains, and in the more easterly-lying Stalon complex of the Middle thrust rocks, very few observations have been made on small-scale folding and other lineation structures.

CHAPTER XXVI

ROCKS AND MINERALS OF ECONOMIC VALUE

More than three hundred years ago the search for silver ore started in some parts of the Swedish Caledonides. More extensive ore prospecting did not, however, begin until the twenties and thirties of this century. In this review deposits of economic importance which have been in production, or will probably be worked, will be described.

Lead deposits of the eastern marginal zone of the Swedish Caledonides. The lowest part of the autochthonous Eastern Cambro-Ordovician consists of sandstones with intercalated shales. Scattered finds of fossils within the basal sediments indicate that they are of Lower Cambrian to latest Proterozoic age. In the Lower thrust rocks of the eastern marginal zone there are similar sandy formations which have been recrystallized into quartzites. During the last thirty years of prospecting activity, extensive disseminations of galena and sphalerite have been discovered in the autochthonous sandstones and thrusted quartzites mentioned. In some places these are concentrated into real ore deposits, and the Boliden Mining Company has carried out prospecting and commenced mining in places. The three most prominent mineralized areas are: the *Laisvall area* in the most southeasterly part of the Norrbotten mountains; the *Dorotea area* in the southernmost part of the marginal zone of the Västerbotten mountains; and the *Vassbo area* in northernmost Dalarne. According to a report by GRIP of the Boliden Company (1960) the Laisvall mine produces about 700 000 metric tons of lead a year; from 1960 Vassbo should produce 150 000 metric tons a year, while Dorotea with its poor ore in the Lower thrust rocks is being reserved for future use. The Laisvall ore and the Vassbo ore lie in autochthonous sandstones. According to the report, the paragenesis comprises galena, sphalerite, pyrite, calcite, barite, fluorite, sericite and accessories. Mineralization occurs particularly where quartzitic beds lie between impounding beds of shale or shaly sandstone. It is of interest to note that the mineralization is often linked to tectonic features such as faults, fissures and crush zones associated with the overthrust movements.

Silver-lead ore occurrences within the high mountains of Norrbotten. In 1634 a Lapp reported an ore deposit on Mt. Nasafjäll, lying near the international border in the extreme southwestern part of the Norrbotten mountains. The discovery of silver-bearing lead ore caused a great sensation and the Lapp was rewarded by the Swedish Queen. The Nasafjäll silver ore was mined mainly during the years 1635–1659. However, the poor economic results and distant position of the ore in a waste area stopped production. From 1635 to 1712 the entire lead production amounted to 270 metric tons and silver production to about 1 metric ton. The lead comprised about 27% and the silver about 0,1% of the mined ore. The main ore minerals present were galena, sphalerite and pyrrhotite. The silver-bearing lead ore occurred as lenses in a quartz vein lying in the boundary zone between granite and gneissic–mylonitic rock. In places a sheet of black graphitic schists lay at the boundaries of the quartz veins and the enclosing feldspar-rich rocks. The Nasafjäll ore deposit is situated within the southern part of the Nasafjäll antiform-shaped Precambrian basement window.

Silver-bearing lead ore has been found in several other places within the southern Norrbotten mountains. Most of the ore impregnations were discovered in the high mountains to the west of Lake Saggat, these mountains used to be called the Kvikkjokk mountains, but the ore impregnations discovered in them are of little value. They occur in quartz veins in high grade schists of the Upper thrust rocks (the Seve–Köli schists) as irregularly formed intermittently occurring bodies. Ore minerals present are silver-bearing galena with pyrrhotite, chalcopyrite, and sparse occurrences of arsenopyrite and boulangerite.

The lead ore impregnations in quartzitic rocks within the eastern border of the Caledonides are younger than the tectonic emplacement of the mountain zone, as also are the silver-bearing lead ore lenses found in quartz veins penetrating the high-grade schists of the Kvikkjokk mountains. The writer is of the opinion that the easterly and westerly situated lead ores are contemporaneous and probably record a period of early post-orogenic hydrothermal activity within the Caledonian mountain chain. A geochemical survey of mineralization from the eastern border in a WNW direction might give valuable scientific results.

Chalcopyrite–sphalerite ore deposits within the western part of the Caledonian mountain schists (within the Upper thrust rocks). A large part of the chalcopyrite–sphalerite deposits within the Caledonian mountain schists are found in a SSW–NNE-striking belt in the most westerly part of the Västerbotten mountains and in an adjacent part of the northwestern Jämtland mountains. Most of the actual ore deposits lie in the Seve–Köli complex. Some northern occurrences are, however, localized in rocks of the Storfjäll nappe complex.

The *Stekenjokk ore.* The largest known occurrence of chalcopyrite–sphalerite ore in the ore belt mentioned is the Stekenjokk ore deposit. This carefully surveyed ore deposit lies in the extreme southwestern part of the Västerbotten mountains, only 7 km from the international border, in the central lower part of the Remdalen syncline. The ore deposit is surrounded by the Stekenjokk quartz keratophyre of the Lasterfjäll group.

The Stekenjokk ore, discovered in 1918 by Högbom of the Geological Survey, soon became the object of a detailed survey. In spite of electrical measurements, attempts to uncover the ore and diamond drilling, the considerable dimensions and high quality of the ore were not verified during the first period (see Högbom, 1925). In 1952 a new prospecting period started in the Stekenjokk ore deposit. The area was remapped on a large scale and as a result of about 30 000 m of diamond drilling the configuration of the ore deposits is now known in detail. Zachrisson has described the Stekenjokk ore resources (unpublished report, 1963) and has kindly permitted the writer to communicate the following facts.

The ore-bearing Stekenjokk quartz keratophyre is composed of light-coloured volcanic rocks with tuffitic portions and slate intercalations. There are all gradations from tuffaceous rocks to slates. The slates, including some tuffitic types, are often graphite bearing. The graphitic slate often lies closely attached to the ore bodies. Lens-shaped gabbro intrusions frequently occur in the vicinity of the ore bodies also.

The metasomatic transformation of rocks in connection with ore mineralization has caused the quartz keratophyres and tuffites immediately around the ores to change into sericitic quartzose rocks, and the basic members to change into chlorite schists. During folding of the rocks, and perhaps also in connection with ore mineralization, the graphitic slates were the sites of gliding. In some places graphitic material has migrated into the surrounding rocks and has caused graphitic slickensides.

The Stekenjokk ore is a compound ore, and the main minerals present are pyrite, chalcopyrite, sphalerite and pyrrhotite. A very low percentage of galena is present. Three main ore types occur: *compact pyrite ore* with 30% S and varying % Zn and Cu; *disseminated Cu ore* with heavy impregnation of chalcopyrite and pyrrhotite in quartzose rocks; and *impregnation ores*, e.g. impregnations in graphitic phyllite bounded to compact pyrite ore, or heavy impregnations in tuffites and sericitic quartzites.

The Stekenjokk ore deposit is ruler-shaped and can be followed for a stretch of about 2200 m. In 1962 the ore deposit was calculated to consist of more than 13 000 000 metric tons.

The *Ankarvattnet ore* lies about 24 km SSW of the Stekenjokk ore, in the same ore-bearing belt. According to Zachrisson's report the Ankarvattnet ore lies in a calcareous phyllite—the Blåsjö phyllite in the description of the Blåsjö area (see Nilsson, 1964). The ore seems to be conformable with the enclosing phyllite and lies near bodies of gabbro, but is never in contact with the intrusive basic rock. Two main types of ore are discernible, namely the *pyrite ore* and the *pyrrhotite ore.* Both of them are generally compact ores but impregnation ore types occur too. The Ankarvattnet ore deposit is calculated to be about 750 000 metric tons.

The *Remdalen ore* is situated about 16 km NNE of the Stekenjokk ore, and lies in the upper part of the Seve–Köli sequence, within the Remdalen greenschist of the Remdalen group. There is a gabbro lens in the vicinity of the ore deposit. The Remdalen ore consists of two main types: *pyrrhotite with chalcopyrite* and *pyrite with sphalerite.*

Following the description above of three occurrences of pyritiferous ore within the Seve–Köli complex two such deposits within the Storfjäll thrust complex will be mentioned. The deposits lie in a northerly continuation of the SSW–NNE directed ore belt mentioned above.

The *Ropen ore.* In the southern part of Mt. Södra Storfjället and west of Lake Ropen, several small occurrences of chalcopyrite–sphalerite have been located in the "Ropen granite" which may be a recrystallized quartz keratophyre, situated in the lowermost part of the Storfjäll nappe complex. The zones of ore impregnation are characterized by the so-called white schists of sericite–quartzitic nature. These white schists were formed during seeping-through of the ore-producing solution, and they indicate the path of the solution.

The *Tjåter ore*. The Tjåter ore deposit lies on the southern slope of Mt. Tjåter, immediately northeast of Mt. Södra Storfjället. The ore consists of pyrite and sphalerite with some chalcopyrite, and lies in the central part of some light-coloured sericitic schists of the same kind as those mentioned in the Ropen area. The sericite schists pass upwards into graphitic slates. Below them occurs phyllitic slate. The stratigraphy of the mountain area—the Gieravardo–Jofjället area, with Mt. Tjåter in the southeastern part—is incompletely known, but the Tjåter ore is probably of little value.

The chalcopyrite–sphalerite deposits of the ore-bearing western belt described are probably representative of the area. From the above description and from a general inspection of other known ore deposits of the belt it appears that the mineralizing solutions precipitated out in different types of rocks and at different levels within the Upper thrust rocks. The ore material might have come from pyritiferous black slates, gabbroid rocks and other rocks with ore minerals, and the solutions then precipitated out not too far from the source of the material. It is of interest to note that east of the ore-bearing belt no pyritiferous ore-impregnation of any importance has been reported within the relatively low grade metamorphic belt with fossiliferous bedrock, or in the even more easterly-lying high grade metamorphic part of the Upper thrust rocks. It may also be mentioned that the rocks of the ore belt, particularly the fine grained ores, are often intensely and finely folded, with more or less pronounced tectonic banding. This type of tectonic alteration of rocks has been studied by the writer in the area west of the central Västerbotten Björkvattnet–Virisen area and has been called the *Grokas facies* (after *Mt. Stora Grokas*).

Apart from the belt of Cu–Zn ores in western Västerbotten and northern Jämtland, there are some scattered Cu ores in the Upper thrust rocks of the Southern Swedish Caledonides. Some of these have been worked from time to time. The *Cu ore deposits* on the slopes of *Mt. Åreskutan* in the western part of Jämtland may be mentioned. The ore bodies consist of pyrrhotite, pyrite and chalcopyrite and sometimes also of sphalerite and bornite; they appear partly as diffuse impregnations, partly as sheets or lenses. The ores lie in quartzose gneissic rocks, in a sequence of gneissic rocks and amphibolites.

Iron ore deposits. No iron ore has been reported in the Upper and Uppermost thrust rocks, but in the thrust basement of the Middle thrust rocks there is a rather large occurrence of titaniferous iron ore—the *Ruoutevare ore*—lying near the eastern border of the Norrbotten mountains about 12 km northwest of Kvikkjokk village. In the Ruoutevare ore region of Precambrian thrust-basement igneous rocks, the iron ore lies enclosed in anorthosite and gabbroic anorthosite. Together with the rocks mentioned and considered as belonging to the same intrusive suite, there are K-rich rocks of monzonitic, syenitic and granitic composition as well as ultrabasic rocks of dunitic, pyroxenitic to hornblenditic composition. The rich part of the titaniferous iron ore is calculated to have an area of 135 000 m^2 and the thickness of the ore-bearing sheet is, according to drillings, some tenths of a metre. The quantity of ore is calculated to be 20 million metric tons.

Magnesite occurs in the middle, eastern part of the Norrbotten mountains as small lens-shaped bodies lying in the so-called Eastern amphibolite of the Upper thrust rocks. During the second world war, the *Äpartjåkko magnesite* on Mt. Äpartjåkko was worked.

Limestone has been quarried mainly in the Storsjö area of Jämtland. The *limestone quarries at Brunflo and Fugelsta*, both lying near the eastern end of Lake Storsjön in autochthonous Ordovician limestone, are also worthy of mention.

Soapstone. Peridotite rocks are metamorphosed to serpentine and soapstone. Some of the soapstone occurrences in the Upper thrust rocks of western Jämtland have been utilized. The *Handöl soapstone quarry* produces, *inter alia*, soapstone stoves.

Roofing slates and flags of good quality seem to be sparse in the Swedish Caledonides and in most parts of the frequently isolated mountain areas, such flags as occur are of little value. In the Offerdal area, lying in the central part of Jämtland to the north of Lake Storsjön, there are plane-foliated schists—the *Offerdal schists*—which are quarried at Trången and Lien. The schists are micaceous and are supposed to be an intensely tectonized lower part of the thrust-basement of the Middle thrust rocks.

CHAPTER XXVII

THE RELATIONSHIP OF THE SWEDISH PART OF THE SCANDINAVIAN CALEDONIDES TO ITS MAINLY NORWEGIAN COUNTERPARTS

The Swedish or middle eastern part of the Scandinavian Caledonides continues to the north, west and southwest, mainly on Norwegian territory. The Finnish and Russian parts of the mountain chain are very small and of limited interest in this brief comparison. The subdivision of the Swedish Caledonides into autochthonous rocks—Lowermost, Lower, Middle, Upper, and Uppermost thrust rocks—is applicable to most of the Scandinavian Caledonides.

The erosional remains of the *autochthonous* sediments are exposed in a very narrow zone along the eastern border of the Swedish Caledonides. This zone can be followed towards the northeast over Finnish and Norwegian territory to southeast of Laksefjord where it thins out. In the other direction—in southern Norway—there is also a narrow zone of the sediments along the eastern border of the thrust rocks, from Trysil in the northeast to Stavanger in the southwest. From the Mjösa area a branch of the autochthonous Cambro–Silurian extends to and encompasses the Oslo fjord area in the south. The autochthonous Cambro–Silurian rock complex just mentioned forms a very thick sedimentary sequence, in striking contrast to the thin remains in all other autochthonous parts of the Caledonian eastern border zone. The broad Cambro–Silurian region of Central Jämtland in Sweden, in several respects a counterpart of the south Norwegian Mjösa–Oslo region, is composed to a very great extent of Lower and Lowermost thrust rocks.

The *Lowermost thrust rocks* or the East Jämtland Cambro–Silurian nappes seem to have no real tectonic counterparts in Norway.

The *Lower thrust rocks* in Sweden, composed of fossiliferous Cambro–Silurian sediments, Vemdal quartzite, tillite, sparagmite and slices of the Precambrian basement rocks are widely distributed in the southern Norwegian Sparagmite region, in the Valdres region and possibly further southwest.

East of the Bergsdal area, in Valdres and Jotunheimen, the Valdres sparagmite covers considerable areas. The Valdres sediments consist of arkoses, conglomerates, feldspar-bearing quartzites and slates, and very much resemble the Late Precambrian sparagmite group rocks of the Sparagmite region. The sheet of Valdres sediments lies in part directly on fossiliferous Ordovician sediments and there is a slab of Precambrian igneous rocks lying between the sparagmite and the fossiliferous Ordovician. The opinion held by Norwegian geologists, and advocated strongly by STRAND during the last decade, is that during Ordovician times the geosynclinal area of southern Norway was transformed into a mountain chain and overthrusting took place. A large nappe from a western root zone with Precambrian igneous rocks in its lower part came to rest in Valdres and adjacent areas. Before the Valdres sparagmite sequence was laid down, partly on erosional remains of the thrust-basement rocks, and partly on reexposed Ordovician strata, most parts of the Ordovician nappe had eroded away. The writer is of the opinion (KULLING, 1955) that the Valdres sparagmite sequence, together with its underlying Precambrian igneous rocks, represents the Valdres thrust complex, lying on thrust rocks with Ordovician sediments in their upper part. During a stay in the Valdres area in 1960 the writer strengthened this view (KULLING, 1961). A roofing slate beneath the Valdres clastic rocks in the Mellene area, lying on finely folded phyllite of the fossiliferous Ordovician with levels of graptolite shales of Arenigian and Llandelian age, was looked upon by the writer as a basal, stressed part of the Valdres thrust rocks. In another area, in which a thin slab of syenite lies between the Valdres arkose and the underlying phyllite, the syenite and the lower part of the arkose have been tectonized in the same way. Over very large areas, the Valdres sparagmite lies directly on the fossiliferous Ordovician, but material of the fossiliferous sedimentary sequence has not been found in the

Valdres clastics—they are not submarine turbidity current material, but mainly supramarine arkosic deposits of Late Precambrian age.

In the northernmost part of Swedish Lapland the Lower thrust rocks continue northwards through Finnish Lapland to the Norwegian Finmark region, where part of the thrust rocks were quite recently described under the name of the *Jerta nappe* (see the Norwegian part of this book). From the area around the southern part of Porsangerfjord going northeast to Varangerfjord, and including the big Varanger peninsula, the broad mountain border zone is designated as Lower thrust rocks. As is clear from the description of their occurrences in Norway these northern Lower thrust rocks are of a composition other than their counterparts in northern Sweden. In the northern part there are primarily very thick sandstone deposits bearing an upper and a lower tillite formation.

The *Middle thrust rocks*, which consist in Sweden of stressed, often mylonitic Precambrian crystalline rocks (granites, syenites, augen gneisses etc.), metamorphic arkoses (sparagmites), and tectonic banded quartzites and schists (hardschist rocks), occur in Norway, but their distribution is still not well known.

The bedrock of the Norwegian border zone in Nord Tröndelag, round the northern part of the Trondheim region, will be considered first. On his geological map of Central Scandinavia (1896) Törnebohm considered the metamorphic Precambrian granite and porphyry complex of the boundary zone, in the district around Lake Torrön etc., as the lower part of his Main thrust nappe. This complex, according to recent geological surveys, surrounds the Lower thrust rocks to the north, west and south; the Lower thrust rocks are composed of fairly well-preserved granite and porphyry; on the Swedish side of the international border it is called the Olden anticlinal complex. The metamorphic "Basal gneiss" of Nord Tröndelag is thought by the writer to be part of the Middle thrust rocks, and so are the basal gneisses of the *Tömmeråsen anticline* in the northeastern part of the Trondheim region.

In this connection attention may be drawn to Ramberg's geological results in the Fosen district which lies at the western border of the Trondheim area, around the inlet of Trondheimfjord (Ramberg, 1943). He concludes that the bedrock of the Fosen district—both the old Precambrian basement rocks and the Trondheim region rocks downfolded into them—belong to one and the same tectonic complex. He emphasizes that "the Vestrand is none other than a continuance of the Caledonian chain developed in the Trondheim district".

The geological map and monograph of Törnebohm (1896) give useful information about the distribution of the Middle thrust rocks of southern Norway. The lowest part of the Middle thrust rocks in the southern Swedish mountains is the augen gneiss unit. From the western part of Härjedalen this characteristic rock complex can, with interruptions, be followed on the Norwegian side, first along the southeastern border of the Trondheim region to its southwestern corner and then along its western side. Southeast of the Trondheim region, in the Sparagmite region, Törnebohm's map also shows some tectonic klippen with augen gneiss in their lower parts and metamorphic sparagmites constituting their upper parts. Other klippen are composed only of metamorphic sparagmites and metamorphic sediments of other composition. All the klippen southeast of the "Trondheim region"—the *Kvitvola klippe*, the *Koppang klippe* and others—belong to the writer's Middle thrust rocks.

The augen gneiss rocks along the western border of the Trondheim region occur together with porphyroblastic and banded schists. This rock zone may also be looked upon as belonging to the Middle thrust rocks. The composition and tectonic position of the "basal gneiss" to the west of the western augen gneiss zone is, however, still rather little known. The "basal gneiss" probably contains complexes with different tectonic positions.

Wegmann (1959), who surveyed the Driva area in the western border zone of the Trondheim region and paid particular attention to the distribution of the augen gneiss along the Trondheim region borders, called the zone the "zone de déformation intense à la base du complexe de Trondheim (avec les gneiss oeillés et roches analogues)". He called the southwest Norwegian "basal gneiss" region to the west of the augen gneiss zone the "Complexe des vieux gneiss avec zones de granite". In the same complex Wegmann also included the Olden granite area in northwestern Jämtland and the Precambrian basement rocks east of the Scandinavian Caledonides. He thus put together different Precambrian crystalline rock complexes, which according to the previous account are distributed among Middle thrust rocks, Lower thrust rocks and autochthonous basement rocks.

It is very doubtful whether the tectonic subdivision of the Scandinavian Caledonides, arrived at for the eastern part of the chain, is valid for the extreme southwesterly parts of Norway. What rock complex in the Bergsdalen area is, for instance, a likely counterpart of the Middle thrust rocks? The Bergsdalen area lies off the southern part of the basal gneiss area mentioned above. According to Kvale (1948) there are three nappes—the uppermost or the Jotun nappe; the Upper Bergsdal nappe; and the Lower Bergsdal nappe. The Jotun nappe probably belongs to the Middle thrust rocks. Both of the Bergsdal nappes have been subdivided into sheets, all of which are composed of about the same kind of rocks. But in contradiction to what is the rule for lower and upper nappes in the eastern part of the Caledonian orogen, the Lower Bergsdal nappe shows evidence of higher regional metamorphism than does the Upper Bergsdal nappe. A possible solution to the problem of metamorphism may be that the nappes come from areas west of the axis of major metamorphism, and that the Upper Bergsdal nappe started from a root zone with lower rock alteration than did the Lower Bergsdal nappe. Below the lower of the two nappes mentioned comes the so-called basal gneiss; this basement rock complex shows migmatitic transformation. To the west and above this highly altered rock complex there are no western counterparts of the Bergsdal nappes, but in the Bergen area there occurs a rock sequence of eugeosynclinal deposits with fossiliferous strata and volcanic rocks of true Upper thrust rock facies.

Information about the Caledonides of the extreme southwestern part of Norway lends support to the writer's view that the tectonic picture of the region cannot yet be fitted into that given for the eastern parts of the Caledonides; in particular no real western equivalent of the Middle thrust rocks is distinguishable.

Turning to the northern Scandinavian part of the chain, the Middle thrust rocks of Swedish Lapland cross the international border west of the Akkajaure area, and north of the Torneträsk and Pältsa areas. Little need be said about the northern continuation of the Middle thrust rocks. According to PADGET (1955) the northern continuation of the Middle thrust rocks in northernmost Swedish Lapland consists of overthrusted sparagmitic schists with mylonites, superposed by metamorphic eugeosynclinal rocks with the collective name of the *Birtavarre Series.* In more northerly parts of northern Scandinavia, Norwegian geological maps show vast areas of mainly gneissic overthrust rocks which have been surveyed very little. They probably consist of both Middle and Upper thrust rocks.

The *Upper thrust rocks* of Jämtland and Swedish Lapland extend across the international border. The Trondheim region with its southern continuation, the Otta region, is the largest Upper thrust rock area in Norway. A very large part of the thrust rocks of Northern Norway, called the "Mica schist–marble group" on the 1960 geological map of Norway, may be Upper thrust rocks. In the southwest Norwegian fjord area the metamorphic "Cambro–Silurian rocks" of eugeosynclinal facies are, as stated before, assigned to western areas of Upper thrust rocks by the writer. The relationship of the Upper thrust rocks in the Bergen arc to the migmatitic basal gneiss and the Bergsdal nappe rocks in the east has been discussed in the preceding pages.

The *Uppermost thrust rocks* seem to be restricted to the most westerly parts of Swedish Lapland and to the high Norwegian mountains west of Lapland. The Uppermost thrust rocks or the Rödingsfjäll nappe (KULLING, 1948) extends in northern Norway from Namdal in the south to south of Vestfjorden in the north. On the Norwegian geological map of 1960, the thrust rocks show an abundance of "igneous rocks in the Caledonian zone" mainly of granitic composition. The distribution of thrust-basement rocks and the Caledonian intrusive rocks among them has been only partially determined. Probably most of the acid igneous rocks are of thrust-basement type.

SELECTED BIBLIOGRAPHY

ABBREVIATIONS USED IN SELECTED BIBLIOGRAPHY

Geol. Fören. Stockholm Förh.	= Geologiska Föreningens i Stockholm Förhandlingar
Norsk Geol. Tidssk.	= Norsk Geologisk Tidsskrift
Norg. Geol. Undersøkelse	= Norges Geologiske Undersøkelse
Sveriges Geol. Undersökn.	= Sveriges Geologiska Undersökning
Kgl. Vetenskapsakad.	= Kungliga Svenska Vetenskapsakademien

Asklund, B. (1933). Vemdalskvartsitens ålder. (The age of the Vemdal quartzite.) *Sveriges Geol. Undersökn.*, Ser. C382.

– (1938). Hauptzuge der Tektonik und Stratigraphie der mittleren Kaledoniden in Schweden. *Sveriges Geol. Undersökn. Årsbok, Ser. C: Avhandl. och Uppsat.*, **417**.

– (1960). *See* Magnusson and others (1960).

Asklund, B. and N. Marklund (1954). Aktuella Skandinaviska fjällproblem. (Actual Scandinavian Caledonian problems.) *Geol. Fören. Stockholm Förh.*, **76**, 103–167.

Bailey, E. B. and W. J. McCallien (1950). The Ankara mélange and the Anatolian thrust. *Nature*, **166**.

– (1952). Ballantrae igneous problems: Historical review. *Trans. Edinburgh Geol. Soc.*, **15**.

Beskow, G. (1929). Södra Storfjället im südlichen Lappland. *Sveriges Geol. Undersökn. Årsbok, Ser. C: Avhandl. och Uppsat.*, **350**.

De Munck, V. C. E. A. (1951). *Geology and petrology of the northwestern Artfjäll. Swedish Lapland.* Diss, Amsterdam.

Du Rietz, T. (1935). Peridotites, serpentines and soapstones of northern Sweden etc. *Geol. Fören. Stockholm Förh.*, **57**, 133–260.

– (1938). The injection metamorphism of the Muruhatten region. *Sveriges Geol. Undersökn. Årsbok, Ser. C: Avhandl. och Uppsat.*, **416**.

– (1960). Tectonic conditions in the Front Range of the Swedish Caledonian in Central Norrland. *Sveriges Geol. Undersökn. Årsbok, Ser. C: Avhandl. och Uppsat.*, **568**.

Faddegon, J. M. (1940). *Geologische en petrologische onderzoekingen in het Rivovardogebied en omgeving.* (Zusammenfassung) Diss, Amsterdam.

Foslie, S. (1941). Tysfjords geologi. *Norg. Geol. Undersøkelse.*, **149**.

– (1942). Hellemobotn og Linnajavrre. *Norg. Geol. Undersøkelse.*, **150**.

– (1949). Håfjellsmulden i Ofoten og dens sedimentaere jern-manganmalmer. (The geology of the Håfjellsmulden/synform/ in Ofoten and its Fe-Mg-ores of sedimentary origin.) *Norg. Geol. Undersøkelse.*, **174**.

Föyn, S. (1937). The Eo-Cambrian series of the Tana district, northern Norway. *Norsk Geol. Tidsskr.*, **17**, 65–164.

– (1964). Den tillitförende formasionsgruppe i Alta—En jevnföring med Öst-Finnmark og indre Finnmark. (The tillite-bearing group of formations in Alta—A comparison with Öst-Finnmark and the inner part of Finnmark.) *Norg. Geol. Undersøkelse.*, **228**, 139–150.

Frödin, G. (1921). Über die Geologie der zentralschwedischen Hochgebirge. *Bull. Geol. Inst. Univ. Uppsala.*, **18**, 57–197.

Grip, E. (1960). The Skellefte district and the Laisvall area. Part of guide to excursion Nos. A27 and C22. *Intern. Geol. Congr. 21st Copenhagen, Rep. Session, Norden*, 3–14.

Hamberg, A. (1910). Gesteine und Tektonik des Sarekgebirges nebst einem Überblick der skandinavischen Gebirgskette. *Geol. Fören. Stockholm Förh.*, **32**, 681–724.

Hedström, H. (1930). Om ordoviciska fossil från Ottadalen i det centrala Norge. (On Ordovician fossils from Otta in central Norway.) *Det Norske Vidensk.-Akad. i Oslo. Avh. I. Mat.-Nat. Kl.*, **1930**, 10.

Högbom, A. (1925). De geologiska förhållandena inom Stekenjokk–Remdalens malmtrakt. (The geology of the Stekenjokk–Remdalen ore district.) *Sveriges Geol. Undersökn. Årsbok, Ser. C: Avhandl. och Uppsat.*, **329**.

– (1941). Nyare undersökningar inom Remdalens malmtrakt och dess omgivningar. (New investigations in the Remdalen ore district and its environs.) *Sveriges Geol. Undersökn. Årsbok, Ser. C: Avhandl. och Uppsat.*, **439**.

Högbom, A. G. (1894). Geologisk beskrifning öfver Jemtlands län. (Geology of Jämtland county.) *Sveriges Geol. Undersökn. Årsbok, Ser. C: Avhandl. och Uppsat.*, **140**.

– (1920). Edn. 2.

KAUTSKY, F. (1940). Das Fenster von Gautojaure in Kirchspiele Arjeplog, Lappland. *Geol. Fören. Stockholm Förh.*, **62**, 121–147.
– (1945). Die unterkambrische Fauna vom Aistjakk in Lappland. *Geol. Fören. Stockholm Förh.*, **67**, 129–208.
KAUTSKY, G. (1953). Der geologische Bau des Sulitelma-Salojaure Gebietes in den nordskandinavischen Kaledoniden. *Sveriges Geol. Undersökn. Årsbok, Ser. C: Avhandl. och Uppsat.*, **528**.
KIAER, J. (1929). Den fossilförende ordovicisk–siluriske lagrekke på Stord. (The fossiliferous Ordovician–Silurian sequence of Stord.) *Arbok Univ. Bergen, Mat.-Nat. Ser.*, **11**.
KULLING, O. (1930). Stratigraphic studies of the geology of northeast Greenland. *Medd. Groenland*, **74**, 317–346.
– (1930). Studier över den kaledoniska fjällkedjans stratigrafi och tektonik inom norra delen av svenska Lappland. (On the Caledonian stratigraphy and tectonics in the northern part of Swedish Lapland.) *Geol. Fören. Stockholm Förh.*, **52**, 647–673.
– (1933). Bergbyggnaden inom Björkvattnet–Virisen-området i Västerbottensfjällens centrala del. (Geology of the Björkvattnet–Virisen area in the central part of the Västerbotten mountains.) *Geol. Fören. Stockholm Förh.*, **55**, 167–422.
– (1934a). Något om åldersförhållandena inom Lapplandsfjällens östra randområde. (On the age-relations of rocks in the eastern border zone of the Caledonides of Lapland.) *Geol. Fören. Stockholm Förh.*, **56**, 60–64.
– (1934b). The "Hecla Hoek formation" round Hinlopenstredet. Scientific results of the Swedish–Norwegian arctic expedition in the summer of 1931. Part XI. *Geogr. Ann., Stockholm*, **1934**, 161–254.
– (1938a). Grönstenarnas placering inom Västerbottensfjällens kambrosilurstratigrafi. (On the stratigraphical positions of the greenstones in the eugeosynclinical deposits of the Västerbotten mountains.) *Geol. Fören. Stockholm Förh.*, **60**, 153–176.
– (1938b). Notes on varved boulder-bearing mudstone in Eocambrian glacials in the mountains of northern Sweden. *Geol. Fören. Stockholm Förh.*, **60**, 392–396.
– (1942). Grunddragen av fjällkedjerandens bergbyggnad inom Västerbottens län. (The geology of the marginal part [the eastern part] of the Caledonides of Västerbotten county.) *Sveriges Geol. Undersökn. Årsbok, Ser. C: Avhandl. och Uppsat.*, **445**, 1–320.
– (1947a). Några fjällgeologiska frågor. (Some Caledonian problems.) *Geol. Fören. Stockholm Förh.*, **69**, 127–135.
– (1947b). Aktuella fjällfrågor. (Actual Caledonian questions.) *Geol. Fören. Stockholm Förh.*, **69**, 475–485.
– (1955). Den kaledoniska fjällkedjans berggrund inom Västerbottens län. Beskrivning till berggrundskarta över Västerbottens län 2. (The Caledonian mountain range of Västerbotten county.) *Sveriges Geol. Undersökn. Ser. Ca*, **37**, 101–296.
– (1960). The Caledonian mountain chain in the Torneträsk area, northern Scandinavia. Guide to excursion Nos. A25 and C20. *Intern. Geol. Congr. 21st Copenhagen, Rep. Session, Norden*, 18–76.
– (1961). On the age and tectonic position of the Valdres Sparagmite. *Geol. Fören. Stockholm Förh.*, **83**, 210–214.
– (1964). Översikt över Norra Norrbottensfjällens berggrund. (Kartor 1965.) (The geology of the Caledonian rocks of the northern Norrbotten mountains.) *Sveriges Geol. Undersökn. Ser. Ba*, **19**, 1–166.
KVALE, A. (1948). Petrologic and structural studies in the Bergsdalen quadrangle, western Norway. Part 2. Structural Geology. *Arbok Univ. Bergen, Mat.-Nat. Ser.*, **1**.
LAKEMAN, R. (1952). *Geology and petrology of the Brantsfjäll region, Swedish Lapland.* Diss, Amsterdam.
LINDSTRÖM, M. (1958). Tectonic transports in the Caledonides of northern Scandinavia east and south of the Rombak–Sjangeli window. *Publ. Inst. Mineral. Paleontol. Quaternary Geol. Univ. Lund*, **43**.
LJUNGNER, E. (1948). Kölen och Skanderna. (The Kjöl and the Scandes.) *Svensk geogr. Årsb.*, 190–199.
MCCALLIEN, W. J. (1950 and 1952). *See* BAILEY and MCCALLIEN (1950 and 1952).
MAGNUSSON, N. H., P. THORSLUND, F. BROTZEN, B. ASKLUND and O. KULLING (1960). Description to accompany the map of the pre-quaternary rocks of Sweden. *Sveriges Geol. Undersökn. Ser. Ba*, **16**.
MARKLUND, N. (1952a). A study in a Caledonian major thrust plane. *Bull. Geol. Inst. Univ. Uppsala*, **34**, 205–221.
– (1952b). A Cambro–Ordovician type-section in the Sarvas region SE of Nasafjäll. *Geol. Fören. Stockholm Förh.*, **74**, 353–384.
– (1954). *See* ASKLUND and MARKLUND (1954).
MICHEL, H. (1950). *Geology and petrology of the Borkafjäll region (southern Swedish Lapland).* Diss, Amsterdam.
MULDER, C. J. (1951). *Geology and petrology of the region between Lake Överuman and Tärnasjön (southern Swedish Lapland).* Diss, Amsterdam.
MURRIS, R. J. (1957). *Geology and petrology of the Gieravardo–Jofjället region, Västerbotten, Sweden.* Diss, Amsterdam.
NILSSON, G. (1964). The geology of the Caledonides of the Blåsjö area, Jämtland, Sweden. *Sveriges Geol. Undersökn. Årsbok, Ser. C: Avhandl. och Uppsat.*, **595**.
Norges geologiske undersøkelse (1960). Geologisk kart over Norge. (Geological map of Norway.) Ed. 2. *Norg. Geol. Undersøkelse.*, **164**.

PADGET, P. (1955). The geology of the Caledonides of the Birtavarre region, Troms, northern Norway. *Norg. Geol. Undersøkelse.*, **192**.

PARASNIS, D. S. (1962). *Principles of Applied Geophysics*. London. Methuen's monographs on physical subjects.

POULSON, CHR. (1930). Preliminary notes on the stratigraphy of the Cambro-Ordovician formations of East Greenland. *Medd. Groenland*, **74**, 297–316.

QUENSEL, P. (1960). Beskrivning till karta över berggrunden inom Västerbottens fjällområde. (Description to the bedrock map of the main part of the Västerbotten mountains.) *Sveriges Geol. Undersökn., Ser. Ba*, **21**.

RAMBERG, H. (1943). En undersökning av vestrandens regionalmetamorfe bergarter. (A survey of the regional-metamorphic rocks of the Vestrand [west of the Trondheim field].) *Norsk Geol. Tidsskr.*, **23**, 1–174.

RUTLAND, R. W. R. and R. NICHOLSON (1965). Tectonics of the Caledonides in part of Nordland, Norway. *Quart. J. Geol. Soc.*, **121**, 73–109.

SJÖGREN, HJ. (1900). Enkrinitfynd i fjellskiffrarne vid Sulitelma. (Find of Crinoidéan fossils in the mountain schists at Mt. Sulitelma.) *Geol. Fören. Stockholm Förh.*, 22, 105–115.

STAARGAARD, J. A. (1955). *Geology and petrology of the southeastern Borgafjäll, Jämtland, Sweden*. Diss, Amsterdam.

STRÖMBERG, A. G. B. (1955). Zum Gebirgsbau der Skanden im mittleren Härjedalen. *Bull. Geol. Inst. Univ. Uppsala*, **35**, 199–243.

– (1961). On the tectonics of the Caledonides in the south-western part of the county of Jämtland, Sweden. *Bull. Geol. Inst. Univ. Upsala*, **39**, 1–92.

STÅLHÖS, G. (1956). The sparagmite series and Vemdal quartzite of the Hede region, Härjedalen. *Bull. Geol. Inst. Univ. Uppsala*, **36**, 297–368.

Sveriges geologiska undersökning (1958). Karta över Sveriges berggrund. (Pre-quaternary rocks of Sweden.) *Ser. Ba*, **16**.

TEGENGREN, F. R. (1962). The Vassbo lead ore deposit in Idre, western Sweden. *Sveriges Geol. Undersökn. Årsbok, Ser. C: Avhandl. och Uppsat.*, **586**.

THORSLUND, P. (1940). On the Chasmops series of Jemtland and Östergötland (Tvären). *Sveriges Geol. Undersökn. Årsbok, Ser. C: Avhandl. och Uppsat.*, **436**.

– (1960). *See* MAGNUSSON and others (1960).

TÖRNEBOHM, A. E. (1872). En geognostisk profil över den skandinaviska fjällryggen mellan Östersund och Levanger. (A geological section across the Scandinavian Caledonides between Östersund and Levanger.) *Sveriges Geol. Undersökn. Ser. C: Avhandl. och Uppsat.*, **6**.

– (1888). Om fjällproblemet. (About the Caledonian mountain problem.) *Geol. Fören. Stockholm Förh.*, **10**, 328–336.

– (1896). Grunddragen af det centrala Skandinaviens bergbyggnad. (Geology of the central Scandinavian part of the Caledonides.) *Kgl. Svenska Vetenskapsakad. Handl.*, **28**:5.

UYTENBOGAARDT, W. (1953). On the opaque mineral constituents in a series of amphibolitic rocks from Norra Storfjället, Västerbotten, Sweden. *Kgl. Svenska Vetenskapsakad. Arkiv. Mineral. Geol.*, **1**:19.

VAN DER HARST, T. H. (1965). *Geology and petrology of the Borga region, Västerbotten, Sweden*. Diss, Amsterdam.

VOGT, TH. (1927). Sulitelmafeltets Geologi og Petrografi. *Norg. Geol. Undersøkelse.*, **121**.

– (1945). The geology of part of the Hölonda-Horg district, a type area in the Trondheim region. *Norsk Geol. Tidsskr.*, **25**, 449–528.

– (1950). The geological map Narvik. *Norg. Geol. Undersøkelse.*

WEGMANN, E. (1959). La flexure axiale de la Driva et quelques problemes structuraux des Caledonides Scandinaves. *Norsk Geol. Tidsskr.*, **39**, 25–74.

WIMAN, C. (1893). Über die Silurformation in Jemtland. *Bull. Geol. Inst. Univ. Uppsala*, **1**, 256–276.

YOCHELSON, E. L. (1963). Gastropods from the Otta conglomerate. *Norsk Geol. Tidsskr.*, **43**, 75–81.

ZACHRISSON, E. (1964). The Remdalen syncline. *Sveriges Geol. Undersökn. Årsbok, Ser. C: Avhandl. och Uppsat.*, **596**.

ZENZÉN, N. (1932). Nya geologiska notiser från Idre. (New geological notes from the Idre area.) *Geol. Fören. Stockholm Förh.*, **54**, 533–534.

INDEXES

AUTHOR INDEX

Page numbers in italic denote entries in the bibliographies

SUBJECT INDEX